Opening the Ozarks

A Historical Geography of Missouri's
Ste. Genevieve District
1760–1830

WALTER A. SCHROEDER

University of Missouri Press
COLUMBIA

Copyright © 2002 by
The Curators of the University of Missouri
University of Missouri Press, Columbia, Missouri 65211
Printed and bound in the United States of America
All rights reserved
First paperback printing, 2016

Library of Congress Cataloging-in-Publication Data

Schroeder, Walter A., 1934–
   Opening the Ozarks : a historical geography of Missouri's Ste. Genevieve District, 1760–1830 / Walter A. Schroeder.
     p.   cm.
   Includes bibliographical references and index.
   ISBN 978-0-8262-2071-4 (alk. paper)
   1. Sainte Genevieve County (Mo.)–Historical geography. 2. Land settlement–Missouri–Sainte Genevieve County–History. 3. Frontier and pioneer life–Missouri–Sainte Genevieve County. 4. Human geography–Missouri–Sainte Genevieve County–History. 5. Sainte Genevieve County (Mo.)–History. 6. Sainte Genevieve County (Mo.)–History, Local. I. Title: Historical geography of the Sainte Genvieve District, 1760-1830. II. Title.

F472.S33 S37   2002
911'.778692–dc21

                                                      2002023147

♾™ This paper meets the requirements of the
American National Standard for Permanence of Paper
for Printed Library Materials, Z39.48, 1984.

Jacket Designer: Susan Ferber
Text Designer: Elizabeth K. Young
Typefaces: Usherwood, Oberon

The University of Missouri Press offers its grateful acknowledgment to the Ozark Natural Resources Foundation for a generous contribution in support of the publication of this volume.

to Pat

Contents

List of Illustrations, xi

List of Tables, xiii

Foreword by Terry G. Jordan, xv

Preface, xvii

Acknowledgments, xix

Introduction

1. The Ste. Genevieve District, 3
   The Place, 4
   The Time, 11
   Interpretation, 18

Part I *The Environment of Settlement*

2. "A Country Thus Rich by Nature," 27
   Resources, 28
   Natural Regions, 63
     Mississippi Alluvial Plain, 63
     Bluff Lands and Karst Plain, 66
     Rough Hills, 68
     Mining Country, 70

Part II *Regulating Settlement*

3. French and Spanish Periods to 1795, 75
   French Period before 1770, 77
   Spanish Period, 1770–1795, 80

4. Spanish Period, 1796–1804, 97
5. American Period, 1804–1818, 114
   Confirming Land Claims, 117
   Results of the Land-Confirmation Process, 146
   Surveying Private Lands, 152
   The Landless, 159
   Attempts to Limit Settlement, 164
6. American Period, 1818–1830, 174
   U.S. Rectangular Survey and Land Sales, 174
   Lead Lands, 183
   Indian Lands, 193
   Recapitulation, 194

Part III *Building Communities*

7. Settlement Forms, 201
   Planned French Village, 202
   Unplanned French Village, 212
   Planned American Town, 214
   Linear Valley Settlement, 216
   Discontinuous Valley Settlement, 219
   Compact Settlement, 220
   Large Tracts, 221
8. The French Villages, 225
   Old Ste. Genevieve, 225
   Colluvial Slopes, 234
   New Ste. Genevieve, 235
   New Bourbon, 244
   La Saline, 255
   Dispersed Settlements, 264
9. A Traditional French Village in the Hills, 267
   St. Michel, 267

10. The Mining Country, 284
    Mine à Breton, 284
    Potosi, 295
    Old Mines, 303
    Herculaneum, 320
    Big River Bend, 327
    Farmington Plain, 336

11. Valley Settlements in the Hills, 340
    Plattin Valley, 340
    Lower St. Francis River Valley, 345

12. Private Settlement Colony, 351

13. An Exclusive American Community, 359
    The Bellevue Valley and Caledonia, 359

14. Perry County: A Settlement Mosaic, 368
    Shawnee-Delaware Indian Grant, 371
    Bois Brule, 382
    Fenwick Settlement, 383
    The Barrens and Perryville, 388
    Brazeau and Abernathy Settlements, 396

Part IV *Organizing Space*

15. "Connecting Tissues," 401
    The Administrative Perspective, 402
    A Restless Population, 412
    Pathways of Integration, 415
    The Economic Perspective, 429

Conclusion

16. The Ste. Genevieve District as a Settlement Frontier, 447
    Cultural Diversity on the Frontier, 447
    The Environmental Encounter, 454

Administrative Settlement Policy and
   Traditional Practices, 457
Settlement Pattern, 459
Settlement Systems, 461

Appendices

A. Spanish Governors-General of Louisiana, 467

B. Lieutenant Governors of Upper Louisiana during the Spanish Period (in St. Louis), 469

C. U.S. Government Officials Involved in the Land-Confirmation Process in Missouri, 471

Bibliography, 473

Index, 521

Illustrations

1. France in the Interior of North America before 1763 5
2. Ste. Genevieve District in the Spanish Period 8
3. Model of Partial and Complete Flooding of a Mississippi River Bottom 40
4. Natural Regions of the Eastern Ozarks 64
5. Landownership, 1804, Ste. Genevieve District 118
6. François Tayon Survey, February 6, 1804; date of concession given as October 15, 1799 129
7. Solomon Morgan Survey, Bois Brule Bottom 154
8. Israel Dodge Survey, February 27, 1798 158
9. Limit of Settlement of Missouri, Moses Austin, 1806 168
10. Limit of Settlement of Missouri, Gov. Meriwether Lewis, 1808 169
11. Limit of Settlement of Missouri, Amos Stoddard, 1807 170
12. Limit of Settlement of Missouri, William Russell, 1814 172
13. Public Domain Offered for Sale, 1818–1851 178
14. Landownership in the Historic Ste. Genevieve District, 1830 181
15. Mineral Land Reserved from Sale, 1824 185
16. Types of Settlement Forms in the Ste. Genevieve District 203
17. Villages and Towns of the Ste. Genevieve District, ca. 1823 215
18. Village and Town Establishment, Historic District of Ste. Genevieve 216
19. Plans of Villages and Towns, Ste. Genevieve District 217
20. Possible Sites for the Relocation of Ste. Genevieve, 1785–1795 229
21. Ste. Genevieve 236
22. Ste. Genevieve, Common Fields and Other Tracts 242
23. Connected Plat of New Bourbon 249
24. Henri Peyroux Concession at La Saline, February 22, 1806 260
25. St. Michel Survey by Thomas Maddin, April 27, 1805 274
26. St. Michel Concession, 1799 276
27. Moses Austin Survey, June 2, 1800 290
28. Connected Plat of Mine à Breton 299
29. View of Potosi 301

Illustrations

30. Old Mines Survey, Original Exterior Survey, February 3, 1804 311
31. Old Mines Survey, Interior Survey, December 20–22, 1805 312
32. Herculaneum, Platted 1808 by Moses Austin 322
33. View of Herculaneum, Sketched by Charles LeSueur, April 9, 1826 324
34. Farmington Plain, Ste. Genevieve District 328
35. Lands Owned for Lead-Mining Purposes, 1818, Big River Bend, Ste. Genevieve District 335
36. Plat of Maxwell Grant on the Black River, Surveyed February 2, 1816 354
37. Maxwell Grant on Black River, 1799, Ste. Genevieve District 355
38. Perry County Settlement Areas, Ste. Genevieve District 369
39. Middle Mississippi Valley Settlement, 1770 404
40. Part of "Lewis and Clark Map of 1804" 405
41. Ste. Genevieve District Settlement, 1804 406
42. Ste. Genevieve District with New Bourbon District and Cantons, 1803 407
43. 1804 Population, Ste. Genevieve District 408
44. 1820 Population, Missouri 410
45. Historic Ste. Genevieve District, Counties and Townships, State of Missouri, December 19, 1821 411
46. Land Routes, French and Spanish Periods 420
47. Land Routes, American Period, 1804–1820s 424
48. Zones of Economic Development, ca. 1820 436
49. Missouri Post Office Receipts, 1828 441

Tables

1. Population, Ste. Genevieve District, 1769–1830 15
2. Population, Upper Louisiana, 1799–1804 16
3. Classification of Settlement Types by Neighborhoods, Historic Ste. Genevieve District, Early Nineteenth Century 204
4. Population by Neighborhoods, Ste. Genevieve District, 1804 205
5. Village and Town Population, Ste. Genevieve District, 1769–1830 206

Foreword

Terry G. Jordan

Eight decades ago, in an issue of *National Geographic,* Frederick Simpich described Missouri as the "mother of the West." The truth of his assertion remains evident today in the dominantly heartland cultural character of the far greater part of the American West. His is an assertion I personally came to accept after a decade of research on the origins of the cattle-ranching business, a project that came into print as *North American Cattle Ranching Frontiers.* For me, a sixth-generation Texan, to discover that Missouri exerted a greater influence than Texas upon the West in general and ranching in particular ranks as a major-league epiphany. Missouri served as the migratory funnel through which heartland pioneers from the Old Northwest and upland South passed west, often after a residential stopover. The Oregon, Mormon, California, and Santa Fe Trails all had their roots in Missouri. Lesser-known but equally well-beaten paths led from Missouri to Texas. The American West is a giant triangle with apexes in California, the Pacific Northwest, and Missouri.

If Missouri mothered the West, then the Ste. Genevieve District was parent to Missouri. What Walter Schroeder has to tell us about this small district, then, has profound relevance not just for the Ozarks, but for half a continent as well. A cultural prototype formed in the Ste. Genevieve area, made from diverse national and regional raw materials. Even my Texas would feel its influence, conveyed by the Austins, DeWitts, and other Anglo-American founding families.

I, like Walter Schroeder, am a geographer. In our academic discipline, we tend to view matters such as sectionalism rather differently than historians do, though in a complementary way. Walter Schroeder's geographical approach is revealed on almost every page of this wonderful and important book. Geographers emphasize the character of the natural habitat and the manner in which people adapt to and change the land, in the process creating distinctive ways of life and unique humanized landscapes. The explanation for a way of life also leads the geographer to consider cultural diffusion—the spread of people, ideas, and traditions across geographical space. When these several strands—habitat, adaptation, landscape, diffusion, and way of life—are drawn together, the result is an explanatory portrait of a region. This is precisely what Walter Schroeder has so masterfully accomplished for the Ste. Genevieve District, an achievement hugely magnified by the pivotal role this corner of Missouri played in American westward expansion.

Opening the Ozarks, in my estimation, is one among perhaps one hundred books that can be called essential to our understanding of the Eur-american settlement frontier. Pay close attention, dear reader. This book possesses far greater importance than its modest title suggests.

Preface

The inspirational text for my book is taken from the first seven paragraphs in Clarence J. Glacken's introduction to *Traces on the Rhodian Shore*. In that passage of secular scripture for a geographer, we read that Western civilization has persistently posed three questions concerning the relationship of humans and the earth: Is there purposeful design and order? How has the earth affected humans as individuals and as cultures? How have humans changed the earth from its hypothetical pristine condition? My book entertains all three, but the first receives central attention.

Opening the Ozarks is a search for order in the settlement of the Ste. Genevieve District, a portion of the eastern Ozarks of Missouri. The processes of invading new lands, acquiring and occupying property, and creating landscapes of communities in all their comprehensiveness and complexities must have purpose and order behind them, if we reject, as Western civilization has, a hypothesis of randomness. It is my mission to explain how purposeful order came about.

Acknowledgments

In the spring of 1980, an interdisciplinary group at the University of Missouri–Columbia assembled under Susan Flader's leadership to study the community of Ste. Genevieve, Missouri, from its founding to the first decades of the nineteenth century. Members included Flader and Susan Boyle of history, Osmund Overby of art history and archaeology, Chad McDaniel of anthropology, myself of geography, and several research assistants, including David Denman and Anna Price. The Ste. Genevieve Project, as it was called, received two consecutive grants from the National Endowment for the Humanities, 1981–1985, with Flader as principal investigator, which provided support to accumulate, organize, translate, and computerize documents from the Ste. Genevieve Archives, a microfilm collection at the Western Historical Manuscripts Collection at the University of Missouri–Columbia, and other repositories. Later, with support from other funding sources, selected records from the Spanish Archivo General de Indias were added. The several participants, individually and as a group, presented papers on their work at professional meetings and at special seminars in the town of Ste. Genevieve, most of the latter arranged by the Foundation for the Restoration of Ste. Genevieve. Several graduate theses and publications resulted from the sustained, scholarly focus on Ste. Genevieve.

I am deeply indebted to all the above persons and organizations for their ideas and support during the life of the Ste. Genevieve community study and afterward. Susan Flader's continuing encouragement for me to produce a full-length monograph in my particular area of interest, Ste. Genevieve's relationship with its larger district, is, to a large extent, the reason that this effort finally found closure. I am deeply grateful for her support and excellent counsel over this extended period of time. I also thank her more specifically for her careful reading of the manuscript as a historian and for suggesting numerous changes in the organization and detail. I thank the others who read the manuscript and made valuable comments from their special areas of interest: John Bullion, Richard Bienvenu, John Wigger, and Osmund Overby of the University of Missouri–Columbia; William E. Foley of Central Missouri State University; and Terry G. Jordan of the University of Texas at Austin.

Over the years, many residents in the historic Ste. Genevieve District have helped in various ways. Among them are townspeople in Ste. Genevieve, Fredericktown, Farmington, St. Mary, the Bellevue Valley, Potosi, Herculaneum, and Perryville. The hospitable residents of the dispersed community of Old Mines, most of them members of the Old Mines Area Historical Society, deserve special thanks for making each visit to Les Vieilles Mines a grand learning experience and uniquely memorable. County assessors and clerks in Washington, Madison, and Ste. Genevieve Counties dug deep into big, old books to answer questions of early-nineteenth-century landownership.

Librarians were the key to discovering documents to supplement those amassed by the Ste. Genevieve Project. In the State Historical Society of Missouri at Columbia, Laurel Boeckman, Peggy Platner, Marie Concannon, Linda Brown-Kubisch, Fae Sotham, and Elizabeth Bailey graciously satisfied my frequent requests for help. Before them, it was Goldena Howard, also my neighbor, who introduced me to the society's storehouse of materials on Missouri history. James W. Goodrich, director of the society, took interest in my research and made sure that the resources of the society were at my disposal. In the university's Ellis Library, Sally Schilling, Esther Fetterhoff, and Geoffrey Swindells never failed to locate other early-nineteenth-century government documents.

In the Missouri State Archives in Jefferson City, Patsy Luebbert and Gary Beahan helped me from the start. It is unthinkable to explore seriously the complex land issues of territorial and early-statehood Missouri without introduction by Luebbert to the great variety of pertinent materials in the state archives. It was my good fortune to be searching there when many documents important to this study, especially the beautiful Soulard land surveys, were being cataloged and organized for microfilming. Kenneth Winn, director of the Missouri State Archives, and Lynn Morrow receive my thanks for making sure that my days in Jefferson City were pleasant and productive. In a similar way, Michael Flowers and Norman Brown in the Division of Geology and Land Survey (at Rolla) of the Missouri Department of Natural Resources enabled my work with public and private land surveys to be productive.

Documents of all sorts about Ste. Genevieve and its district found their way over the years to St. Louis and ultimately into the collections of the Missouri Historical Society. Librarians and scholars at the society helped expedite finding and copying various materials relating to Ste. Genevieve.

Since cartography has become technologically based, I have turned to geographic information specialists for producing the numerous maps and diagrams without which a spatially oriented study would be incomplete. My colleague James D. Harlan, senior research specialist in the Geographic Resources Center of the Department of Geography, deserves my and the readers' thanks for his outstanding maps of high resolution that involved manipulating various layers of information. Adrianne Nold Black and Jeffrey Thomas constructed most of the other maps. Susan Ferber helped with graphic materials, composed the tables, and otherwise advised in matters of production.

Jesse H. Wheeler Jr., my former colleague and department chair, took great interest in this research in its earlier stages, and I recognize his encouragement posthumously. Kit Salter, my colleague and present chair, has supported my work in all ways possible to him, and I deeply appreciate his unflagging support. Mary Wilson did much of the typing of the early manuscript, and she merits my huge thanks for such a splendid job, done so pleasantly. The College of Arts and Science provided financial support for cartography. Annette Wenda skillfully edited the lengthy manuscript, and readers will be as grateful as I am for her excellent work. Jane Lago and Clair Willcox at the University of Missouri Press transformed words, maps, and tables into final book form.

The nearly two decades of off-and-on work on this research involved my family. Paul and Julie grew up with "Ste. Genevieve" as if she were an invisible member of the family. During the final stages of this work, Julie helped me in ways too numerous to mention. Pat needs no formal expression of thanks here for enduring my years of thinking about Ste. Genevieve, but she will get it: "Thanks, Pat!"

Opening the Ozarks

Introduction

1

The Ste. Genevieve District

Into this wonderful land, at the base of the Ozark Mountains . . .
— Longfellow, *Evangeline*

Potosi, Thursday, 5th Nov. 1818
I begin my tour where other travellers have ended theirs, on the confines of the wilderness, and at the last village of white inhabitants, between the Mississippi river and the Pacific Ocean.
— Henry Rowe Schoolcraft,
Journal of a Tour into the Interior of Missouri . . .

Evangeline's search for her beloved Gabriel led her north from New Orleans into "this wonderful land, at the base of the Ozark Mountains," where Gabriel was reportedly living among other French Canadians. Longfellow set his fictional account of exiled Acadians in the years around 1760 when Canadians were, in fact, establishing the village of Ste. Genevieve at the eastern base of the Ozarks. There they intended to establish a traditional French society and cultural landscape.

In 1818, more than a half century later, American Henry Rowe Schoolcraft stood at Potosi, the new American center of lead mining in the eastern Ozarks, and fancied himself standing on the brink of civilization as he prepared to leave on an expedition to explore the wilderness that stretched to the Pacific Ocean. During that half century, the young United States had annexed and Americanized the eastern Ozarks. Settlement had proceeded from French Ste. Genevieve westward into the interior as far as American Potosi, barely sixty miles from the Mississippi River, where, according to Schoolcraft, it stopped, leaving the wilderness beyond.

This book is an account of how the people who entered the Ozarks from the 1760s to the 1820s acquired and occupied land. In an agricultural society, supplemented by the extractive activities of mining and hunting, land is wealth; control of land is necessary for economic gain and the creation of viable communities.

When people entered the eastern Ozarks—a term, incidentally, not in use for this region until the mid-nineteenth century—they had a multitude of choices of how and where to select land, how to use it, and whether to stay or move on. Consequently, an account of initial land occupance asks such questions as: Who were the people who settled the land, and why did they come to it? What was the role of government in directing people onto the land and managing them after their settlement? How did government policies interfere with the vernacular practices of acquiring and using land? How did the transfer of government from Spain to the United States affect the process of settling the land? What kinds of land did the new settlers look for, and where did they find them? What was the role of extended families in the geography of settlement? How attached were settlers to the lands they settled and made productive? How were their neighborhoods and communities formed, and how stable were they? How did the network of routes develop that linked the communities with each other and with the rest of the world? Why did this district, the first one well settled in Missouri, fade in relative importance after 1820?

This account begins in the 1760s, more than a half century after the initial emplacement of French villages in the middle Mississippi Valley. Therefore, the focus is not on the arrival and establishment of the French Canadian presence in the valley, but on the spread of settlement into the historic Ste. Genevieve District of the eastern Ozarks.

The Place

The Ste. Genevieve District was one small piece of the vast arc of French-claimed territory from New Orleans to the Saint Lawrence, much of which lies in the drainage basin of the Mississippi River (fig. 1). This entire basin, the extent of which was more hypothesized than known, was claimed for France and named Louisiane in 1682 by Robert Cavelier, sieur de La Salle, upon his descent of the Mississippi River. The terms *Haute Louisiane* and *Basse Louisiane* date at least from the 1730s. The boundary between the two divisions shifted from Natchez northward to the Ohio River, then in 1799 back southward far enough to place New Madrid within Upper Louisiana. Upper Louisiana, as originally applied by the French, included the entire Ohio River basin, but after the 1760s its usage was restricted to the Mississippi River drainage basin west of the

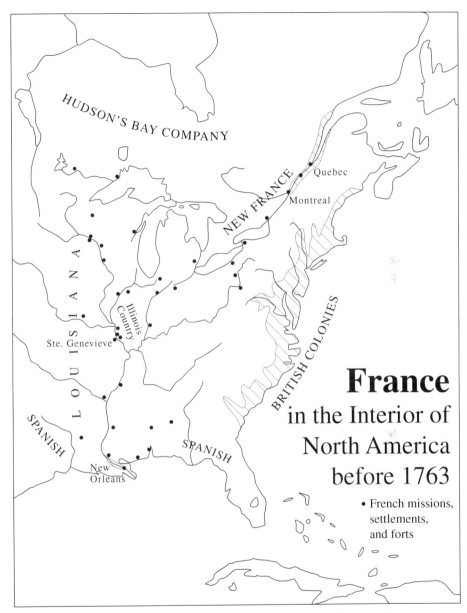

1. *France in the Interior of North America before 1763.* The French presence in the Illinois Country and Upper Louisiana consisted only of scattered settlements along waterways in the midst of a vast expanse of Indian territory. (Based on Map U-11D in *United States History Atlas* [Maplewood, N.J.: Hammond, 1997])

river. Louisiana continued in official usage throughout the Spanish regime as Luisiana and gave its name to the American purchase of the territory in 1803.[1]

The term *pays des Illinois* (Illinois Country) had been in use since the French explored the lands inhabited by the Illinois group of Indians in the seventeenth century. Illinois had a variety of Spanish spellings (Ilineses, Ylinese, Ylinoa, and so on) until stabilized as Illinois around 1800 and accepted by English-speaking Americans. The limits of the Illinois Country are imprecise, because the various Illinois tribes moved widely during the eighteenth century after their defeat by Indians to their north. The Illinois Country, however, always included both sides of the middle Mississippi River from the Missouri River almost to the Ohio River. Officials designated the Illinois Country together with the Wabash Valley as one of the nine administrative subdivisions of the Province of Louisiana very early, in the 1720s, with its own commandant and judge. With the transfer of the east bank to Britain and the west bank to Spain in 1770, the Mississippi River became an international boundary, and the western side of the Mississippi Valley became known as the Spanish Illinois and the eastern side as the British or, later, American Illinois, or simply as Illinois. The alluvial land on the eastern side of the river in the Illinois Country is still known as the "American Bottoms."[2]

Thus the Illinois Country, a term referring to Indian space, lay within the larger Upper Louisiana, a term referring to hydrographic space, as long as both terms included land on both sides of the Mississippi River. Illinois Country was a geographic designation made from Canada. Upper Louisiana was a geographic designation originating from New Orleans.[3] After the American Revolution, Upper Louisiana dropped out of usage for the eastern side of the valley, which then became the Northwest Territory of the United States.

The Spanish administration used the terms *Upper Louisiana (Alta Luisiana)* and *Illinois Country (Ylinoa)* interchangeably for the European-settled region on its side of the valley from the mouth of the Missouri River as far south as the Ohio River junction. Both St. Louis and Ste. Genevieve carried the geographic identifications "of the Illinois Country" and "in Upper Louisiana." Lieutenant Governor Trudeau's official report

---

1. Antoine Le Page du Pratz, *Histoire de la Louisiane*, 1:162–65.
2. François-Xavier Martin, *The History of Louisiana*, 148; Clarence W. Alvord, *The Illinois Country, 1673–1818*, 150–51, 169–70.
3. Charles J. Balesi speaks of the "transformation" of the Illinois Country to Upper Louisiana when political control shifted to New Orleans (*Time of the French in the Heart of North America*, 246).

to his Spanish superiors in New Orleans, dated January 15, 1798, defines the terms: "The settlements of His Catholic Majesty in Ylinoa form part of Upper Luisiana and are located on the western bank of the Misisipy River . . . between the two confluents which the above river forms with the Ohio and the Misoury." In the report he used the term *American Ylinoa* for the eastern bank. He signed his report from "San Luis de Ylinoa."[4]

Although *Illinois* and *Louisiana* were interchangeable terms in the European-settled area, *Spanish Illinois* was totally inappropriate for referring to the rest of Upper Louisiana. Upper Louisiana extended far beyond Spanish Illinois, far into the Indian-occupied and -controlled Missouri and upper Mississippi River regions, the western and northern limits of which no one knew. Maps presumptuously showed the whole drainage basin as Spanish territory, although Spain and predecessor France exerted absolutely no control over it and usually ventured into it at peril. It was blatant European imperial arrogance displayed cartographically. If maps showed actual conditions, the geopolitical area of Spanish and French in Upper Louisiana would be restricted to narrow belts, if not mere clusters, along the Mississippi and Missouri Rivers and not much farther upstream than St. Louis. During the Spanish administration, *Missouri,* as a regional term, referred to lands along the Missouri River only, as "in the Missouri country."

After 1803, the use of Spanish Illinois and Upper Louisiana for the western side of the river abruptly ceased, and that side, including its constituent Ste. Genevieve District, became successively the Territory of Indiana (1804), the Territory of Louisiana (1804), the Territory of Missouri (1812), and then the State of Missouri from 1821 on.

The Ste. Genevieve District was an administrative subdivision created by the Spanish shortly after their arrival in Upper Louisiana in 1770 (fig. 2). The district focused on the village of Ste. Genevieve, its "post" where the district commandant lived. Its original boundaries passed through unoccupied land: on the north, the Meramec River was a demarcator visible in the landscape and clearly evident on maps; on the south, Apple Creek served the same purpose for the same reason. Because colonial authorities never set a western limit to the Ste. Genevieve District, real administrative authority went as far as French and American settlement went.

The Ste. Genevieve District continued into the American era as an ad-

---

4. "Trudeau's Report Concerning the Settlements of the Spanish Illinois Country—1798," in *The Spanish Regime in Missouri,* ed. Louis Houck, 2:247–58. Usage of *Illinois* and *Upper Louisiana,* in their various languages and spellings, can be traced in the documents in Houck's *Spanish Regime.*

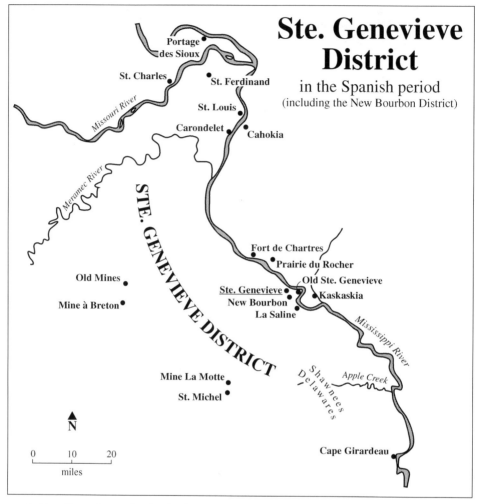

2. *Ste. Genevieve District in the Spanish Period.* The district lay along the west side of the Mississippi River, south of St. Louis, between the Meramec River and Apple Creek. Ste. Genevieve village was its administrative center.

ministrative unit. Renamed a county, it was diminished to its present extent by 1821 through the creation of new counties. The term *historic Ste. Genevieve District* refers to the original area created by the Spanish, disregarding subsequent detachments from it.

During French and Spanish regimes, geographic linkages—administrative, economic, religious, familial—followed the continental arc of French settlement, either southward to New Orleans or northeastward to the Saint Lawrence. In the 1790s, however, linkages began to be reoriented eastward through the Ohio Valley and then accelerated after the American acquisition of Louisiana. The reorientation was not complete by any means, especially not economically, as some middle Mississippi Valley products still took the line of least effort down the river to New Orleans.

The significance of the valley's location changed when its political control changed. Before 1804, the middle Mississippi Valley was a colonial outpost at the far ends of empires. France, Spain, and Britain vied in the peripheral continental interior and sought accommodation with resident Indians. European rulers generally neglected its development or, at best, treated it as a buffer to protect colonies elsewhere or as a pawn to be used in larger imperial strategies. Despite pronouncements from the very highest levels that the valley would be developed for its grains, lead, and furs, it received only desultory high-level action to go with the words of intention.

But the Americans considered the valley, certainly the Ste. Genevieve District because of its lead, an integral part of their continuously expanding national domain, land that the United States was destined by nature to effectively control and develop. Indians need not be accommodated; they could be removed. The valley lay contiguous to the expanding national core, not separated by an ocean. With such a mission and energetic commitment of national resources (roads, postal service, administrative and military services, and serious attention to Indian affairs), and with the seemingly inexhaustible demographic power of American immigrants to carry it out, Americans would convert the region into part of their geopolitical state. This was not just a hope, an expression of words; Americans firmly believed in it and brought it about.[5]

The conception of the valley also changed after the purchase in yet another fundamental way. European colonials thought geographically in lines. The Mississippi River, perceived as a north-south line, was their

5. The Illinois Country is analyzed by the geographic concept of core and periphery in Donald W. Meinig, *The Shaping of America: A Geographical Perspective on 500 Years of History*, 1:193–202.

baseline, their line of reference, and its tributaries were east-west lines. All properties, except mining concessions and cattle ranches, had a "front" to them. Lands "fronted" on the Mississippi River or other lines in the landscape. The implication is that settlers would build their houses on the fronts of the attenuated properties. It was a linear, riverine conception of spatial organization. Neither the French nor their Spanish administrators ever thought that they would occupy the total area of the landscape. Theirs was a linear world.[6]

In contrast, Americans thought in areas. Their division of space, and how it was to be settled and organized, was in squares and fat rectangles. No one side of a property was intended to be more important than other sides. The area each family claimed extended in all directions from the farmstead. Their system of land division partitioned off all space and foresaw the day when the total land surface would be owned and occupied, not just stream-border tracts with houses and villages in lines. Americans had a land-survey system that tied everything together into a coherent whole in contrast to the French-Spanish system of separately located properties, each with a plat but without overall coordination. The French and Spanish had never made a map showing how their hundreds of grants or properties fit together spatially, but the Americans began the official settlement process with a coordinated, detailed map of the entire area.[7]

The middle Mississippi Valley, by virtue of being in the center of the vast network of the Mississippi River system, held great potential to be a center of economic activity. Noted very early, its centrality continued as a geographic advantage until the railroad era rendered it partially obsolescent. The Ste. Genevieve District lay along the most critical segment of the hydrologic system, the stretch between the mouths of the Missouri and Ohio Rivers. People and products funneled into this stretch from the west, north, and east, including the Great Lakes via easy portages. The Ohio connected to the east, and its role increased in time, moving products in both directions and, extremely important, concentrating thousands of westering settlers into that stretch. Thus, in national terms, the middle Mississippi Valley between the mouths of these major tributaries

---

6. Martin, *The History of Louisiana,* 152. D. Aidan McQuillan notes that the French were unable or unwilling to move into eastern (British) townships of Quebec because of an alien system of land subdivision (*Prevailing over Time: Ethnic Adjustment on the Kansas Prairies, 1875–1925,* 31, 37).

7. A full treatment of the differences among colonial land-division systems is in Edward T. Price, *Dividing the Land: Early American Beginnings of Our Private Property Mosaic.*

functioned historically as a settlement and transportation focus, a function that culminated in the rise of St. Louis at its northern end as the dominant city of the pre-railroad American interior.[8]

All the while, Ste. Genevieve lay astride this water highway that connected it to all major parts of the French and Anglo-American settled continent, and even to the Caribbean and Europe. Though distant from imperial and national capitals, Ste. Genevieve's location was far from isolated, for it witnessed currents of people and products for more than a century.

The Time

Having claimed the Mississippi basin as a French possession by virtue of exploration, the French entered the middle Mississippi Valley from their base in Canada for trade and settlement in the first years of the eighteenth century. At the time, Indians occupied the valley with varying degrees of control. Illinois tribes, such as the Peoria, Weas, Kaskaskias, Mitchigamis, and Cahokias, lived and operated primarily on the eastern side of the river, and the Canadians established their first trading posts, missions, and villages among them.[9] By midcentury, French Canadians had villages at Cahokia, St. Philippe–Fort de Chartres, Prairie du Rocher, and Kaskaskia, the last named serving as the principal economic and administrative center and Fort de Chartres as the military center for the Illinois Country settlements.

Farther west of the Mississippi River were the Missouris along the river bearing their name, the Osages in the Ozarks south of the Missouri River, and the Chickasaw south of them. The eastern Ozarks held few, if any, permanent Indian villages, but Osage parties visited the region frequently enough for it to be considered within their domain. Illinois tribes and the French shared land, resources, and marriage partners, but friction between the Osages and the French continued throughout the century.[10]

---

8. For an account of the geographic factors of centrality and the hydrologic network in the rise of St. Louis, see Ellen Churchill Semple, "Geographic Influences in the Development of St. Louis."

9. For Illinois Indians, see Wayne C. Temple, *Indian Villages of the Illinois Country* and the accompanying *Atlas* by Sara J. Tucker; and James Scott, *The Illinois Nation*. For French interaction with the Illinois, see Joseph Zitomersky, *French Americans–Native Americans in Eighteenth-Century French Colonial Louisiana* . . . ; and Natalia M. Belting, *Kaskaskia under the French Regime*.

10. For Osage Indians, see John J. Mathews, *The Osages: Children of the Middle Waters;* Willard H. Rollings, *The Osage: An Ethnohistorical Study of Hegemony on the Prairie-Plains;* Francis La Flesche, "The Osage Tribe"; and Carl H. Chapman and Eleanor F. Chapman, *Indians and Archaeology of Missouri,* 91–111.

French miners entered the eastern Ozarks around 1720, anticipating finding silver; they found lead instead. Coming upriver from the south, these French introduced black slaves into the Illinois Country, and Canadian French already resident apparently accepted black slavery without much question.[11] Indeed, Indian slaves were already present among the Indians and Canadian French in the Illinois Country. The first titles to land in the eastern Ozarks were awarded in the 1720s for mining, but the mining camps did not evolve into permanent settlements. Villagers on the eastern side of the Mississippi directed the mining operations.[12]

In the late 1740s—the year has never been established, if indeed one year can serve such a designation—Kaskaskians of Canadian and Indian ancestry began to cultivate land and live on it across the Mississippi River in the bottomland called the Pointe Basse. This settlement, founded spontaneously and informally, became known in the 1750s as Ste. Genevieve and is the first permanent European settlement within the present state of Missouri. Ste. Genevieve soon took on a life of its own with a church and resident priest. In 1763–1764, Pierre de Laclède and Auguste Chouteau, who came north from New Orleans, formally founded St. Louis as a trading post, with agriculture relegated to a support function.[13]

With the Treaty of Paris of 1763, which closed the French and Indian War, France surrendered control of the entire Mississippi drainage basin.

---

11. The French, Spanish, and Americans shied away from using the words *esclave, esclavo,* and *slave,* except in official documents. The French community used the terms *nègre* and *noir,* and Americans used *negro, black,* or *colored.* The Catholic church in the Ste. Genevieve District, which owned slaves, referred to them as servants. Mulattos, those with partial African ancestry, were sometimes separately recognized, but usually they are classified with Africans, especially by Americans. *Métis,* those of mixed Indian and French ancestry, are not generally identified in the middle Mississippi Valley by that term. If they lived in the French communities, they were apparently considered French.

12. Martin, *The History of Louisiana,* 117, 130.

13. On the founding of Ste. Genevieve, see Carl J. Ekberg, *Colonial Ste. Genevieve: An Adventure on the Mississippi Frontier,* 2–47; and Charles E. Peterson, "Early Ste. Genevieve and Its Architecture." Much has been written on the founding of St. Louis. See the cofounder's account: Auguste Chouteau, "Narrative of the Settlement of St. Louis." Other standard sources include Charles E. Peterson, *Colonial St. Louis: Building a Creole Capital;* James Neal Primm, *Lion of the Valley: St. Louis, Missouri,* 9–38; William E. Foley and C. David Rice, *The First Chouteaus: River Barons of Early St. Louis,* 1–35; J. Thomas Scharf, *History of Saint Louis City and County, from the earliest periods to the present day: including biographical sketches of representative men,* 1:62–78; John Francis McDermott, "Myths and Realities Concerning the Founding of St. Louis"; and Charles van Ravenswaay, *St. Louis: An Informal History of the City and Its People, 1764–1865,* 10–47.

It turned over control of the eastern part and Canada to the United Kingdom, and the western part, by virtue of the Treaty of Fontainebleau of the preceding year, to Spain. Though now an international boundary, the Mississippi River still unified the Indian and French-settled lands bordering it. International power politics played out in Europe violated human geography in America, but, in the time of empires, no one thought it inappropriate if boundaries divided communities and one nation governed people of another nation and ethnicity.

Though the treaty is dated 1763, it took some time before representatives of the new owners could reach the middle valley. The British occupied the villages on the eastern side in 1765, but the Spanish did not occupy St. Louis until 1770. During these interim years of anticipation of new local authorities and then for some years afterward, French Creoles, most of whom were in villages on the eastern side, had to decide whether their lot would be happier under the Protestant and enemy British or under the Catholic Spanish, whose king was blood-related to their French king. Many, certainly a majority as the years unfolded, chose the Spanish alternative, and the British east-side villages lost population as Spanish Ste. Genevieve and St. Louis gained. Thus, the basic "French" stock of the Ste. Genevieve District derived mainly from families of French Canadian origin, many intermarried with Illinois, that had been in the valley and in America for some time. They were rightly Creoles, and they called and considered themselves Creole or French. They did not usually refer to themselves as Canadians.[14]

14. The term *French* refers broadly to the entire French-culture group, easily identified by language, although it excludes French-speaking black slaves. *French* is the standard term for dichotomizing between French and Americans in the New World (Meinig, *Shaping of America*, 2:18). The word *Creole* is customarily used for those born in the subtropical and tropical portions of the New World of French or Spanish ancestry. It includes those of French ancestry with mixed African and Indian blood. The term came up the Mississippi Valley in the middle of the eighteenth century from Lower Louisiana and was applied to the middle Mississippi Valley French, the majority of whom were of Canadian origin. *Creole,* however, is not used for French of Canadian origin who remained in Canada. One French visitor in the late eighteenth century implied that *Creole* carried a negative connotation when he said that French Canadian settlers in the valley were changed into Creoles by isolation, solitude, hunting, and contact with Indians (Nicolas de Finiels, *An Account of Upper Louisiana by Nicolas de Finiels,* 107–18). By this reasoning, Creoles are culturally derivative from French immigrants to North America in the same way that Americans are culturally derivative from British immigrants. Though most of the valley French had Canadian ancestry, they no longer referred to themselves as Canadians by the last half of the nineteenth century (John Francis McDermott, *A Glossary of Mississippi Valley French, 1673–1850,* 60–62; Ward A. Dorrance, *The Survival of French in the Old District of Sainte Genevieve,* 5; Joseph Tregle Jr., "On That Word 'Creole' Again: A Note").

Spanish administrators did not interfere with local French affairs and concentrated on external affairs, especially defense, the fur trade, and Indian affairs. Spanish policy restricted immigration to select groups, and only a few immigrants came. This policy preserved French culture and kept it from appreciable dilution or modification from others, with the important exception of Africans and Indians.[15]

The village of Ste. Genevieve grew from twenty-three persons in 1752 to six hundred in 1769, most of the increase coming in the years immediately after the arrival of the British at Kaskaskia. Afterward, population grew more slowly and through the early 1780s came largely from natural growth. Slaves accounted for approximately 40 percent of the population of six or seven hundred (tables 1 and 2).[16]

This stable, if not stagnant, settlement geography of more than thirty years abruptly changed in the late 1780s after unabated riverbank erosion and a series of floods, climaxing in the one of 1785, memorialized as *l'année des grandes eaux*. Residents of the village of Ste. Genevieve, up to then located in the Pointe Basse along the Mississippi River, abandoned their houses progressively over a period of years until nothing was left. Such freeing of people from their attachment to a place enabled them to make choices of where to go and which lands to use for farms and resources. The new village of Ste. Genevieve, nestled in a break in low bluffs, was the primary choice for relocation, but others chose to move elsewhere in the uplands. New finds of lead ore in the interior and a (falsely) perceived lessening of Osage dangers supported the dispersal.

Two other events in the early 1790s influenced these initial years of settlement dispersal. One was the arrival of Pierre-Charles Delassus de Luzières, a nobleman refugee from the French Revolution, who received his own administrative district, the New Bourbon District, which was carved out of the Ste. Genevieve District. The other event was the invitation and encouragement of the Shawnee and Delaware to settle in the southern edges of the district.

By far the most radical change in the settlement geography of the Ste. Genevieve District came in the mid-1790s when the Spanish government

15. Multicultural relations (French-Indian-African) in the middle Mississippi Valley are addressed in Belting, *Kaskaskia under the French;* Zitomersky, *French Americans–Native Americans;* Ekberg, *Colonial Ste. Genevieve,* 86–124, 197–239; Balesi, *Time of the French,* 232–53; Tanis C. Thorne, *The Many Hands of My Relations: French and Indians on the Lower Missouri;* Arvarh E. Strickland, "Aspects of Slavery in Missouri, 1821"; and John Mack Faragher, "'More Motley than Mackinaw': From Ethnic Mixing to Ethnic Cleansing on the Frontier of the Lower Missouri, 1783–1833."

16. On the growth of Ste. Genevieve, see Ekberg, *Colonial Ste. Genevieve;* and Lucille Basler, *The District of Ste. Genevieve, 1725–1980.*

Table 1

POPULATION
STE. GENEVIEVE DISTRICT, 1769–1830

		Total Population[a]	Ste. Genevieve		Year	White Male	White Female	Mulatto, Black		Indian[b]
1769	Spanish Censuses	600	600							
1772		691	691		1772	264	140	287		
1773		676	676		1773	251	149	276		
1785		594	594							50[c]
1787		657	657		1787	259	132	266		
1788		896	896	New Bourbon						
1791		973	973		1791	388	241	344		
1795		1,155	1,002	153	1795	412	275	315		
1796		1,539	1,156	383	1796	499	289	368		1,800[d]
1799		2,069	1,509	560				428		
1800		2,421	1,792	629	1800	1,077	767	577		1,500
1804 (Stoddard)		2,870	2,870							
1804 (Schroeder)[e]		3,610	3,610							1,540[f]

	Total Population[a]	Ste. Genevieve	Washington	Madison[g]	Jefferson[h]	Wayne[i]	Perry	St. Francois	Indian[b]
1810 U.S. Census	4,620	4,620							
1814 Estimate[j]	6,726	4,217	2,509						
1817 Estimate[j]	7,947	2,750	1,709	1,248	1,208	1,034			(2,400)[k]
1820 U.S. Census	13,056	4,962	2,769	2,047	1,835	1,443			
1822 Mo. Census[l]	14,883	3,181	3,744	1,907	1,838	1,614	1,599	(1,000)[m]	
1824 Mo. Census[l]	16,947	2,025	4,803	1,647	1,934	2,527	2,340	1,671	
1828 Mo. Census[l]	20,366	1,705	6,236	2,276	2,367	3,009	2,743	2,030	
1830 U.S. Census	22,912	2,186	6,784	2,371	2,592	3,264	3,349	2,366	few

Note: Figures are not strictly comparable because the areas of districts and counties changed.

a. Totals do not include Indians in the final column.
b. Not included in totals, because not counted in censuses. Figures do not include any transient or "roaming" Indians.
c. Chiefly Peorias living in a village near Ste. Genevieve.
d. Shawnees and Delawares; includes a small number in swamps south of Cape Girardeau.
e. See footnote 4, page 403 for this calculation.
f. Shawnees and Delawares; includes Shawnees on the Meramec River.
g. All of Madison County is included in this table, although small parts had earlier been transferred to the historic Cape Girardeau District.
h. All of Jefferson County is included in this table, although parts had earlier been transferred to the historic St. Louis District.
i. All of Wayne County is included in this table, although all of the county had earlier been transferred to the historic Cape Girardeau District.
j. Estimates based on percentage of population that was white male in the censuses of 1814 and 1817 for the territorial legislature.
k. Includes Indians in the Mississippi swamps south of Cape Girardeau.
l. Counts made in Missouri for reapportionment of the state legislature.
m. Population figure missing in the census; figure roughly estimated so that a district total may be determined.

Table 2

**POPULATION
UPPER LOUISIANA, 1799–1804**

Total Population

	Total	St. Charles	St. Louis	Ste. Genevieve	Cape Girardeau	New Madrid
1799 Spanish Census	6,028	895	2,272	1,509	521	831
1800 Spanish Census	7,215	1,110	2,467	1,792	740	1,106
1804 Stoddard Estimate	10,340	1,550	2,780	2,870	1,650	1,500

Percent Increase

	Total	St. Charles	St. Louis	Ste. Genevieve	Cape Girardeau	New Madrid
1799–1800	20	24	9	19	42	33
1800–1804	43	40	13	60	123	36
1800–1804 Annualized	11	10	3	15	31	9

Note: Stoddard derived his 1804 population estimates by extrapolating from the last Spanish complete census of 1800. Though he did not give the rates of increase he used (which varied among the five districts), they may be calculated from his population estimates for 1804.

opened its Illinois Country to immigration from the young United States. Americans entered the district in ever-increasing numbers.[17] Coming predominantly from the upland southern states of Kentucky, Tennessee, and Appalachian Virginia and Carolina, these timber-clearing woodsmen, stockmen, and farmers with slaves scattered into all the valleys of the district. Though businessmen and speculators were drawn more to emerging St. Louis, the mining areas of the Ste. Genevieve District attracted developers, beginning with Moses Austin in the Spanish period. Americans revolutionized the lead industry and added new ingredients of planned towns, roads, and economic networks to the settlement pattern. Few newcomers reinforced the French Creole stock. By 1803, Spanish Illinois was at least one-half American, although the proportion in the

17. "Americans" are those people living in and coming into Upper Louisiana from the United States. That is the way the term was used at the time they came. French in the valley never referred to themselves as Americans before 1804 (Lawrence Kinnaird, "American Penetration into Spanish Louisiana," 211). "American" does not include Canadians, Indians, or Africans. Persons born in England, Ireland, Scotland, or Europe and coming more or less directly to the Mississippi Valley were called English, Irish, Scots, or the like. If they lived long enough in the British colonies or in the United States, they became, of course, Americans.

Ste. Genevieve District was still slightly in favor of the French cohort, including their numerous slaves. Total district population, including slaves but excluding Indians, was approximately thirty-six hundred, of whom fifteen hundred lived in villages and twenty-one hundred in outlying neighborhoods. To this total should be added about fifteen hundred resident Indians.

Retrocession of Louisiana to France by the Treaty of San Ildefonso of 1800 and recurring rumors of American interest in the land across the Mississippi prompted maneuvering in land titles on the part of opportunistic residents who understood what was going on. It also prompted more Americans to enter to get free land before transfer to the United States.

Though the Louisiana Purchase was signed on May 2, 1803, and approved by Congress that October, Americans first established a military presence in Upper Louisiana when Capt. Amos Stoddard took charge in St. Louis on March 10, 1804, from the last Spanish lieutenant governor, Charles de Hault Delassus. During 1800–1803, France, titular administrator, had never sent anyone to take charge in St. Louis from the Spanish. Their absence did not make any difference to the settlement process; the last two lieutenant governors of Spanish Upper Louisiana were of French culture (from Flanders in the Spanish Netherlands), and the local Ste. Genevieve commandant was thoroughly French in culture.

Great changes in the settlement process were expected from the change in government, and great changes occurred. The political process devolved into factions arguing and bickering in public and private over broad policies and petty details. This was a far cry from the authoritarian regimes of France and Spain, which lacked public debate, and from the basic political culture of the French segment of the population, which was not to question authority.

Issues involving land and settlers were in the forefront both of local concern and of national policy. National policy for the new territory was unambiguous. It was to recognize landownership under the prior regime, as agreed to by terms of the Louisiana Purchase, and to facilitate the further settlement of the land. But to carry it out at the local level was another matter. Issues raised during the process to confirm existing land titles and claims seemed endless, complicated by a constant stream of settlers seeking new land but unable to acquire title until those existing claims were settled. In the midst of this near chaos over land, the lead industry was booming, but that only brought on more land issues, as nabobs set themselves up to control land and mines by legal maneuvering and by violence.

Following the War of 1812, the settlement process became more orderly. The confirmation process eventually resolved most land claims but left enough unresolved so that the issue continued to fester. Land went on sale in 1818.[18] As population of the historic district increased to 6,726 in 1814, to 13,056 in 1820, and to 16,947 in 1824, six new counties were erected. They required county seats, a completely new element in the expanding settlement system, and new roads interconnected them and linked them with the outside world. By the end of the 1820s, the basic settlement pattern and system was in place in the district. The historic district had reached the end of the frontier stage in the ongoing settlement process.[19]

Interpretation

The frontier is a persistent theme in American history and well it should be. No one questions that the experience of people spreading geographically into lands new to them has been a major force in creating the nation. In recent decades, studies of the American frontier have expanded beyond their traditional focus on a westering people from the Atlantic into the interior to include frontiers in the Southwest, the Gulf region, and the Mississippi Valley. When westering Americans reached the middle Mississippi Valley at the beginning of the nineteenth century, they encountered not only the expected resident Indians but also an unexpectedly well-functioning French society into which they had to merge. Historians and historical geographers are beginning to explore this confluence of different ways of life. Thus, our fascination with the frontier finds new geographies to explore.

Recent attention by historians to the trans-Appalachian region, including the Mississippi Valley, focuses on its interpretation as a multicultural frontier. Jay Gitlin, for example, believes that local community studies of borderlands are necessary to discover the distinctive dynamic of each of the several cultures of the Mississippi Valley. He argues that local perspectives allow us to understand the multicultural settlement process

---

18. General histories of territorial Missouri include William E. Foley, *The Genesis of Missouri: From Wilderness Outpost to Statehood*, 131–302; Floyd C. Shoemaker, *Missouri's Struggle for Statehood, 1804–1821*; and Louis Houck, *History of Missouri, from the Earliest Explorations and Settlements until the Admission of the State into the Union*, vol. 3.

19. A general history of the first ten years of statehood is in Perry McCandless, *A History of Missouri: Volume 2, 1820 to 1860*. Other historians have used 1830 as the end of the early settlement period of Missouri (Eugene M. Violette, "Early Settlements in Missouri"). Hattie M. Anderson chose 1828 (an election year) and 1830 (a census year) to end her respected study of early settlement and politics ("Missouri, 1804–1828: Peopling a Frontier State").

more insightfully than the imperial perspectives of Madrid, Paris, and Washington, through which historians have traditionally approached the history of the valley. John Mack Faragher also emphasizes the multicultural frontier of the region, noting variations within the usual categories of Indians, French, Spanish, and Anglos, and asks for comparative histories of frontier societies. He proposes focusing on communities as units rather than on the frontier individualism that has underpinned standard histories of the American frontier. Faragher illustrates his ideas in a study of intercultural interactions in the lower Missouri River region near St. Louis, a frontier that he describes as having a long history of generally good relations among Creoles, emigrant Indians, and Americans; it was a "frontier of inclusion." Events following the War of 1812 and surrounding the drive for statehood converted the Missouri frontier to a "frontier of exclusion." Faragher notes that traditional American history has forgotten the frontier period of cultural inclusion. The terms *frontier of inclusion* and *frontier of exclusion* come from geographer Marvin Mikesell, who pointed out the significance of multicultural frontiers in American frontier history a generation earlier.[20]

Stephen Aron traces the transformation of frontier Kentucky from an eighteenth-century "hunters' world" of reasonably good relations between Indians and American hunters, which was itself a legacy of prior French presence in the Ohio Valley, to a "civilized world" in the early nineteenth century in which the privatization of land for speculation and economic development won out over "rights in the woods." The transformation, Aron says, was one of lost opportunities to continue a viable multicultural society. Similarly, Daniel H. Usner Jr. approaches the eighteenth-century history of the lower Mississippi Valley as a history of cross-cultural interaction and cooperation. His focus is on a vigorous exchange economy among Indians, Europeans, and African slaves. He says that examination of local communities, even in matters as elementary as face-to-face episodes of exchange, is the key to uncovering the nature of intercultural cooperation. The multicultural society of the lower Mississippi Valley began to collapse when an export-directed economy supplanted the frontier-exchange economy among its various constituent groups.[21]

---

20. Gitlin, "On the Boundaries of Empire"; Faragher, "Americans, Mexicans, Métis: A Community Approach to the Comparative Study of North American Frontiers"; Faragher, "'More Motley than Mackinaw'"; Mikesell, "Comparative Studies in Frontier History."

21. Aron, *How the West Was Lost: The Transformation of Kentucky from Daniel Boone to Henry Clay;* Usner, *Indians, Settlers, and Slaves in a Frontier Exchange Economy: The Lower Mississippi Valley before 1783.*

As historians turn their attention to multicultural frontiers through community studies, the settlements of the middle Mississippi Valley do indeed hold promise. Although broader interpretations are still pending, work proceeds on subjects of narrower focus. Susan L. Flader, Susan Boyle, Kevin J. Koehler, David D. Denman, Steven R. Call, and Lucille Basler explore aspects of the social and community history of the villages of Ste. Genevieve and New Bourbon, all of them largely restricted to colonial time. Tora Lorraine Williamsen analyzes the town plan of Ste. Genevieve. The remarkable collection of historic buildings that centers on Ste. Genevieve, considered the largest and best-preserved collection of French colonial architecture in the United States but including later architectural traditions, holds clues to the interpretation of community history, as Charles E. Peterson and Osmund Overby point out. Carl Ekberg provides the most comprehensive account of Ste. Genevieve as a functioning community, including its cultural diversity, but his impressive and highly informative account does not include the geographer's eye on the landscape and resources and necessarily ends abruptly at the Louisiana Purchase.[22]

Unfortunately, no one has seriously continued Ste. Genevieve's interesting story into the American period to explore the integration of French, Creole, and French-culture black elements with American, American-culture black, and immigrant-German population elements.[23] It is as if the community's history stopped in the first decade of the nineteenth century. Neither has anyone united the Illinois and Missouri sides of the middle Mississippi Valley for historical study to show that they in fact constitute a single, interactive society of multicultural elements. Clarence Alvord and more recently Winstanley Briggs, Charles Balesi, and Carl Ekberg focus on the Illinois Country as a single settlement region for the

22. Flader, "Settlement History of Old Ste. Genevieve"; Boyle, "Did She Generally Decide? Women in Ste. Genevieve, 1750–1805"; Koehler, "French New Bourbon: A Study of a Spanish Missouri Community during the 1790s"; Denman, "French Peasant Society in Flux and Stress: The Reintegration of Traditional Village Communal Activity in Ste. Genevieve, 1703–1830"; Call, "French Slaves, Indian Slaves: Slavery and the Cultural Frontier in the Illinois Country, 1703–1756"; Basler, *District of Ste. Genevieve;* Williamsen, "An Analysis of Colonial Town Planning in North America and Its Influence on Ste. Genevieve, Missouri"; Peterson, "Early Ste. Genevieve"; Overby, Ste. Genevieve Project of the Historic American Buildings Survey, 1985; Ekberg, *Colonial Ste. Genevieve.* Forthcoming is Ekberg's *François Vallé and His World: Upper Louisiana before Lewis and Clark* (Columbia: University of Missouri Press.)

23. The best example in the Ste. Genevieve District of the integration of different groups into a multiethnic community built around the exploitation of a resource is a detailed archaeological study by Michael K. Trimble et al., "Frontier Colonization of the Saline Creek Valley."

French, but their emphasis is also chiefly restricted to colonial time. Joseph Zitomersky's meticulous population study of French and Indian settlements in the Illinois Country ends in the 1760s. Stuart Banner analyzes the different conceptions of law and property of contrasting French-Spanish and Anglo-American legal cultures and traces the tensions and compromises that occurred between them during the transition period after the purchase. Standard community histories of Kaskaskia, Cahokia, and early St. Louis are heavy on families, personalities, and these communities' "Frenchness," although Jay Gitlin and others seek new interpretations of the Americanization of French Creoles of St. Louis. Ward Dorrance accounted for the survival of the French language in the Ozark mining community of Old Mines and in so doing presented an account of French cultural practices in the historic Ste. Genevieve District. Archaeological study has contributed significantly to the understanding of some French colonial communities, especially the Saline Valley and Ste. Genevieve. Daniel Usner and John Mack Faragher incorporate the westering Indians into their preliminary studies of the multicultural frontier of St. Louis and the lower Missouri River Valley, but emigrant Indians' more substantial presence in the Ste. Genevieve District and contribution to its life and that of the neighboring Cape Girardeau District remain utterly unexplored. Resident Osages and Illinois are the subjects of several studies, but these studies are geographically peripheral to the Ste. Genevieve District and the eastern Ozarks.[24]

The middle Mississippi Valley, as it centers on St. Louis, begs for an overall interpretation by integrating geographic concepts of spatial organization, in the vein of William Cronon's interpretation of Chicago. The few short studies of this portion of early Missouri that examine how people used space are Lynn Morrow's short but perceptive studies on New Madrid and its hinterland and on traders operating out of Ste. Genevieve, David Denman's work on La Saline and its salt-marketing connections, and Milan J. Kedro's brief study of how Ste. Genevieve used roads as connectors to establish an incipient mining hinterland around it. Morris Arnold's book on the Arkansas Post far to the south examines, among other topics,

24. Alvord, *Illinois Country;* Briggs, "Le Pays des Illinois"; Balesi, *Time of the French;* Ekberg, *French Roots in the Illinois Country;* Zitomersky, *French Americans–Native Americans;* Banner, *Legal Systems in Conflict: Property and Sovereignty in Missouri, 1750–1860;* Gitlin, "'Avec bien du regret': The Americanization of Creole St. Louis"; Dorrance, *Survival of French;* John A. Walthall, ed., *French Colonial Archaeology: The Illinois Country and the Western Great Lakes;* Usner, "An American Indian Gateway: Some Thoughts on the Migration and Settlement of Eastern Indians around Early St. Louis"; Faragher, "'More Motley than Mackinaw'"; Mathews, *Osages;* Rollings, *Osage: Ethnohistorical Study.*

how that post related to its larger region and explains why the French never geographically expanded their Arkansas settlement beyond it.[25]

The historical geography of the eastern Ozarks—as an American-settled region, excluding the French villages—remains unwritten for the American period of the first decades of the nineteenth century except for a few site-specific studies. Robert Flanders presents thoughtful research on Caledonia as a community in the Bellevue Valley and its distinctive "high church" culture. Richard Joseph Janet's study of St. Mary's Seminary at Perryville uncovers the uniqueness of the Anglo Catholic settlement there and its persistence in a region of contrasting settlement strategies. Several accounts of mining activity exist, but they emphasize economic and legal history, personalities, and control of the mines, rather than a history of environmental exploitation and a settlement geography of mines and mining. The best understanding of the historical geography of lead mining and the mineral district in general comes through biographies of Moses Austin, William Henry Ashley, and John Smith T. A century ago, Arthur Winslow provided a valuable outline of the long history of lead mining in the eastern Ozarks, with reference to specific mines, geographic locations, and changing technology. Harbert Clendenen's impressive doctoral dissertation on settlement morphology of the Current River country, 1820–1850, focuses on only one type of settlement process and pattern in the eastern Ozarks, the one created by upland southerners. Explanation is by way of tracing cultural diffusion from Kentucky and Tennessee into the Ozarks and by analyzing an environment of deeply entrenched valleys and narrow ridges.[26]

The middle Mississippi Valley and eastern Ozarks also beg for more attention to the role of the natural environment, especially its variety of re-

---

25. Cronon, *Nature's Metropolis: Chicago and the Great West;* Morrow, "New Madrid and Its Hinterland, 1783–1826"; Morrow, "Trader William Gillis and Delaware Migration in Southern Missouri"; Denman, "History of 'La Saline': Salt Manufacturing Site, 1675–1825"; Kedro, "The Three-Notch Road Frontier: A Century of Social and Economic Change in the Ste. Genevieve District"; Arnold, *Colonial Arkansas, 1686–1804: A Social and Cultural History.*

26. Flanders, "Caledonia, an Ozarks Village: History, Geography, Architecture"; Janet, "St. Mary's of the Barrens Seminary and the Vincentians in Southeast Missouri, 1818–1843"; Steven J. Bellovich, "The Establishment and Development of the Main Colonial Territorial Routes of Lead Movement in Eastern Missouri, 1700–1965"; Lucy Elizabeth Hanley, "Lead Mining in the Mississippi Valley during the Colonial Period"; Welton Lyle Willms, "Lead Mining in Missouri, 1700–1811"; David B. Gracy II, *Moses Austin: His Life;* James Alexander Gardner, *Lead King: Moses Austin;* Clokey, *William H. Ashley: Enterprise and Politics in the Trans-Mississippi West;* Dick Steward, *Frontier Swashbuckler: The Life and Legend of John Smith T;* Winslow, *Lead and Zinc Deposits;* Clendenen, "Settlement Morphology of the Southern Courtois Hills, Missouri, 1820–1860."

sources, in settlement history. Apart from the work done on lead mining and salt already mentioned, no work has treated the basic resources of water, soil, grass, and timber and how they made possible the initial European entrance into trans-Mississippian territory in the valley. Neither has any work comprehensively treated how the environment has undergone change due to more than two centuries of European-style settlement, despite the large number of site-specific investigations.

Little has been done on the topic of land acquisition and landholdings in the eastern Ozarks except in respect to the founding of old Ste. Genevieve. Allen Henry Rose chronicled in all their complexity the official acts that brought the U.S. land system to Missouri, but his history was written without any sense of geography: where the lands were and how the acts affected the settlement of the region.[27]

The above bibliographic essay for the Ste. Genevieve District discloses a need for geographical perspectives, which this book addresses. One perspective is the history of land, in contrast to the history of political, economic, and social institutions. The story of settlement is told as one of assessing land, acquiring it, occupying it, and forming communities from it. A tract of land changes in significance as the district passes through years of continuing settlement. With a focus on the land, the shift in 1804 from Spanish-French to American occupation, which usually is the ending and starting point for histories of the region, is easily transcended to show continuity of settlement across that time divide. Focus on acquiring and occupying land also puts attention on natural resources and their exploitation. There is an environmental history of the district to be told that complements its social and community history. How did district residents perceive the district's resources? How did they transform the natural landscape into a "civilized one"? How did the exploited resources contribute to economic success and social integration of the various communities in the district?

Attention to details of geographical differentiation within the district provides another geographical perspective. The district was far from spatially homogeneous in its opportunities to settlers. Local variations allow, if not require, variety in settlement strategies and patterns that can result in a mosaic of different imprints on the landscape. Attention to microgeographic detail reveals a more complex settlement frontier than heretofore visualized.

27. A. H. Rose, "The Extension of the United States Land System to Missouri, 1804–1817."

Finally, modern geographical perspectives provide insight into how the district, as an identifiable settlement region, was spatially organized. To understand how communities form and grow requires examination into how they are interconnected with each other by family and social ties, by economic activity, and by administrative governance. For the region to develop integrity requires investigation into how it is interconnected with other parts of the country.

The great French historian Fernand Braudel reminds us that geography has as much to tell us about the human experience as the richest documents in the archives.[28] Those who study the Ste. Genevieve District are fortunate to have an exceptionally rich documentary base. To that archival base we will now add geography.

The basic organization of the book is simple. Part I lays out the natural environment as to its meaning for settlement. Part II is a history of governmental laws, policies, and other actions that were to direct and influence the course of settlement. This discussion proceeds chronologically through four chapters, beginning with the pre-1770 French regime, continuing through the Spanish regime, and concluding with the administrative complexities of the American federal administration. A special concern is how the various governmental land laws and policies were actually carried out and the extent to which they were superseded by customs and traditions of settlers. Part III, the geographical and longest part, explores the various settlement patterns that resulted from different people entering the eastern Ozarks under different administrative laws and policies and occupying different environmental niches. Settlement neighborhoods are grouped into seven basic patterns. Detailed historical accounts illustrate each of the patterns, while retaining the distinctiveness of each neighborhood. Discussion begins at old Ste. Genevieve and proceeds through the district in eight chapters organized geographically. Part IV analyzes the evolving administrative framework, population movement, and transportation network of the district, then shows how the numerous neighborhoods and settlements were integrated into systems of interaction and connected to places outside the district. Thus, the historic Ste. Genevieve District becomes a dynamic, functioning settlement system, not a static distribution of unlinked people and neighborhoods that the chapters on settlement patterns may portray. The Conclusion considers the Ste. Genevieve District in the context of a settlement frontier and offers comments on its legacy two centuries later.

28. Braudel, *The Identity of France*, 32.

Part I
The Environment of Settlement

2

"A Country Thus Rich by Nature"

When the Virginian Moses Austin was examining Spanish Upper Louisiana in mid-January 1797 as a place to establish a lead business, he extolled its great potential:

> The Illinois Country is perhaps one of the most Beautifull and fertile in America and has the perculiar advantage of beeing interspersed with large plains or prairies and Wood Lands, where a Crop can be made the first year, without the trouble and Expence of falling the timber, which in every other part of America exhaust the strength and purse of a New Settler.... Nature has undoubtedly intended this Country to be not onely the most agreeable and pleaseing in the World, but the Richest also.... She has the Richest Lead Mines in the World.... [S]he has Salt Springs on Each side of the River, and also Iron Ore in great quantities. These Minerals are more usefull in a Country then Gold or Silver. A Country thus Rich by Nature cannot be otherwise then Wealthey with a moderate shere of Industry. Its also to be remember.d that all the Wealth of this extensive World may be warfted to a Market at any time of the year Down the Missisipi at an easy expence.[1]

The resources of the middle Mississippi Valley appealed to all people who ventured into it, and by the close of the eighteenth century, they fair-

---

1. Moses Austin, "'A Memorandum of M. Austin's Journey from the Lead Mines in the County of Wythe in the State of Virginia to the Lead Mines in the Province of Louisiana West of the Mississippi,' 1796–1797," 539–40. There is no general treatment of this natural region of the Missouri Ozarks. A detailed treatment is in preparation by Tim Nigh and Walter A. Schroeder under the working title of "Atlas of Missouri Ecoregions," sponsored by the Missouri Department of Conservation, Jefferson City. The Ozarks in general is discussed as an ecoregion in Tim Nigh et al., *The Biodiversity of Missouri*, 28–32. Clendenen examined Tennesseeans' assessment of the environment of the neighboring Current River hills in his "Settlement Morphology," 37–53. Carl O. Sauer draws contrasts between French, German, and American approaches to the Ozark environment in *The Geography of the Ozark Highland of Missouri*, thereby demonstrating that peoples of different cultures occupy the same natural environment differently.

ly well understood what it had to offer. To the French entering from Canada, the environment was climatically milder, biologically richer, and more encouraging. To the Americans arriving from Kentucky, Tennessee, and Appalachia, the environment was quite familiar, a major reason so many of these upland southerners entered the eastern Ozarks.[2]

Settlers assessed their new location in terms of climate, rock, soil, plants, wildlife, rivers and water, minerals, and general healthfulness. They were keenly aware of distance from the Mississippi River because of the high cost of overland transport. They also conceptually organized the new land into four basic natural regions that geographically defined how they were going to use its resources.

Resources

"The climate is almost the same as in France," wrote a French captain who passed three summers and two winters at Fort de Chartres in the mid-eighteenth century.[3] The location was latitudinally midway between Canada and Lower Louisiana, and using the latitudinal reckoning of climate then prevailing, the French reasoned that winters would not be as rigorous as in Canada nor summers as long as along the Gulf, although summer days could be as hot. There would never be a shortage of food due to this temperate and benign climate. Here indeed was *America felix*.

The French found the climate better for their staple wheat than that of either Canada or the Gulf Coast, and both French and Spanish administrators wanted to develop the Illinois Country into a wheat granary. Wheat, which was pressing its climatic limits in Laurentian Canada, did well in the nearly two hundred–day growing season in the Illinois Country, and its yield was higher than in the wetter summers farther south.

---

2. Both Thomas Jefferson and Count Volney agreed that the "interior" was warmer by three degrees of latitude than the Atlantic Coast, that is, a place in the interior had a climate of a location three degrees closer to the equator than a place at a corresponding latitude on the Atlantic Coast. That was incorrect, argued Daniel Drake, who maintained that the two places had substantially the same climate. Drake believed the misperception of a warmer climate arose because people also went south while traveling westward into the interior, chiefly down the Ohio drainage, and did not realize it (*Systematic treatise, historical, etiological, and practical, on the principal diseases of the interior valley of North America, as they appear in the Caucasian, African, Indian, and Esquimaux varieties of its population*, 476–77).

3. René Cruchet, "La Vie en Louisiane de 1752 à 1756, d'après un manuscrit bordelais inédit," 66. Georges-Victor Collot, who visited the valley in 1796, reported that the Illinois Country had intense summer heat, just as hot as summers in New Orleans and St. Domingue, and that winters were colder than the same latitude in Europe ("Victor Collot: A Journey in North America," 125).

The French who came from Canada grew spring wheat, sown in the spring and harvested in autumn, because Canadian winters were too severe for winter wheat. Autumn sowing of wheat, as practiced in France, had been introduced into the Illinois Country before 1725, but the farmers from Canada, apparently fearing winters too cold, doggedly continued with spring wheat until wheat was sown in fall 1789 in the bluff lands and produced the "most beautiful wheat that one could see even in Europe," according to commandant Henri Peyroux.[4] Americans finally introduced winter wheat on a large scale later in the decade, and it did remarkably well in the more moderate winters of the middle Mississippi Valley. Perhaps the explanation for Creole preference for spring wheat lies in how the seasonal rhythm of spring wheat fitted into the seasonal operation of lead mines. Work in the mines, lead hauling to the river, and fur trapping were done when labor was not needed for spring wheat planting and fall harvesting. An alternate explanation is that the French let the seasonal operation of lead mines and fur trapping, which were more lucrative enterprises, determine the seasons of planting and harvesting wheat. Yet another factor was the French practice of opening the fences around their plowed fields after autumn harvest and allowing cattle and "herds of pigs" to come inside during the winter. They would have trampled to death any emerging autumn-sowed wheat next to fields of harvested corn.[5]

Although corn produced bountifully and without failure in this climate, the French persisted in their preference for wheat and accepted corn grudgingly. When the wheat crop was short or failed, as when floods prevented its sowing early enough in the bottomlands to get a crop that season, the French were "reduced" to corn for their primary grain. When assessing the quality of land, it was rated in terms of wheat: *bonne terre à ble.*[6] Corn, everyone knew, could grow virtually anywhere, and the Americans generations earlier had made it their staple. In general, because Americans were moving zonally, from east to west, in contrast to the southward-migrating Canadians, their crops and agricultural systems could be brought into the new land virtually intact.

4. Peyroux de la Coudrenière to the Governor, September 2, 1790, PC 203-402 (see the Bibliography for explanation of abbreviations). Also see Paul de Saintpierre to the Governor, December 29, 1790, PC 203-507; Marcel Giraud, *A History of French Louisiana*, 5:459.

5. De Hault Delassus to Carondelet, May 1, 1793, PC 214-4.

6. *Régistre d'Arpentage*, 68, 71, 73, 83, 210, 212, 214, 215. Corn was reported to be the chief crop in the Big Field at Ste. Genevieve in 1808 (Zadok Cramer, *The Navigator, containing directions for navigating the Monongahela, Allegheny, Ohio, and Mississippi waters*, . . . 135).

The long summers tempted French and Americans to try crops more at home farther south. The French grew cotton successfully at Ste. Genevieve, and so did the Americans in creek bottoms throughout the district. Farmers as far north as St. Louis harvested a cotton crop in three successive years, but it still was not dependable in enough years to think about commercial production. Cotton disappeared from farms soon after cotton cloth was available, except in the remotest regions. Indigo was tried. Tobacco and hemp were eminently successful and were important cash crops, especially for Americans.[7]

Carbonate bedrock, both limestones (calcium carbonate) and dolomites (magnesium carbonate), dominates the eastern Ozarks.[8] Because of the tremendous length of geologic time that bedrock has been weathering in the Ozarks, well over a quarter-billion years, the mantle of decomposed and disintegrated rock material above solid bedrock is generally very thick, and solid bedrock exposures at the surface are not at all common. Early settlers, therefore, had only limited knowledge of the underlying substrate, and its nature and significance came largely through observing residuum and soils at the surface.

The carbonates are everywhere soluble, and a quarter-billion years of water moving through them have allowed vast amounts of the rock mass to be dissolved and carried away. Interbedded with the carbonates is chert, or flint, a microcrystalline quartz of silicon dioxide. It constitutes only a minor fraction of the solid rock mass, but, in contrast to the carbonates, it is insoluble. As the carbonate portion dissolves and is removed, the chert portion of the rock mass remains and accumulates on the surface. Ridge tops, hillsides, and creek channels are all distinguished by an abundance of sharp-edged chert fragments, as all travelers in the eastern Ozarks noted. Great gravel bars of rock fragments, mostly chert, choke Ozark stream channels. Most of the Ozark soils contain a high proportion of chert and limestone rock fragments, and settlers prized the pockets of soils that do not.

Where conditions are appropriate, the rock-solution process led to the development of sinkholes, small and large, deep and shallow. In places, these sinks were true funnels (*entonnoirs* to the French), cup-shaped concavities up to thirty or forty feet deep where one could see down into the

---

7. "Report of Don Pedro Piernas to Gov. O'Reilly, Describing the Spanish Illinois Country, Dated October 31, 1769," in *Spanish Regime,* ed. Houck, 1:69; LCC 163; "Trudeau's Report," in *Spanish Regime,* ed. Houck, 2:256.

8. A detailed geology of Ste. Genevieve County is in Stuart Weller and Stuart St. Clair, *Geology of Ste. Genevieve County, Missouri.*

bedrock beneath, even into the underground void and cave stream.[9] At other sinks, fine soil had washed in, plugging the subterranean outlet with impermeable clay that held a pond *(marais)*.

Some forty miles west of Ste. Genevieve are igneous knobs of granite and rhyolite now pretentiously called the St. Francois Mountains. Surprisingly, neither the French nor the Americans before 1820 paid much attention to these eye-catching pink and purple rocks, although the early miner Renaudière noted in his 1723 description of Mine la Motte that "on the other side of the St. Francis river a large number of mountains may be seen the color of whose stone gives strong indication of mineral wealth."[10]

Soil was the most important environmental discriminant for selecting specific sites for settlement in the Ste. Genevieve District.[11] Good-soil land would be settled by farmers; poor-soil land would not. It was that simple. Water, grass, and timber were universally available and therefore less important than soil in the selection of a specific site to occupy.

Good soil *(bonne terre,* or "first-rate land" in the American land survey) was of three origins: alluvium, loess, and chert-free limestone.[12] Alluvial loams covered the bottoms of the Mississippi River. Their productivity was legendary. Desultorily managed, they still produced well, but better managed, they yielded much more abundantly. When villagers decided to move the built village of Ste. Genevieve out of the floodplain, they continued to use the high-yielding alluvial soils of the bottoms for their crop lands, although they continued to be flooded no less frequently than before. Not every piece of the floodplain had soil good for staple grains. In places, it was too sandy and in others too clayey and wet *(terre humide)*.

---

9. Carbonate solution and karst terrain in the Ozarks is described in A. G. Unklesbay and J. D. Vineyard, *Geology of Missouri,* 53–74.

10. Rev. John Rothensteiner, "Earliest History of Mine La Motte," 208.

11. Modern county soil surveys are available for most of the counties of the historic Ste. Genevieve District. See, for example, Burton L. Brown and James D. Childress, *Soil Survey of Ste. Genevieve County, Missouri;* and Dorris F. Festervand, *Soil Survey of Perry County, Missouri.* Though their soil terminology is outdated, Curtis Fletcher Marbut *(Soils of the Ozark Region)* and Carl O. Sauer *(Ozark Highland,* 36–45) are better than the modern surveys in relating soils to choice of settlement and human use of the land.

12. The plats of conceded land (1796–1806) and Spanish concession requests (1760s–1804) are replete with notations on soil, which indicates that soil quality mattered when evaluating land for settlement. Likewise, the field notes of the U.S. land-office survey, beginning in 1815, include comments on the suitability of the soil for agriculture for every mile of survey line. From these documents one can glean a detailed understanding of the local geography of soil as viewed at the time.

Neither the French nor the Americans did much to mitigate these poorer soils of the floodplain, although they talked about draining the wet swales. Sandy soils might not be good for grain, but they yielded quantities of pumpkins, squash, melons, and other vegetables. Alluvial soils also occurred in small, disconnected pieces along all the creek valleys of the interior. Loamy and well drained, they were prized locations for American farms with small crop acreages.

Loess, the fine silt blown off the Mississippi alluvial plain during waning stages of the Ice Age, mantles all uplands within a few miles of the Mississippi River bluff line. It is a loose, friable, minimally leached deposit, up to ten or fifteen feet thick but thinning rapidly with increasing distance from the bluff. The soils that formed at the top of the loess deposit are wonderfully productive and remarkably easy to work with the simplest farm implements, such as a wooden plow and hoe. Loessial soils are much better for wheat than the heavier and wetter alluvial soils of the bottoms. However, they rapidly erode into rills and ravines when the vegetative cover is removed.

In the interior, farmers identified a productive residual soil formed on chert-free limestones in basins among the hills. Loose, deep, and reasonably free of rock fragments, this limestone soil holds nutrients and responds well to natural manuring and simple management. When settlement spread into the interior, these small basins were the first lands to be occupied and become organized communities.

Poor soils (*terres mauvaises,* or second-rate land in the American survey) covered about two-thirds of the total area of the district and thus served as the great restrictor of agricultural settlement. These soils were judged poor because of their excessive rockiness, steep slopes, and droughtiness.[13] Rock fragments constituted as much as 50 to 90 percent of the soil mass, but 20 or 30 percent was more common. Actually, the matrix of clays and silts that embodied the rock fragments was by itself reasonably fertile and productive for corn, but after a couple of plowings and heavy rains, it rapidly washed away from the ridges and hillsides, leaving behind an even greater concentration of immobile rock fragments. In other places, where the land was less broken, a fragipan, or dense clay layer, impeded water movement up and down in the soil, which made the soil prone to excessive droughtiness during dry spells when subsoil moisture could not readily rise by capillary action into the

---

13. The French used the term *terre aride* for soils that dry out excessively during dry spells, a condition now referred to as pedologic or soil drought as opposed to atmospheric drought.

root zone of crops. Settlers discounted this soil environment, locally known as post-oak country *(terres aux chênes à pieux)*, for farming.

The French had special terms for some distinctive soil surfaces. *Glaise* or *salée* was a patch of white, sticky clay with a "greasy and adhesive feeling." The white color was a salt efflorescence, which Americans called a lick. *Glaises* were often shallow depressions in a limestone terrain that held water temporarily after rains. Both salt and water attracted animals, and *glaises* were special sites for hunting and trapping. The French used *vase* for the soft, yielding mud of creek beds and banks, especially for the lowermost portions of tributaries to the Mississippi River.[14] *Marais* was a shallow pond of open water, either perennial or seasonal. It would likely be fringed by cattails, sedges, cordgrass, or shrubby vegetation so that a person would approach it through a marsh.

It did not take many annual plantings of the same crop on a piece of land to reduce crop yields on all but the alluvial soils. Constant cropping, especially for demanding crops such as tobacco and corn, rapidly removed nutrients, and the land became *aride* (barren). Because cattle and hogs ran loose on open range most of the year, natural manuring occurred only when the French congregated their animals on fenced cropland after harvest and rarely on American cropland. Soil eroded to a surprising degree on upland sites, according to requests by farmers to move to other lands because of excessive soil loss.[15] But the alluvial soils kept producing. The Mississippi River bottom at Ste. Genevieve continued as the most productive agricultural tract of the district, even though it began crop production decades before any other piece of land in the district.

The woods formed a "closed" forest only in those selected places where tall trees were so closely spaced that little sunlight reached the

14. Timothy Flint, *A Condensed Geography and History of the Western States, or the Mississippi Valley,* 2:66; John Francis McDermott, *Glossary of French,* 81. The term *glaise* has sometimes been translated as marl. In the English-language edition of Collot's "Journey in North America," published the same year (1824) as the French-language edition, *vase* is regularly rendered as mud (282).

15. William Cronon points out a similar ecological experience in the early settlement of New England: "Lands cleared for crops frequently had to be turned back to pasture or woods less than a decade after their first planting." Cronon quoted one planter who wrote in 1637 that the soil "after five or six years . . . grows barren beyond belief; and . . . puts on the face of winter in the time of summer." As in the Mississippi Valley, animals ranged in open fields, and a supply of manure was lost to the cropped fields (*Changes in the Land: Indians, Colonists, and the Ecology of New England,* 150–51).

ground. "Gloomy forests of pine" cloaked the hills along Establishment River through which the road to the mining country passed.[16] In creek bottoms, the tangle of vines and undergrowth with huge sycamores, cottonwoods, elms, and walnuts presented a daunting forest that became dark early in the evening when the sun dropped behind the enclosing ridges.

But a closed forest was very much the exception for the district. In fact, the French did not use their word, *forêt,* in the Illinois Country, although they used it in Canada, the Ohio Valley, and Lower Louisiana. For the Ozarks, the French used *bois* (woods). When the Spanish were promoting the Ste. Genevieve District for settlement by Kentuckians, their advertisement written in English called attention to the "great abundance of range, both in the prairies and woods" and to "the woods [that] are well stocked with timber." No mention of any forests. Neither did Americans refer to the treed landscape as forest, but rather as woods or woodland, as Moses Austin did in his 1797 account of the Illinois Country quoted earlier. Surveyors and others used "timber" to indicate the presence of trees of a size suitable for lumber and construction purposes, an essential resource on a frontier, but "timber" does not necessarily imply trees dense enough to constitute a forest.[17]

Almost everywhere away from river and creek bottoms the woods was "open," a woods with mature trees spaced far enough apart that sunlight reached the ground and grasses thrived. It was a generous mixture of trees and grasses in a wide range of ratios and, in most places, with little or no woody underbrush. "The high grounds are seldom so thickly covered with wood as to prevent the growth of grass," observed Amos Stoddard. The high grounds, he continued, "exhibit more the appearance of extensive meadows than of rude and gloomy forest." In some places on

---

16. Henry Marie Brackenridge, *Recollections of Persons and Places in the West,* 210. The Establishment River probably was so called because it empties into the Mississippi River across from Fort de Chartres and the French *établissement* there. Finiels called it Rivière de Chartres (*Account of Upper Louisiana,* 51).

17. Compare *bois-debout* and *grand-bois* used in the Mississippi Valley for virgin timber (woods), but not necessarily closed forest (McDermott, *Glossary French,* 27, 83). Pierre-Charles Delassus de Luzières, "An Epoch in Missouri history; being a facsimile of a brochure printed in Lexington, Kentucky in the year 1796," 2–3. Recent studies confirm that the Ozark "forest" was more open than in most "natural stands" today and that fire was a significant factor in keeping the woodlands open (Doug Ladd, "Re-examination of the Role of Fire in Missouri Oak Woodlands"; Richard P. Guyette and E. E. Cutter, "Tree-Ring Analysis of Fire History of a Post Oak Savanna in the Missouri Ozarks"). See Michael Batek for a detailed study of the spacing and size of trees in the Ozark forest from 1815 to 1845 in the Current River region ("Presettlement Vegetation of the Current River Watershed in the Missouri Ozarks").

the ridges and upper slopes, trees were so far apart as to create a true savanna or parklike landscape. When American surveyors ran the Fifth Principal Meridian through the eastern Ozarks in 1815, they sometimes could not find trees convenient for blazing and had to raise piles of stones for mile markers. No one had trouble traveling through these open woods by horse, cart, or wagon. Open-range grazing animals, whether domestic or wild, had no trouble finding enough forage in these woods, and, as Schoolcraft observed: "All the hay collected by the inhabitants, for their winter's stock, is cut either in these open oak woods, or in the glades. This renders these elevated open tracts, which in other respects, may be considered unfavourable for agriculture, extremely advantageous for raising cattle."[18]

Red cedar *(Juniperus virginiana)* was not at all as prevalent in the natural landscape of the eastern Ozarks as it is today. Tolerant of dry soil and needing sunlight, it thrived on bluffs and other rocky outcrops. Woodland fires held its distribution in check, so it did not spread so readily onto the dry-soil uplands and south-facing slopes otherwise suitable to it. The French prized cedar. It rotted less quickly in contact with the earth and was the wood of choice for houses whose walls were built by vertical posts sunk into the ground and for the miles of fences of upright stakes that enclosed village lots and common fields. According to residents, cedar posts lasted for some twenty-five to thirty years, whereas other woods rotted in five years.[19] Though they did not sink so many posts into the soil, Americans also prized cedar, especially for furniture.

Shortleaf pine *(Pinus echinata)* was another special wood used by both French and Americans for sawed lumber and pitch and tar.[20] It often occurred in large, almost pure stands, called pineries *(pinières, piniaires),* and this concentration aided its exploitation. A few persons with the capital to invest in sawmills and who could secure large land concessions for pineries could monopolize its commercial exploitation. Thus, wealthier people of the French community early grabbed up the pineries close to Ste. Genevieve, and speculators and developers of the American community got control of the pine stands farther away somewhat later. Where pines were just another tree in the woodland mix, they were less

---

18. Stoddard, *Sketches, Historical and Descriptive, of Louisiana,* 229; Schoolcraft, *Travels in the Central Portions of the Mississippi Valley,* 239.

19. Brackenridge, *Recollections of Persons and Places,* 21; "Fortifications of St. Louis—Report of Lieut.-Gov. Perez, Dated 1788," in *Spanish Regime,* ed. Houck, 1:271.

20. The early-nineteenth-century distribution of pine in the Ozarks has been mapped from U.S. land surveyors' field notes by F. G. Liming (*The Range and Distribution of Shortleaf Pine in Missouri*).

likely to have been cut unless sawmills were operating in the locality for hardwood lumber as well. The demand for pine boards for construction and furniture was so great in Upper Louisiana, which made their cost "exorbitant" by 1804,[21] that pineries accessible by water a hundred miles away in the Meramec and Gasconade drainages were being exploited in the 1820s.

French found the woods and grasslands abounding in natural foods. They enjoyed pecans *(pacanne)* and walnuts *(noyer)*, the native plum *(prune)*, paw paw *(acimine)* and persimmon *(piakimine, plaquemine)*, mulberry *(alises)*, and a grape something like the French *pinot noir*, smaller and less sweet from which the *habitants* made wine. Similarly, native plants served medicinal purposes. Among them were St. John's wort *(millepertuis)*, centaury, *viperine* (for treating bites), juniper berries *(genièvre)*, and sassafras.[22] Americans also took advantage of plants of the wild, as they were often the same as those in the lands they last lived in. Americans brought with them their rich lore of simples, or medicinal plants, in the natural environment.

The woods and grasslands provided game for food, furs, and skins. French and Americans hunted and trapped animals themselves, or traded Indians for them. They included bison in the earlier years, bear, deer, elk, beaver, fox *(argenté* and *noir)*, otter, wildcat, squirrel, rabbit, wildcat, turkeys, and mink.[23] Like many settlement frontiers, the eastern Ozarks was a hunter's paradise.

Trees were the first resource to be degraded to the point of requiring protection. Already in 1797, the New Bourbon commandant had commented on the elimination of trees close to the villages and the increasing distances that villagers had to go to get wood for fencing and domestic use. But it was around the saltworks and lead smelters that the most extensive degradation took place. The need for prodigious quantities of wood for the continuous fires for numerous log furnaces over the years eliminated trees for some distance around and kept them from growing back. Elimination was so thorough that mine operators soon demanded large concessions, up to a league square (9.4 square miles), to ensure enough fuel, then found them insufficient and asked for more. The Mine la Motte neighborhood was reported denuded by 1770, the Mine à Bre-

---

21. Stoddard, *Sketches of Louisiana*, 217.
22. Cruchet, "La Vie en Louisiane," 68, 67. Wine at Ste. Genevieve was made from dwarf vines of the "poor hills" west of Ste. Genevieve, not from the giant vines along the Gabouri and Mississippi, which did not bear much fruit (Brackenridge, *Recollections of Persons and Places*, 203).
23. Cruchet, "La Vie en Louisiane," 66.

ton and Old Mines neighborhoods shortly after 1800, and the Big River area by 1824. Around Mine à Breton, land was so repeatedly depleted of its timber that it was expected "not to be re-timbered for ages by a new growth." Depletion of timber caused the value of the mines themselves to decline noticeably. Travelers described the landscape around the mines as "barren," meaning deficient in timber.[24] Two centuries later, these mining landscapes still have not fully recovered from their early and repeated detimbering.

The landscape opened up to prairies in various locations, including some on the Mississippi alluvial plain and some in Ozark hollows and ridge tops. The largest, however, were on the limestone plain that stretched from Ste. Genevieve southeastward all the way beyond Apple Creek into the Cape Girardeau District.[25]

Grasslands were the most valued environments for the placement of French villages and croplands. Without exception, every French village in the middle Mississippi Valley was set in a natural prairie or adjacent to one.[26] The French were not woodsmen; they did not have forest-clearing axes.[27] They did not go through the arduous labor of felling mature trees to create fields and pastures, if a *prairie naturelle* was available. It was common sense to take advantage of luxuriant, nutritious natural grass for pasturage as well as for cut hay. The French also placed their village croplands in natural prairies. Even the mining settlements at Mine la Motte and Mine à Breton were set in small prairies. The first land grant at Mine la Motte was a square centered on a natural prairie as described in 1723.[28]

Americans also took advantage of natural grass openings in the woodland. Amos Stoddard, the American captain who first administered the territory and had an excellent working knowledge of its geography, noted: "When practicable, prairie lands have been included in plantation

---

24. On New Bourbon: [Luzières], "Observations on the abuses which are opposed to the progress and perfect success of the welfare of the Illinois," Census of New Bourbon, 1797, PC 2365-345; on Mine la Motte: MN 15; on Mine à Breton, Old Mines, and Big River area: *ASP-PL* 3:579–80; for quote about Mine à Breton: U.S. General Land Office, Field Notes, Missouri, 17 December 1815, 1:273; for travelers' description: *ASP-PL* 3:613.

25. Walter A. Schroeder, *Presettlement Prairie of Missouri,* 13–14.

26. Giraud noted that the French villages of the Illinois Country in 1723 were in prairies, but he offered no reason (*History of French Louisiana,* 457–62).

27. Terry G. Jordan and Matti Kaups, *The American Backwoods Frontier,* 95–96. The European ax had a thick wedge suitable for splitting, but it would get stuck when chopping at a trunk. A woodsman's ax had a thin blade and a helve that could bounce away from the tree with each swing.

28. Rothensteiner, "Earliest History," 208.

surveys. They afford an early grass for cattle, and produce an abundance of hay of no very inferior quality. The soil of them is generally stronger than that of the circumjacent grounds; and almost the only labor required to convert them into tillage fields, is making the necessary fences round them."[29]

The cost of converting prairies to fields was much less than the cost of clearing forested land in the American East, and this differential, fully recognized at the time, hastened settlement of these new lands. In the 1820s, every American settlement neighborhood in the Ste. Genevieve District, no less than the French villages that preceded them, was associated with a prairie or, at the absolute least, with naturally open, grassy woodlands.

There is no question that fires created and maintained the grasslands of the Ozarks, both the more pure stands of bluestem prairies and the stands of grasses among the trees of woodlands. Contemporary accounts describe near-annual fires.[30] Set by lightning and by Indians, French, and Americans, they kept the woods open to improve hunting, both by improving sight of game and by increasing the supply of grasses. Frequent ground fires did not allow much combustible material to accumulate and were not intense enough to kill mature trees but only the tree seedlings that sprung up annually. Thus, through fire, the open woodland lacked the dense, brushy undergrowth that would characterize it in later decades, after fires were suppressed and reduced in number and frequency. The desire to prevent fires came early, not only for protection of wooden buildings and fences as one would expect, once these constructions spread beyond the compact villages in the 1790s, but also because fires began to be seen as destructive of resources. The New Bourbon commandant, in his thoughtful report of 1797 on the status of his district, recommended that burning be stopped at least in the vicinity of the villages, if not everywhere. He lamented that the fires, which he said smokers and hunters set, destroyed fences and burned up wood that could have been fuel for the villages or the lead and salt industries. The commandant's recommendations went unheeded.[31]

At the beginning of the nineteenth century, Americans introduced the term *barrens* for landscapes in the eastern Ozarks that were deficient in trees large enough to be useful for timber. Some barrens were "a denuded surface [with] stunted growth of oaks," and their soil could be exces-

29. Stoddard, *Sketches of Louisiana*, 213.
30. Ibid.
31. [Luzières], "Observations on the abuses," Census of New Bourbon, 1797, PC 2365-345.

sively cherty and unfit for agriculture.[32] A "barren ridge" was avoided. However, other treeless barrens were grass-covered loam lands, and the same landscape was a prairie. The central third of Perry County was covered with tall grass and a prairie to the French, but to its initial Kentucky settlers, who did not yet know the word *prairie,* it was a barrens. Soils were superb and readily adapted to plowing and bountiful hay harvests. In this case, barrens connoted nothing negative for settlement. In fact, the Barrens community rapidly developed into one of the district's most prosperous.[33]

Despite its immense size, the Mississippi River was not the barrier to human geography that it is today. People and goods moved across the channel in simple canoes, bateaux, and barges, and cattle and horses swam it. British captain Philip Pittman, who was stationed at Fort de Chartres in the 1760s, noted, "The communication of this village [Ste. Genevieve] with Cascasquias is very short and easy, it being only to cross the Missisippi [sic], which is about three quarters of a mile broad at this place."[34] It was, in fact, just as easy for people and goods to move across a mile of water as a mile of land. To travel overland between St. Louis and Ste. Genevieve, it was actually easier to use the east bank by crossing the river twice, once at both ends of the journey, than to avoid crossing the river by staying on the west bank for the entire trip. To think of the river primarily as a political boundary separating different administrations, which is how it was thought of by distant officials in London, Madrid, and New Orleans, is to fail to appreciate its unifying force in the everyday lives of the people on opposite sides of it.

All alluvial rivers are constantly shifting their banks and reconfiguring their beds, and the Mississippi was no exception. In 1796, engineer Victor Collot noted the ever changing nature of the channels and the uselessness of constructing "a faithful chart" of the river between the mouths of the Ohio and Missouri Rivers.[35]

It is necessary to understand some principles of how alluvial rivers

32. *Senate Journal,* 11th General Assembly of Missouri, 1st sess., 1840, 513.
33. Whereas barrens may carry a negative connotation in the twentieth century, it most certainly did not in the early nineteenth century. See Carl O. Sauer's discussion of the term as used at the time in Kentucky, where many Perry County settlers came from (*Geography of the Pennyroyal,* 123–30).
34. Pittman, *The Present State of the European Settlements on the Mississippi,* 50. According to Stoddard in 1804, the Mississippi River between St. Louis and the Ohio River was fifteen hundred yards wide with a channel depth of twenty-five to thirty-five feet (*Sketches of Louisiana,* 372).
35. Collot, "Journey in North America," 285.

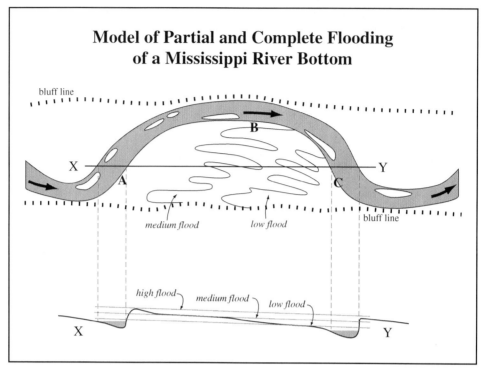

3. *Model of Partial and Complete Flooding of a Mississippi River Bottom.* Frequent low floods cover only the downstream part of the bottom, which is the last part of the bottom to be occupied by settlers. As floodwater heights increase, the last part of the bottom to be inundated is its upstream part, where settlement occurred earliest and villages were located.

change their channels and how the alluvial plain is continuously destroyed and created in order to appreciate Ste. Genevieve's changing relationship with the river and explain the pattern of occupation and abandonment of cultivated land on the river bottom. Figure 3 shows a model of a typical alluvial "bottom" or "point" like the one at Ste. Genevieve.[36] In its course around a bottom, the river's swiftest current tends to impinge upon the bank at *A*, the cut bank, and destabilize it by wetting the silts and clays and destroying their cohesion. Riverbanks are mostly eroded by large blocks of earth slumping in. Higher water enhances the slumping process because the bank is saturated higher up. At *B* and *C* on

36. The model is adapted from Theodore H. Schmudde, "Some Aspects of Land Forms of the Lower Missouri River Floodplain." For a thorough discussion on floodplain formation, see Luna Leopold et al., *Fluvial Processes in Geomorphology,* 317-28.

the outer and lower ends of the point, water moves more slowly along the bank, which results in a tendency to deposit, rather than erode. Material eroded from cut banks upstream is deposited at these places and creates a point bar. Deposition occurs at water level or just below it, so the newly accreted land appears first as a bar when the river is low, but grows in elevation as emergent vegetation traps more sediment above average water level. Therefore, the lower end of a bottom always tends to be closer to the water elevation of the river than its upper end. As years go by, the upper end of the bottom progressively disappears by bank slumping, the lower end grows by point-bar accretion, and the entire bottom translocates downvalley as a unit. This process requires no floods.[37]

As the river elevation rises gradually above bank-full stage into flood, water submerges the lower bottom first, because it is lowest in elevation, shown as "low flood" in Figure 3. Many annual floods rise no higher than a submergence of the lower bottom. As the flood worsens, submergence gradually extends backward up the bottom, contrary to the flow of the river water. This "backflooding" has no current. The water creeps into the minor swales of the surface and leaves the slightly higher ridges uninundated ("medium flood" in Figure 3). Because only the highest floods submerge the entire bottom, people and animals can take refuge on the minor ridges and the highest ground at the upper end of the bottom. This gradual submergence of the bottom without current does little erosive damage to the surface and tends to allow fine sediments such as silts and clays to settle out as deposits.

The rare high flood submerges the entire bottom. When river water elevations are high enough, water spills out of the channel directly onto the bottom at its upper end at A and takes a straight course across the bottom. This water has a strong current to it, because it actually flows down a steeper slope than if it had taken the longer channel curve around the bottom (cross section *XY* in Figure 3). The deeper water with its cur-

---

37. *Pointe* (French) and *point* (English) were used by both French and Americans for the convex river bends around which one had to travel. They continued in use for a long time by river men, although *bottom* gradually took over. A concave bank was called an *anse* or *côte*. Concurrently, *pointe* was also used for an extension of woods into a prairie, and *anse* for a prairie extension into woods (McDermott, *Glossary of French*, 14, 123; Helmut Blume, *The German Coast during the Colonial Era, 1722–1803*, 21). *Alluvial plain* and *floodplain* are often used interchangeably. *Alluvial plain* refers generally to the entire plain created by processes of river deposition and constructed upon river deposits, or alluvium. *Floodplain* is more precisely that portion of the alluvial plain that can reasonably be expected to be inundated by present overbank flows, or floods. Portions of the alluvial plain that are not floodplain are terraces, or second bottoms.

rent can do more erosive scouring, which helps to create or deepen the swales of the bottom and deposits coarser sands instead of fine silts. Water spilling out of the channel at *A* also tends to deposit materials directly on the riverbank at that place, because the water velocity drops abruptly when water leaves the deep river channel. This deposition creates a natural levee that is responsible in part for the high ground at the upper end of the bottom. The original village of old Ste. Genevieve was most likely located on a low natural levee and escaped the minor and moderate floods, although the croplands did not. The built village of Kaskaskia also occupied slightly higher ground and was inundated only in the higher floods.

Thus, two different kinds of river hazards occur at two different places on these alluvial bottoms. One is the upper end of the bottom, subject to slumping of the banks year in and year out and a continuous, progressive loss of land to the river. The lower part of the bottom—lower both by elevation and by being farther downriver—is most subject to flooding and lingering standing water after floods.

Therefore, settlement of the Ste. Genevieve bottom, both by building and by farming land, logically began on its upper end. New properties were added to the Ste. Genevieve common field progressively toward the less desirable, more flood-prone lower end, but the built village never did extend onto them. However, the first people to abandon the Ste. Genevieve bottom were those who lost land and buildings to bank caving at the upper end.[38] The same evolving human use of a bottom also occurred in the American-settled bottom, next downstream. The whole surface of both bottoms became severely hazardous during the rare high-magnitude floods and forced complete evacuations.

Villagers expected some flooding of their lands each year from the Mississippi; floods were "as periodical as those of the Nile."[39] Almost on cue, the river rose in May from the annual sharp rise of its major tributary, the Missouri, and began to fall in June. As a large river, the Mississippi was slow to rise, slow to crest, and slow to fall. In most years, only the point bar and swales of the lower bottom would flood, and in occasional years, floods would reach the upper bottom, but even then the low ridges were rarely inundated. Even when portions of the upper bottom flooded, corn could still be planted if the water receded by early June, but

---

38. See the reconstruction of Old Ste. Genevieve by David Denman and Susan L. Flader in Flader, "Ste. Genevieve: An Interdisciplinary Community Study," app. E.

39. Thomas Ashe, *Travels in America, performed in . . . 1806, for the purpose of exploring the rivers Alleghany, Monongahela, Ohio, and Mississippi, and ascertaining the produce and condition of their banks and vicinity,* 282.

spring wheat, which was usually growing in the fields when the floods came, was less sure. A flood of several days' duration ruined it, and the chances of replanting it after a late flood for autumn harvest were slim. Wooden fences, whether the upright stakes of the French or the rail fences of Americans, would become buoyant even in currentless water and float away. Log and sawed-timber buildings would also become buoyant and float or be dislodged from their foundations.

The French who came to the valley from Canada had no experience with large-river floods. The Saint Lawrence, a tidal estuary, does not flood as alluvial rivers do. The French who came up from Lower Louisiana had experienced river floods, but of a different type. There, the channel was more fixed in place, and the higher natural levees that were the sites of houses and gardens were hardly ever submerged, whereas floodwaters spread over extensive backswamps and fields behind the levees. Thus, the French who settled in the Illinois Country, wherever their provenance, had to learn how to live with a large river of a different nature. The fact that they chose the floodplain for their villages and fields and persisted in living there throughout most of the eighteenth century suggests that they did not fully understand the hazards of living on a floodplain and were unable or unwilling to adjust to them, if we apply late-twentieth-century notions of how to live with floods.

An alternative and more likely interpretation is that the Creoles did learn the Mississippi's vagaries and accepted them as a part of life. Throughout the eighteenth century, they continued to live on bank-caving, flood-prone bottoms, losing buildings and crops, but apparently few, if any, lives directly from floods. The lieutenant governor observed in 1798 that residents of Ste. Genevieve "are accustomed to lose two out of five harvests regularly," although by this time they had moved most of their houses from the vulnerable river's edge. The New Bourbon commandant proposed levees, but like so many of the improvements proposed by local officials, they were not built.[40]

Bank caving did not keep the Mississippi Valley French from their riverine locations. At Fort de Chartres, they watched the Mississippi creep closer and closer to the fort by bank caving. In 1756, the Mississippi channel was a distant half mile from the new stone fort, but in just nine years it had reached the walls, a recession of the banks at an average rate of 250 feet per year, highly observable to anyone. The river began under-

40. "Trudeau's Report," in *Spanish Regime*, ed. Houck, 2:248; [Luzières], "Observations on the means of increasing promptly and with success the progress of agriculture of the district of New Bourbon and of that of Ste. Genevieve," Census of New Bourbon, 1797, PC 2365-345.

mining the fort in December 1765, and portions began falling into the river.[41] The French who had been awarded land at Fort de Chartres and nearby St. Philippe in 1722 and 1734 lost half or more of it from bank caving. Was not something learned from these and similar experiences along the river?

The lowest, almost annual, Mississippi River floods went uncommented. The bottom at Ste. Genevieve was flooded "once in about ten or twelve years," according to Captain Stoddard, but this observation likely refers to severe flooding during which virtually all of the bottom was affected, even reaching the houses of the village.[42] The highest floods elicited major comments for their destructive effects. They include the following, from the founding of Ste. Genevieve until 1820:

1772   Caused abandonment of Fort de Chartres.
1777   Severe bank caving threatened to destroy the church and residences at old Ste. Genevieve.
1785   The celebrated *année des grandes eaux*, which was both deep and sustained, lasted from late April into August; preceded by lesser floods in the preceding years; all bottoms flooded from above Cahokia to below Cape Girardeau; considerable damage and loss of crops and animals; direct cause for the relocation of the built village of Ste. Genevieve.
1788   Severe bank caving; flood washed away fences; wheat harvest lost; illness.
1794   Two-thirds of the Ste. Genevieve bottom under five and six feet of water; wheat lost; residents resorted to corn; fences gone.
1797   Harvest lost, animals lost; food shortage; illness; "great damage at Ste. Genevieve"; "children starving"; "famine at New Bourbon."
1802   Bottoms flooded; cabins carried away.
1811   Highest flood to date at Ste. Genevieve; all crops totally destroyed; too late to sow late crops.
1815   At Herculaneum.[43]

41. Pittman, *Present State,* lxiii, 46; Belting, *Kaskaskia under the French,* 52; Clarence Walworth Alvord and Clarence Edwin Carter, eds., *The New Regime, 1765–1767,* 125, 177, 340–41. Stoddard described the collapse of silt and sand banks, which reportedly sunk boats (*Sketches of Louisiana,* 370–71). Bank caving was also on the minds of the Spanish when looking for fort sites in 1766. They passed by a location in the Bois Brule bottoms, which they were inspecting for a possible fort to protect Ste. Genevieve, because the banks there caved in (St. Ange to Ulloa[?], August 4, 1766, PC 2357-22). For a graphic account of bank caving at the old village of Ste. Genevieve, see "Peyroux de la Coudrenière to Esteban Miró, March 8, 1788," in *Spain in the Mississippi Valley, 1765–1794,* ed. Lawrence Kinnaird, 3:245–46.

42. Stoddard, *Sketches of Louisiana,* 216.

43. The St. Louis District of the U.S. Army Corps of Engineers maintains a complete record of historic flooding by the Mississippi River. In addition, major floods be-

The Mississippi River was subject to winter icing. River ice was less due to the local climate than to ice floating down from colder regions to the north, although an unusually cold winter at Ste. Genevieve would enhance icing. When water was naturally low in winter, ice would form in slow-moving chutes and build out from banks. Large, thick ice floes coming downstream jammed as they dragged on the shallow river bottom or piled up in river bends.

New arrivals to the valley or visitors traveling on the river frequently described river ice, as though it was an unexpected phenomenon worthy of comment, whereas locals seemed to accept icing in stride without comment. Ice interfered with cordelling and poling along the banks. When the first Spanish lieutenant governor came up the river in November 1768 to take charge for Spain, ice was already so bad that his group had to abandon their bateaux not far above the Ohio and walk overland to Ste. Genevieve. People crossed the Mississippi on ice between Kaskaskia and Ste. Genevieve in January 1766. Amos Stoddard reported that for three consecutive winters, 1803–1805, ice completely covered the Mississippi River at St. Louis from bank to bank, and in January 1805, it reportedly exceeded twenty-two inches in thickness.[44]

A type of river environment of special interest for settlement was the mouths of the small tributaries to the Mississippi. All of these tributaries, for a couple hundred yards or so up from their junction with the Mississippi channel, held calm backwater from the big river. Because the bot-

---

fore 1830 have been recorded in various documents. They are the subject of correspondence in the Papeles de Cuba from 1777 to 1797 (e.g., for 1785: 11-319, 512, 581, 117b-191, 193, 195; for 1788: 201-941, 959; for 1794: 208a-480, 209-598, 643; for 1797: 204-717, 214-874, 879, 957). Specific floods of interest are 1766 (Alvord and Carter, *New Regime,* 227); 1785 ("Report of Gov. Miró to Conde de Galvez of the Great Overflow of the Mississippi in 1885 [sic] in Upper Louisiana," July 10, 1785, *Spanish Regime,* ed. Houck, 1:235–36; Clarence Walworth Alvord, ed., *Kaskaskia Records, 1778-1790,* 375; Ekberg, *Colonial Ste. Genevieve,* 421–22); 1811 (Firman A. Rozier, *Rozier's History of the Early Settlement of the Mississippi Valley,* 318–19; Flint, *Condensed Geography,* 2:79). St. Louis newspapers carried accounts of floods beginning in 1808. Scharf summarizes Mississippi River flooding to 1883 (*St. Louis City and County,* 2:1060–68). For floods before the founding of Ste. Genevieve, see Belting, *Kaskaskia under the French,* 18, 65–66. The effect of flooding in general is discussed in Alvord and Carter, *New Regime,* 438–39. Ekberg gives accounts of the effects of floods of the 1780s and 1790s on community life of Ste. Genevieve (*Colonial Ste. Genevieve,* passim).

44. Don Pedro Piernas to Gov. O'Reilly, in *Spanish Regime,* ed. Houck, 1:66–67; Alvord and Carter, *New Regime,* 362; Stoddard, *Sketches of Louisiana,* 236. Twenty-two inches would probably not be possible in running water. Perhaps this exceptional thickness, if correct, was in ice floes stacked on top of each other or achieved in quiet-water ponds.

toms of their channels had to be as low in elevation as the bed of the Mississippi where they joined it, water in these mouths was as deep as in the adjacent Mississippi channel, but lacked any current and thereby provided safe harborages in which to pull boats off the open Mississippi. Banks falling in along the big river, logs and trees coming downriver with force to batter and capsize any boat tied up along the bank, and rapid bankside currents made the Mississippi banks unsafe. Thus, Captain Pittman discovered that the mouth of the Kaskaskia River was a "secure port for large batteaux [sic], which can . . . load and unload without the least trouble."[45] La Petite Rivière served as harborage for old Ste. Genevieve, and the Gabouri mouth for new Ste. Genevieve. Likewise, the Saline River mouth protected boats for the La Saline saltworks, the muddy mouth of the Joachim served Herculaneum, and Cahokia Creek served Cahokia. The precise site of villages and shipping points on the Mississippi River depended on the presence of a tributary mouth of deep, calm water.

The bed and banks of these tributary mouths were composed of fine silts and clays deposited in the calm backwater. The soft mud (*vase* to the French) presented a problem for poling boats up these channels and usually required cordelling from the bank. The mud also presented problems when attempting to build bridges over these creeks because it was so difficult to find a solid footing for timbers sunk vertically into the bed. Some distance up the creek valley, where the channel became shallow and gravel bottomed, was the usual place for crossing, whether by ford or by bridge.

People drank directly from the Mississippi River. River water was "salubrious," and even "alterative and medicinal," according to physician Daniel Drake, and was "preferably drunk immediately from the river, without clarification by deposition, or by any artificial process." Others, however, let river water sit for a while to let sediment settle.[46] The numerous perennial streams of the district also furnished convenient drinking water, but most people in the interior preferred to get water from small springs that headed each hollow. Larger springs issued forth in the bluff lands where sinkholes diverted large amounts of runoff into underground drainage systems. The more regular flow of these springs powered mills more dependably than streams whose flows were derived primarily from runoff from irregular rainfall. The mill that long served

45. Pittman, *Present State,* 42.
46. Drake, *Systematic treatise,* 662; Berquin-Duvallon, *Vue de la colonie espagnole du Mississippi, ou des provinces de Louisiane et Floride Occidentale; en l'année 1802, par un observateur résident sur les lieux,* 18–19.

both Ste. Genevieve and New Bourbon was powered by springwater.[47] Brackish groundwater issued forth both naturally as salt springs and in dug pits in the valleys of the lower Saline and Meramec Rivers.

In the towns and villages, hand-dug community wells, probably only slightly deeper than the static water table at ten to twenty feet deep, provided dependable sources of water. Ste. Genevieve, New Bourbon, St. Michel, and Potosi had community wells according to records; probably other villages also had them. Wells and springs in limestone produced a "salubrious beverage," but water from wells in alluvium (as on the floodplain at old Ste. Genevieve) "may abound in decaying vegetable remains" and be "impregnated with insalubrious matters" and were avoided in preference for rock-lined cisterns and rain barrels.[48]

The district's rivers and streams also supplied fish. Indians fished more in ponds and quiet waters, and the French and Americans more in the Mississippi and other flowing waters. Did these latter avoid the ponds, which were probably more accessible and simpler to fish, because of their negative association with fevers and diseases? The Mississippi held catfish (allegedly up to nine and ten feet long!), pike, carp, buffalo, and eel, and smaller streams provided perch, trout, and sunfish.[49]

The eastern Ozarks was known to be rich in useful minerals from the beginning of European entry into the Mississippi Valley. In fact, it was the quest for minerals that excited early exploration of the lands along the river. Resident Indians gave Father Marquette information on iron and other metals in the region; they had dug for lead and extracted salt from the region's saline waters long before Europeans entered. The useful minerals of the region known by 1830 include the metals lead, iron, copper, cobalt, and manganese and the other minerals of saltpeter (niter), salt, and limestone. Silver was an elusive metal. Recurrent reports of it fueled European interest in the Ozarks and may have been responsible for the earliest explorations that turned up lead instead.[50] Indeed

---

47. Ekberg, *French Roots,* 271. The stream is now called Vallé Spring Branch on the south side of Ste. Genevieve.

48. Drake, *Systematic treatise,* 662.

49. Malcolm Comeaux, "An Archaeological Perspective on Animal Exploitation at French Sites in the Illinois Country," 94; Cruchet, "La Vie en Louisiane," 66; Luzières, "An Epoch in Missouri history," 6.

50. Elmer Cave Barrow, "The Early History of Iron Mining in Southeast Missouri (1815–1861)," 1–2. The best general account of known mineralization in the early nineteenth century is that given by Henry Rowe Schoolcraft (*A View of the Lead Mines of Missouri*), supplemented by Moses Austin's comments on lead (*ASP- PL* 1:188–91). Tiff, or *tuf blanc,* as a sign of lead ore appears in documents as early as the 1740s

the shiny, silver-colored cubes of galena may have been mistaken for silver.

Lead was the most exploited mineral. All of the lead mines of the pre-American eastern Ozarks lay in the Ste. Genevieve District and were the direct responsibility of its commandant. The residents of Ste. Genevieve (earlier from Fort de Chartres and Kaskaskia) were responsible for developing lead-mining activities, but later in the American period, as the industry further developed and expanded, St. Louis and East Coast interests entered the lead business of the eastern Ozarks.

The geography of the lead ore—it occurred at the surface over thousands of acres—determined how it was to be exploited before 1830. The lead ore, which included pure lead crystals of weights up to several pounds, was in a weathered, surficial material of varying depth, covered only by a thin veneer of soil and vegetation. Most often the lead ore was in "a red clay mixed with gravel, forming together a hard compact body, or else attached to, or combined with, . . . tiff."[51] Miners used tiff *(tuf blanc)*, or barite, also called a vitreous spar, as a sign of lead ore. Lead ore in solid bedrock deep below the surface was difficult and expensive to dig and went unexploited largely because of the abundance and ease of extraction of surface deposits, which could be dug by any man, woman, or child with a shovel and strong back. Surface digging required no capital investment, no technology, no education in the mining business. In this respect, it was like trapping for furs or shooting deer: an exploitative resource activity that one could independently engage in. Young men could easily be lured to dig ore just as they could be to trap beaver. The call of the lead pits affected pursuit of agriculture and the social and economic stability of villages and towns, both French and American, by

---

(Hanley, "Lead Mining in the Colonial Period," 25), but it was not exploited as a useful mineral until the beginning of the twentieth century when it came to dominate mining in Washington County. Though the French did not find silver (Foley, *Genesis of Missouri*, 13–14), Stoddard reported a silver mine on the Meramec River (*Sketches of Louisiana*, 216), and a silver mine operated on the St. Francis River in Madison County in the 1870s and 1880s; the place-name "Silver Mines" commemorates the site (Thomas R. Beveridge, *Geologic Wonders and Curiosities of Missouri*, 50; Winslow, *Lead and Zinc Deposits*, 7:701–2). Silver is currently a coproduct of lead mining in the Ozarks; Missouri ranked sixth in the nation in the early 1990s among silver-producing states.

51. John Rice Jones, "Salt Springs and Lead and Copper Mines," November 6, 1816, *ASP-PL* 3:604. Moses Austin noted that lead was found in all the small creeks "washed down from the hills, and it is not uncommon to find in the draughts leading to creeks and rivers, and in gullies made by the spring rains, mineral in pieces from ten to fifty pounds weight brought down by the torrents: Some hundreds have been collected in this way" (*ASP-PL* 1:190).

draining off young men no less than did the call of the hunt for furs and skins. Toward the end of the 1700s, the American Moses Austin sank a shaft eighty feet deep into bedrock, but this innovation to get at an expanded supply of ore or to higher-grade ore did not entice others to do likewise.[52] Apparently, there was still enough ore at the surface, and it was just too easy to dig pits.

Because the lead ore occurred over tens of thousands of acres of land, no person could monopolize mining. Should someone seize a miner's pits, the latter could go elsewhere and find more lead. There would always be new finds discovered over the next hill by some fortunate searcher, although hopes were always greater than reality. Naturalist John Bradbury reported that out of forty sites dug in the Richwoods area, thirty-eight contained lead ore.[53] The widespread occurrence of lead ore at the surface resulted in numerous short-lived mining camps scattered across the landscape and worked against the development of a single primary mining center and settlement focus.

The French smelted lead ore in the simplest way by piling it on top of logs that were then set afire. One firing did not recover much lead, and even after two firings, the lead yield still was only 40–60 percent. These simple log *fourneaux,* which hardly merit the English translation as "furnaces," could be built anywhere using local wood. In 1798, Austin introduced a reverberatory furnace, which recovered a much larger percentage of the lead. A huge stone structure fixed in place and requiring a rather continuous supply of ore, it was the most powerful force in concentrating the industry, both geographically and organizationally. Diggers from hundreds of pits abandoned their individual log furnaces and hauled their ore to Austin's furnace or later to others that copied his. Whereas mining continued to be geographically dispersed, smelting became more concentrated and fixed, although numerous log furnaces persisted in remoter locations. Manufacturing lead into sheets or into gunshot other than for one family's needs also was a concentrated activity. It required investment capital that few had, and it required a minimum amount of lead to be profitable. Americans initiated lead manufacturing.[54] It is noteworthy that the French who warehoused and shipped lead

52. *Missouri Gazette,* June 20, 1811.
53. Bradbury, *Travels in the Interior of America in the Years 1809, 1810, and 1811,* 250.
54. French Creole lead-smelting methods are described in Schoolcraft, *View of the Lead Mines,* 93–112; Moses Austin, "Description of the Lead Mines in Upper Louisiana," *ASP-PL* 1:189; Carl J. Ekberg, "Antoine Valentin de Gruy: Early Missouri Explorer," 147–48; and Winslow, *Lead and Zinc Deposits,* 6:199–200. The early lead mining and smelting industry is treated in Hanley, "Lead Mining in the Colonial Peri-

at Ste. Genevieve for a half century did not develop any significant processing activity that would have added value to their product. Wealthier French, however, did join the capitalization of American enterprises.

Transporting lead ore, smelted lead, or lead products was a separate function, and for a long time, individual miners or farmers hauled lead during the farming off-season. Marketing was in the hands of a select few. Individual miners sold their ore locally, and they and lead haulers were captives to a pricing system beyond their control. Local marketers had connections at St. Louis, New Orleans, Philadelphia, and elsewhere to sell the product, and were subject to price fluctuations and shifting demands for lead governed by events as far away as Europe.[55] Through all the economic vicissitudes of the time, including the American and French Revolutions, change of government in Louisiana, and the Napoleonic wars and their aftermath, lead continued to be produced in the Ste. Genevieve District and sustained the ambitions and hopes of its residents, poor and wealthy, French and American. Lead also helped powerfully to direct the course of settlement in the eastern Ozarks, establish the central places, and determine road networks.

The lead mines and environs were deemed unfit to raise and maintain cattle. Horses, cattle, dogs, cats, and chickens reportedly dropped dead from licking lead slag or breathing fumes of arsenic and sulphur.[56] Trees were stunted around the lead mines, and observers interpreted this simple spatial association to mean that lead ore caused diminutive trees, although stunted trees such as post oaks occurred elsewhere on cherty surfaces not known to contain lead. People and animals had to stay away from the "poisonous effluvia" produced by smelting;[57] perhaps that is why black slaves did most of the smelting. Runoff that percolated through the residue from smelting fouled the water, rendering it undrinkable. It is not known whether knowledge of the deleterious effects of lead on human and animal health came from lead mining in Spain and the Spanish Netherlands or was learned from long experience in Upper Louisiana, or possibly from both sources. Later research has proved the toxicity of lead, and these historic mining areas continue to bedevil Missouri public agen-

---

od"; Willms, "Lead Mining, 1700–1811"; and summarized in Winslow, *Lead and Zinc Deposits*, 6:267–71.

55. Two biographies of Moses Austin put the lead industry of the Ste. Genevieve District in the context of larger regional and national events: Gracy, *Moses Austin: His Life;* and Gardner, *Lead King: Moses Austin.*

56. Flint, *Condensed Geography,* 2:86; Henry Marie Brackenridge, *Views of Louisiana, together with a Journal of a Voyage up the Missouri River, in 1811,* 150.

57. To Richard Bates, December 17, 1807, *Bates Papers,* 1:244.

cies charged with the responsibility of maintaining environments healthful for humans and wildlife.

Recognizing the serious human and animal health problems associated with lead production, the first Missouri state legislature enacted a law in 1821 requiring the fencing of furnaces. Fencing was to be at least ten yards from furnaces and kept in repair for six months after furnaces were last used.[58] Actually, French miners had as a custom fenced mines and furnaces long before, because it was their own crops and animals that suffered. The 1821 law was necessary, because by then speculators and absentee owners of mines disregarded the environmental health of the miners and local residents and had to be coerced to take the commonsense precautions.

Iron also was present as red hematite in partially decomposed rock near or at the surface, sometimes as "iron banks" on valley sides. Soils throughout the eastern Ozarks are typically red from iron oxides derived from the weathering of iron-bearing rocks. Some of the mountain knobs in the headwaters of the St. Francis River have an iron content as high as 65 percent, and they became the focus for exploitation of iron. Early "furnaces" consisted of logs and rock with a clay lining. At iron forges, similar to a blacksmith's simple forge, iron was mixed with charcoal to remove impurities, resulting in "blooms" of approximately eighty pounds. Successful ironworks began in 1815–1816. All iron operations used natural resources similarly. All depended on hematite wagonned in not more than five miles. Local timber was the fuel for roasting the ore, with local charcoal and limestone for mixing. Springs and streams provided waterpower for hammering and machinery and thereby determined the sites of all furnaces and forges.

Although the district contained all of the basic natural resources for ironworks and usually in great abundance, cheap transportation was lacking. Hauling from interior locations by wagon in the absence of regularly navigable streams increased the cost of products so much (excluding the Maramec Iron Works) that they could not compete with ironware brought in on the Mississippi River. Therefore, these early ironworks, at best, produced bar iron for local blacksmiths, or articles for local settlers only, such as horseshoes, grubbing hoes, plow shares, ax heads, mattocks, wagon-wheel rims, kettles, and skillets. Exploitation of iron also chronically lacked capital for investment until the end of the 1820s when Thomas James invested heavily in the Maramec Iron Works.[59] That it

58. *LPGN* 1:794–95.
59. James D. Norris gives a full account of the Maramec Iron Works in *Frontier Iron: The Maramec Iron Works, 1826–1876*. Other accounts of the nineteenth-centu-

took a personal fortune to develop a successful frontier ironworks is analogous to the personal fortune that Moses Austin put into his frontier lead operation to make it successful.

Settlers on all frontiers needed gunpowder (black powder), but the eastern Ozarks was better endowed with the basic resources for making it than most. Such an obvious environmental advantage "ought not to be passed over unnoticed," wrote natural scientist John Bradbury when assessing the Ozarks for settlement.[60] Gunpowder is a combination of varying amounts of saltpeter, or niter, with charcoal and sulphur. Potassium nitrate was available in the form of abundant bat guano in the district's numerous, large limestone caves. To concentrate it into saltpeter was a simple process that backwoodsmen could readily do. Local trees provided charcoal. Sulphur was an otherwise wasted byproduct of lead smelting of even the crudest fashion. Thus, because all the necessary ingredients were present locally as well as the demand, it is possible that from the 1720s, when the first lead was produced in the Ste. Genevieve District, musket-toting hunters and settlers were making gunpowder.

Typically, men made the saltpeter right at the caves and sometimes the gunpowder also. In the latter case, it was a subsistence activity, for one's personal use, just as a family would boil its own supply of salt at a salt spring. More commonly, the miners hauled the saltpeter to another place, a "factory," where others made the gunpowder. Gunpowder making was an exceptionally dangerous activity with spectacular accidental explosions and loss of life. Though the abundance of saltpeter was common knowledge early in the district, gunpowder making was slow in being commercially developed during the colonial period. Early in the American period, Frederick Bates, a territorial leader, asserted, "The Country abounds with . . . caves of Salt Petre; but none of them are yet wrought to advantage."[61] That was an overstatement, but it pointed out the continuing potential for commercial development of a frontier necessity.

It was after the arrival of Americans, who owned many more guns and used them much more often than the French, that gunpowder making developed into a commercial activity in the eastern Ozarks. In 1810, a

---

ry iron industry in southeastern Missouri are Frank L. Nason, *A Report on the Iron Ores of Missouri;* Arthur B. Cozzens, "The Iron Industry of Missouri"; Barrow, "History of Iron Mining"; and James M. Swank, *History of the Manufacture of Iron in All Ages,* 332–37.

60. Bradbury, *Interior of America,* 247.

61. To Richard Bates, December 17, 1807, *Bates Papers,* 1:244; Stoddard, *Sketches of Louisiana,* 392.

man and his two sons journeyed to caves on the Gasconade River to make saltpeter, returning in just three weeks to St. Louis with three thousand pounds for making black powder.[62]

The most celebrated commercial saltpeter and gunpowder operation of the Ste. Genevieve District was William Ashley's. Ashley, involved in lead mining at Mine à Breton–Potosi, reportedly discovered a cave near the head of the Current River. He began making saltpeter at the cave in 1813 or 1814 and hauled it to Shibboleth lead mines east of Old Mines where he and Lionel Browne manufactured it into gunpowder. They hauled sixty thousand pounds of saltpeter, an amount sufficient to make thirty thousand dollars worth of gunpowder, from the cave in an eighteen-month period. The fact that saltpeter was carted over rough Ozark ridges and valleys such a long distance (eighty-six miles) indicates its high value. Gunpowder was also sometimes used to blast open lead-mining pits.[63]

As with lead smelting, reduction of cave guano into saltpeter was a process harmful to the free-ranging cattle, hogs, and wild animals of the Ozark woods. In 1815, a half decade before the state law requiring fencing of lead operations, the Missouri territorial legislature passed a law requiring the fencing of saltpeter works and the stream water draining from them so that animals could not reach them.[64] It is an open question whether such fencing was actually done considering the remoteness of most of the Ozark caves.

Salt for meat preservation, skin tanning, and fur processing was manufactured at La Saline near the mouth of the Saline River and on the Meramec, although after the turn of the century, these local sources waned as supplies came increasingly from the Ohio River region.[65] The salt at La Saline, directly on the Mississippi River and so easily accessible, was exploited well before the French arrived in the valley.

The Ste. Genevieve District lies within a large region subject to strong earthquakes that center upon New Madrid, just below the junction of the

---

62. William Clark Breckenridge, "Early Gunpowder Making in Missouri," 89. The U.S. government awarded its first lease for saltpeter mining in a cave on the upper Saline River in 1813 (*Bates Papers,* 2:249–50).

63. Clokey, *William H. Ashley,* 29–32; G. W. Featherstonhaugh, *Geological report of an examination made in 1834 of the elevated country between the Missouri and Red rivers,* 46.

64. Breckenridge, "Early Gunpowder Making"; *Glimpses of the Past,* 4:58; *LPGN* 1:341.

65. Production of salt on the Saline River has been examined by Denman, "History of 'La Saline'"; and Trimble et al., "Frontier Colonization."

Ohio and Mississippi Rivers. Records indicate earthquakes large enough to comment on occurred in the continental interior before 1830 in 1699, 1776, 1791–1792, 1795, 1804, 1805, 1811–1812, 1816, 1818, 1819, and 1820, and most likely in other years for which records are lacking. The 1795 tremor was reasonably severe, and the ones in 1804 disjointed buildings in Kaskaskia and caused American soldiers to abandon their quarters there. The exceptionally strong series of earthquakes during the winter of 1811–1812 at New Madrid, celebrated today as possibly the strongest to have occurred in the conterminous United States in the last two centuries, felled chimneys as far north as St. Louis and probably damaged structures at Ste. Genevieve, as it lay much closer to the epicenter. However, land documents for the Ste. Genevieve District at the time make no mention of those earthquakes.[66]

As they did with other natural hazards, people seem to have accepted earthquakes as uncontrollable natural events and let them pass by as such. "Shakes . . . are in this part of the country extremely frequent and are spoken of as matters of every day occurrences," proffered a resident of Cape Girardeau in 1819. The 1811–1812 earthquakes of a magnitude of 8.4 on the Richter scale received little notice in the St. Louis newspaper, despite widespread damage and change in the physical landscape, and Moses Austin's records and correspondence contain no mention of these earthquakes that rang bells in Charleston, South Carolina, and cracked masonry in Washington, D.C.[67] Apparently Austin's lead furnace and establishment at Mine à Breton, 110 miles from the epicenter, suffered no damage.

The year 1811 had begun with the spectacular passage of Halley's comet that was taken in the middle Mississippi Valley as an ominous sign. Those who believed this, and there were many, were not disappointed.

66. Basic accounts of the New Madrid earthquakes of 1811–1812 are by Myron L. Fuller, *The New Madrid Earthquake;* and James Lal Penick Jr., *The New Madrid Earthquakes.* Of interest also is B. C. Moneymaker, "Some Early Earthquakes in Tennessee and Adjacent States, 1699 to 1850."

67. Stephen H. Long, *Account of an Expedition from Pittsburgh to the Rocky Mountains in the Years 1819, 1820,* 88. The 1811–1812 earthquakes received simple notices of their occurrence in the *Missouri Gazette* issues of December 16 and 21, 1811; and February 7, 23, and 29, 1812.

Territorial secretary Frederick Bates did not have anything to say about the earthquakes in his copious correspondence about the territory *(Bates Papers)*. Compare this blasé attitude toward earthquake hazards with that of colonial South Carolinians toward hurricane hazards: "Presumably settlers simply accepted hurricanes as occasional hazards over which they had no control" (Roy Merrens, "The Physical Environment of Early America: Images and Image Makers in Colonial South Carolina," 551).

All summer, the highest flood of record for the time ravaged the bottoms, destroyed the harvest, and prevented the sowing of late crops. Then came the incredibly strong earthquakes that began in November and December and continued into the next spring. During Christmas week of 1811, the Osages, possibly also disturbed by the comet and tremors, arrived in Ste. Genevieve after an absence of many years and raised great anxieties among the villagers. The 1811 comet was more than just an astronomical curiosity to Ste. Genevieve residents, and the earthquakes and other unusual events of the year were placed in a broader context with religious implications, according to Fermin Rozier, a longtime resident and chronicler of the village's history.[68] Yet, even if the extraordinary natural events of 1811–1812 affected religion and personal behavior, they apparently had little direct effect on the settlement process of the Ste. Genevieve District.

Settlers assessed places for their healthfulness. Because the most prevalent health concern was the variety of fevers, they defined the environmental healthfulness, or "salubrity" *(salubrité)*, of a place. Other diseases such as smallpox, diphtheria, and whooping cough were also present, but fevers affected nearly everyone and were related to characteristics of a specific location and thus entered in decisions of where to locate.[69]

"The fevers" encompassed a wide variety of types, including what we know today as mosquito-borne malaria and yellow fever.[70] Though now known to be caused by mosquitoes that have previously bitten another human, malaria and other fevers were not directly associated with mos-

---

68. Rozier, *Rozier's History*, 319–20.
69. The classical nineteenth-century account of malaria and other diseases in the Mississippi Valley is physician Daniel Drake's *Systematic treatise*. Darrett B. Rutman and Anita H. Rutman explored the effect of malaria on economy, social structure, and culture in early Maryland ("Of Agues and Fevers: Malaria in the Early Chesapeake"), and medical geographer Melinda Meade has helpful comments on the effect of malaria on settlement of the American South ("The Rise and Demise of Malaria: Some Reflections on Southern Settlement and Landscape"). Chad McDaniel introduced the issue of malaria at Ste. Genevieve through analysis of death records in 1981 in his presentation on preliminary findings from a demographic study of Ste. Genevieve, later cited in Flader, "Ste. Genevieve" (6–8). Ekberg also described illnesses, including malaria, and their treatment at Ste. Genevieve (*Colonial Ste. Genevieve*, 247–67).
70. Terms included *autumnal* (perhaps the most common), *bilious, intermittent, remittent, congestive, miasmatic, malarial, marshy,* and *malignant fevers,* and *chill-fever, ague,* or *dumb ague*. For simplicity, *malaria* or *the fever* referred to them as a broad category. *Malaria* was not used in the early decades of the nineteenth century in the middle Mississippi Valley.

quitoes before the middle nineteenth century. Malaria causes fever alternating with chills, anemia, and enlargement of the spleen. Though usually not fatal by itself, it often leads to fatal complications, hence the necessity for care and treatment. Its onset in the middle Mississippi Valley was late summer and autumn, for which reason it was called "autumnal fever." It drained people of physical and mental energy, and, when endemic in a community, gave the impression of a population of dullards and slackers. Observations by visitors from August through November should be interpreted with this in mind. The fever often reoccurred in infected people the second year and perhaps in later years, but subsequent bouts were much less virulent and, indeed, a kind of natural immunity set in thereafter. Thus, a newcomer underwent a "seasoning" period.

Although everyone was subject to fevers, the perceived degree to which Indians, blacks, French, and Americans were subject to and suffered from fevers differed significantly. Whites thought Indians had some immunity and instead attributed fever-induced lassitude in Indians to their race and lifestyle. In general, entrants into the valley ignored the experience of Indians in assessing the healthfulness of the environment for European settlement.

Europeans and Americans thought that black-skinned persons were relatively immune to autumnal fever, and, when they did get the disease, it was milder and less enervating. Their origin from a hot, wet environment in Africa or the Caribbean had supposedly already seasoned them or given them a natural immunity. A major reason for the importation of Africans to Louisiana was their perceived natural ability to work harder and longer in summer heat and humidity, and a good part of this perceived ability was reduced susceptibility to fevers. It was much later that African resistance or natural immunity to malaria, which indeed is the case, was determined to be due to sicklemenia, or the "sickle-cell trait," among some Africans. Thus, the fact that black-skinned persons were less likely to get fevers probably played a role in their being brought in bondage to the Illinois Country for labor, whether in mines or river work.[71]

---

71. African slaves' resistance to malaria is discussed generally in Meade, "Rise and Demise of Malaria." Rutman and Rutman have comments on the difference in resistance to malaria between black and white populations ("Of Agues and Fevers," 34–35). Peter H. Wood also discussed the comparative susceptibility of Africans and Europeans in coastal colonial Carolina and concluded that it played a significant role in settlement and economy. He further noted that the sickle-cell trait is not a "racial" trait, but a trait of populations that have lived for generations with malaria, like Si-

Observers, if they noted a difference between the prevalence of fevers between French and American populations, generally concluded that Americans in the bottoms were more susceptible than the French, but Americans in interior locations away from big rivers were most healthy of all. According to Amos Stoddard, who was on the scene in the first decade of the nineteenth century and observed populations on both sides of the river, "The endemics in Upper Louisiana are almost exclusively confined to English Americans, who were born and educated in more northern climates; and even these after a residence of one or two years in the country, generally enjoy a good degree of health." Stoddard further noted a difference between the states of origin of Americans: "Those from the eastern and middle states are subject to these first diseases.... Those from the southern states enjoy good health."[72]

It is quite possible that the perceived difference between French and Americans was simply one of seasoning and length of residence in the valley. In earlier years, the French had a high incidence of fevers, but by the time the Americans began arriving and contracted the disease, often by whole family units, the French, without new immigrant infusions to their population, had already passed through the disease and acquired some immunity. A constant flow of Americans into the valley kept the disease prevalent among their cohort. Nevertheless, Dr. Drake considered the French population of the middle Mississippi bottomlands "sickly," and implied that their short stature and even their character had something to do with their long history of living in a fever-ridden environment.[73]

---

cilians (*Black Majority: Negroes in Colonial South Carolina from 1670 through the Stono Rebellion*, 76–91).

The 1804 "Remonstrance of the People of Louisiana against the Political System Adopted by Congress for Them" includes the assertion that African laborers are necessary because of the climate. Their "natural constitution and habits of labor enable them to resist the combined effects of a deleterious moisture, and a degree of heat intolerable to whites" (ASP-Misc 1:399). On the subject of Africans' being more capable of hard work than Europeans in Louisiana, see Giraud, *History of French Louisiana* 5:325; and Martin, *The History of Louisiana*, 130. Usner points out that even if Africans had a substantial hereditary immunity to malaria that made them useful for labor in the New World, they were susceptible to other new diseases contracted from Europeans, but this susceptibility was probably worked out by the time they reached the Illinois Country (*Indians, Settlers, and Slaves*, 34). Among recent studies that demonstrate how New World colonizers took advantage of known African resistance to malaria to import slaves is David Adams, "Malaria, Labor, and Population Distribution in Costa Rica: A Biohistorical Perspective."

72. Stoddard, *Sketches of Louisiana*, 237.

73. Constantin François Chasseboeuf, comte de Volney, *A View of the Soil and Climate of the United States of America*, 337; Drake, *Systematic treatise*, 138.

In general, the whole middle Mississippi Valley had a reputation of being fever ridden, but that reputation could be differently understood. For example, the French who came upriver from Lower Louisiana or who traveled to and from Lower Louisiana evaluated the Illinois Country in comparison to Lower Louisiana and clearly considered it more healthful. The fevers that enveloped the Illinois Country were less virulent than those that ravaged Lower Louisiana, and they occurred during a shorter season of the year. Yet, the fevers of the Illinois Country probably affected as great a percentage of the population as fevers in other similar regions open to European settlement.

The whole middle Mississippi Valley experienced fevers; no region was immune to them. Population until the 1780s was almost exclusively riverine, and one would expect that all of it was exposed to mosquito-borne diseases. Thus, there was no way, so it appeared, to avoid fevers altogether in choice of settlement location. The American George Morgan, who examined the valley in the mid-1760s, reported that ague and fever were "remarkably prevalent" from June to October, and that "of the Garrison & Inhabitants of Fort Chartres & Kaskaskia few have escaped being more or less afflicted therewith. . . . But the Climate very well agrees with Cannadians & Europeans, except the first Year or two." Morgan's statement, however, disregards the unknown numbers of people, whether French Canadians or Americans, who might have chosen not to come because of the valley's reputation for fevers. According to at least one credible observer in the 1790s, Henry Marie Brackenridge, the reputation of the disease prevented "many" from coming and caused "many" to return to the East.[74] Thus, the perceived unhealthfulness of the valley as a region probably slowed its settlement.

As population extended into interior locations some distance from the Mississippi bottoms, the reputation of the valley as fever ridden changed to incorporate more geographic precision, but a precision that was confounded by the constant movement of people. Those who had contracted the disease in Kentucky, for example, and brought it with them when they moved into the more healthful hill country of the eastern Ozarks helped to identify their new settlement as fever ridden. Not knowing how the disease was acquired or spread, an observer passing through that location would have concluded that it was unhealthful. Thus, relatively disease-free locations may have gotten a bad reputation undeservedly.

Experience had taught settlers that the first clearing in the woods was

---

74. "Voyage down the Mississippi, November 21, 1766–December 18, 1766," in *New Regime,* ed. Alvord and Carter, 11:439; Brackenridge, *Views of Louisiana,* 110.

fever free, but when the settlement enlarged and other families came and cleared land, fevers appeared. Hunters, trappers, explorers—all basically solitary folk—experienced far fewer cases than cultivators of the soil. Human carriers are needed to transmit the disease. The observation that solitary cabins were disease free may have been a reason some families constantly sought remoteness and "moved on" when others joined them in the neighborhood. Yet, even though fevers became rife when a neighborhood was established, disease did decline as population increased and the land became more civilized and tamed, that is, more deforested, cleared of brush, and laid bare.[75] The well-recorded strong desire to populate a new settlement neighborhood as fast as possible and to remove trees may be associated with a desire to rid the neighborhood of fevers. Expectations that fevers would eventually be overcome strongly sustained settlers through their period of seasoning. In their process of changing the environment from wilderness to one of cleared farms, they would free it from the scourge of febrile diseases, and the period of seasoning was therefore to be stoically endured when conquering new lands.

As experience accumulated, settlers developed a keener relationship between fevers and specific locations. They recognized several common characteristics of a fever-prone site: water, heat, vegetable decomposition, and darkness. These factors converged in late summer and autumn in shaded, standing water with vegetation growing in it and rotting. These favorable conditions occurred on the Mississippi River bottomlands, where floods left pools of water and marshes throughout the summer, although just as important were wetlands from local rains and high water tables. The conditions also occurred in creek valleys where quiet pools of water darkly shaded by tall trees during the sultry days of late summer produced ideal "miasmatic" conditions. Favorable conditions also occurred on the upland prairies where wet winters and springs left water standing throughout the summer in solution depressions with vegetation rotting from mats of the previous year's dead grasses.

By all accounts, the Mississippi River floodplain, including the village of Ste. Genevieve, was the most fever-ridden environment of the district. Chad McDaniel's unpublished study on mortality rates at Ste. Genevieve village strongly supports the notion that fevers now called malaria, or complications from them, brought on deaths, especially of infants, in late

---

75. Erwin Heinz Ackerknecht, *Malaria in the Upper Mississippi Valley, 1760–1900*, 75; Drake, *Systematic treatise*, 710, 727. See also Benjamin Rush's comments on the effect of land clearing and cultivation on fevers in the American East ("An Inquiry into the Causes of the Increase of Bilious and Intermitting Fevers in Pennsylvania").

summer and autumn. The European-educated commandant of New Bourbon recognized that illnesses in his district came from "putrid unhealthy exhalations of . . . stagnant and putrescent waters" and acknowledged the necessity to drain the alluvial marshes at the base of the bluff on which he lived. A few years later, American captain Amos Stoddard concluded decisively that diseases were "superinduced by the pestilential vapors, which arose from the rivers, and from the decayed vegetable substances, produced in great abundance on the bottoms and along the borders of them." An inspection of Spanish records indicates that years with general illness in the Illinois Country, severe enough to comment on in official records, quite often followed major spring and summer river floods: 1784, 1785, 1788, 1794, and 1797.[76]

Conversely, sites perceived to be less fever prone were open lands on high ground from which water drained away rapidly: cleared fields and bare ground, even dead prairie grass or forest litter not in water, and all sunny places. These conditions were met on bluff tops, such as at New Bourbon, but when breezes blew from the bottoms in summer, people at New Bourbon came down with fevers like anyone else. In 1826, the United States Army built Jefferson Barracks, the military point of control for the American West, high on a Mississippi River bluff top just south of St. Louis with intentions to get away from the fever-ridden land along the river, but again, when winds blew from the river bottoms, the site was no longer fever free. Many a young American soldier from the East was initiated into the hazards of western life by contracting enervating fevers at Jefferson Barracks.[77] Settlers favored the sunnier, less wooded north sides of valleys over shadier south sides, but when summer south winds blew over the valleys, they too succumbed to fevers. Thus, the perception of relatively disease-free locations often proved false.

Perhaps the most fever-free environment in the Ste. Genevieve District was the mining country that was cleared of timber and distant from al-

76. Chad McDaniel, "Aspects of the Demography of Ste. Genevieve"; [Luzières], "Observations on the means," Census of New Bourbon, 1797, PC 2365-345. The commandant, his wife, his daughter and son-in-law, his youngest son, and his hired men (presumably also newly arrived) all had fevers shortly after arrival in Upper Louisiana (Dehault de Lassus to Baron de Carondelet, October 17, 1794, PC 209-666). Stoddard, *Sketches of Louisiana,* 237; PC 201-959, 204-717, 208a-480, 209-666.

77. Drake, *Systematic treatise,* 138. In a similar manner, Thomas Jefferson earlier drew the plan for the new town of Jeffersonville, Indiana, in 1802 explicitly to prevent "miasmatic conditions that lead to yellow fever" (John Reps, *Town Planning in Frontier America,* 194–96).

luvial rivers. Rainwater drained rapidly through the coarse-textured soils. Everyone acknowledged that the relative absence of fevers contributed to the general healthfulness of the mining country. (The health hazard from lead mining and smelting was a separate matter.) The French and Americans attributed the salubrity of the mining region, and of the interior in general, to the abundance of carbonate rocks. Among them was Captain Stoddard, who noted that the presence of calcareous rock in the interior "neutralizes the putrid exhalations or prevents them."[78] Because carbonate rocks underlie the interior, its residents breathed "neutralized air" and presumably seldom succumbed to endemics, following the contemporary reasoning that fevers came from breathing "bad air" rather than from mosquitoes. Limestone was a purifier of water. This belief not only was held by educated persons in the district, but also was a basic tenet of folk wisdom among ordinary farmers as well.

If settlers thought high ground and interior locations were more healthful, then why did they continue to live on unhealthful low ground? One answer is the perception that fevers were temporary and one would overcome them after seasoning. Another answer is that settlers decided that the advantages of more productive land and better accessibility to markets from locations in the bottoms overcame their fear of endemics, however strong. The French visitor Count Volney concluded that in the great American interior, "Health is never allowed to enter into competition with profit, and the more fertile plains and hollows one therefore chose . . . though at the expense of health, and the hazard of life."[79]

Besides fevers, another contemporary concern about the linkage between the healthfulness of the natural environment and settlement revolved around the belief that winds from different directions have a direct effect on human health and temperament. Evidence of this is found among French, American, and Spanish populations; Indians and African Americans might have held similar beliefs.

The notion that the classical four humors were associated with climate, through winds, was an entrenched belief in eighteenth-century France. The four elements of air, fire, water, and earth were matched with the four humors of the body respectively: blood (hot-moist), yellow bile (hot-dry), phlegm (cold-moist), and black bile (cold-dry). Climate influenced, if not controlled, these humors. Cold, dry north winds produced people who were physically robust, but gloomy and not very intelligent,

78. Stoddard, *Sketches of Louisiana,* 238.
79. Volney, *View of Soil and Climate,* 230. Rutman and Rutman similarly note that settlers took malaria in stride ("Of Agues and Fevers," 54).

and south winds produced people who were passionate and intelligent, but lazy. Crime increased in summer because of the nature of winds. Climate and winds also affected human health in periods of time shorter than seasons. For example, a sudden change from warm, moist south winds to cold, dry air from the northwest, as with the passage of a cold front, could not only change one's mood and attitude by upsetting the balance of humors in the body but also bring on physical ailments. If one humor dominated over the others, mental and physical sickness resulted.[80]

Both French and Americans were sensitive to the effect of winds on human health. Count Volney wrote observations during his sojourn in the Ohio Valley in 1795–1798 such as "northeast winds in the United States oppress the brain and produce torpor and head-ache." Amos Stoddard asserted with conviction that western winds in the middle Mississippi Valley are "more easy for respiration" and "brace the system," and south winds "create lassitude and sluggishness."[81] South winds prevail in the valley in summer, so any unquestioned belief in their effect on the human mind and body could actually bring about less work and effort, as a self-fulfilling prophecy. Was belief in the effect of winds responsible in part for the reputed languor of the Mississippi Valley French Creoles and black slaves in summertime?

It was the Spanish who brought the relationship of winds and human health into official settlement planning. Their planned cities were to be laid out so that winds had an equal opportunity to enter the city from all directions to keep the four humors in balance and to keep harmful winds from entering. Thus, the city grid was to be positioned with the four corners of the square pointing in the four cardinal directions and the grid tilted at an angle of forty-five degrees from north so that harmful winds could not easily enter the city and flow down streets to "cause much inconvenience." Buildings were to be placed in such a manner as to enjoy the better north and south winds.[82] The village grid of new Ste. Genevieve is laid out at a large angle away from north. Was this sixteenth-century notion of town planning, sensitive to wind direction and incorporated into the Law of the Indies, brought to the middle Mississippi Valley by a Spanish administration in the 1780s?

---

80. Clarence J. Glacken, *Traces on the Rhodian Shore: Nature and Culture in Western Thought from Ancient Times to the End of the Eighteenth Century,* 10–12, 80–82, 557.

81. Volney, *View of Soil and Climate,* 137; Stoddard, *Sketches of Louisiana,* 235.

82. Dora P. Crouch, Daniel J. Garr, and Axel I. Mundigo, *Spanish City Planning in North America,* 34, 36.

Natural Regions

At the beginning of the nineteenth century, when people walked more than rode and their radius of travel was less, their division of space into natural regions was at a local scale. To them, meaningful environmental differentiation was between field and woods, ridge and valley, marsh and bluff, and between useful and "useless" country. These perceived microregions, or environmental niches, numbered into the hundreds for the Ste. Genevieve District. However, to make sense of their land on a broader scale, settlers mentally aggregated these small niches into four larger regions of greater generalization, defined by perceived usefulness of the region's resources rather than its inherent natural characteristics. Theirs was a largely extractive economy that developed directly from exploiting the resources of the environment and gave little thought toward understanding and preserving it.

The four regions lie roughly in belts parallel to the Mississippi River at Ste. Genevieve (fig. 4). In sequence, they are the alluvial plain or bottoms along the Mississippi River, the karst plain and bluff lands, the rough hills, and the mining country. Beyond the mining country lay a seemingly endless country of more rough hills, poorly known and largely unused before 1820.

Mississippi Alluvial Plain

To the traveler on the middle Mississippi River, the alluvial plain appeared as a five-mile-wide flat valley bounded by two parallel chains of hills or bluffs.[83] The Mississippi River meanders through the plain, leaving segments of it first on one side, then on the other.

In the Ste. Genevieve District, the Mississippi alluvial plain occurs in two major disconnected segments. The Ste. Genevieve bottom, or Pointe Basse, the smaller of the two, lies between the river and the villages of new Ste. Genevieve and New Bourbon. About ten square miles large in the middle eighteenth century, it contained a common field known as Le Grand Champ, or Big Field, and the old village of Ste. Genevieve. The larger segment is the Bois Brule bottom below the mouth of the Saline River in Perry County, three miles wide by fifteen miles long. The alluvial plain in the middle Mississippi Valley is much more extensive on the Illinois side where it came to be known as the American Bottom before the Louisiana Purchase. The former point of land in the American Bottom at Kaskaskia, now called Kaskaskia Island but physically continuous with the Pointe Basse and the Bois Brule bottom, was separated from the

---

83. Collot, "Journey in North America," 287.

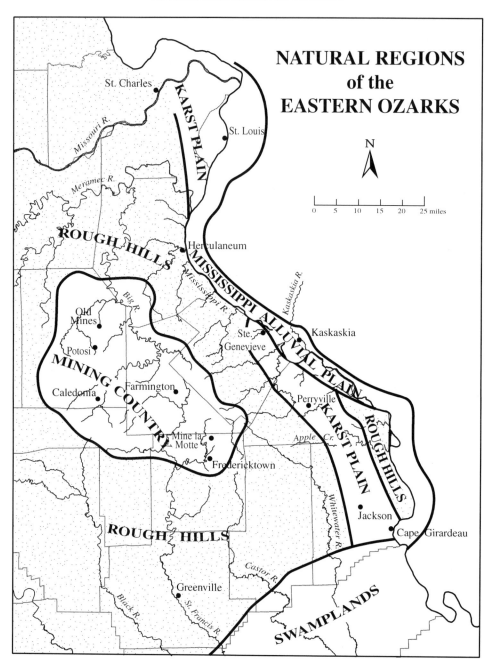

4. *Natural Regions of the Eastern Ozarks.* Settlers conceived the Ste. Genevieve District as consisting of four natural regions.

American Bottom by an avulsion of the Mississippi River in 1881.[84] Except for this spectacular, radical channel change, changes in the geography of the channel have come from the ordinary daily erosion and accretion of banks and islands.

The alluvial plain in the Illinois Country was primarily a natural prairie with trees forming walls along riverbanks and colonizing islands. Grasses and sedges differed from those in the drier upland prairies and required the special designation of *prairie basse* or *prairie humide*.[85] Standing water *(marais)* and marshes characterized the alluvial plain next to the bluffs. A long *marais* lay at the base of the bluffs at New Bourbon, and an even longer lake, a *grand marais,* was also against the bluffs of the Bois Brule bottom.[86]

An alluvial-plain environment was familiar to the Ste. Genevieve French Creoles from their earlier experiences in eastern-bank villages. At Kaskaskia, Fort de Chartres, and Prairie du Rocher, they had become familiar with alluvial land, floods, bank caving, and fevers. They knew how to pick out slightly higher ground for structures and crops and leave the marshy swales for natural hay meadows. They avoided the lowest-lying land at the lower end of a bottom when laying out their long-lot properties. They adjusted their crops to the soils that varied within short distances. They persisted with their staple wheat, but the rich soils were quite suitable for roots, vegetables, grass, hemp, flax, and fruit trees. The long-lot property subdivision of the French cut across the curvilinear swales and ridges and tended to even out soil differences among landowners. These alluvial soils sustained yields year after year. Captain Stoddard noted in the early nineteenth century that the American Bottom had been under cultivation for more than 120 years but was still as fertile and productive as land just put into cultivation.[87]

---

84. J. H. Burnham, "Destruction of Kaskaskia by the Mississippi River."

85. See the 1734 Broutin map in Tucker, *Atlas,* pl. xxii and Collot's 1796 map in "Journey in North America" (287). A model of natural vegetation of the Ste. Genevieve bottom done for archaeological purposes and based on soils and other data contains six different "floral associations." Four of the associations, accounting for more than three-quarters of the bottom, are grass dominated: bottomland (tall grass) prairie, bottomland (wet) prairie, bottomland grasses-forbs, and bottomland marsh (Eric E. Voight and Michael J. O'Brien, "The Use and Misuse of Soils-Related Data in Mapping and Modeling Past Environments: An Example from the Central Mississippi River Valley").

86. *Régistre d'Arpentage,* 223.

87. Stoddard, *Sketches of Louisiana,* 229. A measure of how much the alluvial land was prized is the relative value of the Bolduc landholdings when the estate was sold in 1817. Land in the Pointe Basse alluvial plain went for $8.90 per square arpent, and croplands in the uplands went for a paltry $2.08 per square arpent ([Partition and sale of estate of Louis Bolduc], July 1, 1817, LRB 561).

An alluvial-plain environment was also reasonably familiar to Americans. Those Americans who first settled the Bois Brule bottom with their square properties wisely selected the highest ground and choicest soils, leaving the wet clay marshes for latecomers to cope with. Plats surveyed before 1804 for Americans in the Bois Brule bottom meticulously distinguish its microenvironments: *bonne terre à maïs* (good soil for corn) on the higher ground, *partie qui noie* (part that floods), and the large *marais du Bois Brulé*. Even the slight irregularities of the alluvial surface, which were so important for differentiating soil wetness, appeared as *terrein bas inégal et humide* (wetland, low and uneven).[88]

Other useful resources were cane *(Arundinaria gigantea)*, which grew in almost impenetrable thickets of cane brakes along some riverbanks and backwaters, and rushes *(Equisetum hyemale)*, which provided winter feed for animals. Wintering of animals on Mississippi River islands, such as Horse Island at the northern end of the Bois Brule bottom, was a common practice.[89] Low water in winter facilitated coaxing animals to swim to the islands, but the water was still high enough to keep them contained.

The alluvial plain was the earliest natural region to be occupied by farmers who began to put the Pointe Basse under the plow in the late 1740s, and it was the most thoroughly occupied, with no usable land unclaimed in 1820. It was also the most thoroughly humanized landscape, being transformed from wet prairies, woods, and riverbank timber into neat rows of field crops and pastures. It was, to the early-nineteenth-century eye, a civilized, and therefore beautiful, landscape.

Bluff Lands and Karst Plain

Behind the alluvial plain lie the bluff lands and karst plain, a belt from six to twelve miles wide and a distinct natural region to both French and Americans. Neither group conferred a name on the region, except as a prairie or barrens for its open landscape, but it was special in their minds. By the end of the Spanish period, it was thoroughly occupied in stark contrast to the next belt inward, the rough hills, and this contrast persisted through the 1820s. The inner, or western, boundary of the karst plain served therefore as the sharp boundary of contiguously occupied land in the district.

When viewed from the curving Mississippi, which brings the viewer alternately closer and farther from it, the straightness of the bluff line

---

88. *Régistre d'Arpentage*, 229.
89. John Reynolds, *The Pioneer History of Illinois*, 195.

may not be readily apparent, and the Finiels map shows the bluff lines as a series of cusps. But when sighted along their base or their crest, their straightness is unmistakable. The French called the bluffs at new Ste. Genevieve, where at sixty feet they are lowest, the *petites côtes;* where the bluffs are slightly higher at New Bourbon by twenty feet, the sharp French eye distinguished the *grandes côtes,* but this term has meaning relative only to the even lower bluffs at Ste. Genevieve.[90]

Limestone ledges crop out here and there along the bluff face, although debris washed down from above obscures most bedrock faces. The limestone ledges, readily separable into convenient-size slabs and large blocks, provided stone for buildings and paving. A colluvial apron of loessial soil wash spreads out at the base of the bluffs and merges into the soils of the alluvial plain.

Ravines and small creek valleys interrupt the bluff line, so ingress into the upland interior was possible without mounting the bluff face directly. A prominent ravine likely determined the precise site of New Bourbon village. The widest break in the bluffs behind the Pointe Basse was the double entry of the two forks of Gabouri Creek onto the alluvial plain. Here there is no bluff at all, and the low terrace between the two small forks presented a logical choice for the site of new Ste. Genevieve, just off the Mississippi alluvial plain but still adjacent to the village's croplands and the river.

The plain that lies atop the bluffs extends from the mouth of Establishment River southward through Perry County and past Apple Creek into Cape Girardeau County. The plain is six miles wide at Ste. Genevieve and broadens to nearly fifteen miles at Apple Creek. Its best expression as a smooth to rolling surface is behind Ste. Genevieve and New Bourbon and again in central Perry County from Perryville to Apple Creek. Where streams cross the plain, as along the River Aux Vases and the branches of the Saline, the plain loses its smoothness and becomes distinctly hilly, but not nearly so rough and rocky as the hill country to the west.

The underlying highly soluble limestone is responsible for a karst surface pitted with hundreds of sinkholes and solution basins into which surface runoff sinks and then moves underground through caverns. A surveyor wrote on his plat of a concession southwest of New Bourbon, "point ou les 3. sources s'engoufrait et disparaissent" (place where the

---

90. *Côte* (Spanish *cuesta;* the English cognate is coast) implies hills with a linearity, as in a coastline (e.g., the *Côtes d'Or* of landlocked Burgundy). The term was used, and still is today, for linear bluffs bordering a river floodplain and not as a general term for hills (Henri Baulig, *Vocabulaire Franco-Anglo-Allemand de Géomorphologie,* 32, 440).

three springs are swallowed up and disappear).[91] Shallow ponds spangled the karst plain, often several on a single property of which the largest was the Grand Marais, a solution basin of three hundred acres. The ponds provided water for villagers' livestock, and deer and other wild animals congregated at them, making them choice spots for hunting and trapping. Plats show numerous springs *(sources* and *fontaines)* along the Gabouri Creek, the River Aux Vases, and the Saline River and smaller streams.

Settlers rapidly claimed this natural region, once the bonds that kept them on the alluvial plain had been broken. The French moved onto the karst plain behind the Pointe Basse in the mid-1780s and the Americans onto the barrens of Perry County in the late 1790s. Under human use, it was a more fragile environment than the alluvial plain. The loess and limestone soils readily eroded, once their grass cover was broken even from the shallow scratchings of a wooden plow, and soil erosion and loss of fertility became serious problems on some fields after only ten years of use.

Depletion of wood was another environmental concern. The karst plain, meagerly timbered to begin with and rapidly carved into farms after 1790, could supply neither the Ste. Genevieve village blacksmith with enough wood for his charcoal furnace in 1799 nor the skin tanner who worked two miles away from the village.[92] By 1811, just twenty years after the establishment of the new village of Ste. Genevieve, the usable timber on the karst plain west of the village was nearly gone.

In the barrens south of the Saline River, abundant grass, sufficient but not abundant timber, adequate water, and friable, fertile soils were the foundation for a highly successful farming community of Kentuckians. The same resources characterized the karst plain south beyond Cinq Hommes Creek occupied by agricultural Shawnee and Delaware who farmed the land not much differently from Americans and the French.

Rough Hills

Westward from the karst plain are the rough hills, a thoroughly dissected landscape of a never-ending succession of wooded hills and valleys that characterize literally thousands of square miles of the Ozarks. In the Ste. Genevieve District, the natural region stretches from the Meramec River in the north to beyond Apple Creek in the south. The belt fronts directly on the Mississippi River from the Meramec to the Establishment River and in that reach creates high, pinnacled bluffs, the high-

---

91. *Régistre d'Arpentage,* 191.
92. LPC 54, LCS 15; LCS 202.

est and boldest rock cliffs anywhere along the Mississippi River between St. Louis and New Orleans. Near the mouths of the Joachim and Plattin Rivers, the bluff tops rise higher than three hundred feet above the river water, although the sheer rock face is much less than that. In this reach, the river washed the base of the rock-faced cliffs to form subsurface ledges or shoals, called *platin,* most likely the origin of name of the Plattin River.[93] Another disconnected tract of rough hills front the Mississippi much farther south, from Cinq Hommes Creek to Cape Girardeau, and here too the cliffs are bold and rise directly from the river.

The rough hills were a landscape of "mostly oaks and hickories—much dispersed—with a vigorous growth of native grasses beneath," as Schoolcraft described his thirty-four-mile route from Herculaneum to Potosi in 1821. Fourteen miles of the distance was along a stony, "arid ridge" without water, "firm enough for an excellent road, but the growth of timber was not sufficiently dense to protect the traveller from the rays of the sun." In places the oaks were scrawny post oaks.[94] Tree cover was denser in the deep, narrow valleys.

The French left the sterile hills mostly vacant, occupying only a few locations for special resources such as sugar maples, pines, and cedars, and using both the creek bottoms and grassy woods for grazing. In fact, they drove their cattle up narrow valleys in winter for protection and for winter feeding on creek-bottom cane.[95] But all these various activities in the rough hills were widely spaced and seasonal. Surveyors described lands on the Establishment *as terre généralement montueuse couverte de pins et d'un sol peu fertile* (mostly hilly and pine covered, with infertile soil), and even more bluntly as *mauvaise terre* (bad land). Land along the River Aux Vases was hardly any better: *montagneuse et pierreuse* (mountainous and rocky) and *terre maigre* (unproductive land).[96]

To the woodsman Americans, however, the same rough hills were desirable and inhabitable. An economy built around miniplots of corn and open-range stock raising was adaptable to this environment, but it required considerable acreage per family. It was easy to raise cattle in summer on the native grass of the open oak woods of the ridges, and to winter them in the valley bottoms on cane and rushes. Hay, necessary for stabled horses and cows, could be gotten from the native grasses. Open-

---

93. *Plateen* was the term for the rock ledge at that place (Z. Cramer, *Navigator,* 134), although John Francis McDermott defines *platin* as a marshy place, from Lower Louisiana usage (*Glossary of French,* 120).
94. Schoolcraft, *Travels in the Mississippi Valley,* 239, 252; *Régistre d'Arpentage,* 85.
95. LCS 78, LCS 236.
96. *Régistre d'Arpentage,* 65, 75, 76, 80, 88, 189, 218.

range swine subsisted on the abundant mast of the oak woods with hardly any care. The Americans strung out their solitary farmsteads in the valleys, finding and taking advantage of the small pieces of good alluvial loams in the creek valleys. U.S. land-office surveyors without fail reported creek-bottom land as first-quality land and fit for cultivation as interpreted from the presence of walnut, elm, ash, and other tree indicators, but they reported the rocky slopes and ridges as unfit for cultivation.[97]

The agricultural Shawnee and Delaware, like the French, also avoided the steep, rough hills when occupying land along Apple Creek. Instead of settling right along the Mississippi, which might have been expected for the superb access it provided, they passed over ten miles of rough hills and placed their villages and fields in the fertile karst plain. Perhaps they wanted to get away from the river for other reasons, but it made sense for these Indian farmers to live and farm on the best-soil land of their large Spanish grant.

Most of the rough hills remained *domaine royale* under the Spanish and then renamed public domain under the Americans. As such, everyone could use the hills usufructually for hunting, timber cutting, and grazing. A surveyor for the United States argued in 1816 against spending government money to survey the hills into sections for settlement, because he thought little of it would ever be bought by incoming settlers and the government would lose money on its survey. The rough hills region was the natural region of the Ste. Genevieve District least changed by human use of it.[98]

Mining Country

Beyond the rough hills lay a fourth region with immense amounts of lead ore and known simply as the mineral district, or mining country. Written on French maps as early as 1718 were such designations as *pays plein de mines* and *terrain rempli de mines*. Early American maps similarly have "lead mines" written on them in this area.[99]

---

97. U.S. General Land Office, Field Notes, Missouri, vols. 206–8.
98. William Russell to William Rector, April 20, 1814, *TP* 14:753, 758. Where the rough hills are loess-capped, as behind Ste. Genevieve, they were settled by German immigrants beginning in the 1830s.
99. Tucker, *Atlas,* pl. 15, 24; Gary E. Moulton, ed., *The Journals of the Lewis and Clark Expedition,* maps 6, 123, 125. The French used *mine* to refer to a deposit, whether worked or unworked. Thus, the designation on maps of a region *rempli de mines* does not mean that the lead ore was being mined, but merely indicates the presence of lead. Therefore, *Mine à Joe* may be translated either as Joe's discovery or Joe's working mine. The American "diggings," however, always implied that ore was being produced or had been produced in the past.

The French exploited lead ore first around 1720 at Mine la Motte in the St. Francis River drainage and Mine à Renault in the Meramec River drainage. At the turn of the nineteenth century, eight decades later, miners were still digging by pickax in those two drainage basins, and they continued digging into the 1820s and beyond until the bounds of surface ore became known. Miners dug into valleys, slopes, and ridges. They cut timber for log furnaces from whatever source was nearest.

Mixed among the several mining localities were smooth limestone- and sandstone-floored basins with good soil that Americans occupied around 1800 despite their distance from the Mississippi. Partially timbered, partially in grass, and well watered, they offered a highly desirable spatial combination of the basic frontier resources on any square mile of land. The good residual soils of each basin contrasted sharply with the uncultivable land of surrounding igneous knobs and chert hills. The Bellevue Valley was judged by Spanish government surveyors as *terre de la premiaire qualité,* and at Cook's and Murphy's, the land was uniformly *bonne terre à bled et à maÿs.* The U.S. land surveyors similarly judged all basins as "first rate" for agriculture.[100] Although the basins did not contain lead ore, their farm products fed the nearby miners, and the farmers themselves found seasonal employment in mining neighborhoods.

More than half of the total area of the mining country, however, lay unoccupied on account of its roughness and absence of arable soil. If the hills contained any settlement at all, it was because miners had no choice but to live on the poor land at their mines. Lead miners at Old Mines lived on land described as *terre montueuse et escarpée* (hilly and broken land) with numerous springs.[101]

The Ste. Genevieve District lacked none of the natural resources for rapid and successful settlement by Europeans and Americans. They were abundant, widespread, and conducive to productive and prosperous settlement. But the simple presence of basic natural resources is not sufficient to explain their utilization by settlers. Immigrants brought with them different notions of which land to occupy and how to occupy it, and governmental land policies and regulations, both Spanish and American, affected resource exploitation and the spread of settlement in general. Who settled in the district and where, how they acquired land, and the multiple issues concerning land titles are topics that need next to be addressed in the story of the settlement of the Ste. Genevieve District.

100. *Régistre d'Arpentage,* 100, 101, 239, 240, 246, 247, 248, 250; U.S. General Land Office Surveys, Missouri, Field Notes, vols. 204–7.
101. *Régistre d'Arpentage,* 121–22.

Part II
Regulating Settlement

3
French and Spanish Periods to 1795

Choice of land is critical for success in the settlement process. Where and why do settlers choose to locate in new lands? All persons arriving bring with them practices and customs from their earlier locations and, with adjustments to any new circumstances, would be expected to continue as before in the new country. Their basic culture instructs them how to lay out their farms, how to settle in kinship networks, how to regard the healthfulness of sites, and how to make a multitude of other decisions. Thus, the Ste. Genevieve District received the vernacular practices of settlers from Canada; monarchical France; Africa; the Gulf Coast and Caribbean; the Ohio Indian Country; tidewater Maryland; and Appalachian Kentucky, Tennessee, Virginia, and Carolina. The result is an exceptional mix of landscapes, each one with its unique settlement history.

But the vernacular settlement process of making land decisions by culture transfer from one region to another does not tell the whole story. The settlement process in the Ste. Genevieve District, as in other places undergoing the invasion of a people into new territory, was overseen by governments that had goals for the new land and developed strategies to achieve them. Situated aloofly in distant places such as Paris, Madrid, New Orleans, or Washington, administrators had plans that fitted into the larger schemes of central governments than the geographical peculiarities of a local frontier situation.

The Spanish government viewed Upper Louisiana broadly in defensive terms, as a front line against the British and then the Americans. The Spanish developed strategies for settlement, such as immigration policy and settlement clustering, to meet the goal of defense against invasion. Later, the American government, President Jefferson in particular, had plans to use trans-Mississippi Upper Louisiana to solve a national problem—the relocation of displaced eastern Indians—although it abandoned these plans before detailed strategies could be implemented. The

American government then viewed the trans-Mississippi West, in general, as a region for continuing settlement of westering Americans. It developed numerous strategies for the orderly settlement and occupation of the land, but those strategies, formulated from an eastern perspective and, like those of the Spanish, often inappropriate given the geographical peculiarities of a local frontier situation, did not always harmonize with the settlement process under way.

The vernacular process brought to the Ste. Genevieve District by settlers, therefore, took place under an omnipresent veneer, sometimes thick, sometimes thin, of high-level governmental goals and strategies. To a large degree, the two converged. For example, the French custom of living in small villages with surrounding fields and grazing lands, which may have had ancient European roots in defense purposes, fitted nicely with the Spanish administrative strategy of clustering for defense and populating the province with a peasantry. The American custom brought from Kentucky and Appalachia of living dispersed on separate farmsteads fitted nicely with the government's land-division system of large squares and rectangles. The government's role was to establish a broad framework and allow individuals to develop their parcels as they saw fit.

Yet, in other situations, the vernacular and governmental approaches to settlement were disharmonious, and larger-scale, longer-term government perspectives clashed with the way people wanted to occupy and use the land, which was usually with smaller-scale, shorter-term perspectives. Examples of disharmony are: a Spanish immigration policy that prohibited settlement by certain people who would have populated and developed the land; Spanish land-granting regulations that favored the rich with more land and more valuable land than they could gainfully occupy and use while restricting other productive entrepreneurs; attempts of American officials to geographically limit settlement in Missouri of people who defied such limitations; an American mineral land–leasing policy unacceptable to local miners; and American policies to prohibit white settlement on Indian land and the unsurveyed public domain, both of which were contrary to settlers' practices.

The first stage of settlement of the Ste. Genevieve District ended in 1770, the end of the French regime in the valley, by which time the initial imprint of European culture had been made on the landscape by the planting of the villages of Ste. Genevieve and St. Louis and the establishment of an economy based on agriculture, lead and salt, and trade in furs.

The second stage ended in 1795, after a quarter century of Spanish rule, during which population grew only slightly and with little expansion of the settled area.

The third stage, the final decade of Spanish rule, was one of rapid and radical changes in governmental policies and settlement processes, resulting in significant population growth and spread of settlement. The Americanization of the district began, even though Spanish governmental policies for settlement still officially prevailed. Toward the end of that period, the anticipated change to an American government enfeebled administrative control of the settlement process and fostered manipulation of the law for personal gain.

The fourth stage was the American period from 1804 to 1818, a period of confusion in governmental strategies to direct settlement while the flow of settlers who occupied land through the vernacular process continued. It was a time of great disequilibrium between policies on paper and processes on the land, a time of uncertainty and insecurity in settlement activity.

The fifth and concluding stage began with land sales in 1818 and ended quite arbitrarily at the end of the 1820s. Introduction of land sales restored order to the land-settlement process, and other land issues of the previous stage found eventual resolution. During this period, counties and towns took form, and an incipient spatial system of central places emerged. By the end of the 1820s, administrative rules for acquiring land were fast converging with the local vernacular processes of occupying it.

French Period before 1770

Canadian French moving through the Great Lakes region established the first French settlements in the Mississippi Valley at the beginning of the eighteenth century. Cahokia, founded in 1699 as a missionary outpost, also served as a base for fur traders and other itinerants. Kaskaskia, settled by Canadians in 1703, became the most important of the string of French and Indian settlements in the alluvial lands along the Mississippi. During these early years, the French introduced black slaves into the Illinois Country for boat rowing, mining, building, and only incidentally for agriculture.[1]

The French experience of moving onto Indian lands in Canada was

---

1. Balesi, *Time of the French*, 246–47; Daniel H. Usner Jr., "From African Captivity to American Slavery: The Introduction of Black Laborers to Colonial Louisiana." Before black slavery was introduced into the Illinois Country and concurrent with it for some decades thereafter, the French had Indian slaves. See Call, "French Slaves, Indian Slaves."

The French regime in the middle Mississippi Valley in general is discussed by Belting, *Kaskaskia under the French;* Alvord, *Illinois Country;* and Zitomersky, *French Americans–Native Americans,* whose chap. 1 (pp. 3–47) is a historiography of French-colonial Louisiana.

transferred to Cahokia and Kaskaskia and from there extended to new settlements on the western side of the Mississippi. In its broadest conception, French policy was to embrace the Indians and transform them into French, or *la mission civilisatrice*.[2] By the time it reached the Illinois Country, French policy was basically one of friendly cooperation, founded on trade, gifts, and living among the Indians and sharing their lives. French colonial administration needed no procedure for extinguishing Indian title to the land, because the two peoples shared it. Even though the French themselves could acquire "official" ownership of the land from their own colonial administrators by land grants, they recognized no "official" Indian title to the land in the Ste. Genevieve District.[3]

During the French period, the king or his designate in New Orleans had the authority to grant land and permission to exploit resources. However, according to Winstanley Briggs, local administrators in the Illinois Country ended up registering land locally for whoever requested it, acting like modern county recorders and not like the officious administrators in Lower Louisiana and Quebec.[4] In the Illinois Country, a select few received large grants, among whom was Philippe François Renaut, who received six grants in the Illinois Country in 1723. Each was an immense tract of $1\frac{1}{2}$ leagues of frontage, usually on a river, by a depth of six leagues, or approximately eighty-four square miles each, or five hundred total square miles. Two were in the historic Ste. Genevieve District, one at Mine la Motte and the other at Mine à Renault, both clearly for lead mining. Renaut's grant at Fort de Chartres was specifically for food supply for miners. Renaut came to the Illinois Country, worked the mines of his grants, then returned to France in 1742, thereby letting his grants lapse.[5] The Mine la Motte tract stayed reasonably intact and undivided, passing through several ownerships, numerous leases, and considerable litigation, and ended up in a square. It is Missouri's oldest continuously European-titled tract of land. Mine à Renault, however, reverted to *domaine royale* upon abandonment.[6]

Though lead mining was the earliest instigator of governmental land granting in the historic Ste. Genevieve District, other settlement activities had also commenced before 1770. The French occupied the mouth of

2. Olive Patricia Dickason, "A Legend Reconsidered: The French and Indians in the New World," 21–22.
3. Charles C. Royce, comp., *Indian Land Cessions in the United States*, 545–49.
4. Briggs, "Le Pays des Illinois," 33–34, 44–45.
5. Martin, *The History of Louisiana*, 152; Rothensteiner, "Earliest History." Renaut signed his name without an *l*, but place-names deriving from his name are spelled Renault (Ekberg, *French Roots*, 35).
6. Ekberg, "Antoine Valentin de Gruy."

Saline Creek to make salt and the Pointe Basse alluvial bottom for agriculture. Ste. Genevieve village came into existence in the late 1740s, probably by informal Creole practices, but the settlement was soon formalized and grew by land concessions issued by the French commandant at Kaskaskia.[7]

St. Louis offers a sharply contrasting experience during its initial years in the French period, one of congruence of governmental policy and the customs of the people. Pierre de Laclède obtained a grant in 1763 from the director general of Louisiana to trade with Indians on the Missouri and upper Mississippi Rivers. Presumably, this grant also permitted him to establish a trading-post village at a location of his choice. He and his younger associate Auguste Chouteau selected the site of St. Louis in 1764 and, prior to its occupation, laid out a traditional French village, which they clearly intended to be more than a simple Indian trading post.[8] Although the village anatomies of Ste. Genevieve and St. Louis have similarities, the processes of initial establishment contrasted: the former apparently without formal permission and supervision, the latter clearly with. French settlement tradition was the common denominator that made them similar.

By the Treaty of Fontainebleau, signed on November 3, 1762, the western side of the valley passed from French to Spanish hands. Then, by the Treaty of Paris, in 1763, the eastern side passed from French to British. The shift was neither sudden nor smooth in Upper Louisiana. After learning that French authority would be effaced from Upper Louisiana in favor of Spanish and British, many French on the older-settled eastern side moved to the Spanish side. Aware of the earlier fate of the Acadians under British rule, they preferred a Spanish Catholic rule to a British. In addition, Indians harassed the French in their east-bank villages before the British arrived, driving some French and Peoria to the west bank.[9]

Louis St. Ange de Bellerive, the French commandant in the Illinois Country, was one of the many French who preferred to relocate to just-founded St. Louis on the Spanish side. He became, apparently on his own initiative and with Spanish knowledge, the first local commandant on the west side of the river, although merchants usurped his authority in trade matters.[10] It was he who took the initiative to grant lands to fellow settlers moving across the river, and he continued to exercise this function

---

7. Denman, "French Peasant Society"; Flader, "Settlement History"; Ekberg, *Colonial Ste. Genevieve*, 2–25.
  8. Peterson, *Colonial St. Louis*, 6–23.
  9. Alvord and Carter, *New Regime*, 108, 145, 299.
  10. Don Pedro Piernas to Gov. O'Reilly, in *Spanish Regime*, ed. Houck, 1:73.

from 1766 until Spanish authorities arrived in 1770. St. Ange granted eighty-one town lots in St. Louis from 1766 to 1770 and possibly some at Ste. Genevieve. Similarly, Philippe-François Rastel de Rocheblave, a Kaskaskian, crossed the Mississippi to become local commandant of Ste. Genevieve, probably appointed by St. Ange and subordinate to him. Rocheblave issued at least one land concession before 1770. Most of the town lots and agricultural properties during these four transitional years of growth most likely became private property simply through the establishment of residence and mutual recognition by fellow villagers. The total amount of lands conceded by St. Ange in this transition period from 1766 to 1770 was at least 5,710.60 arpents.[11] How much of this land was at Ste. Genevieve is not known.

In 1770, Ste. Genevieve and St. Louis were the only villages of consequence on the west side, although a few people were also living off and on in tiny settlements at La Saline and Mine la Motte. The incoming Spanish placed their administrative center at St. Louis in 1770 despite more people at Ste. Genevieve. In 1770, Ste. Genevieve had grown to a population of about 460 plus 370 African slaves, including the fluctuating population of La Saline and the lead mines.[12] Ste. Genevieve, La Saline, and Mine la Motte were standard toponyms on maps by 1770 and bore witness to the world that the French had opened the eastern Ozarks.

Spanish Period, 1770–1795

Because few Spanish came to settle in Upper Louisiana during the twenty-five years of Spanish rule from 1770 to 1795, Spanish presence basically amounted to governance, which consisted of only a few administrators in St. Louis and district posts and a small number of soldiers. Most of the latter, if they remained, married into the resident Creole population. Spanish were all but absent away from the two post villages.

Although officers of Spain arriving in Upper Louisiana brought a long tradition of colonial administration and expectations to carry out land regulations and decrees issued at New Orleans, they generally did not enforce the regulations; as a result, local custom of the French settlers prevailed. There were exceptions, but, in general, acquiring and occupying land in the Spanish Illinois went on as it would have without the Span-

---

11. As reported in Floyd C. Shoemaker, *Missouri and Missourians*, 1:195. Shoemaker gives no source for his figure of eighty-one town lots. Ekberg, *Colonial Ste. Genevieve*, 336–40. Summary of the quantity of land granted in the district of Louisiana, November 2, 1804, ASP-Misc 1:405–6.

12. Ekberg, *Colonial Ste. Genevieve*, 42, citing Thomas Hutchins, *A Historical and Topographical Description of Louisiana and West Florida*.

ish presence. Only in immigration policy did Spain exert strong influence in the settlement process. The kinds of people who entered and who did not enter highly affected the patterns of settlement that developed within Upper Louisiana.[13]

Spanish land laws and policies for Louisiana had roots in the preexisting fundamental Law of the Indies in force throughout the Spanish lands of the New World. Spanish settlement mostly occurred in areas populated by non-Christian Indians, and the Law of the Indies addressed this situation. In Louisiana, however, the Spanish Empire spread into land already settled by other kindred Europeans who were also Catholic and whose organization of space for economic, administrative, and cultural purposes was essentially the same as their own. The settlement provisions of the Law of the Indies did not address the Louisiana situation. What good would be served by changing the status quo of the functioning French settlements? Why disrupt the way of living? The Spanish would accomplish nothing by introducing the Law of the Indies, especially here at the tenuously held geographical limits of their empire. Motivation to enforce the official settlement policy was lacking.

Because its interest in the Illinois Country was as a defensive buffer against colonial rivals, Spain directed its administrative attention to such matters as immigration, militias, control of the Mississippi River, and the fur trade, which it contested with the British. It was necessary to control immigration in this colonial borderland, lest disloyal settlers enter and detach the sparsely settled territory from Spanish allegiance. Consequently, Spain's strict immigration policy prohibited the entrance of many potential settlers and thereby appreciably slowed the settlement process until 1795. Constantly worried about British designs on their territory, the Spanish concentrated defense at St. Louis, which became a fortified town. The fur trade, in which the Spanish had little experience, es-

---

13. An excellent review of Spanish and French legal practices, emphasizing land, is in Banner, *Legal Systems in Conflict,* 11–81. Spanish immigration and settlement policies in Upper Louisiana are treated in Kinnaird, *Spain in the Mississippi Valley;* Arthur Preston Whitaker, *The Spanish-American Frontier, 1783–1795: The Westward Movement and the Spanish Retreat in the Mississippi Valley* (primarily below the mouth of the Ohio River); David J. Weber, *The Spanish Frontier in North America;* Louis Pelzer, "The Spanish Land Grants of Upper Louisiana"; Gilbert C. Din, "Spain's Immigration Policy and Efforts in Louisiana during the American Revolution"; Abraham P. Nasatir, ed., *Before Lewis and Clark: Documents Illustrating the History of Missouri, 1785–1804;* and Houck, *History of Missouri,* 1:287–387, 2:1–328. Spanish settlement policies of Upper Louisiana are also treated in basic histories of colonial Louisiana, such as Alcée Fortier, *A History of Louisiana;* Charles Gayarré, *History of Louisiana;* Giraud, *History of French Louisiana;* and Martin, *The History of Louisiana.*

pecially commanded Spain's attention, but it remained in the hands of local French with whom administrators found ways to work and skim a percentage of the profits. On the other hand, the lead business lay safely within the Ste. Genevieve District and was geographically inaccessible to the British, but the business also remained in the hands of local French. If the Spanish presence was weak in Upper Louisiana as a whole and focused on defense at St. Louis, it was virtually nonexistent in the Ste. Genevieve District.

Though detailed regulations directed Spanish administrators as to where people were to settle in the Illinois Country, how they clustered or dispersed, or even how much land a family was to be awarded, by and large the Spanish left these matters to the local commandants and local customs. The first Spanish officer to reach Upper Louisiana, Captain Francisco Ríu, did not interfere with civil or local affairs, which remained in the hands of St. Ange.[14] One exception to the Spanish hands-off policy in land matters at the local level was any land request of a square league or more, which required referral to the governor at New Orleans.

Spain accepted without question existing French titles to land in villages and the mining country. Spain also accepted the French *pied,* arpent, and *lieue* as units of land measure, not insisting on their *vara*.[15] The entire French system of land description, division, and organization continued without interruption. The most obvious change was use of the Spanish language in documents prepared by the provincial lieutenant governor; local commandants generally retained French.[16]

Spanish also accepted the decades-long French practice of using African slaves in Upper Louisiana. Although the Spanish had their own laws for the regulation of slaves, they did not introduce them into Upper Louisiana and let the French *code noir* continue.[17] In general, the new Spanish government accepted the societal practices of their fellow Catholic Europeans.

14. Houck, *History of Missouri,* 1:288–89; "Ulloa Sends an Expedition to the (Spanish) Illinois Country to Establish a Fort and Settlement and His Rules for the Government of the Same," March 14, 1767, in *Spanish Regime,* ed. Houck, 1:1–19; Gilbert C. Din, "Captain Francisco Ríu y Morales and the Beginnings of Spanish Rule in Missouri." In a more general way, Carlos III instructed Governor Ulloa upon arriving at New Orleans not to make any innovation in the government (Weber, *Spanish Frontier,* 200).

15. Peterson, *Colonial St. Louis,* 6–8, 19.

16. "Religious Condition of Louisiana, 1772," in *Spanish Regime,* ed. Houck, 1:120.

17. Banner, *Legal Systems in Conflict,* 43, 95; Blume, *German Coast,* 117–19; Leslie B. Rout Jr., *The African Experience in Spanish America, 1502 to the Present Day,* 80–87. The *code noir* for Ste. Genevieve is examined in Ekberg, *Colonial Ste. Genevieve,* 197–239.

France and Spain were sister European nations. In addition to sharing the same Catholic religion and opposition to Protestant England at a time when such matters dominated interimperial affairs, their blood-related monarchies were joined in a "Family Compact" between cousins Carlos III of Spain and Louis XV of France.[18] Yet, even more important than royal bonds, their basic cultures, when viewed from the process of settling land in eighteenth-century America, were similar in types of agriculture, the form of settlements, community structure, and land law. Spanish and French agricultural and settlement systems had more in common with each other than either had with English, American, or any of the indigenous Indian systems. To a large degree, the settlement forms laid out by the Spanish Law of the Indies differed only in detail from French practice in the Mississippi Valley. Both had grid-pattern villages, and both had the tripartite village system of town lot, arable field strip, and commons for grazing and wood supply. Settlement process and pattern in Upper Louisiana during the Spanish regime are understandable through French cultural practices without invoking the Spanish Law of the Indies or other Spanish settlement regulations except those concerning immigration.

The superficial overlay of Spanish law upon French custom in Upper Louisiana, in regards to settlement, was exemplified by the Spanish choice of local officers. Men of basic French culture administered the local districts. These commandants, such as the Creole François Vallé II at Ste. Genevieve (1794–1804), Canadian-born Louis Lorimier at Cape Girardeau (1793–1804), and Charles Tayon at St. Charles (1792–1804), enjoyed the respect and support of the local *habitants*. It was literally on their doorsteps where the land-granting process began and the local administration of justice took place. Although these commandants worked under a Spanish-appointed lieutenant governor, they continued to operate locally with considerable freedom. Later, even the appointed lieutenant governors at St. Louis, Zenon Trudeau (1792–1799) and Charles de Hault Delassus (1799–1804), were French in culture, and a governor-general himself at New Orleans, the baron de Carondelet (1792–1797), was of French culture. One wonders whether these administrators made their everyday decisions more from their ingrained French culture and for the benefit of the local Creole French than by enforcing Spanish settlement regulations that may not have made good sense to them.[19]

18. *Encyclopaedia Britannica,* 11th ed., s.v. "Spain."
19. Zenon Trudeau was a French Creole born in New Orleans in 1748 (Houck, *History of Missouri,* 2:58). Both Charles de Hault Delassus and Francisco Luis Hector, baron de Carondelet, were from Hainault in French-culture Wallonia, then ruled by Spain as part of the Spanish Netherlands (1:365, 2:135). Of course, imperial admin-

Another major problem with using Spanish law to interpret the settlement of Upper Louisiana arises from the fact that the Spanish policy makers thought almost exclusively in terms of better-populated and more economically developed Lower Louisiana when issuing regulations and policies for the Province of "Louisiana." Many of the land regulations issued at New Orleans were utterly inapplicable to the different physical geography of Upper Louisiana, and other regulations that would have been helpful for the specific conditions of Upper Louisiana were lacking. For example, there were neither regulations for lead mining and mineral-land ownership, which occurred only in Upper Louisiana, nor any attention to persons in animal-product activities (furs, skins, pelts, tallow, grease). To interpret the settlement process in Upper Louisiana from 1770 to 1804 through Spanish regulations and policies made for agricultural Lower Louisiana is both difficult and unfair.

A root cause of Spain's weak influence on settlement in Upper Louisiana is that it did not deploy an adequate number of people or the necessary resources to enforce details of its land regulations. For example, Spain did not appoint a surveyor for conceded land beyond the villages, a person who could have exercised great influence in the assignment of land, until 1795, even though regulations required all conceded land to be surveyed. Neither did the Spanish instruct any person to keep an aggregate list of land concessions in Upper Louisiana, and they could not know whether a grant being made overlapped onto one already made or how compact a settlement was.

Finally, the sheer distance from New Orleans to Upper Louisiana—a thousand-mile, three-month boat trip—daunted the Spanish from doing what they knew was necessary to carry out their intentions, unless some crisis developed. Lacking a real crisis in Upper Louisiana, except for the attack on St. Louis in 1780, the Spanish deployed their limited supply of civil servants and resources into other places of their vast empire more needy of attention. When viewed in the broadest context, it was there-

---

istrators did not have to be of the same ethnicity as the government for which they worked. A significant example is Alejandro O'Reilly, an Irishman by birth and education, who was an effective administrator and enforcer of Spanish law and policies in Lower Louisiana. The point is that administrators of a different ethnicity from the government, but of the same as the people they govern, may understand the basic culture of the people whom they govern better than administrators of a different ethnicity, and that, of course, figured in Spain's appointment of French-culture administrators for Louisiana. Spain administered Upper Louisiana more like an occupier than a colonizer, much as Spain administered the Catholic peasantry of Hainault in the Flemish Netherlands. Of course, appointment as lieutenant governor of Upper Louisiana required knowledge of the French language (Frederic L. Billon, *Annals of St. Louis in Its Early Days under the French and Spanish Dominations, 1764–1804,* 76–77).

fore logical for the Spanish to allow this smoothly functioning, French-inhabited farthest corner of their empire to continue to operate as before, by local commandants following tradition, certainly in regard to settlement process and patterns. The Spanish imprint on settlement was insubstantial.[20]

Although the Law of the Indies contained standardized rules for the spread of Spanish settlement onto Indian lands in the Americas, rules to honor and protect Indian settlements were essentially moot in Upper Louisiana because the tiny areas of European settlement along the Mississippi River did not impinge onto land on which Indians were agriculturally engaged in permanent villages. The Osage villages lay far to the west, and relations between them and European populations were not of a daily, interactive nature. More important, the Osages were not a high-culture nation like those of central Mexico or the upper Rio Grande, and settlement policies worked out for those lands would hardly be appropriate for the lands of the seminomadic Osages and remnant Illinois groups who lacked an accumulation of wealth and provided no labor supply. In reality, the Spanish left Indian affairs that dealt with land to the resident French commandants.[21] The Spanish acknowledged that French in the valley had a working system with the Indians based on trade, and they openly adopted it. In this respect, one should note that the forts the Spanish built and garrisoned in the valley were not for protection against Indians, as elsewhere on the Spanish northern-American frontier.

Spanish policy did make a distinction between "baptized" or Christian Indians, who could own land inherently, and all other Indians, who had to obtain a written title from Spanish authorities. Indians, once landowners, could sell their land. In Upper Louisiana, no Indians as a tribe or nation acquired written title to the land before 1793, and if individual Indians acquired title, it was most likely because they were mixed blood and obtained it by virtue of their white parentage. In such cases, the *pardos* (Métis) operated within the European community.[22]

Spanish regulations and policies concerning settlement in Louisiana came in five major issuances: the Law of the Indies, O'Reilly's regulations

---

20. The writer of a broadside of 1772, supposed to be Daniel Blouin, an important French entrepreneur in the valley, sought to unify and promote the economic development of the valley by emulating British economic ideas, thereby dismissing the ineffective Spanish presence in the valley (Clarence Walworth Alvord and Clarence Edwin Carter, eds., "Invitation Serieuse aux Habitants des Illinois").

21. Weber, *Spanish Frontier,* 229–31.

22. Francis P. Burns, "The Spanish Land Laws of Louisiana," 564–65; *ASP-PL* 5: 763.

of 1770, Galvez's policies of 1778, Gayoso's instructions of 1797, and Morales's regulations of 1799. They treated such basic matters as the authority to grant lands; immigration and eligibility for land; the procedure for granting; and stipulations on sizes, shapes, and location. Specific orders and edicts directly from Spain or New Orleans supplemented these major issuances. American legal authorities later reviewed all these statements in excruciating detail when confirming land claims, because the U.S. government logically believed that Spanish authorities should have awarded land in compliance with them. Our understanding of Spanish land regulations has come from review of them during the land-confirmation process.[23]

The Law of the Indies was a comprehensive statement. It was first promulgated in the sixteenth century and since then formed the basic law for administration of Spanish territories in the New World. The governors in New Orleans and lieutenant governors in St. Louis had copies and were instructed in it, including those parts that treated immigration and settlement. As important as it should have been for administration of Upper Louisiana, lieutenant governors and district commandants did not mention the law either in the receiving of new settlers in the colony or in the allocation of land. The law was not mentioned in the development of Laclède's plan of St. Louis, the laying out of new Ste. Genevieve, or the platting of the villages of New Bourbon and St. Michel, all of which occurred during the Spanish years. It was not mentioned in the granting of lands away from the villages. Ironically, American authorities, who assumed the law influenced Spanish land-awarding procedures, may have studied it more thoroughly than any Spanish administrator in Upper Louisiana.

The Law of the Indies is a lengthy, detailed document that was frequently amended over its long life. Provisions required tripartite villages, consistent with French practice, and permitted larger amounts of land according to settlers' resources, social status, and governmental service, also consistent with French practice.[24] Another provision assumed that

---

23. The second Board of Land Confirmation repeatedly used the phrase "laws, usages, customs, and practices" to refer to Spanish administration in Upper Louisiana (*ASP-PL* 5:218–315). Also see Burns, "Spanish Land Laws"; A. H. Rose, "Extension of the Land System"; and Banner, *Legal Systems in Conflict*.

24. The best translation of those portions of the total 148 ordinances of the Law of the Indies that deal with towns and settlement, as issued by Philip II in 1573, is in Crouch, Garr, and Mundigo, *Spanish City Planning*, 6–22, followed by a commentary on them, 23–65. The authors interpret three American cities, of which St. Louis is one, on the basis of the law. Also see Reps, *Town Planning*, 24–46, for the effect of the provisions of the law on town planning.

settlers held their lands as individual properties, and that land was not held communally, also consistent with French practice. The law, in general, indicated that the standard settlement was a clustered village, although it also allowed individuals to occupy lands away from villages.

Shortly after his arrival and establishment of order at rebellious New Orleans, Gov. Alejandro O'Reilly issued a set of laws for Louisiana that included numerous provisions for settlement.[25] The laws were made public on February 18, 1770, barely three months before Lt. Gov. Pedro Piernas arrived in St. Louis to administer Upper Louisiana. (Spanish governors-general of Louisiana are listed in Appendix A and Spanish lieutenant governors of Upper Louisiana are listed in Appendix B.) The first laws designed specifically for the Province of Louisiana, they did not significantly change settlement procedures from the existing French practices in Louisiana. In fact, O'Reilly might have wanted to let the basic French practices continue in order not to further agitate the discomposed citizens of Lower Louisiana. Upper Louisiana is not referred to anywhere.

One significant change introduced by O'Reilly that was extended to Upper Louisiana in later years was the introduction of grants of one square league for cattle raising, to be located away from existing river-border long lots. Cattle raising was a primary land use in Spanish northern America, and the river-fronting long lot was ill-suited for large herds of cattle that the French were beginning to accumulate. Designed to be in natural grasslands, these square tracts, later referred to as *sitios,* became popular in Lower Louisiana, and, in the 1790s, in Upper Louisiana. In Upper Louisiana, the French term *vacherie* (cattle ranch) was used.[26] *Vacheries* in Upper Louisiana did not front on anything and took various compass orientations. Concessions for *vacheries* came later to Upper Louisiana in part because the smaller herds of Upper Louisiana grazed on harvested common fields (not present in Lower Louisiana) or in open,

25. O'Reilly had been made governor of Louisiana after Antonio de Ulloa had been ousted by the residents of New Orleans in 1766. O'Reilly's laws are in *ASP-PL* 5:289–90.

26. John W. Hall, "Sitios in Northwestern Louisiana"; John W. Hall, "Louisiana Survey Systems: Their Antecedents, Distribution, and Characteristics," 98–111. According to O'Reilly's laws, in order to get land beyond the river in tracts one league square, a settler had to have two Negroes, one hundred cattle, and some sheep (Charles Gayarré, *Histoire de la Louisiane,* 2:370). In his comprehensive work on cattle ranching in North America, Terry G. Jordan speculates that the term *vacherie* may have been a borrowing, at least in meaning, of the Spanish *vaqueria,* because in France it meant a cowshed or transhumant dairy herd. Jordan also speculates that cattle ranching in Upper Louisiana derived from Lower Louisiana, but that stock from the Illinois Country was later sent to the Gulf Coast to make herds larger (*North American Cattle-Ranching Frontiers,* 120–22).

grassy woods of the royal domain. The idea of private property a square league or larger also extended in Upper Louisiana beyond *vacheries* to encompass mining and timberland uses. The huge Mine la Motte tract, a holdover from French days, acquired its square shape, with corners pointing in the four cardinal directions and fronting on nothing, during the Spanish regime.

During O'Reilly's tenure, Spain established the basic process for a settler to acquire title to land that stayed in force until the American takeover. The settler was first to make a request of the local district commandant who then was to certify that the land was vacant and did not conflict with other owned tracts. The request then went with the commandant's recommendation to the lieutenant governor of Upper Louisiana who ordered a survey and put the applicant in possession of the conceded land, at that point called a *concession*. The lieutenant governor then forwarded the application to the governor-general, later the intendant, in New Orleans for approval in the name of the king of Spain, an act that perfected the concession and converted it into an official grant. This formal process was cumbersome and lengthy in time for a region so distant from New Orleans, and few concessions were actually forwarded to New Orleans for perfection. Consequently, a simple concession was treated as legal ownership during the whole Spanish period.[27]

Spain's efforts to attract immigrants into Louisiana bore hardly any results during O'Reilly's tenure, from 1769 to 1772, or during the tenure of his successor, Luis Unzaga, from 1772 to 1779. Hopes to attract Spanish Catholics were never much more than that. Spain tried to induce the settlement of Acadians, and then of Irish, German, and French Catholics, but few of the Europeans came.[28] In Upper Louisiana, French Creoles from the British-side villages continued to relocate on the Spanish side, but not at the rate that they had moved earlier, because the population pool that provided them had dwindled. A few Americans entered in these pre–American Revolution years, but they were conspicuous by their rarity. Among them was a John Hildebrand and family, who settled on the Meramec, midway between St. Louis and Ste. Genevieve, in 1774 and gained a concession from Lt. Gov. Fernando DeLeyba on November 24, 1779. This family may claim to be the first successful establishment of an American family in Upper Louisiana away from the villages.[29]

---

27. A. H. Rose traces the official process of acquiring land and perfecting its title ("Extension of the Land System," 17–18). Also see the full account in the second Board of Land Commissioners' report (*ASP-PL* 5:702–8).

28. Din, "Spain's Immigration Policy," 242, 247–48.

29. Houck, *History of Missouri*, 2:73–76; *Régistre d'Arpentage*, 565. Houck identifies Hildebrand's farm on Saline Creek just above its junction with the Meramec as

For the Spanish, settlement of distant Upper Louisiana was always of secondary concern to their presence at New Orleans and Lower Louisiana. Nevertheless, the next governor, Bernardo Galvez (1778–1787), described by biographers as energetic, decisive, and pragmatic, strongly advocated the growth and development of the entire Louisiana Province and gave special attention to Upper Louisiana.[30] Galvez clearly wanted to encourage its population growth, especially by immigration of Catholics, except those from countries with which Spain was not friendly, such as England and the Protestant Dutch Netherlands. Galvez did not issue a unified set of instructions for immigration and settlement as O'Reilly had done, but a series of acts and orders during his tenure served the same purpose.

Galvez acted upon Lt. Gov. Francisco Cruzat's recommendation of December 8, 1777, to recruit French Canadians to Upper Louisiana by ordering him to do so. The Spanish had heard that the Catholic Canadians languished in wretched condition under British rule, oppressed by being forced to bear arms against rebellious New Englanders. In a general letter of instructions dated February 19, 1778, to all post commandants in Upper Louisiana, Galvez intended to provide for their resettlement among Creoles of Canadian origin in the Spanish Illinois Country by awarding them long lots with a frontage of five arpents and a depth of forty arpents. Galvez ordered administrators to show kindness and gentleness to the new colonists, because "the source and origin of all empires has [sic] been the refuge and kind usage which men find in the gentleness of the laws."[31]

Galvez's more open and aggressive immigration policy still did not result in appreciably more settlers into Upper Louisiana. After twelve years, during which the American Revolution took place, the population of Upper Louisiana had risen from 1,313 in 1773 to only 1,591 in 1785, and most of this meager addition consisted of natural increase and removal from the east bank. Settlers crossed over during the civil disorder that followed General Clark's seizure of Kaskaskia in 1778 and again after 1782 when the American garrison withdrew, leaving the east-bank settlements unprotected.[32]

As the American Revolution drew to a successful close, a new situation

---

the first American settlement in Upper Louisiana. Peterson identifies American Robert Owen, who settled on the River des Pères just west of St. Louis village in 1779 as "perhaps the first of his nationality to take up lands in Mo" (*Colonial St. Louis,* 23). Houck says that Owen came into the country in 1789 (77).

30. Din, "Spain's Immigration Policy," 243–44.

31. "Immigration to Upper Louisiana to Be Encouraged—1777, 1778," in *Spanish Regime,* ed. Houck, 1:152–54; "Decree of the Governor [Bernardo de Galvez], February 19, 1778," in ibid., 155–57.

32. Austin, "'Memorandum of M. Austin's Journey,'" 534, 537.

confronted the Spanish in the valley: Americans were westering at a faster rate than theretofore imagined. In addition to the British threat from the North, which was a threat to the control of the fur trade as much as to the seizure of territory, and in addition to keeping relations smooth with Indians, the Spanish now faced a growing "American problem."

In 1782, Governor Galvez, in a bold policy change, recommended that Americans be allowed to enter in small numbers. Otherwise, he believed, they would "pile up" on the banks of the Mississippi and become a bigger, more unmanageable problem in the longer term.[33] Galvez recognized the unifying character of the river. Although it might have appeared to royal advisers across the ocean as a convenient line of territorial demarcation, to those who were on the scene, the river could serve that function only with military monitoring and enforcement in a populated area, and Galvez recognized Spain's powerlessness to do that.

Galvez's pragmatic strategy to let Americans in gradually and in small numbers would presumably allow them to be assimilated into Catholic peasantry. These woodsmen would not be allowed to have firearms, for which they were notorious, and would thus not be able to hunt. Galvez, with his European mind, apparently thought Americans would revert to village-dwelling peasants, European style. The expenses of bringing real Spanish Catholic peasants to America were prohibitive, but the assimilation and conversion of Americans to a European-type peasantry might not be.

The crown did not take Galvez's recommendation. Other issues dominated the fluid relations between Spain and the new United States. Spain was involved with Indians and Americans in the new American Southwest (Mississippi and Alabama), where all three contested territory. Also, the Mississippi River was fast becoming a troublesome issue. In fact, in 1784, just one year after the Treaty of Paris between the United States and Britain ended their war, Spain closed the river to free use by Americans, where Spain controlled both banks, and charged taxes or deposit fees for transshipment at New Orleans. In addition, Spain was encouraging Indians who were being displaced from the western United States to enter Upper Louisiana. Spain's diplomatic initiatives were to restrain the growth of the American West. Indeed, Galvez reversed his earlier recommendation and now, considering the larger issue of international diplomacy, prohibited American immigration into Upper Louisiana.[34]

33. Din, "Spain's Immigration Policy," 255–56.
34. On taxes and deposit fees: Whitaker, *Spanish-American Frontier,* 3–14; on Indians being displaced: Foley, *Genesis of Missouri,* 64; on Galvez's reversal: Din, "Spain's Immigration Policy," 256.

This stance was continued by his successor, Esteban Miró, who wrote to Lt. Gov. Francisco Cruzat at St. Louis on July 13, 1785, that no Protestants were to be admitted into Upper Louisiana, only French Catholics, and preferably those with slaves. French Creoles from the eastern side continued to be gladly received, and more Kaskaskians did move into the Ste. Genevieve District as a result of the imposition of new U.S. laws, especially the prohibition of slavery in the newly organized Northwest Territory and the new land policy just introduced of selling land to settlers, both occurring in 1787. At Kaskaskia, one now had to buy land by auction in competition with others and deal with an impersonal, politically appointed government representative, whereas the Spanish, a mere ferry crossing away, where relatives already lived, would give land to settlers by simple request to the local commandant, proportionate to one's wealth, and without competition from others. Discouraged with the way things were going and apprehensive of the future, many crossed the river in 1788 and 1789. Among the new arrivals were some of the wealthiest Kaskaskia families, such as the Janis, Ste. Gemme Beauvais, and Caillot dit Lachance families. Along with the Creoles came some American residents from the east bank such as John Dodge and family, the first Americans to take up residence in Ste. Genevieve village. The terrible flood of 1785, which ravaged Kaskaskia no less than old Ste. Genevieve, certainly conditioned the Kaskaskians for relocation. Spain invited the flood-displaced Kaskaskians to relocate, acting "compassionately" when civil authorities on the American side did nothing. All of these newly arrived families participated in the rebuilding of a new Ste. Genevieve, where a new life and business opportunities were more promising. Kaskaskia, it was said, lost 77 percent of its population, whereas Ste. Genevieve village grew from 670 in 1787 to 923 in 1791.[35] This important demographic and economic shift was more the result of local conditions and American land policy and cannot be counted a success of Spanish immigration policy.

On December 1, 1788, by royal *cedula,* Spain remarkably reversed its immigration policy and invited Americans to enter all of Louisiana. Protestants could enter, as long as they took an oath to raise their children to be Catholics, refrained from public worship and preaching, and swore allegiance to the Spanish king. Spain awarded them land in pro-

---

35. Esteban Miró to Franco Cruzat, July 13, 1785, PC 3b-771; Alvord, *Kaskaskia Records, 1778–1790,* 410. Thirteen Kaskaskia families, totaling 135 persons (of whom 69 were slaves), arrived in Ste. Genevieve between December 1, 1787, and the end of December 1789 ("Immigrants from the United States," in *Spain in the Mississippi Valley,* ed. Kinnaird, 3:290). Ekberg, *Colonial Ste. Genevieve,* 431–32, 460, 462–63.

portion to the number of slaves they had. As an example, Miró invited Benjamin Sebastian of Kentucky to settle in Upper Louisiana and promised him and others land up to eight hundred acres each; in cases of "men of influence," he promised up to three thousand acres.[36]

Spanish thinking, as expressed through Governors Galvez and Miró, recognized the unity of the entire Mississippi River drainage basin, including its vast Ohio, Cumberland, and Tennessee tributaries, and that this natural geographic unity would translate eventually into human geographic unity. By being in a different natural region than the Atlantic drainage, the trans-Appalachian American West would, in time, become detached from the East and form with the Spanish territories a single economic and political unit.[37]

The United States, however, did not see things that way. The Republic was working out ways to incorporate the new West with the older East, even though it knew that hydrographic patterns moved the West's products through New Orleans and away from the East. The swiftness of settlement of Kentucky and the Holston Valley and Cumberland basin of Tennessee meant that remembrances and personal ties with the East remained strong. Migrants to Kentucky and Tennessee had not been repelled from the East as much as they had been attracted to lands in the West. Thus, they remembered the East positively and harbored no ill thoughts of the regions they had left that could be translated into a desire to establish a new republic or merge with Spanish territory. The paramount issue for them was to keep the Mississippi open for export of products, and this became the major issue for the American government.[38]

Thomas Jefferson was not fearful of the West's separation. Upon hearing of Galvez's opening of Spanish Louisiana to all Americans, the then secretary of state remarked that this invitation "will be the means of delivering to us peaceably [rights to navigate the Mississippi] what may otherwise cost us a war." Jefferson knew that with Americans on both sides of the Mississippi, the river would be open to them, if not also the eventual control of trans-Mississippi lands. Jefferson could not imagine that

36. Whitaker, *Spanish-American Frontier*, 101–2, 157–62. This new royal order had been preceded earlier in the year by permission from Governor Miró to allow Protestant Americans to settle in the Tywappity prairie bottoms below Cape Girardeau and at New Madrid, as long as they had good reputations, did no preaching, and built no churches (Esteban Miró to Manuel Perez, February 15, 1788, PC 5a-320, 325). Gov. Miró to Benjamin Sebastian, September 16, 1789, *Congressional Globe*, 25th Cong., 2d sess., 1838, app. 398.

37. Whitaker, *Spanish-American Frontier*, 105; Gilbert C. Din, "The Immigration Policy of Governor Esteban Miró in Spanish Louisiana," 174.

38. Meinig, *Shaping of America*, 2:4–7.

Americans would ever revert to a European-type peasantry. Like all earlier Spanish efforts to attract people into Louisiana, whether Catholics or Protestants to be Catholicized, the 1788 overture was a failure in numbers of immigrants. During 1787–1790, only 293 Americans, of whom 106 were slaves, came to the entire Spanish Illinois.[39]

The most prominent settlement in the middle Mississippi Valley that resulted from the 1788 immigration policy was George Morgan's planned American colony at New Madrid on the Mississippi River, just below the mouth of the Ohio River. Although Morgan's scheme appeared to fit well into Spanish plans, Governor Miró found reasons not to approve the project and thwarted the plan.[40] Some French and American settlers eventually drifted into Morgan's carefully planned village, but certainly not to the grand extent envisioned.

Spain failed to attract Americans to settle in Upper Louisiana for several reasons. Kentucky and Tennessee were still receiving settlers in 1788–1790. Good land, with virgin soil, was still plentiful. Markets for Kentucky produce were growing, including the army units fighting Indians north of the Ohio River. Also, one wonders how the widespread reputation of oppressive Spanish rule might have influenced Americans. The Black Legend—*Leyenda Negra*—about an evil Spain that treated its subjects arbitrarily and ruthlessly circulated among Americans and may have caused individual families to think twice before entering a Spanish dominion. The thought of being forced to become Catholic intimidated some potential immigrants, although official policy did not require conversion, only that children must be Catholic. Americans were also intimidated about being required to obtain permission to do just about anything, being forbidden to have local elections and public Protestant worship, and not being able to participate in land speculation. In addition, the Mississippi River was now open to Americans, and the advantage of use of the river to Spanish-side residents was gone.[41] Simply put,

---

39. Jefferson quoted in Whitaker, *Spanish-American Frontier*, 103; Foley, *Genesis of Missouri*, 63. Spanish commandants at St. Louis frequently invited Protestant Americans living in American Illinois to cross over to the Spanish side during 1791–1794, but, according to one account, they "chose to live in a land of Freedom" ("Memorial to Congress by Inhabitants of the Territory," December 18, 1815, *TP* 17:268–69).

40. Din, "Immigration Policy of Miró," 169–70; Max Savelle, *George Morgan: Colony Builder*, 218–28. Among the numerous accounts of George Morgan's colony at New Madrid are Max Savelle, "The Founding of New Madrid, Missouri"; Savelle, *George Morgan: Colony Builder;* Houck, *History of Missouri*, 2:108–29; *Spanish Regime*, ed. Houck, 1:275–321; and John W. Reps, "New Madrid on the Mississippi."

41. On Kentucky and Tennessee: Whitaker, *Spanish-American Frontier*, 160; on the Black Legend: Weber, *Spanish Frontier*, 336; and Banner, *Legal Systems in Conflict*, 92–93; on Americans: Din, "Immigration Policy of Miró," 161–64.

the Spanish Illinois did not yet offer any advantages over the American West but instead presented disadvantages, real and imagined.

Francisco Luis Hector, baron de Carondelet, began his five-year term as governor on the first day of the new year, 1792. The governor, though a French-culture Fleming, was a loyal Spanish administrator. He was the product of European thinking about social stratification and unequal partition of land and did not understand the realities of the New World with open societies and unlimited land. Throughout his tenure, Carondelet persisted in imagining a Catholic, European, docile, and loyal peasantry in Louisiana. He could not imagine gun-bearing Americans, unsubmissive to authority, living in Louisiana. Carondelet opposed outright the policy of unlimited American immigration. For example, he immediately disallowed New Andalusia, a proposed American-settlement colonization scheme on the Meramec between Ste. Genevieve and St. Louis, for which his predecessor Miró had granted permission. Carondelet, in clear words, instructed Lieutenant Governor Trudeau at St. Louis not to admit any American or English settlers. Spain, basing its decision more on the situation in Lower Louisiana and the Yazoo and West Florida borderlands and on unfolding events in revolutionary Europe, rescinded its open-door policy and returned once again to one of strict separation of Spanish and American space.[42] The earlier policy had borne no significant results, and those who had come had not abandoned their religion and language and most certainly had not laid down their guns.

Carondelet resumed an immigration policy to attract Spanish, French, German, Dutch, and Flemish Catholics to Louisiana, with special attention to the latter three. On April 17, 1792, Don Henri Peyroux de la Coudrenière, commandant at Ste. Genevieve, received permission to form a new settlement of Germans, Flemish, and Dutch, and later Jacques Clamorgan of St. Louis proposed to bring Germans from Holland. These intentions were good, but nothing resulted from them.[43]

Carondelet also turned to Indians for settlers. Spain could use its long experience of Christianizing and civilizing Indians, especially with those

---

42. Whitaker, *Spanish-American Frontier*, 153–54; "New Andalusia, an Anglo-American Colonization Project: A Royal Invitation to the Industrious; Carondelet to Trudeau," June 8, 1792, in *Spain in the Mississippi Valley*, ed. Kinnaird, 4:xxv, 46–53; "Carondelet: Instructions to Trudeau, New Orleans," March 28, 1792, in *Before Lewis and Clark*, ed. Nasatir, 1:151–53.

43. Official permission given to Peyroux de la Coudrenière to form a new establishment by Andrés Armesto, April 17, 1792, PC 208a-488; Clamorgan to Carondelet, October 4, 1793, in *Spain in the Mississippi Valley*, ed. Kinnaird, 4:208–15; Baron de Carondelet to Peyroux de la Coudrenière, September 24, 1792, PC 205-675; Baron de Carondelet to Peyroux de la Coudrenière, 1792, PC 205-681.

Indians already partially Europeanized by interaction and those who had been victimized and displaced by American occupation of their lands and who presumably harbored anti-American feelings. The major groups invited into Upper Louisiana were the displaced Shawnee and Delaware from the Ohio Valley. A key figure in movement of these Indians to the Spanish Illinois was Louis Lorimier, a Métis trader with a Shawnee wife, who himself had been forced from his post in Ohio during the Indian-American land struggles in the Northwest Territory. Lorimier received a generous donation of land in the sparsely populated region around Cape Girardeau, specifically as a sponsor and supervisor of the Shawnee and Delaware. In 1793, Governor Carondelet personally awarded the Indians, as a single group, a huge tract centering on Apple Creek in vacant lands between Lorimier's new post at Cape Girardeau and the Cinq Hommes River.[44] The Spanish wanted the west bank of the Mississippi more fully occupied, especially so close to the mouth of the Ohio River, the chief conduit of westward American penetration.

Carondelet's third target for immigration was French of the ancien régime who were fleeing revolutionary France. These Catholics of wealth, it was reasoned, would find a sympathetic milieu in Upper Louisiana, where ideas of revolutionary republicanism had not taken root. One noteworthy immigration concerned the Ste. Genevieve District. In April 1793, Governor Carondelet agreed to a plan to bring French émigré families from Gallipolis in Ohio Territory to the Spanish Illinois.[45] These new settlers were profoundly unhappy with their lot in the malarial wilderness at Gallipolis. Part of the plan to bring them to Louisiana was the establishment of a village, La Nouvelle Bourbon, on the bluff top above the Pointe Basse common field, with refugee nobleman Pierre-Charles Delassus de Luzières as commandant. The Luzières family did settle on the bluff top, but the Gallipolis colony did not follow en masse as hoped. Although a few settled elsewhere in the Spanish Illinois Country, the planned colony of émigrés, despite its publicity and fanfare, did not develop.

The Treaty of San Lorenzo del Escorial, or Pinckney's Treaty, signed by Spain and the United States on October 27, 1795, was the result of forces

---

44. *ASP-PL* 2:551; "Proclamation Issued by Carondelet concerning Lorimier," January 4, 1793, PC 2363-718. Gregory Evans Dowd analyzes the Shawnee and Delaware struggle to maintain their integrity, along with other Indians with whom they sought common cause, in *A Spirited Resistance: The North American Indian Struggle for Unity, 1745–1815*. Spain had more interactive involvement with Indians in their borderlands in Alabama and Mississippi than in Upper Louisiana. See Whitaker, *Spanish-American Frontier*.

45. [Carondelet] to the Duque de la Alcudia, April 26, 1793, PC 2363-320.

and processes well beyond Louisiana and cannot be interpreted by Spanish strategies in the Mississippi Valley.[46] The treaty, however, was a major turning point for immigration and settlement of the Spanish Illinois Country, because it established a new direction for Spanish-American relations. The treaty defined the boundary between the United States and Spanish territory, it opened the Mississippi River to Americans without duties, and it gave Americans the right of deposit (to unload and load) at Spanish New Orleans. Spain had no cards left to force a future American hand. By the treaty, Spain acknowledged failure to achieve its goal of populating Louisiana against American encroachment. It was tacit acknowledgment of the inexorable expansion of Americans into the Mississippi Valley, the American Southwest, and ultimately across the river into Spanish territory. Spain forever abandoned thought of eventual incorporation of Louisiana as an integral part of its empire and began entertaining notions of using the non-Hispanic province as a pawn in international diplomacy. Who would offer the most for this expendable real estate?

After the treaty, Carondelet reversed himself and permitted immigration from the United States, although still officially restricted to Catholics. When he announced the new immigration and settlement policy, this time Americans were listening, heard the message, recognized the advantages of Upper Louisiana, and entered in large numbers. Thus, the year 1795 closed a chapter in the settlement history of the Ste. Genevieve District and of all Upper Louisiana, and 1796 began a new one, the beginning of its Americanization and rapid settlement.

---

46. Whitaker, *Spanish-American Frontier*, 201, 217–22.

4

Spanish Period, 1796–1804

The new immigration policy that opened Upper Louisiana to Americans finally brought in the large numbers of people that Spain had long wanted, but those immigrants were Protestant Americans instead of European peasants. During the remainder of the Spanish period, Spanish Illinois was progressively Americanized in people and in settlement processes. In 1804, three out of five residents were Americans, and the dispersed American farm became a basic settlement form to supplement the enduring French Creole villages.

The American move into Upper Louisiana was part of American immigration into Spanish territory in general. But the American migration into Upper Louisiana was especially significant, because there it overtook a long established and functioning French society, which included blacks and Indians, and there the American and French ways of occupying land and organizing society interacted in 1804 on nearly equal demographic grounds.

Why was the immigration policy successful after 1795? First and primary, conditions had to be present in Kentucky and elsewhere to motivate people to want to leave their homes.[1] Chief among them was the filling up of the best arable lands, some of which were already being abandoned due to decline in fertility. As children reached maturity, their choice of land was dwindling. Virgin land suitable for agriculture lay farther west, not within their midst. In addition, land had become a speculative commodity in Kentucky, Tennessee, and neighboring Appalachia. Land was bought and sold at increasingly higher prices to the disadvantage of ordinary families. Settlers bought land from speculators and other land promoters often without the benefit of a land survey, and titles were not always secure. Squatters who had improved their lands and expected later

1. The situation in Kentucky in the last two decades of the century is presented by Aron in *How the West Was Lost*, 1–50.

to get title to them by usage found themselves evicted by land companies or speculators who had obtained title to the lands. Daniel Boone was the most famous of the many who lost land in this way in Kentucky and sought land in Spanish Illinois under a different land system.[2] To the dispossessed, a free land grant from an authoritarian governor seemed a safer way to hold land. The simple land system of Spanish Illinois seemed fairer than the bully system of land speculators.

By autumn 1796, just a few months after the announcement of the new immigration policy, Lt. Gov. Zenon Trudeau reported a wave of Americans reaching Upper Louisiana. In a revealing letter of October 10, 1796, to Governor Carondelet, Trudeau described the numerous requests of Americans coming down the Ohio to enter Spanish territory. He reported having promised land already to an astonishing four hundred families, who, if they did not gain admission, Trudeau warned, would take up residence right across the river and become enemies. At the rate they were coming, Trudeau forecast ten thousand American families in a few years. He asked the governor for limits to the numbers he should admit. Carondelet, shaken by such numbers of the disliked Americans, instructed the lieutenant governor to admit farmers, but not hunters.[3] Farmers are peaceful peasants; hunters carry guns. But was not the westering American both farmer and hunter?

To make the new policy better known and to publicize the attractiveness of their lands, Spanish Illinois commandants circulated English-language pamphlets and placed advertisements in newspapers in Kentucky. New Bourbon commandant Luzières authored (he signed his statement on May 17, 1796) an eight-page pamphlet extolling the advantages of living in Spanish Illinois. It was printed in English in the *Kentucky Gazette* in Lexington on October 29, 1796, and probably in other issues and in

---

2. The Spanish land-granting process, which required that new grants not overlap existing ones, resulted in remarkably few land conflicts as contrasted to the acquisition of land in Kentucky at the same time. Indiscriminate granting of undeveloped land by Virginia before Kentucky statehood and by Kentucky itself after resulted in a complicated tangle of land claims and led to evictions of settlers by land companies who acquired titles to large tracts of land (Paul Wallace Gates, "Tenants of the Log Cabin"; Kinnaird, "American Penetration," 219–24). Ironically, after leaving his land-title problems behind in Kentucky and receiving a land grant from Spain, Daniel Boone had his Spanish land grant rejected by the American Board of Land Commissioners. To get his title confirmed required a special act of Congress (John Mack Faragher, *Daniel Boone: The Life and Legend of an American Pioneer*, 235–63, 299–300, 307; A. H. Rose, "Extension of the Land System," 26–27).

3. Zenon Trudeau to Baron de Carondelet, October 10, 1796, PC 34-715; *ASP-PL* 6:712.

other papers. In it, the recruiter-commandant described favorably the region's geography in detail, but throughout he emphasized the magnificent benevolence of the Spanish government. The government, Luzières wrote, "grants *gratis* and without any expences whatever (even without any office charges for titles and patents) as much land as those who settle there can cultivate, use and desire, proportionably to the strength of their families, their circumstances, their trade or profession, and in such place as they may choose." The country is "exempt from chicanery & lawyers. . . . [Differences] are decided upon and terminated . . . always without any expence. . . . [T]he Indian nations [are] all friendly to Spain. . . . [T]he greatest and most reasonable liberty of conscience and action is allowed, [so] that every one may go and come and do, without any obstacle or contradiction, whatever is not injurious to the king's interest, or that of any one else." It was a document carefully crafted to address the concerns of cash-short Kentuckians. He spoke to settlers' concerns about unfriendly Indians in Kentucky. He also allayed Americans' apprehension of a Catholic regime that reportedly forbade Protestant worship.[4]

The public recruitment of Americans on American soil by a Spanish official may not have helped Spanish-American relations, but it did produce great results. Numerous Kentuckians, when requesting land in Spanish Illinois, testified to the effectiveness of these advertisements. When William James petitioned for land in early 1798, he related that "in consequence of the encouragements, announced in Kentucky, that should be granted to honest, substantial, and Catholic inhabitants . . . he determined last year to come and settle himself, with his family and slaves." John Dodson, one of the first to settle in the Bois Brule in the spring of 1798, came "as the result of an announcement published in the Gazette of Kentucky." Job Westover, a native of Pittsfield, Massachusetts, also heard of opportunities in Spanish Illinois and arrived in 1797 at age twenty-five. A carpenter and millwright, Westover had skills much in demand. A slave-

---

4. Luzières, "An Epoch in Missouri history." Luzières signed the letter on May 17, 1796. Upper Louisiana had no printing presses in 1796. There was one in Lower Louisiana, but it was reserved for governmental use, and besides it made no sense to have the ad shipped from New Orleans for distribution in Kentucky. Luzières probably paid for the printing expenses himself. Commentary by Frances L. S. Duncan and Jacqueline P. Bull accompanying the republication of the pamphlet in 1958 interprets its significance in terms of intrigues to separate Kentucky from the United States, which, of course, is the reason that Kentucky interests republished the pamphlet. The commentary admits ignorance of its actual effect on settlement of Upper Louisiana (Missouri), which was Luzières's real purpose for printing it.

holding Protestant, he received a concession of 325 arpents on February 10, 1798, on the River Aux Vases just west of newly relocated Ste. Genevieve village.[5]

Yet another example of the effectiveness of the advertisements is the response of thirty-one families on the Red River in Kentucky. The advertisement convinced them to leave a land of "high price" and "very high taxes" for a place of free land and no taxes. They sent a scout, Risin (or Risine or Rissene) Boyer to obtain a concession for them to settle as a group. Their request was granted on May 4, 1797, for fertile land on the unoccupied limestone plain between the Mine à Gabouri and the Prairie à Bollieur (Boyer?) near present Farmington.[6]

A large group of German-American Protestants also sought to emigrate to the Ste. Genevieve District in 1799 in response to Luzières's advertisements. Unable to find enough unconceded good-soil land in the district, they discovered what they were looking for on the karst plain behind Lorimier's post at Cape Girardeau and reserved land for even more German American families on the way. Lorimier enthusiastically recommended these "hardworking and peaceful farmers" to the lieutenant governor. This was the beginning of the prosperous Whitewater Dutch community that formed the American core of the Cape Girardeau District.[7]

The governor instructed the lieutenant governor and local district commandants to be especially proactive in encouraging immigrants who were professionals and developers. New Bourbon commandant Luzières, mindful of Ste. Genevieve's lack of the services of a physician and surgeon, successfully negotiated in 1797 with the Fenwick family, a Catholic family that had moved from Maryland to Kentucky, to settle in the Ste. Genevieve District with ample lands for each member of the family, especially the physician, Dr. Walter Fenwick.[8]

One of the most celebrated Americans to enter as a result of the new immigration policy, and one who also had read the advertisements, was Moses Austin. Austin, in the lead business in western Virginia, had learned about the extensive lead deposits in the Ste. Genevieve District and negotiated directly with Governor Carondelet in New Orleans for a large land

---

5. "Land Claims in Missouri, January 21, 1834," *ASP-PL* 5:748; request for land by Jacques Dodson, April 15, 1798, LCS 168; John Glendower Westover, *Selected Memories*, 1–3.

6. Petition for land by Risin Boyer et al., May 4, 1797, LCS 116.

7. Louis Lorimier to Charles Dehault [sic] Delassus, March 20, 1800, PC 218-352; Louis Lorimier to Charles Dehault [sic] Delassus, January 15, 1800, PC 218-356.

8. Dehault Delassus [sic] to Carondelet, April 18, 1797, PC 213-903.

grant and certain favored conditions for himself and a large entourage of skilled workers. Austin came and quickly rose to a leading position in the mining district.⁹

In contrast to the stream of Americans, French and Creole immigration during these eight years dwindled to a trickle. By the mid-1790s the east bank had fairly well exhausted itself of Creoles willing to cross the river. Ste. Genevieve received hardly any more persons of French culture, whether from Canada, Lower Louisiana, the Caribbean, Flanders, or revolutionary France.¹⁰

Most of the immigrants into Upper Louisiana after 1795 were ordinary American farmers. After the initial wave that responded to public advertisements, chain migration took over. As letters and people moved back and forth, potential immigrants heard good things about the Spanish territory. Families in the Holston Valley of eastern Tennessee heard from their former neighbors, the Byrds, who were some of the first to go to Upper Louisiana. According to one of the emigrant families, the Byrds and others who had been out to see the country "put the people on fire to move," and a "mania" swept over the valley. Whole family clans packed up and left the Holston Valley for Spanish Illinois.¹¹

By now it was clear that the Americans were discarding preconceived notions that the Spanish authorities were cruel, repressive, and arbitrary. Americans already in Upper Louisiana could vouch for a benign and quite superficial government. Mistreatment and arbitrariness were certainly not part of their experience. Once land had been conceded from the abundant reservoir of royal domain, a new settler probably had no more contact with a Spanish administrator. Surveyors, syndics, and local *commissaires de police* were French and, in the American-settled neighborhoods, American.

In the late 1790s, acquisition of land was far more important to westering Americans than any perceived difference between distant Spanish and American governments. Thus, the geographical unity of the valley

---

9. Gracy, *Moses Austin: His Life,* 53–54. According to Gracy, Austin had earlier heard reports about Upper Louisiana lead, but the firsthand account by the New Bourbon commandant, which he read, cemented his decision to move.

10. One of the few French émigrés of revolutionary France who made it to the Ste. Genevieve District after 1796 was Jean-René Guiho, sieur de Kerlegand, whose experience illustrated the difficulties that French haute culture had in adjusting to frontier conditions. Kerlegand settled at New Bourbon and disturbed the village's harmony so much that he was forced to flee to New Orleans (Walter A. Schroeder, "Opening the Ozarks: Historical Geography of the Ste. Genevieve District [Missouri], 1760–1830," 132–33).

11. Reynolds, *Pioneer History of Illinois,* 249–51.

was fast becoming a reality constructed around a people who valued land above political systems. In that context, one must realize that Americans who were entering Upper Louisiana in 1799 and 1800 suspected that it might one day become part of the United States. "It is but just to the memories of these people," noted a Tennessee emigrant recalling those years, "to state that a presentiment existed in their minds that the country would come under the American government, and they, or at least their children, would enjoy equal rights."[12] So many rumors were circulating about the future status of this Spanish-administered but not Spanish-settled land that such presentiments divulged a good understanding of the implications of the settlement changes put in motion in 1795.

To meet the land needs of the increased number of settlers in Upper Louisiana, Spain in February 1795 appointed its first full-time surveyor, Antoine Soulard, so that tracts could be properly demarcated on the ground and plats made. Soulard employed deputies, both French and American, to help handle the huge volume of accumulated, unsurveyed concessions and the large number of new concessions being awarded weekly. (Arrival of Americans in their midst had impelled the resident French to seek formal concessions for land they had long occupied without documentation.) Thomas Maddin, an American, was Soulard's deputy surveyor for the Ste. Genevieve District.[13] The initiation of land surveying was visible evidence that both settlers and administrators were taking matters of land occupation more seriously. *Bornes de pierre* and notched trees began appearing in the landscape to mark property lines and corners.

The change in Spanish immigration policy necessitated a rethinking

12. Ibid., 220. In 1796, rumors were rampant of possible retrocession to France, made more believable by Victor Collot's mission to the Mississippi River, but these had little direct effect on the settlement of Upper Louisiana (Baron de Carondelet to the Principe de la Paz, November 1, 1796, PC 2364-144; Durand Echeverria, "General Collot's Plan for a Reconnaissance of the Ohio and Mississippi Valleys, 1796"; Houck, *Spanish Regime*, 1:xxiii).

13. Houck, *History of Missouri*, 2:58, 225. Surveys had been made of St. Louis town lots in the late 1760s by Martin Duralde and others. Duralde was appointed official surveyor of Upper Louisiana by Lieutenant Governor Piernas in 1770 (Peterson, *Colonial St. Louis*, 9 n. 23). St. Louis surveys were entered in the *Livres Terreins,* which are in the Missouri Historical Society, St. Louis. It is believed that no surveys were made of town lots in Ste. Genevieve or other villages until the American period. Soulard employed standard French surveying practices. For example, his surveys and plats indicate he used a *témoin* (witness) for marking corners and other points on his survey lines. Often the *témoin,* buried under the point, was *mâche fer* (or *fer mâche* [chewed iron]), a piece of metal, possibly slag from lead smelting. Undoubtedly, many of these fragments of buried metal are still in place two centuries later.

of the procedures for land distribution in Louisiana, and on September 9, 1797, Gov. Gayoso de Lemos issued new instructions to commandants in regard to granting land for new settlers. Because these instructions were in force when the great bulk of Americans arrived and when the great bulk of land concessions were made to both French and Americans, they later assumed critical importance in the technicalities of the American land-confirmation process. They were of great significance to the Illinois Country, because they pointedly closed the Illinois Country to Protestants and returned to a Catholic-only immigration policy. When Lieutenant Governor Trudeau questioned his superior on this reversal of policy (Trudeau was strongly against it), Gayoso replied on July 9, 1798, that his instructions "must be followed to the letter" and that Trudeau was "not given any latitude" in this matter. "Americans are not the most suitable subjects as prospective colonists and . . . the intention is that the western part of the Illinois is to be settled primarily by Canadians."[14] There was little Trudeau could do about such clear language from his superior, but there was still less he could do about the Protestant Americans who were entering Spanish Illinois either ignorant of or ignoring the governor's instructions.

Gayoso's major innovation was to standardize the amount of land awarded to settlers by a formula of two hundred arpents per husband and wife, fifty arpents for each child, and twenty arpents for each slave, not to exceed eight hundred arpents for one family. These figures became ingrained in the thinking of commandants and lieutenant governors for the rest of the Spanish period. The figures also implied an acceptance of more compact parcels that Americans wanted and a rejection of the size and shape of traditional French long lots for future concessions.

Thus, while Gayoso might have wanted to populate Upper Louisiana with a Catholic peasantry, he worded his instructions to address the large number of new American immigrants. The commandants and lieutenant governors dutifully applied them, but only to the common American farmer. A lieutenant governor could still award however much land he wanted to, to whomever, including those who could pay him for his generosity. After all, Upper Louisiana was ruled by a monarchy, and not everyone needed equal treatment under the law. The "law" was for common people. Thus, whereas an immigrant family of four received only three hundred arpents according to the formula, other immigrant families received handsome awards of one thousand or more arpents. Nowhere

14. *ASP-PL* 5:290–91. A date of September 3 instead of September 9, 1797, is used in "Instructions to commandants in regards to new colonists," PC 2365-378. Gayoso to Zenon Trudeau, July 9, 1798, PC 133-400.

in any official pronouncement was it specifically stated that authorities had this freedom to award land as they wished. It was simply understood to be a prerogative of an officer of an authoritarian government subject to review only by his superiors.[15]

One prominent example shows how Gayoso's instructions for awarding land to new settlers were applied selectively. On March 15, 1797, mining entrepreneur Moses Austin negotiated with Governor Carondelet for a grant of four square leagues. The skilled workers he brought with him to build the lead works were also guaranteed land by the governor's contract with Austin, but their quantities were in strict accordance with Gayoso's formula. Brothers Samuel and Jacob Neal, for example, asked for land in the creek valleys west of Austin's grant. Samuel and family received 240 arpents, and Jacob, with a much larger family, received the maximum, 800 arpents. Both concessions (dated November 22 and November 20, 1799, respectively) carried the precise language of the formula for granting land, including instructions to get a perfected title from the intendant.[16] The Neals were Protestants, like Austin.

Gayoso's formula was also applied to French-culture Catholic immigrants. François Berthiaume, a gunsmith associated with Louis Lorimier and the Shawnee and presumably a Catholic, received a concession inside the Indian tract at the mouth of Apple Creek. The size was 420 arpents, for himself and wife (200 arpents), four children (200 arpents), and one slave (20 arpents). In fact, the concession from Lieutenant Governor Delassus, dated December 28, 1799, specifically states that the quantity is "proportionate to the number composing his family, conformably to the regulations of the governor general of this province."[17]

The wave of American immigrants into Louisiana was so large within only three years that the worst fears of the Spanish government and of a certain segment of the French community in Lower Louisiana were coming true. In Lower Louisiana, civil and church authorities expressed concern that the Americans were not being culturally assimilated, as long-standing Spanish immigration policy insisted. Even worse, they were affecting the morals and lifestyles of the French. In other words, the French

---

15. This difference caused great consternation later during the land-confirmation process, as the board looked in vain for written authority that gave Spanish officials the power to grant lands of unlimited size, sometimes in numerous parcels, to whomever they wished.

16. Gracy, *Moses Austin: His Life,* 63–65; LCS 227; on Samuel Neal: LCS 226; on Jacob Neal: LRA 220. Also see wording in concessions LCS 231 (September 1, 1799) and 234 (January 3, 1800).

17. *ASP-PL* 5:801. Also see wording in the concessions LCS 18, LCS 18a, LCS 72, and LCS 74.

were being Americanized, rather than vice versa. "Americans were to incorporate with us, not we with them," lamented one French officer in Lower Louisiana.[18] Spanish goals of Hispanicizing, Gallicizing, or simply Catholicizing the Americans were now more remote than ever.

The situation in Upper Louisiana, however, was different. American moral character and its influence on the residents were not the primary issues; economics was. Americans were industrious and contributed to a theretofore undeveloped economy. They invigorated the mining industry and brought trade skills in carpentry, masonry, tanning, milling, wagon making, boatbuilding, and gunsmithing and the professional skills of doctors, surveyors, and engineers. They brought business entrepreneurship and ties to the American East to supplement the commerce of the French. Moses Austin achieved fame as a real Spanish *empresario*. The Americans' reputation as productive farmers was no less complimentary. They not only raised hemp, which French Creoles could not or would not without slaves, but even processed it into ropes, which reduced imports. Lieutenant Governor Trudeau had earlier noted that Americans "work hard, develop mills, build better houses, and spur economic development." "We should seek Catholics," he continued, "but the Americans are industrious and make good farmers."[19]

In 1800, David Wilcox, a Kentuckian, discovered iron on the Meramec and sought to operate a forge and smelter with ten workers from Kentucky as well as ten to fifteen families. Frankly admitting that he could not promise that they would become Catholic upon taking up residence in the Illinois Country, he did vouch for their becoming honest and faithful subjects of the king of Spain. Lieutenant Governor Delassus recognized the enormous economic advantages that this initial iron manufactory would bring and forwarded Wilcox's application with his approval to New Orleans. The reply was an abrupt no; no Protestant Americans could establish themselves in Upper Louisiana, despite glowing recommendations for their entrance by the lieutenant governor and the district commandants.[20] The iron forge and smelter remained unbuilt. To veto-

---

18. Victor Collot, Général de Brigade, to Citizen Talleyrand, Minister of Foreign Relations, Paris, 15 Brumaire, An X, November 6, 1801, in *Before Lewis and Clark*, ed. Nasatir, 2:670.

19. On reputation: Santiago Mackay to Gayoso de Lemos, November 28, 1798, PC 2365-297; on hemp: LCS 241; "Trudeau's Report," in *Spanish Regime*, ed. Houck, 2: 256.

20. Carlos Dehault [sic] Delassus to Ramon de Lopez y Angulo, November 28, 1800, PC 260-800; petition from David Wilcox to Carlos Dehault [sic] Delassus, November 8, 1800, PC 260-1801; Lopez y Angulo to Carlos Dehault [sic] Delassus, January 24, 1801, PC 260-1803.

ing authorities in New Orleans, cultural correctness took precedence over economic development.

Americans who were considering relocating in Spanish Illinois were getting mixed messages. The territory was ruled by a Catholic monarchy that preferred Catholic immigrants, but Protestant Americans could enter if they would raise their children as Catholics and not participate in Protestant worship. How would the Spanish government enforce this? Could they enforce it? Many Protestants were already in Upper Louisiana, demonstrably successful and satisfied. Tennessee families had entered, received permission to settle, and were content in their own growing American Protestant neighborhoods. But the official policy daunted some. "We were destined for the Murphy's settlement, on the St. Francis river," wrote one potential immigrant, but upon learning that Spain required the father "to raise his children Catholics, [he] determined not to live under such a government."[21] Some Protestants called the Spanish government's bluff, whereas others took the Spanish at their word.

Spain appointed Juan Bonaventure Morales the intendant general of the Province of Louisiana on October 22, 1798, and he took over power to grant lands from Gayoso de Lemos, who continued as governor.[22] Shocked at the huge size of some of the grants of his king's land that governors and lieutenant governors had awarded, especially to Americans, Intendant Morales issued a new set of land ordinances on July 17, 1799. They did not replace Gayoso's regulations in any substantial way (the eight hundred–arpent limit continued) but made them more explicit and comprehensive.[23] Like those of Gayoso, they figured prominently in the land-confirmation process of the American government because they were the last set issued and were in force when the last Americans arrived under the Spanish regime.

Though both Gayoso's and Morales's regulations laid out clear procedures for obtaining lands, the procedures, at least in Upper Louisiana, already seemed to be breaking down, whether willfully from local administrators who disagreed with them and ignored them or from administrative ignorance of them. One certain factor was the large influx of so many immigrants and so few authorities motivated and able to enforce official procedures.

21. Reynolds, *Pioneer History of Illinois,* 250.
22. *ASP-PL* 5:735.
23. *ASP-PL* 3:432–35, 5:291–94; Pelzer, "Spanish Land Grants," 9; A. H. Rose, "Extension of the Land System," 22–24; Burns, "Spanish Land Laws," 570–74. Burns also describes the "acrimonious dispute" between Morales and Gayoso over land granting (576–79).

According to Joseph Pratte, a Ste. Genevieve land speculator and lead miner, after Morales took over land granting for Upper Louisiana the lieutenant governor could merely make recommendations to the intendant, but those recommendations, even when not returned approved by Morales, were uniformly considered valid by residents and commandants in Upper Louisiana and were "transferred from hand to hand as such."[24] Thus, Lieutenant Governor Delassus continued uninterruptedly to award concessions, although they were officially only recommendations to the intendant. Even neighborhood syndics were now granting permissions to settle, and no one questioned their validity.[25] Several years earlier, in 1795, the lieutenant governor had surrendered his prerogative to grant village lots *(emplacements)* and authorized local commandants to grant them in their district villages.[26] Apparently, the commandants did this casually and orally, because few seem to have been recorded and sent up the line for approval. These procedural irregularities were not as serious as were the first indications that land-seeking Americans were blatantly ignoring Spanish authority and entering without seeking permission from anyone and without requesting concessions from anyone.

It is not clear whether those arriving after 1796 in Upper Louisiana were always made aware of the land regulations of Gayoso and Morales. It is not clear, in fact, whether they were even publicly posted in the various districts of Upper Louisiana as regularly done for official statements.[27] Charles de Hault Delassus, the last lieutenant governor of Upper Louisiana, acknowledged receipt of Morales's regulations, but he admitted not publishing them and not passing them on to his subordinates and commandants, because quite frankly he did not intend to observe them himself. In fact, he had openly "remonstrated" against them, and asserted with some pride that the regulations of Morales were never in force in his jurisdiction.[28] Yet, in one concession dated December 28, 1799, Lieutenant Governor Delassus wrote that the size of the grant would be made "conformably to the regulations of the governor general of this province; and after this [concession] is executed, the said party shall have to solicit the title of concession in due form from the intendant general of these provinces, to whom alone corresponds, by royal or-

---

24. Testimony by Joseph Pratte, December 14, 1832, LPC 24c.
25. LCC 149; LCC 178s.
26. LCS 81½. The practice occurred as early as 1792 with or without the lieutenant governor's approval (LCS 93).
27. W. A. Schroeder, "Opening the Ozarks," 140.
28. *ASP-PL* 8:796, 801.

der, the granting of lands and town lots."[29] Evidence indicates that the authorities in Upper Louisiana were indeed aware of the regulations emanating from New Orleans, but they ignored them because they disagreed with them or they applied them selectively.

After the retrocessive Treaty of San Ildefonso of October 1, 1800, Spanish authorities in Louisiana continued to act in their official capacities until France fulfilled European terms of the treaty. When France did so in 1802, Spain issued a royal order on October 15, 1802, to turn the province over to France, but no French functionary came to New Orleans, and Spanish officials continued to govern Louisiana as though it was still theirs. Meanwhile, on December 1, 1802, due to the death of the assessor in New Orleans, Intendant Morales called a halt to all land granting in the province, and in the same year the king of Spain issued an edict forbidding the granting of any land to Americans.[30]

However, the royal edict to stop American immigration made no difference at all in Upper Louisiana. It was far too late to dam the migration river now flowing for more than five years. People were entering without obtaining permissions to settle, and there was no way to enforce the law, even if the increasingly hapless administrators wanted to. When Delassus in 1803 received the orders from New Orleans not to award any more land, he must have been amused at the futility of it all.[31]

After residents of Upper Louisiana learned that the province had been retroceded to France, neither the lieutenant governor nor the local commandants tried to prevent Americans from establishing their way of life in the midst of a European culture. Instead, these officials and other French community leaders put their own personal economic gain over Gallicization of Americans and cast their futures in the hands of Americans either as partners or by learning their ways. Both the lead business and fur trading, the two economic mainstays, underwent rapid hybridization. The French of Upper Louisiana also learned that land could be a speculative commodity and welcomed more and more Americans who would buy their extensive landholdings that they had obtained free but now could sell at handsome prices to Americans.

---

29. *ASP-PL* 5:801.

30. Royal order: *ASP-PL* 5:274–76. Intendant Morales ordered a halt to the acceptance of requests and their granting because of the death of the assessor, which created a vacancy in the chain of land-granting authority (*ASP-PL* 5:735; Martin, *The History of Louisiana*, 564).

31. *ASP-PL* 2:389. The fact that officials and residents of Upper Louisiana operated to a large degree under their own procedures for awarding land and recognizing its ownership constitutes evidence that Upper Louisiana had begun functioning independently from authorities in Lower Louisiana.

Realizing from 1800 onward that they would be leaving office and had little to lose and potentially much to gain by selling their signatures, the Spanish administrators conceded land generously.[32] And when the United States showed interest in purchasing New Orleans in March 1803, more intrigue entered the realm of private ownership of land and the settlement process in general. The logical course of action to residents and immigrants was to establish claim to land. Enterprising residents, both French and American, tried to get as much land as possible by concession from the lieutenant governor; the others, the ordinary settlers, both resident French and entering Americans, tried to occupy and establish claim to land by settlement rights before the old regime was replaced.

Despite orders from New Orleans not to award any more land, Lieutenant Governor Delassus continued to concede land and order surveys through the winter of 1803–1804, even after the formal surrender of Lower Louisiana to the United States at New Orleans on December 20, 1803. The late concessions and surveys went to relatives, friends, and all others who paid Delassus for this last act of aristocratic largesse with a monetary douceur, or "sweetening." The lieutenant governor had absolutely nothing to lose as retiring administrator for a retiring regime—to whom was he accountable?—and he owed absolutely nothing to the incoming regime. Land was appreciating fast and expected to rise in value even further, and he could help his friends profit substantially. But Delassus had another motive: he was heavily in debt, and his father, the New Bourbon commandant, even more so. Delassus had received hardly any stipend from Spain, and now the "sale" of land concessions presented itself as a convenient way to pay debts and thereby restore honor to his family name.[33] Because the territory had been retroceded to France in 1800, then sold to the United States in 1803, and the governor and intendant had issued orders not to award any more land, Delassus lacked land-awarding authority and was obliged to backdate his actions to a date before December 20, 1803, or better, to 1799, before the Treaty of San Ildefonso, and enter the name of his predecessor, Trudeau.[34] Unauthorized and unethical, backdating had to be done quickly and discreetly. It also needed the connivance of surveyor Soulard to backdate new surveys and to erase and write new, earlier dates on surveys already completed. Soulard may have solicited douceurs for himself for surveys in the ab-

---

32. Burns, "Spanish Land Laws," 579–80.
33. Robert R. Archibald, "Honor and Family: The Career of Lt. Gov. Carlos de Hault de Lassus [sic]." Lieutenant Governor Trudeau reported that the commandants at Ste. Genevieve, New Bourbon, and Cape Girardeau received one hundred pesos a year, and the others nothing ("Trudeau's Report," in *Spanish Regime,* ed. Houck, 1:257).
34. Houck, *History of Missouri,* 2:226–30.

sence of written concessions. In December 1803 and January 1804, inclement months for fieldwork, he and his deputies completed an unusually large number of surveys, then again a remarkable seventy-seven in February 1804. The majority bore concession dates of 1797, 1798, 1799, and 1800, before the Treaty of San Ildefonso and before the requirement that concessions pass through Intendant Morales's hands.[35]

Upper Louisiana heard news of the sale of Louisiana, signed in Paris on May 2, 1803, toward the end of June.[36] The date that the United States took formal possession of New Orleans, December 20, 1803, became the critical cutoff date for carrying over landownership in Louisiana from pre-American regimes to the American regime. The newly arrived and arriving settlers of Upper Louisiana recognized fully the importance of the date and scrambled to lay claim to their lands. In the Bellevue Valley, for example, families were arriving that fall and winter. One settler later testified:

> Thomas Baker told him that his father and Benjamin Crow notified all those who claimed places to be on them by a certain day, and accordingly, on the day thus appointed, Thomas Baker and many others did go and remain on certain pieces of land, and that the father of Thomas Baker and Crow did go around, and saw that Thomas and others thus in possession or occupancy of certain lands, and report says that they afterwards went to St. Louis as witnesses to prove up their claims to land thus occupied as aforesaid.[37]

These assertive and independent Americans, in French-Spanish Upper Louisiana without permission to enter, could honestly swear before God that they were "on the land" and "occupying it" in order to obtain it through settlement rights.

Thus, the period from mid-1802 to March 1804 was a time of intense maneuvering to claim landownership in Upper Louisiana. The lieutenant governor offered backroom deals, and Americans rushed in to claim land by standing on it with a witness present. Capt. Amos Stoddard's arrival in St. Louis on February 24, 1804, and assumption of U.S. authority in a formal ceremony on March 10, 1804, brought an abrupt end to all this

---

35. Untitled and undated opinion by Judge J. H. Peck on land fraud in "Opinions and Arguments Respecting Land Claims in Missouri," June 3, 1836, *ASP-PL* 8:832–47; "Capt. Amos Stoddard to the Secretary of War about the Spurious Grants of Land in Upper Louisiana," January 10, 1804, in *Annals of St. Louis in Its Territorial Days, from 1804–1821,* by Frederic L. Billon, 371–73. Judge Peck's opinion is probably one of the most complete judicial statements on land fraud involving private claims from the Spanish period. Also see A. H. Rose, "Extension of the Land System," 55–62.

36. Houck, *History of Missouri,* 2:36.

37. *ASP-PL* 7:829–30.

fast-paced and quasi-legal activity. A new order to the settlement process was beginning.

If the Spanish land-granting officers in Upper Louisiana did not always adhere to the numerous laws and regulations designed for them, they did follow general procedures, which, according to an American administrative review of them in 1834, "seemed to have grown into gradual though silent operation, originating from the circumstances of the country, and accommodating itself to the necessities and condition of the people." The "silent operation" gave virtually complete discretionary power to the land-granting officers in admitting persons as settlers and awarding land. Amos Stoddard later noted that "the laws, rules of justice, and the forms of proceeding, were almost wholly arbitrary[,] for each successive Lieut. Governor has totally changed or abrogated those established by his predecessor." And, it might be added, discretionary power mutated into disregard of regulations toward the end of the Spanish period. Some historians have interpreted this discretionary power as "chaotic and systemless" Spanish policy, whereas others believe that it exhibited Spanish administrative "flexibility and sensibilities" to frontier conditions.[38]

Among the discretionary powers of all lieutenant governors was granting land for services to the king as *mercedes* (graces).[39] The Spanish began this practice early in the Americas, as did the other colonial powers, and the practice was never questioned during Spain's long history in administering its dominions. Lieutenant governors in Upper Louisiana presumed to have the authority to grant *mercedes* in the name of the king, whose land they were giving away.

Other examples of Spanish authorities using their own discretion included disregarding the requirement to obtain a perfected title from the

---

38. "On Land Claims in Missouri, with Translations of Sundry Spanish Laws and Customs Relating Thereto," February 19, 1834, *ASP-PL* 6:923; Amos Stoddard, "Transfer of Upper Louisiana: Papers of Captain Amos Stoddard," 95–97. Also see Stoddard, *Sketches of Louisiana,* 243–67; and *ASP-PL* 6:924. Pelzer, "Spanish Land Grants," 6; Charles R. Cutter, *The Legal Culture of Northern New Spain, 1700–1810.* Banner interprets the discretionary power as acting according to the local customs of the people (*Legal Systems in Conflict*).

39. *ASP-PL* 6:924; Stoddard, *Sketches of Louisiana,* 257. An example of a *merced* is a land concession of sixteen hundred arpents on the River des Pères to François Tayon in 1786 in gratitude for the valuable services that he rendered as the St. Charles commandant (*ASP-PL* 5:779–80; Houck, *History of Missouri,* 2:82). As a governmental reward for services, the Spanish *merced* is comparable to American land grants to veterans of the Revolutionary War and land grants to Meriwether Lewis and William Clark for their expedition to the Pacific.

governor. They allowed inchoate concessions to be treated as legal assets; such concessions could be transferred, inherited, and mortgaged, and no one else could claim the land. Lack of survey, or even lack of warrant or order of survey, did not render a concession any less valid. Neither did lack of gainful occupation of land by cultivating so many arpents or erecting buildings. There was no reversion to the royal domain of lands left uncultivated, unoccupied, or unsurveyed at the end of one or four years, as stipulated. A concession did not even have to be located on the ground.

The lack of congruence between written law and actual settlement process became more prevalent in the closing years of the Spanish regime, when settlers disregarded immigration law. Until 1796, comparison between law and actual immigration lacks much significance, as so few entered Upper Louisiana and those who did enter were mostly French Catholics and the very kinds of immigrants Spain recruited. After 1796, however, Spanish immigration policy vacillated in how to address the relentless westward march of Americans until no plan became workable. Americans were after land, and no immigration policy would likely contain them any longer, certainly not in the face of an impotent administration.[40]

The buffer location of Louisiana between the expansive United States on one side and numerous, formidable Indian nations on the other contributed to the circumstances that liberalized or expanded the discretionary powers of Upper Louisiana authorities in matters of settlement. At this extreme northern geographical projection of Spanish rule into the continental interior, the exposure and vulnerability of the Spanish presence required the maximum flexibility of administrative action in the enforcement of constricting laws and rules.[41]

Thus, the Spanish administration in Upper Louisiana ended without leaving much of an imprint on its settlement. The goal of defense brought forts and soldiers, but their role in local settlement was inconsequential. Few Spanish actually settled and stayed. Spanish policy was successful by impeding American settlement until 1795, but after then Americans forced Spain to let them enter with their own settlement practices. The Spanish chiefly left a settlement legacy of perplexing laws that poorly addressed the settlement needs of Upper Louisiana and led to years of confusion in the land-confirmation process.

---

40. The relentless pressure of Americans moving outward from the United States in 1783 is an organizing theme of Meinig's second volume of *Shaping of America*, 3–77. The pressure could not be effectively controlled by laws, either those of the United States or those of territories being entered.

41. *ASP-PL* 6:924–25.

When the events of the 1760s brought about the partitioning of the interior of North America into Spanish and British spheres with utterly no regard for the various Indian nations that continued their full and active control of most of that immense space, the Mississippi River provided the easy, natural line of demarcation for European empires. It was, as the Spanish minister, the count of Aranda, suggested, an obvious line on a map and an obvious feature in the landscape that all could see and observe.[42] Yet, dividing political space by a major river violates the principle of human geography that rivers unite, creating the geographical paradox of a region united hydrologically for its social and economic life but divided for administrative reasons. During the Spanish period, the middle Mississippi Valley grew increasingly into shared space where people came together, from both the east and west sides, down from the north and up from the south. Yet, in the minds of distant government officials, the river became increasingly fixed as a psychological boundary between different imperial jurisdictions with strikingly different ways to occupy and administer space.

Toward the end of the Spanish period, Louisiana became a pawn in the complex and ever-changing power politics in revolutionary Europe. The United States was drawn into the picture because of the necessity to have unthreatened use of the river and its mouth. In the end, in 1803, the United States found itself not only with the use of the river but also with ownership of the vast drainage basin of the entire Mississippi system west of its trunk. Thus, the natural physical and human geographic unity of the basin was returned to political unity. No longer would settlement policy be made by European minds using European models. No longer would European solutions be used for American situations. No longer would European wars and power politics determine immigration into the valley. Policy for the entire Mississippi basin would now be made by Americans for Americans. This did not mean that the discordance between administrative actions and the vernacular settlement process immediately faded away. The discordance continued, but it would now be expressed as one between federal land policies made in Washington for the settlement of its distant western territory and the vernacular land-occupying practices of westering Americans.

---

42. Weber, *Spanish Frontier*, 199, 203. Also see comments on the artificiality of the Mississippi River as a boundary by Usner (*Indians, Settlers, and Slaves,* 121) and Gitlin ("Boundaries of Empire," 81–84).

5
American Period, 1804–1818

A single sentence in Article 3 of the Treaty of Cession of April 30, 1803, stipulated that "the inhabitants of the ceded territory shall be maintained and protected in the free enjoyment of their . . . property."[1] This was to be expected; property rights are normally respected with transfers of government. What was different for Upper Louisiana was the contrasting method of obtaining and occupying land. The Spanish government had given it away upon request; the American government sold it. The Spanish system in practice was direct and simple; the American system was complex and bureaucratic. Residents who understood this contrast and anticipated its implications for appreciation of land values managed to take advantage of the change before the transfer of authority.

Because the people who speculated in land and contrived to secure their properties were important and influential and their lands were some of the most strategically placed in the territory, the fate of their claims dominated public debate over land during the transition years.[2] On the other hand, the transfer presented a completely different situation to the ordinary French Creoles, and they responded differently. Simply put, many of them did not or could not understand the new American system.

The French had a paternalistic institutional system whereby a commandant or cleric oversaw the individual's welfare. The village settlement system, made cohesive through kinship ties and close spatial contact, promoted group solidarity and looked to tradition as the way to do things. The commandant arbitrated differences. The public and legal institutions of the Americans, in which responsibility lay with the individ-

---

1. *LPGN* 1:2; *ASP-PL* 5:276–77.
2. To put the land issue within the broader context of territorial politics, see Foley, *Genesis of Missouri*, 170–74, 210–13, 249–53; William E. Foley, "The American Territorial System: Missouri's Experience"; van Ravenswaay, *St. Louis*, 122–45; David D. March, *The History of Missouri*, 1:158–261.

ual and his initiative, were difficult to understand for these French. Consequently, many failed to present themselves before American authority and apply for titles to their lands, even though they held concessions to them. Ignorant and in dread of new American laws to a greater degree than Americans had been cowed by Spanish rulers, they deferred from the litigation required for getting their property titles confirmed. In addition, a certain complement of the French had never bothered to get a concession, because their commandants had told them that the "ax and the plow" are the best proofs of landownership.[3]

How, indeed, would these naive Creoles of a different culture stand and plead their case for land before an impersonal, imperious, and alien board of judgment that lacked the paternalism of the French-Spanish system? Daunted by the new American laws with which they felt threatened, if not assaulted, they distanced themselves from the new system and eventually became officially landless squatters on land they had previously gainfully occupied for years. They were gullible to incoming Americans who bought or took over their improvements and then got them confirmed. These folk, of unknown numbers, have not been treated well if at all by historians because they lie near the bottom of the socioeconomic scale and left little written record. Probably, many merged inconspicuously into the general population or moved on. Among this group were the hundreds of persons engaged in various ways in fur trading, hunting, and mining. John B. C. Lucas, of French birth and Pennsylvania residence, who had visited Upper Louisiana in the 1790s and later played a major role in the new American administration of the territory, commented in 1807, "[T]hese people are poor but remarkably honnest harmless and unsuspecting[;] they dislike the Americans in generall because those with whom they had intercourse have taken more or less advantage of them." He continued, "[T]he greatest Number of those french inhabitants, have Little or No Land."[4]

---

3. Judge Lucas to the Secretary of the Treasury, January 29, 1807, *TP* 14:88–90; Recorder Bates and William Rector to Josiah Meigs, November 25, 1816, *TP* 15:210–11; Flint, *Condensed Geography,* 1:311; David Francis McMahon, "Tradition and Change in an Ozark Mining Community." Dorrance also comments on the dispossession of Creoles' land (*Survival of French,* 43). Harvey Wish shows that the experience of the poorer French Creoles with the incoming Americans sharply contrasted with that of the wealthier Creoles who cooperated and cast their fortunes with the Americans ("The French of Old Missouri, 1804–1821: A Study in Assimilation").

4. Lucas to Secretary of Treasury, January 29, 1807, *TP* 14:90; Judge Lucas to the Secretary of War, February 9, 1807, *TP* 14:95. The numerous French Creoles who lost their land after 1804 are known because of their presence in the baptism and marriage records but their absence from landownership records after 1804.

American farmers already in the territory generally knew what the transfer meant in terms of landownership. Having learned the value of land titles from their Kentucky experiences, they participated in the process of registering their claims, whether concessions or settlement rights. Other Americans who had moved into St. Louis and other villages understood the implications of the transfer for land appreciation and played the speculative game of land claims just as entrepreneurial French did, often in collaboration.

Unlike the government transition from France to Spain, which took nearly seven years to accomplish in the 1760s, the American government took charge rapidly. Within days after his arrival, Captain Stoddard issued various proclamations on how the territory would be administered. The United States quickly erected a new judicial system in both personnel and process to arbitrate its new laws, and a strong military presence assured residents that the Americans meant what they said.[5]

The flow of Americans across the Mississippi continued after American authority arrived in St. Louis. These land seekers expected to be able to claim and get clear title to it through the system of land sales that had been worked out during the previous decade in the Northwest Territory. However, this was not to be. Numerous delays for various causes prevented the inauguration of land sales for fourteen long years. Thus, the period from 1804 to 1818 was characterized by a pronounced population growth from immigration that spread into unoccupied lands, but without any method to get title to those lands. The story is one of continuous delays, confusion, and frustration. Most of the problems centered on the process of the confirmation of pre-American property.

People who were arriving almost daily to settle on the west side of the Mississippi had only two options, if they chose not to live in a village. One could buy land from a landowner at an inflated price and trust that the land claim from pre-American years, whether by Spanish concession or by settlement right, would eventually be confirmed. Or, one could occupy the public domain as an illegal squatter until land went on sale and trust that the parcel would be offered to you first through as yet undefined and unallowed preemption rights. Newcomers' insecurities and frustrations at obtaining land were accompanied by the frustrations of those Americans who were resident before 1804. Those who had already claimed land, whether by concession or by settlement rights, could not confidently move, sell, or divide their lands while titles stayed inchoate

---

5. Banner details the arrival of the American legal system in *Legal Systems in Conflict*, 85–100.

and unconfirmed. Settlers could and did sell their lands in whole or part, but it was often with a proviso that the buyer bought it knowing that title was not confirmed.

The U.S. treaty agreement to respect the property rights of its new citizens necessitated that it determine who owned land, how much, where it was, and whether the land claim was valid. These were unanticipatedly tough questions to answer, and they require a full account of the chief issues that affected the complex process of land confirmation of pre-1804 residents and the equally complex process of getting post-1804 immigrants onto the land.

Confirming Land Claims

On March 2, 1805, almost a year after the United States took control at St. Louis, Congress created a Board of Land Commissioners in St. Louis to handle the examination and confirmation of private land claims in Upper Louisiana.[6] Figure 5 shows the location of all land claims in the Ste. Genevieve District that were finally confirmed. Also provided for at the same time in St. Louis was a territorial agent, an officer of the U.S. secretary of the treasury, to represent the land interests of the United States in the confirmation process. To act on pre-American land claims was not new to the U.S. government; it had earlier set up a procedure for processing French land claims at Vincennes, Kaskaskia, and Detroit in the Northwest Territory, as they had passed to the United States through in-

---

6. *Stats. at Large of USA* 2 (1805): 324–29. Many have investigated the land-confirmation process in Missouri, but the topic still lacks a comprehensive account. Among the more helpful accounts are A. H. Rose, "Extension of the Land System"; Pelzer, "Spanish Land Grants"; Ada Paris Klein, ed., "Ownership of the Land under France, Spain, and United States"; Eugene M. Violette, *Spanish Land Claims in Missouri;* Shoemaker, *Missouri and Missourians,* 1:194–201; March, *The History of Missouri,* 1:210–61, which is exceptionally clear and readable; Lemont K. Richardson, "Private Land Claims in Missouri"; and Thomas Maitland Marshall's introduction to *Bates Papers,* 1:23–24. Paul Wallace Gates reviews the struggle with private land claims in Missouri within the context of private land claims throughout American history in his authoritative *History of Public Land Law Development,* 93–108.

The U.S. Public Land Commission compiled the various land laws for the sale and disposition of the public domain to 1883. Thomas Donaldson, the compiler, concluded in his chapter on the confirmation of private land claims that there is "no one branch of jurisprudence where greater research and extent of legal erudition have been displayed than in the discussion and determination by the judicial tribunals of the intricate questions which in this connection have arisen" (*The Public Domain, Its History with Statistics . . . ,* 366). Similarly, Gates, eighty-five years later, concluded that "no problem caused Congress, officials of the General Land Office, and Federal courts more difficulty or took up as much time as the private land claims, that is the grants of land made by predecessor governments" (*Public Land Law Development,* 87).

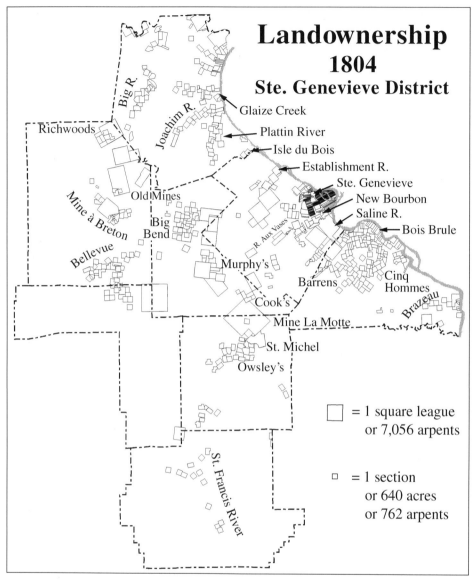

5. *Landownership, 1804, Ste. Genevieve District.* The map shows land claimed in 1804 and eventually confirmed for private ownership. Also shown are twenty-four named neighborhoods in 1804.

tervening British rule. What was new in Louisiana was the introduction of Spanish land law and the difficulty of its interpretation and application, especially to Upper Louisiana.[7] Nevertheless, given clear instructions and the authority to make decisions, subject to final confirmation by Congress, Congress and the board confidently expected the process to last a few years at most.

The board, with its accessories of clerk and translator, was composed of strong-willed and principled individuals, a contentious threesome who often found it difficult to cooperate. (Board members are identified in Appendix C.) Members absented themselves from meetings, abstained from voting, and used other tactics that delayed or prevented decisions. Meetings were frequently turbulent and physically abusive.[8] Some members took the sides of political allies and influential claimants and participated in local land speculation. Commissioner Frederick Bates, who also served as territorial secretary, was noted for his narrow interpretation of the law and instructions, but always with an impartiality uncommon in frontier politics. Confrontational attitudes caused board members to clash over small matters that grew into major matters, such as an argument over adequate documentation proving that the kings of France and Spain delegated to provincial governors the power to grant land. At one point, a frustrated commissioner reported to Washington that "the dignity of the Board is reduced to the Lowest ebb."[9] Clearly, the board itself was part of the land-confirmation difficulties.

At the outset, one must understand that land may be considered private property both by having received title to it through an established official and legal process, which permits land speculation, absentee own-

---

7. Payson Jackson Treat concludes that private land claims in the Louisiana Territory were more difficult to adjudicate than private claims in the Northwest Territory, and the experience in the latter was not much help in the former (*The National Land System, 1785-1820*, 226-29). Meinig also emphasizes differences in the legal philosophies of French, Spanish, and American governments in the process of merging the Louisiana Purchase into the existing federation (*Shaping of America*, 2:16-17).

In Peoria, Illinois, the French were deported (or driven) from their homes during the War of 1812 and their lands confiscated. They had to be compensated for them, and that involved the reestablishment of landownership (Judith A. Franke, *French Peoria and the Illinois Country, 1673-1846*, 46-49).

8. Houck, *History of Missouri*, 3:48-49.

9. Lucas to Secretary of Treasury, January 29, 1807, *TP* 14:85-86; *ASP-PL* 3:586-87. The basic division was between the "junto," which supported the large grants issued by Trudeau and Delassus to both French and opportunistic Americans, and the newly arriving Americans who opposed such favoritism (Foley, *Genesis of Missouri*, 287-90; March, *The History of Missouri*, 1:218-28).

ership, and the accumulation of large amounts of land, and by having lived on it and gainfully used it for some time, a practice called soil or settlement rights, which is a custom of the people of many cultures. The Americans who constituted the board were caught between these two basic concepts of landownership. They fully understood the concept of settlement rights and would have liked to accommodate it, but they were obliged to operate under rules laid down by Congress, which ordered judging the validity of claims on the basis of official Spanish land-granting processes. Yet, the board and everyone else knew that written Spanish land regulations could not cover all the cases of land claims, and some proposed overlooking them and proceeding with "customs and practices." Thus, while the board may have earnestly wanted to be fair and do justice in confirming properties, it was initially limited to using the rules and regulations of the Spanish government and to disregarding the custom and practice of settlement rights.

Therefore, the board, obliged to refer to the written laws and regulations of Governors O'Reilly and Gayoso and Intendant Morales and apply them to Upper Louisiana, spent much time trying to understand Spanish law, because it held inconsistencies, generalizations, and vague instructions, and left unaddressed some central issues. One board member expected the entire land-titling process from initial request to the perfection of the title by the governor at New Orleans to be fully documented, which would have resulted in only a handful of properties confirmed. The board strictly applied the Gayoso 200–50–20 formula to claims dated after its promulgation. Any hope that procedures developed to confirm private French lands in the Northwest Territory could have been transferred to the Louisiana Territory quickly evaporated when confronted with the complex nature of Spanish laws, mostly designed for other regions of Spanish America, and their selective application and enforcement.

In addition, Congress and the board set other stipulations.[10] The board accepted only one claim per claimant. Land had to be inhabited and cultivated. Claim size was limited to 640 acres. These stipulations were not grounded in Spanish regulations but derived from the American practice

---

10. *Stats. at Large of USA* 2 (1804): 283–89; (1805): 324–29. A question not yet examined is how the Kentucky experience with land-title issues, which predated Missouri's by a decade or so, influenced federal directives on private claims in Missouri and the board's interpretations of them. Many of the Americans involved in the land-confirmation process had come from Kentucky, and all were aware of Kentucky's muddled situation. For example, one can postulate that the board was biased against large land claims by the enormous problems that had resulted from land speculation in Kentucky in the 1780s and 1790s.

that settlers inhabited the land they used and that 640 acres was at that time the maximum size for a yeoman farmer of the Republic to purchase lands from the public domain. One of the basic principles of the new republic's own land policy was to prevent huge amounts of land going into the hands of individuals for speculative purposes and the development of a class of landed gentry. American policy strove for an egalitarian land distribution, and this principle was extended into the confirmation of trans-Mississippi properties.

By announcing that it would act strictly according to instructions, including size limitations and actual occupancy, the board immediately alienated the large land claimants. Many withheld filing their claims for fear of having them rejected. Fifteen of some of the most influential St. Louisans, all of whom stood to lose a lot of land, immediately remonstrated by a letter to Congress, dated February 1, 1806. In the lengthy, two thousand-word grievance, they described how land conceding actually went on. They bluntly stated, for example, that "occupancy & Cultivation have not always been the Condition on which lands were granted." Capt. Amos Stoddard, who was present in Upper Louisiana during the last year of the Spanish regime and witnessed the last Spanish lieutenant governor at work, admitted that the Spanish land laws were "general rules." It was, Stoddard insisted, always the lieutenant governor's intention to ignore them. If authorities in New Orleans did not like what the lieutenant governor was doing in the distant north of the colony, they said nothing and by their silence must have accepted what was going on.[11]

Desiring fair treatment for its new citizens, on March 3, 1807, Congress relented and gave the board authority to pass on claims according to "usage and customs" of the prior French and Spanish governments. This was a significant change in instructions, but it posed a new problem for the board: what exactly were the "usage and customs"? As the board uncovered these unwritten practices in succeeding years, it asked Congress to pass special legislation for certain categories of claims that the board considered worthy but felt unauthorized to act upon.[12]

---

11. Foley, *Genesis of Missouri,* 158, 170–71; "Petition to Congress by Inhabitants of the Territory," February 1, 1806, *TP* 13:425–30; Stoddard, *Sketches of Louisiana,* 243.

12. *Stats. at Large of USA* 2 (1807): 440–42. For the United States as a whole, see *General Public Acts of Congress, Respecting the Sale and Disposition of the Public Lands, with Instructions Issued, from Time to Time, by the Secretary of the Treasury and Commissioner of the General Land Office, and Official Opinions of the Attorney General on Questions Arising under the Land Laws.* A list of acts of Congress and other actions regarding land claims in Missouri from 1803 to 1874 is in W. A. Schroeder, "Opening the Ozarks," 699–708.

The immense size of some claims bothered the board. In some instances, a claim was presented for a single tract of enormous size, such as Governor Carondelet's grant to Ste. Genevieve priest James Maxwell for 16 square leagues (150 square miles) in the Ste. Genevieve District. In other instances, large size was due to the aggregate of multiple tracts of land, sometimes obtained under one open-ended grant, other times under separate grants to the same person. For example, the total arpentage claimed in the district in the name of François Vallé was in excess of 23,000 arpents (30 square miles) for seven tracts, and the aggregate arpentages of some claimants in the St. Louis area were even greater.[13] Several issues arose for these large claims. Could a claimant legally claim more than one tract? Does a large tract need to be inhabited by the claimant? Does it need evidence of productive use?

Concessions of 1 square league and larger were common throughout Louisiana, and the practice of awarding huge grants was time honored, well known, and accepted in the late Spanish period. But to judge the validity of these large claims on the basis of Spanish practices meant that the imposed 640-acre limit for a family farm had to be compromised. In 1807, Congress moved the maximum size for confirmation to 1 square league, or 6,002.5 acres, and instructed the board to use it.[14]

Although the American standard for farm property as a single tract of land for residence, garden, cropland, woodlot, and grazing prevailed in the American-settled portions of Upper Louisiana, many French lived in tripartite villages of three noncontiguous properties for each resident, only one of which contained the owner's residence. The board decided these properties lay outside its jurisdiction. The issues the board faced with the French villages were multiple: claimants claimed at least two tracts of land, in the village and in the common field; claimants claimed tracts they did not actually reside on; some property was owned in common by the village; many of the constituent land parts lacked concessions; and properties lacked surveys.

---

13. Rev. John Rothensteiner, "Father James Maxwell of Ste. Genevieve," 147–48; LCS 219; Vallé's arpentage as totaled from his surveyed properties in *Régistre d'Arpentage*. Vallé had other unsurveyed properties (LCS 25). Soulard surveyed 87,951 arpents for the five members of the Luzières family, of which son Jacques St. Vrain accounted for 49,375 arpents. Auguste Chouteau's confirmed claims totaled 23,500 acres, and (Jean) Pierre Chouteau's totaled 60,000 arpents (Foley and Rice, *First Chouteaus*, 179–80). As for claims presented before the board, whether or not confirmed, those of the Chouteau brothers totaled 234,000 arpents; the Reverend James Maxwell's, 106,356; Lieutenant Governor Delassus's, 52,200; Louis Lorimier's, 39,881; John Smith T's, 33,610; and Jacques Clamorgan's, an incredible 1,212,486 arpents (Gates, *Public Land Law Development*, 97).

14. *Stats. at Large of USA* 2 (1807): 440–42.

The French who lived in a village under the tripartite land division, which was the majority of the French population, objected to the board's inattention to their situation and presumed lack of authority in dealing with their lands, some of which had been inhabited for forty years. St. Louisans who held not only town lots whose value was appreciating rapidly but also large arpentages in several common fields and commons behind the expanding village presented a petition to Congress in 1806 in which they explained that common-field strips and usage of the grazing commons were integral parts of the villagers' property. Others who understood the French settlement system supported them and argued that despite the lack of specific provisions for the village system in the Spanish regulations for Louisiana (the tripartite village did not exist in Lower Louisiana), the spatial separation of a property into three separate land parcels was a uniformly accepted "usage and practice." John B. C. Lucas explained in 1807 that "the Land in the field and in the commons were [sic] not granted for the purpose of being inhabited, nor the Lot in the village for the purpose of being cultivated," and that they were all part of a single landholding system. Lucas recommended that Congress acknowledge this village system and pass legislation to confirm these properties because the board thought it lacked authority. Even Moses Austin, who had had a protracted struggle with the French Creole miners over communal property at Mine à Breton, recommended that Congress act to confirm common fields, although they were not inhabited and did not fall within the law outlining confirmation procedures. Other local petitions noted that commons were a part of the basic wood supply for villagers and needed to be confirmed.[15] The omission of village lots, common fields, and commons from the land-confirmation process was hindering town progress. People could not sell or subdivide their properties, inherited property was in question, and reorganization of the common-field strips, which were becoming obsolete as individual farm properties, was delayed.

In the end, the board chose not to act on these hundreds of tracts, even though they clearly fell within the "usage and customs" of the French and Spanish systems and the board had no opposition to their confirmation, but it did recommend that Congress address the issue. Congress did and passed on June 13, 1812, eight long years after the Louisiana Purchase,

---

15. "Petition to Congress," February 1, 1806, *TP* 13:425–30; A. H. Rose, "Extension of the Land System," 74–75. The residents of Ste. Genevieve also petitioned for confirmation of the right to land granted as a common, but much later, on December 30, 1818 (*House Journal*, 15th Cong., 2d sess., 30 December 1818, 144). Lucas to Secretary of Treasury, January 29, 1807, *TP* 14:90; *Austin Papers*, 2:122; "Representation to Congress by Committee of Inhabitants," [October 10, 1809], *TP* 14:323–27.

an act confirming the "village lots, outlots, commonfield lots and commons" in the villages in the territory of Missouri. In the Ste. Genevieve District, Ste. Genevieve and New Bourbon were specified in the act, but St. Michel, Old Mines, and Mine à Breton were not.[16] The latter two villages were not part of the tripartite system, but were simple collections of properties inhabited by miners. St. Michel had a tripartite system, but it alone among the villages had been granted a single concession for the entire village that was separately confirmed. The act of 1812 did not fully resolve the issue of landownership in the tripartite villages, however, because the act confirmed each village as a single unit, and the identification of who owned which village lots and common-field strips and their sizes and locations remained to be determined.

To examine the validity of the claims, the board had to gather evidence, because surprisingly neither the lieutenant governor nor the district commandants had kept records of their acts of conceding land. No compendium of landownership existed, except for books of surveys. Aware of the need for evidence of landownership, Captain Stoddard, on the very day he took charge in St. Louis, March 10, 1804, issued a proclamation that land grants and petitions for survey be deposited with the various district commandants by May 15, 1804, that is, within sixty-six days. Few heeded this directive; many had no such documents. The directive was reissued by the act of March 2, 1805, that set up the board, which stipulated that all land claimants were to deposit their pertinent land documents with the recorder of land titles at St. Louis before March 1, 1806, or forever have their claims barred from consideration. Although it was in everyone's interest to move the process forward as fast as possible, this date of one year thence was still too early. Persons in distant locations could not make the deadline, especially because they needed to get their claims surveyed before presenting evidence to authorities. Thus, the new date was again missed by many, and the filing deadline had to be reextended. Some of the slowness in filing was due to large landholders who held back at first when it was made known that the maximum size to be confirmed was 640 acres, then raised to 1 square league, which was still too small for them.[17] This strategy to hold back

16. *Stats. at Large of USA* 2 (1812): 748–52. Donaldson notes that these villages were the first of a long list of towns and town sites acted upon during the administration of the public domain and set a precedent for the confirmation of existing town holdings in other acquired portions of the United States, especially the Mexican cession (*Public Domain*, 298).

17. On Stoddard: Houck, *History of Missouri*, 2:374; on reissued directive: *Stats. at Large of USA* 2 (1805): 327; on holding back: Judge Lucas to the Secretary of the Treasury, February 13, 1806, *TP* 13:444.

paid off. Those who held larger sizes and who filed promptly had their claims confirmed but reduced to the maximum. Those who held back continued to press for confirmation of their total claim in other ways and usually won in later years.

Others did not present any evidence for their lands. Chief among these were the "old French," whose "simplicity and ignorance" and lack of documentation caused them to forego the process and thereby presumably lose their claims and become legally "landless." An observer six or seven years after the cession reported that some of the French Creoles did not even consider themselves *dans l'Amérique*. They spoke of the United States "as a different country than theirs, on the Mississippi."[18] In a final effort to give all people an opportunity to file claims, the board sent one of its members, recorder Frederick Bates, on a multimonth expedition in 1808 to various parts of the territory to collect evidence and testimony.

Evidence consisted of three kinds. One was the written concession document, which, with the lieutenant governor's signature or that of another authorized person, awarded the person a certain amount of land in a certain place.

The second was the plat of survey. According to Spanish regulations, all concessions were supposed to be surveyed before the land was considered to have been conceded. Marked visibly on the land and recorded on paper, the field survey and written plat constituted evidence that the claimant had physically occupied the concession. Some land had gone unsurveyed, because claimants had to pay surveyors' fees and some could not. Nevertheless, surveys were as important as concessions to the board, because only the survey plats gave specific locations and boundaries of the claim.

Unfortunately, surveys and their plats, as a category of evidence, were in great disarray. Because no one had kept a registry of conceded land and the concession document was given to each landowner, it was only surveyor Antoine Soulard's personal collection of survey plats that composed in any respect a central depository of awarded land.[19] "Official" in the sense of being ordered by the government, they were essentially the work of one man and his deputies, and their accuracy and general credibility rested on Soulard's reputation. Lacking basic locational and cadastral information and otherwise substandard in execution, many of those done in the last few years attest to a harried surveyor's carelessness or his intentional omission of basic information. Nevertheless, a simple sur-

18. Bates and Rector to Meigs, November 25, 1816, *TP* 15:211; Reynolds, *Pioneer History of Illinois,* 248.
19. Stoddard, *Sketches of Louisiana,* 259.

vey plat from this key man virtually guaranteed confirmation of landownership.

The third kind of evidence consisted of supporting statements from witnesses under oath, such as neighbors or officials, that a claimant was indeed the person he pretended to be, that indeed the signature on the document was true, that indeed the claimant occupied the land. How much of this was perjury would be anyone's guess. Testimony of this nature was usually the only evidence offered for land claimed by settlement rights.

The most controversial issue the board faced in the land-confirmation process was the suspicion of numerous fraudulent claims. Fraud could have taken different forms, such as false testimony, but most of the allegations concerned Lieutenant Governor Delassus's late issuance of concessions and backdating them in order to make them appear to have been issued before the legal authority of Spain in Louisiana had expired.[20]

After the cession, it became clear that some incoming Americans openly colluded in the business of backdated concessions. In fact, the first American governor, James A. Wilkinson, retained Soulard as surveyor for unsurveyed claims, despite his known role in backdating during the Spanish regime, and, incredibly, allowed Soulard to keep in his possession all of his surveys, including the alleged backdated and altered ones, so critical for land-confirmation action. Soulard's continued possession of his surveys opened him to more suspicion of altering them. Wilkinson and other Americans were reportedly involved in several ways with shady aspects of the land business, especially in the St. Louis District, where land values were highest.[21]

Moses Austin, in an analysis of conditions in the territory in 1806 prepared for the secretary of the treasury, described Delassus, as soon as he heard of the sale of Louisiana, as having agents throughout Upper Louisiana solicit money for concessions, the size of the concession pro-

---

20. Fraudulent claims have attracted the attention of many historians of the territorial period in Missouri, although no single study has treated them comprehensively. A good, brief overview is in A. H. Rose, "Extension of the Land System," 55–62. Also helpful is Richardson, "Private Land Claims." Houck provides a summary in his *History of Missouri*, 3:34–37. Others have used the term *antedating* for the backdating of land concessions.

21. Lucas to Secretary of Treasury, January 29, 1807, *TP* 14:87. Governor Harrison had retained Soulard as surveyor, but cut his pay in half (it was "exorbitant" under the Spanish), in order to survey settlement-right claims of Americans and other claims that had not been surveyed under the Spanish regime in preparation for the expected land-confirmation process (Governor Harrison to Antoine Soulard, November 8, 1804, *TP* 13:71–72). Also see *ASP-PL* 1:187 and Foley, *Genesis of Missouri*, 166, 172.

portionate to the amount offered. Austin implicated Wilkinson and Soulard for collusion with the French land speculators and participation in the process. Austin himself also witnessed firsthand the backdating practice. Because some of the craftsmen who had come with him had failed to request concessions for their lands, Soulard, according to Austin, sent him blank forms, already backdated, for him to file for his men, but he refused to sign them. Austin also alleged in his 1804 report on the status of the lead mines that from 30,000 to 40,000 acres of mineral land had been surveyed after the purchase, most of it in February 1804, by order of the still-sitting Spanish lieutenant governor, obviously to beat the imminent transfer to the United States. These surveys, Austin noted, "have been made with an intention to include every spot of land supposed to contain mineral."[22] Austin's statements on land fraud greatly influenced the board and Washington officials concerned with the matter.

Amos Stoddard was also well aware of fraud by backdating. He accused Trudeau of having received forty dollars for each 100 or 400 acres of mineral land. Delassus, he said, granted twenty-six concessions of 1 square league or greater near the close of the Spanish regime, but they bear dates of 1799 (thirteen concessions), 1800 (nine), 1801 (two), 1802 (one), and 1803 (one). They totaled 271,752 arpents, or an average larger than 10,000 arpents (13.2 square miles) each. Of these twenty-six, only twelve were actually surveyed; the other fourteen, constituting 150,304 arpents, were not. According to Stoddard, Delassus dated some 1799 so that a compliant Governor Gayoso would complete them at New Orleans, not Intendant Morales, who apparently was not part of any scheme to defraud.[23]

The various schemes to acquire land by fraudulent methods were fully described by Rufus Easton in a lengthy letter of January 17, 1805, to President Jefferson. Easton, a judge of the territorial superior court, was himself deeply involved in land speculation at the time and resented both the backdating practice and its apparent acceptance by some American

---

22. On solicitation: *Austin Papers,* 2:117; on Soulard: Gracy, *Moses Austin: His Life,* 93. The craftsmen did receive lands by settlement rights because of their occupation of the land and because Luzières had promised land to them in writing. "Description of the Lead Mines in Upper Louisiana," November 8, 1804, *ASP-PL* 1:191; *Austin Papers,* 2:117.

23. Stoddard to the Secretary of War about Spurious Grants, January 10, 1804, in *St. Louis in Its Territorial Days,* by Billon, 371–73. This detailed account about how the Spanish lieutenant governors committed fraud was written while Stoddard was still in Kaskaskia, before he crossed the Mississippi to take control of Upper Louisiana (Stoddard, *Sketches of Louisiana,* 256). For the career of Stoddard, see Wilfrid Hibbert, "Major Amos Stoddard: First Governor of Upper Louisiana and Hero of Fort Meigs."

officials. Like Austin and Stoddard, Easton presented hard facts of Delassus's and Soulard's activities. Easton wrote to the president that concessions with orders of survey went for sixty dollars for a 500-acre tract, one hundred dollars for a 1,000-acre tract, and that surveys were backdated to 1799. After Delassus learned of the cession to the United States, he upped his charge to one hundred dollars for a 500-acre tract, and increased tract size to 50,000 acres.[24] Judge Easton's accusations were accepted by officials in Washington who then directed the board to investigate them.

An example of a backdated concession date on a Soulard survey that the board examined is the one for underaged François Tayon, son of Charles Tayon, commandant at St. Charles (fig. 6). Delassus made the award in January 1800, at the same time that father Charles received a large grant *(merced)* for his many services since 1770. François Tayon's grant was for 10,000 arpents in the mining country on the lower Fourche à Renault, where it is joined by Old Mines Creek. Soulard's deputy, Thomas Maddin, did not survey it until February 6, 1804, one month before Delassus surrendered charge, but Soulard entered the fictitious date of October 15, 1799, on the survey.[25]

In 1806, the federal government, convinced of Wilkinson and Soulard's collusion, removed both from their offices, and appointed American Silas Bent as the new territorial surveyor. When reviewing his predecessor's work, which was now deposited in public hands, Bent found the surveys replete with erasures and apparent alterations. The *Régistre d'Arpentage,* Soulard's collection of plats, had "leaves cut out, plats and surveys pasted in . . . names rewritten." Bent noted "striking contrasts in the colors of ink used."[26]

Lieutenant Governor Delassus left St. Louis to live in New Orleans, but Soulard remained a resident in St. Louis and was available to articulate

24. Rufus Easton to Thomas Jefferson, January 17, 1805, Thomas Jefferson Papers, ser. 1, reel 32, folios 25558–25561, microfilm, Ellis Library, University of Missouri–Columbia. Major portions of Easton's letter are in Houck, *History of Missouri,* 3:36–37.

25. *Régistre d'Arpentage,* 120; Houck, *History of Missouri,* 2:81–82; "Land Claims in Missouri (No. 65.—Francis Tayon)," January 21, 1834, *ASP-PL* 5:778–79. The first board twice rejected the claim (May 3, 1806, and the second time under assignee Peter Chouteau, August 18, 1810), but the second board confirmed it (November 7, 1833).

26. Silas Bent to Jared Mansfield, September 22, 1806, *TP* 14:8–9; ibid., October 5, 1806, *TP* 14:13–14; Richardson, "Private Land Claims," 143–44. One can confirm Bent's observations by examining the original *Régistre d'Arpentage* in the Missouri State Archives, Jefferson City. The grounds for Governor Wilkinson's removal were more extensive than questionable land actions.

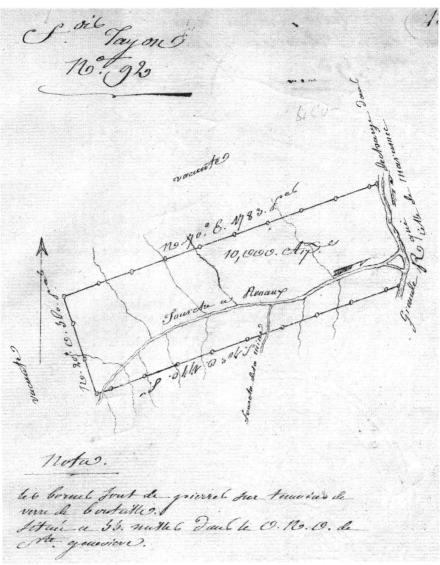

6. *François Tayon Survey, February 6, 1804; date of concession given as October 15, 1799.* Lieutenant Governor Delassus awarded the land that probably contained the historic Renault mines to young François Tayon in January 1800 but backdated the concession to October 15, 1799, before the transfer of Louisiana to France when, supposedly, his authority expired. The tract was hurriedly surveyed on February 6, 1804, one month before the Spanish transferred administration to the Americans. At the bottom, the surveyor notes that the boundary markers are stones on top of pieces of bottle glass. (*Régistre d'Arpentage,* 120)

the Spanish practice of awarding land. Soulard's personal communication to the board in 1806 (he did not testify before it) described how he conducted his business.[27] He accumulated his surveys in several bound volumes, one for each district, labeled A, B, C, and so on, although beginning in late 1805 he put all new plats into a single volume, mixed as to district. He numbered the pages as he added them to the string-bound folios. Each survey sheet included the date of the survey, the date of the lieutenant governor's authorization of the survey, and the date the certificate was issued.

Soulard maintained that any alleged alteration of his documents could be checked against other documents. He admitted to erasures and insertions, but such were made for necessity, not deceit. He maintained that if the erasures were done as fraud, then he could have done better by simply loosening strings at both ends of each spine and substituting a new sheet. He admitted that the Spanish government never bound him by oath. He admitted that sometimes he slightly enlarged properties by surveying, because of rivers with irregular boundaries.

Soulard further noted that toward the end of his employ by territorial governor Wilkinson, applications for surveys of Spanish claims were so numerous that "some involuntary errors may have possibly resulted from the Multiplicity of business, and the registering of the Plots could not be executed with the same degree of Neatness and accuracy," which reads like a weak attempt to explain his alterations and plats made without actual surveys of properties.[28]

Soulard noted that he, himself, claimed only 3,129 arpents, plus 1,600 surveyed for his sons, and that he relinquished several thousand more arpents that the lieutenant governor had awarded him. Soulard explained that if he were involved in giving away land, he could have done much more for himself.

The board received clear instructions from the secretary of the treasury, who was influenced by the reports from Austin, Stoddard, and Easton, to uncover evidence on land fraud and reject claims with such evidence. However, those sympathetic with the land schemers could entertain no possibility of fraud. Consequently, the board, unable to reach conclusions on many claims of alleged fraud by backdating, deferred action on them and left it to the claimants, almost all of whom were among the wealthy and politically active, to pursue their claims through the courts.[29]

27. Antoine Soulard to the Land Commissioners, November 5, 1806, *TP* 14:29–34.
28. Ibid., 32.
29. Questions of fraud that still lingered for the second Board of Land Commissioners were regularly dismissed in favor of the claimants, and concern with the pro-

With respect to Indian lands, however, the new American administration was not so understanding or generous. Americans, who had a concept of strictly separated space, had to extinguish Indian landownership by federal treaty before the public domain could be alienated and Americans could acquire title to the land. Resident Indians did not initially comprehend this new notion of landownership that appeared so abruptly.[30]

The board asked by what authority the Spanish had granted settlers land that had not been formally purchased or otherwise acquired from the Indians. This complex question, the answer to which lay in contrasting approaches to the European colonization of Indian-occupied lands, devolved into a more practical one of whether the board could confirm any land and pass it into private ownership in the United States if the Indian title to the land had not previously been extinguished by treaty or purchase. This was not at all rhetorical; the board had to resolve this general issue for the whole territory before proceeding with individual cases.

John B. C. Lucas recommended to Congress in 1807 that the United States immediately get clear and private title to land in which any land claims had been filed by buying it from Indians. The government had already made a treaty with the Sacs and Foxes on November 3, 1804, by which those Indians surrendered their lands along the Upper Mississippi as far south, on the west side of that river, as the Missouri River. A treaty

---

tection of the interests of the government apparently was no longer important. For example, the second board noted in 1834 that the date of a concession to the village blacksmith was earlier than the date of petition and recommendation, but liberally concluded that since "being written in these three instances in the old French and Spanish way of abbreviating, they are inclined to attribute this discrepancy to a mistake of the recording clerk" rather than to intentional fraud (Minutes, 7:45; *ASP-PL* 7:843).

30. For how Indian title to land was extinguished, see Donaldson, *Public Domain,* 240–42. Especially of note is the prohibition of individuals from purchasing or otherwise receiving lands from Indians.

Stoddard wrote a chapter, "The Aborigines," of fifty-five pages on Indians in his account of the Province of Louisiana, but little of it is specific to a given Indian tribe or nation, although he admitted that there were differences (*Sketches of Louisiana,* 409–63). To Americans, particularly the military, all Indians tended to be considered the same for administrative policy, whether they had been displaced from the East or were resident Indians like the Osages. Americans, of course, never used the term *American* for any Indians. The French *sauvage,* it should be noted, originates from the Latin *silva* and connotes more "forest dweller" than "beast," as the English cognate *savage* does (Peter Moogk, *La Nouvelle France: The Making of French Canada—a Cultural History,* 17–18). The French tended to think of Indians by individual tribe or nation probably because the French interacted with them more at a personal level and recognized and respected their differences. The French literature uses the more specific Shawnee, Delaware, Osages, Sauk, Chickasaw, and so on, rather than *peaux rouges* or *sauvages.*

with the Osages, whose territory included virtually all of Missouri south of the Missouri River, remained to be made when Lucas wrote but was signed the next year, on August 31, 1808.[31] By this treaty, the Osages were to retire west of a due north-south line passing through Fort Osage on the Missouri River, a line far to the west of any land claims.

These two treaties essentially resolved the issue of Indian ownership of the land and officially cleared the way for land to pass into private ownership by the confirmation of Spanish grants and the expected sale of land from the public domain. (At this time, the United States did not recognize any lands belonging to Illinois nations on the west side of the Mississippi River.) It did not solve, of course, the problem of interaction, sometimes violent, between American settlers and Indians. The Osages continued to hunt and otherwise operate in the territory they had surrendered by treaty, although such geographic activity rapidly declined in frequency as the years passed.[32] The Osages and other *grandes nations* continued to frequent St. Louis for trading and gifts specified by treaty.

The Spanish land grant on Apple Creek to the immigrant Shawnee and Delaware, awarded by the governor-general himself, presented a special case for the board. The Indians had received the grant under basically the same provisions as land grants to whites, but did Spanish land regulations permit granting lands to immigrant Indians not native to the land? Were the Indians Catholic, as Spain had required immigrants to be? Was the amount of land in the grant consistent with the numbers and resources of the Indians? Did they inhabit and cultivate it as required by land-granting regulations? Could land be awarded "in common" to a group? The board declined to confirm the grant to immigrant Indians, giving the reason that it exceeded the maximum size allowable. However, that decision was moot because Indian affairs were handled directly by the federal government, which promptly acknowledged the validity of the Indian grant under Article 6 of the Treaty of Cession by which the

---

31. Lucas to Secretary of Treasury, January 29, 1807, *TP* 14:81–82. Also see Governor Lewis's recommendation on acquiring title to Indian lands (Governor Lewis to the President, August 27, 1809, *TP* 14:293–97). Charles J. Kappler, *Indian Affairs: Laws and Treaties*, 74–77, 95–99. The Jeffreon, Jeffron, Jaufflon, Jaustioni, or Jefferson River is thought to be the North River, which empties into the Mississippi River just north of Hannibal, Missouri. Houck misidentified the Jeffreon as the Fabius River (*History of Missouri*, 1:17). The Missouris, who also lived and used land south of the Missouri River, had by this time removed to join their Oto kin in what is now Nebraska. No treaty was made with the Missouris within the present state of Missouri.

32. Royce, *Indian Land Cessions*, 698–701, 714–15, 724–25. Ironically, the federal government moved other Indians from the East onto reservations on land in the Ozarks that they had removed the Osages from only a few years earlier.

United States agreed to respect all agreements that the Spanish government had made with Indian nations. As with the tripartite villages, confirming the Indian tract as a whole did not confirm any land to individual Indians living on the tract.[33] A few French and Americans had already settled on the Indian tract before 1804, some of them in the service of the Indians, and the board confirmed their claims, duly presented. However, having acknowledged the tract's validity, the U.S. government was then obligated to respect Indian title to it and close it to further white settlement.

Two other special cases that came before the board illustrate the nature of Indian landownership as interpreted by the board. In contrast to the Shawnee-Delaware grant, which was made to Indian nations, these two were made to Indian individuals.

One case involved a Métis, Noel Mongrain, son of the aunt of Cheveux Blanc (White Hair, or Paw-Hiu-Skah, meaning white-headed eagle), chief of the Great Osage nation, and thus the chief's first cousin. In 1797, Chief Cheveux Blanc awarded Mongrain one square league on Little Saline Creek, a tributary of the Osage River in present Miller County in central Missouri, for services Mongrain had performed for him and the Osage nation.[34] This reason for granting land is clearly the same as those given by Spanish and American governments for services performed for their governments. In fact, the Osages probably learned the idea of granting land from the Spanish and French, because the Osages did not normally invest parcels of land in individuals.

Mongrain left his Osage kin, went to live with the French community, and publicly renounced any claim as Indian to common ownership of Osage-nation lands, as recorded on June 20, 1797, by interpreter Jacques Sundé and witnessed by Bernard Pratte, a prominent member of the French Creole community. Mongrain claimed he had been hunting on his land for the past ten years, from 1798 to 1808, when he presented his claim to the board for confirmation. When Chief Cheveux Blanc signed the Osage Treaty with the United States in 1808, he intentionally excluded Mongrain's land from the land cession, because it was no longer land of the Osage nation, and he had no authority over it. As far as the Osages were concerned, it was already in the private ownership of Mongrain.

33. Donaldson, *Public Domain,* 96. The federal government, whether in Washington or through a representative in St. Louis, issued numerous proclamations and directives for white persons settled on Indian lands, including the Shawnee-Delaware tract, to leave under threat of being forcibly removed by the militia (*St. Louis Missouri Gazette,* December 9, 1815; *Niles Weekly Register,* October 5, 1816, 11:96).

34. *ASP-PL* 2:494.

On August 24, 1811, the board declined to give any opinion on Mongrain's claim, in effect denying it forever, because Mongrain was one of those nonparticipating French who chose not to pursue his claim further. The board's reason was that the claim was "neither embraced by the law, usages, and customs of the Spanish Government, nor the acts of Congress." In other words, the board accepted Spanish suzerainty over the land throughout all of the Louisiana Territory, including lands inhabited by Indians and far distant from those inhabited by Europeans, and therefore all land transactions, including those done by Indians on Indian land, had to be done under Spanish, or civilized European, law. Only Spain, not Indian nations, had complete dominion over the entire Louisiana Territory.[35]

An associated Indian case illustrates how those wise to the political process and aggressively participating in it received different treatment from the board, although the intentions of the land grantor and grantee were essentially the same as in Mongrain's case. The same Osage chief, Cheveux Blanc, similarly awarded a tract of land on March 19, 1792, to prominent St. Louisan Jean Pierre Chouteau, known as Pierre, also for his considerable services to the Osage nation.[36] The Chouteau grant lay astride the Lamine River near its junction with the Missouri River, mostly in present Cooper County and some 150 miles west of St. Louis. Whereas Mongrain's tract was typical Ozark terrain, undistinguished in any way, Chouteau's tract included valuable salt springs for which he had selected the tract and that were easily exploitable and accessible to Missouri River traffic.

Chouteau, suspecting that Americans might not recognize the validity of an Indian grant, even when formalized by a paper document, petitioned Lieutenant Governor Delassus, with whom he cooperated in many ways, for a separate Spanish grant to the tract. Delassus obliged his friend, and it was on the strength of the Spanish grant giving away Indi-

---

35. Ibid. In the later Osage Treaty of 1825, whereby the Osages further ceded lands west of the present Missouri-Kansas state line, Mongrain received special treatment. He received one section (640 acres) of land, and each of his ten children and four grandchildren, including females, received an equal amount. This land was on the Marais des Cygnes River at the state boundary. Other "half-breeds" with such St. Louis surnames as Chouteau and Tayon also received sections of land by the Treaty of 1825 (Kappler, *Indian Affairs,* 218–19). For Métis in the region, see Thorne, *Many Hands of My Relations* (Mongrain land case, 94–96). A cynical interpretation of Mongrain's experience with land confirmation would be that the board's action shows American bias against, if not outright disdain toward, Indians and "half-breeds."

36. Letter of Instructions: Baron Carondelet, Governor of the Province of Louisiana, to Lieutenant-Colonel Don Carlos Howard, New Orleans, November 26, 1796, *Missouri Historical Society Collections* 3:1 (1908): 87; *ASP-PL* 5:796–97; Clokey, *William H. Ashley,* 37–38; Foley and Rice, *First Chouteaus,* 180–81.

an land, not the original Indian grant giving away Indian land, that the board validated the Chouteau salt spring tract. The Chouteau tract of thirty thousand arpents (just under forty square miles) was confirmed by an act of Congress on July 4, 1836, after Congress removed the reservation of salt springs from confirmation.[37]

The board also wrestled with the issue of floating concessions, which were like wild cards to claim land virtually anywhere. Awardees had located some of their floating concessions before 1804, but the locations of many were still "floating" while the board was collecting evidence.[38] Were they still valid if they remained unlocated after December 20, 1803, and why had they never been located? Could they be sold to others before being located and used by the new owners for claiming land? Did Spanish land regulations allow such practices? Could the holder of a floating concession use it to claim land that was occupied by someone else, but whose title to it had not been acted upon? This latter question was especially acute in the mining region, where landownership was in great confusion, and holders of floating concessions appeared out of nowhere to claim immense acreages of mineral land already gainfully occupied. The board, rightly appalled at such behavior and unfavorably disposed to the idea of wild-card concessions, denied claims based on floating concessions, using as the reason their inconsistency with official Spanish regulations. Claimants would have to pursue their cases through the courts.

No issue that the board faced showed more sharply the divergence between governmental regulations and traditional settlement practices of people than ownership of mines. Although Spanish land regulations made no mention of mineral land at all, the Spanish did grant land for mining in the lead district but usually with the request and award couched in terms of agriculture or cattle raising. This deception reflected the expectation that farmers formed the population base of a society. Only in its closing years, beginning with Moses Austin, did Spain award land explicitly for lead mining. The large mining grants awarded during the earlier French regime to Renaut and others were also eligible for consideration by the board, as they had been acknowledged by Spain.[39]

37. *ASP-PL* 5:796–97.
38. Stoddard, *Sketches of Louisiana,* 245–46, 257.
39. For an example of a concession for a lead mine written in terms of a cattle ranch, see Gabriel Cerré's petition (*ASP-PL* 3:589). Some of the more insightful comments on intruders, leases, and ownership of mineral land are in the *Bates Papers.* Also helpful for administration of the lead mines are Willms, "Lead Mining, 1700–1811"; and Hanley, "Lead Mining in the Colonial Period."

However, by far the most common exploitation of lead mines was by prescriptive rights, or rights of usage.[40] This practice worked because individual miners mutually respected each other's mines. Some litigations did arise, but in the majority of cases, the miners worked out their differences one way or another. Authorities in St. Louis and New Orleans accepted the prescriptive rights system by reasoning that occupance of any single pit was temporary, a few years at the most, and did not contribute to permanent settlement. Authorities did not award land for short-term uses.

Americans introduced a new approach to land occupation in the mineral country, which was to keep lead and saline lands from any private ownership, whether in fee simple or by prescriptive rights. The lead lands would be reserved, that is, put in a kind of federal trust, and miners would lease them and thereby produce revenue for the federal government. With the new American approach, three types of landownership confronted the board for action on mineral lands: prescriptive rights on public or common land, Spanish grants and concessions, and American leases.

In contrast to official Spanish inattention, the lead lands of Upper Louisiana were a primary focus of attention for the American government in Washington, its territorial agent in St. Louis, and its board of land commissioners. Highly coveted lands, they were fought over with fists and guns and in courts, because owners could make great sums of money from them. Even poor people could make a decent living from them. Unlike agriculture, wherein hundreds of valleys throughout the district could satisfy the desire for farmland, active lead mines were site specific and location choices restricted.

Yet, the widespread occurrence of lead over the surface of hundreds of square miles dictated that the settlement of the lead country would be geographically dispersed and that the production of lead would remain fragmented among hundreds of different miners, working individually. This meant that land claims would be numerous and widespread, and the possibility of disagreements over claims and boundaries would be multiplied. It further meant that governmental oversight of individually leased mineral lands would be next to impossible, although the federal government failed to realize that.

Under terms of the Louisiana Purchase, the American government was to guarantee property, but, wondered the board, did prescriptive rights to mines constitute property in this sense? Apparently not, because all pre-

---

40. Prescriptive rights to mines can be traced back to medieval German practices of free mining (Everett Dick, *The Lure of the Land: A Social History of the Public Lands from the Articles of Confederation to the New Deal*, 85; Charles Howard Shinn, *Mining Camps: A Study in American Government*, 24).

1804 unconceded land—the royal domain—automatically became public domain after 1804. Could miners continue to dig under prescriptive rights on the American public domain?

The American mind presumed a sharp division between public and private lands, and the federal government on the East Coast had firm ideas on a policy for the valuable mineral lands that did not allow for continuation of prescriptive rights. One of the first acts of Congress regarding the territory of Louisiana was to order, on March 26, 1804, only sixteen days after the changeover at St. Louis, all persons (called "trespassers" by Congress) off the public domain. Miners who had been working the land for years under traditional prescriptive rights became trespassers overnight. The public domain included all the mines that were being mined by prescriptive rights, including the valuable Mine à Breton commons adjoining Austin's furnace. In fact, Gov. James A. Wilkinson specifically issued a proclamation prohibiting unauthorized mining on the Mine à Breton commons.[41] Immediately upon the posting of the proclamation on September 20, 1805, a loud cry went up from everyone involved in lead mining, most of them in the Ste. Genevieve District and including the former commandants Luzières and Jean-Baptiste Vallé.[42] Faced with such strong reaction from both hardscrabble miners and wealthy lead operators, Wilkinson relented and one month later, on October 28, 1805, allowed miners to continue digging ore on the Mine à Breton commons.[43] The 1804 act of Congress, however, still applied to other mineral lands, even if unenforced.

The federal government, which represented eastern attitudes and not those prevailing in western territories, regarded the public domain and all its resources—lead, timber, soil, and water—as federal resources not to be used without some kind of compensation. Miners should pay for the privilege of digging on the public domain. People in the mining region, however, viewed the public domain as a vacant wilderness that they were converting to gainful usage and making more valuable for society. Instead of degrading and devaluing the public property for future sale, they saw themselves as improving the country. They were, of course, quite willing to buy mineral lands from the government, if only the government would sell them.

---

41. *Stats. at Large of USA* 2 (1804): 283–99; Governor Wilkinson to the Secretary of State, September 7, 1805, *TP* 13:195; Willms, "Lead Mining, 1700–1811," 80.

42. Willms, "Lead Mining, 1700–1811," 81–82, citing "Petition of 135 Inhabitants of Ste. Genevieve," n.d., in the General James Wilkerson manuscripts, Missouri Historical Society, St. Louis.

43. Willms, "Lead Mining, 1700–1811," 82; Houck, *History of Missouri*, 3:45.

But the federal government would not sell. Congress and President Jefferson had become quite aware of the great extent and value of the lead resources of the Ste. Genevieve District. Austin had written glowingly of the productivity of the active mines and promised that many more would soon be discovered. Federal land agent William Carr, mirroring a belief already expressed by Austin, had noted in 1805 the depreciating value of the mines on public property without any benefit to the government, and he suggested a leasing program that would follow a precedent set in 1785 for lands with base metals in the Northwest Territory. But in that case, the mineral lands were "virgin," that is, they had not earlier been put into private ownership or mined by prescriptive rights. The leasing provisions of the 1785 act had not been seriously applied and were an untested model to copy. Nevertheless, Congress passed the federal Lead and Salt Leasing Act on March 3, 1807, by which the government reserved all lead and saline lands from sale in Upper Louisiana and set up a leasing program that was to be a bane for subsequent administrations of the Ste. Genevieve District.[44]

The policy of reserving lead and saline land from sale and leasing it ran headlong into numerous existing claims to mineral land awaiting action by the board. The government decided that mines included within land claims duly filed with the board, such as Austin's and Luzières's and the saltworks at La Saline, could continue in the hands of their claimants, because decisions on those claims were expected shortly. But all land for which claims had not been filed, including the Mine à Breton commons, all new mines opened since 1804, and all individual diggings worked by prescriptive rights would have to cease operations or be leased from the federal government, and such land would officially remain public domain. As it turned out, saltpeter mines, all of them in caves, also fell under the Lead and Salt Leasing Act of 1807.[45]

The leasing program was intended to avoid land and lead speculation

---

44. *ASP-PL* 1:191; William C. Carr to the Secretary of the Treasury, November 14, 1805, *TP* 13:275; Benjamin Horace Hibbard, *A History of the Public Land Policies*, 512–13; *Stats. at Large of USA* 2 (1807): 445–46. The act provided not only for the reservation and leasing of the actual lead-bearing lands but also for sufficient adjacent land as "mine easements" for timber supply and dumping grounds for mine refuse (Donaldson, *Public Domain*, 307).

45. Gov. Meriwether Lewis awarded at least two leases for saltpeter caves, even though there apparently was no provision for leasing minerals other than lead and salt. The first lease was awarded on February 23, 1809, for twelve months for four lots of ground of twenty acres each, to William Mathers to mine saltpeter in caves on the Meramec River. Mathers was permitted to use timber on adjacent U.S. land for firewood ("Lease of Saltpetre Caves to William Mathers," n.d., *Bates Papers*, 2:60–62).

and, second, to make money for the federal treasury off the mining enterprises. Also, so the argument went, by the time lead lands would eventually go on sale, the lands around would have become populated and economically developed, and thus the sale of them would bring far higher prices. Limiting size and time period of leases would prevent monopolies on this important metal. Furthermore, the federal government remembered that a few years earlier, during the undeclared war with France at the end of the 1700s, the United States had experienced a shortage of lead, and consequently the federal government deemed lead a strategic and essential resource in the production and sale of which the federal government should retain a primary role.[46]

The board was instructed by a second act passed the same day, March 3, 1807, to refrain from making any decisions on the numerous claims to land that contained a lead mine or salt spring.[47] Even Moses Austin's perfected grant from the governor could not be confirmed. This was, in one sense, a contradiction, if not a violation, of the Louisiana Treaty to "respect property" of the previous regimes. The claimants could, however, continue to operate their mines to their profit as if they owned them, though action on confirmation was put off indefinitely, which the claimant-owners did.

The first act of March 3, 1807, permitted leases of lengths up to three years for small tracts from 40 to 320 acres. Lessees could have free use of timber for fuel and stone for building from the adjacent public domain. One-tenth of the lead mined would be the leasing fee to the government, a fraction obviously chosen because it was the going rate of payment for lessees on private lands in the district. Frederick Bates, the recorder at St. Louis, awarded the leases and administered the program in the territory, a heavy additional burden for an official soon to be encumbered with the multitude of headaches of the land-confirmation process. Bates awarded the first lease for lead mining on September 26, 1807, to Andrew Miller.[48]

To reserve lead land from public sale required identification of where mines and lead-bearing lands were, but this was not well known. Most of the known tracts were in private land claims and not subject to the leas-

---

46. To Albert Gallatin, October 6, 1807, *Bates Papers,* 1:216; Donald J. Abramoske, "The Federal Lead Leasing System in Missouri." A review of lead-land leasing from the perspective of conservation and management of resources in Missouri is in Richard West Sellars, "Early Promotion and Development of Missouri's Natural Resources," 146–53.
47. *Stats. at Large of USA* 2 (1807): 440–42.
48. "A Lead Mining Contract," *Bates Papers,* 1:196–97.

ing program. But new mines had been opened since 1804, and much more land was expected to become lead producing. This raised a geographical question: should every square mile of land within the general mineral district that was not in a private land claim be reserved from sale and put in the leasing program because any part of it might be lead producing? To answer this question was beyond the board's purview, but the board did need to know whether a tract before it for confirmation lay in the mineral district and therefore was potentially lead bearing.

Appointment of a special federal agent by the act of 1805 to look after federal-land interests was intended specifically for lead interests and reflected the importance that the federal government placed on lead mining in the Ste. Genevieve District. President Jefferson appointed Lt. William C. Carr to this potentially influential and politically sensitive position that touched upon the livelihoods of so many influential persons as well as ordinary miners making a living for their families. Carr arrived in St. Louis in 1808 to begin his work, but his impartiality suffered from his becoming personally involved in the lead-mining debates. When Carr finally submitted a report on September 28, 1811, after three full years of investigation, he listed a mere twenty-eight private claims in mineral areas, only fifteen of which contained lead. Board commissioner John B. C. Lucas blasted the report as both too late to be helpful to the board and totally inadequate in its coverage. Lucas noted that the board, whose business was public knowledge, had before it three claims containing lead ore that Carr did not discover or intentionally ignored.[49]

More helpful in locating land to reserve was the report on the lead mines that the federal government had requested from Moses Austin in 1804.[50] The single most detailed account of lead mines, it described briefly the history of all active and inactive mines and who owned or claimed them. It gave their locations relatively by direction and miles from Mine à Breton because no map of the district yet existed.

Other than Austin's and Carr's reports, identification of lead-bearing lands was based on the knowledge of locals who reported to the board in random manner. Much was hearsay. Trespassers, speculators, and local nabobs manipulated knowledge of lead mines for their own interests. Nevertheless, by 1814, the secretary of the treasury had reserved a total of 150,000 acres, or about six and one-half townships, from sale and confirmation.[51] When making its final report to the federal government in

---

49. *ASP-PL* 3:602–3; Willms, "Lead Mining, 1700–1811," 78–79; *Stats. at Large of USA* 2 (1805): 331–32; *ASP-PL* 3:586–87.
50. Austin, "Description of Lead Mines," *ASP-PL* 1:189–90.
51. Abramoske, "Federal Leasing System," 35–36.

1812, with Lieutenant Carr's defective report appended, the board had no choice but to put the land claims to mineral land in a special category because it had no authority to act on any lands known or even thought to be lead bearing and officially reserved from sale by the act of 1807.[52] Thus, all of the claimed mineral lands were still lying in legal limbo eight years after the Spanish left.

The twin actions of not confirming private claims in the lead district and leasing the public lead-bearing lands were, to put it mildly, extremely unpopular in the Ste. Genevieve District and the territory in general. The wealthy land claimants protested vigorously their deferred situations, although almost all continued to hold on to their valuable properties through these land-contentious times. Some who had finagled these large concessions in the final years of the Spanish administration with the collusion of the local lieutenant governor were all the more defiant to the new and distant government's treatment of their lands, potentially so extremely valuable.

The leasing program staggered from the very start. Although Recorder Bates dutifully continued to issue leases and to receive payments in lead for some of them, enforcement was lacking. Mine owners openly flaunted the program. "All the troops in service would not be able to guard this treasure [of the mineral lands]. Those disposed to purloin it will laugh at legal restraints," Amos Stoddard prophetically observed as soon as he heard of the provisions of the act of 1807.[53] Those few who went to the trouble to abide by the law and get leases often found their mining land occupied illegally by others who could not be dislodged.

Notorious among the latter was John Smith T, who appeared and claimed the land at new lead discoveries, often with a floating Spanish concession that he demanded be recognized, and threw everything into seemingly endless litigation while his henchmen forcibly took over the land, operated the mines, and reaped the profits. Henry Dodge, N. Wilson, and Alexander Craighead, operating as Dodge and Company, had purchased Shibboleth Mine from its discoverer for $6,000 and had received a lease to operate it according to procedures of the federal act. They sent miners onto the property, with equipment, to begin mining. John Smith T and his armed men showed up, claimed the land under a floating concession, and prevented the miners from working. News of Smith T's illegal and shameful seizure of lead mines reached Washington and outraged principled men of the federal government, but the

---

52. *ASP-PL* 2:551–56.
53. Stoddard, *Sketches of Louisiana*, 399.

brazen land-grabber could not be ejected and turned the rich Shibboleth Mine into his base of operations. In such a way, unscrupulous persons gained undisturbed possession of some of the most valuable mines. Frederick Bates accused Samuel Perry, a leading mining entrepreneur at Mine à Breton, of "bold Stakery," that is, of furtively moving his delimiting property stakes outward to aggrandize his surveyed tracts.[54]

The leases amounted to only a small fraction of the total amount of public land being mined, and an even smaller fraction of all mined land. One of the largest lessees was Amable Partenais, who had eight mines leased in 1816, aggregating more than 2,700 acres.[55] Partenais was dutifully paying the government its 10 percent royalty, amounting to $3,979.67 for that single year. The size of this acreage and royalty suggest that large-scale operations were possible under the leasing program.

The larger private operators such as Partenais and Austin had the manpower and means to keep intruders off their lands, although Austin struggled constantly with his nemesis, Smith T. However, the administrators of the lead-leasing program at St. Louis had absolutely no means to ensure the integrity of leases or to keep intruders off public land. Published official statements that threatened intruders with expulsion and punishment met with ridicule.

Just a few years after the lead-leasing act of 1804, Bates, Lucas, and other territorial leaders were already questioning the wisdom of leasing. They thought that Congress should have let the board decide claims to mineral lands, because the board was on the scene, examining the individual merits of all claims, and was in a much better position to make decisions on specific land questions than Washington general policy makers and a single federal agent. They called for repeal of the law and recommended the outright sale of lead land. Yet, the board, caught in the middle of a federal-land issue over which it had no authority, and being attentive to every jot and tittle of the law, could do nothing but act as representatives of Washington-made law. The act of March 3, 1811, which authorized the sale of public lands in Missouri (although actual sale was delayed until 1818), reasserted federal lead-land policy by specifically reserving lead mines from sale.[56] Lead-land leasing was such a mess that most pseudo-owners and would-be owners of lead land continued to op-

54. Dodge, Wilson, and Craighead to Recorder Bates, January 23, 1815, *TP* 15:5–7; Bates to Albert Gallatin, October 6, 1807, *Bates Papers,* 1:217.
55. To the Commissioner of the Land Office, Washington, D.C., November 10, 1816, *Bates Papers,* 2:302–4.
56. To Seth Hunt, February 7, 1808, *Bates Papers,* 1:277–78; *Stats. at Large of USA* 2 (1811): 662–66.

erate as if the land were indeed theirs. The board, disagreeing with federal policy but unable to change it, essentially washed its hands of lead-country concerns and let those matters drift, but Recorder Bates and others continued to offer their opinions on the matter, which were, in brief, that the leasing program was not working.

In addition to land claims based on Spanish land concessions and grants, the board had to deal with hundreds of claims based on possession rights of settlers who were in Upper Louisiana before the transfer of government. Such claim to land, called "settlement rights," was strongly entrenched in the French, Spanish, and American practice that use of land not claimed by any others confers ownership to it after a certain period of years. The practice had been a recognized method of acquiring landownership during the American colonial experience, and it continued in the trans-Appalachian lands of the new republic. Because this large category of land claims from Upper Louisiana had to pass through the land-confirmation process at the same time as Spanish concessions and grants, and because many of these properties that had already been surveyed on the ground were set off in varying compass directions no different from the Spanish concessions and grants, historians and others have sometimes lumped them together with actual Spanish concessions and grants and erroneously called them "Spanish land grants."[57]

The board found no help in how to handle settlement rights by researching Spanish land regulations and had to act purely by congressional direction. Authority for the board to pass on settlement-rights claims was in the act of March 2, 1805, which provided that every person who, prior to December 20, 1803, had the permission of the proper Spanish officer and made a farm and cultivated it was to be awarded a square mile of land and that only one such tract was to be awarded to any person. *Permission* was a key word, and the board diligently looked for it in claims. Though some Americans did have written permissions, most of the American settlers who came in the closing years of the Spanish regime did not bother to secure them. Permission could also have been given orally and have left no documentary evidence. According to Lucas, "[S]everal commandants under the [S]panish government could neither read nor write. . . . [S]everal of them granted permissions [to settle] verbally." Furthermore, Lucas continued, there was the presumption of permission to settle in the absence of documentary evidence, "because the act of settlement being a publick act and that an actual settler could not

57. Misapplication of the term *Spanish grant* for any private land that predates the American land survey continues today. See, for example, Russel L. Gerlach, *Immigrants in the Ozarks*, 15–17; and Duane G. Meyer, *The Heritage of Missouri*, 147–48.

conceal his act of settling nor carry [off] the Benefits of it with him, he is not to be supposed to have done these acts without a previous assurance that he should not work in vain." Lucas concluded that "corporal possession is much respected" by the commandants as well as settlers. The issue of permission was resolved by the act of April 21, 1806, which formally recognized that persons who had cultivated land for three years before December 20, 1803, were assumed to have had the permission of the Spanish to settle, or they would have been evicted.[58]

Speaking for the Americans, Austin proffered that all Americans who came after the purchase and before the transfer of authority in March 1804, as well as all those who were on the way, believed that the new American government would validate their lands.[59] It might also be presumed that arriving Americans expected that their own government would want Americans to populate and Americanize the new territory and would reward them with land titles.

Enabling the board to decide settlement rights by the acts of 1805 and 1806 opened the door to a large number of cases of persons who claimed they were living on a tract on December 20, 1803. It would not be difficult for a claimant to find neighbors, relatives, and friends to testify in support. Indeed, it was almost impossible in the close-knit Bellevue community to find a testifier who was not a friend of long standing, if not a relative. Benjamin Crow testified for brothers James and Thomas McLaughlin, and Thomas returned the favor for Walter Crow, Benjamin's brother. Robert Reed and Benjamin Crow mutually testified for each other. Joseph Reed testified for his cousin William Reed Jr., and the latter testified for his wife's cousin John Lewis.[60] Was there fraud by perjury? Most likely. For example, claims may have been made for land only visited, not inhabited, to which a settler was on his way; many persons were going back and forth between Kentucky and Missouri as they moved during the winter of 1803–1804. Persons who arrived after December 20, 1803, but be-

---

58. *Stats. at Large of USA* 2 (1805): 324–29. Washington had plenty of reports from Americans on the scene at the time of the transfer, such as Stoddard's comment that later arrivers in 1802–1804 applied for concessions but, unable to pay for them or their surveys, were told by Spanish officials to go out, settle, earn money, then pay for them ("Fragment of Letter from Stoddard to Dearborn, June 3, 1804," in *Glimpses of the Past*, 2:107). Lucas to Secretary of Treasury, January 29, 1807, *TP* 14:84; *Stats. at Large of USA* 2 (1806): 391–95. According to Flint, Americans could acquire a permission to settle for a "trifling *douceur*" to the commandant (*Condensed Geography*, 1:313).

59. Moses Austin to [Albert Gallatin?], [August 1806?], *Austin Papers*, 2:116, 120.

60. LCC 38, 39, 99, 104, 105, 144, 146a.

fore the 1808 deadline for filing probably filed claims. Congress, recognizing the difficulty of determining when settlers were actually on the land during the transition years and probably wanting to give the new lands an American population, provided in the act of March 3, 1807, for the confirmation of lands initially occupied during the ten years before December 20, 1803, and whose claimants continued in possession of them at the time of filing.[61] This category of claims became known as the "ten-year claims." It would have been extremely difficult for the board to verify the testimony of settlement-rights claimants by inspection of sites to see how old cabins and fences were, because almost all were in dispersed American-settled neighborhoods, distant from towns and villages.

In the end, the board confirmed most of the settlement-rights claims that had no lead mine or did not conflict with other claims. For most of the claims that it confirmed, the board awarded a standard 640 acres, or 1 square-mile section of land, although in a few it worked out an amount of land according to Gayoso's formula. In still others, the board added the amount of land according to the formula to the standard 640 acres.

In a general but fundamental way, the work of the board was hindered by its distance from Congress and by unrelated events in Washington. Although Congress delegated considerable decision-making authority to the board, the board's decisions were only opinions or recommendations and required final congressional approval. The divided board, unable to reach decisions in some cases, kept referring questions it could not resolve back to the secretary of the treasury, who, in the best bureaucratic tradition, studied issues thoroughly before responding, too often enigmatically and tardily, to the board. Thus, frequent and repetitive communication with Washington further delayed the slow and cautious work done in the territory.

In fairness, Washington had other serious matters to attend. Events leading up to the War of 1812 diverted Congress's attention from the accumulating list of land questions coming to it from western territories. The anticipation and then occurrence of overt hostilities from 1812 to 1815 in the territory of Missouri itself interrupted the process of settling the land. In fact, settlers withdrew from some areas along the Missouri River and the Mississippi River above St. Louis, and others left their farms to participate in military actions. Certain key officials left St. Louis periodically to fight Indians in the decade after 1804, and their absence slowed action on pending land titles and associated land matters.

61. LCC 69; *Stats. at Large of USA* 2 (1807): 440–42.

Results of the Land-Confirmation Process

For six years after its creation in 1805, the Board of Land Commissioners played a central role in the settlement of the Ste. Genevieve District and of the entire territory of Missouri. It was the on-the-scene arm of the federal government that made decisions and offered opinions on matters affecting confirmation of land titles. Public life of the territory focused on the board, both its decisions and the persons who were making them, as land was the most important public issue facing both settlers and the territorial government.

The board signed its final report on January 20, 1812, and sent it to Congress on April 22, 1812, with accompanying letters by Commissioner Clement Biddle Penrose and Clerk Thomas Riddick dated March 24 and 26, 1812, respectively. The report accounted for every claim that had come before it: 3,340 for the former five Spanish administrative districts of the territory of Missouri, including Arkansas. It had, beginning with its first decisions in December 1808, confirmed a total of 1,340 of them, or only 2 out of 5. Concessions accounted for 712, or about half, of those confirmed; orders of survey, 80; settlement rights, 425, or about one-third of those confirmed; and the ten-year provision for settlement rights, 123. Following strictly the instructions and authority given them, the board limited the acreages for confirmed claims. All but 18 were for 800 arpents French measure or 640 acres American, or less.[62]

In the Ste. Genevieve District, which included the New Bourbon quasi-district, the board confirmed 169 claims, of which 103 were by concession, 53 by settlement rights, and 13 by the ten-year provision. A minuscule 6 of those confirmed in the Ste. Genevieve District were for lots in villages, where most people lived, and only 1 was in a common field. Many of the claims in the Ste. Genevieve District that were filed but not confirmed involved lead mines and therefore constituted some of the most prized and notable claims from the Spanish era. William Carr's investigative report addressing the issue of lead mines was included in the report.[63]

The board rejected some claims outright. It also realized that many of

---

62. *ASP-PL* 2:377–79, 388–603. Totals are from Shoemaker, *Missouri and Missourians,* 1:200. However, Shoemaker erred when he noted that only 5 approved claims measured more than 500 acres, and others have repeated this error. Actually, 14 approved claims measured more than 1,000 arpents, another measured 803, and another measured 1,031 acres. Two others of larger sizes were to the "inhabitants" of the villages of St. Ferdinand and Marais des Liards. Six of these 18 claims were in the Ste. Genevieve District.

63. *ASP-PL* 2:551–56, 3:580–601.

the 2,000 unconfirmed claims "had merit," but considered itself unable to act on them because either it found no authority for them in Spanish laws and ordinances or it was not empowered to act on them by instructions and interpretations from the federal government. The board, Commissioner Penrose writing on its behalf, grouped the 2,000 unconfirmed claims with merit into no fewer than forty-nine classes depending on legal circumstances, to which Penrose then surprisingly added, "It is probable the classification may not embrace all the species of claims." The board recommended that Congress recognize most of them as worthy and confirm them, noting that the Spanish government would have recognized them as valid ownerships. For example, village lots and common-field properties accounted for nearly one-quarter of the number of claims, although a much smaller fraction in area. The board recommended that each village be confirmed as a unit with a demarcated outer boundary, including their commons and common fields, because the "United States has no claim to land within, except a few vacant lots which go for support of schools."[64] In essence, the board wanted Congress to recognize the integrity of the unique tripartite village of the Illinois Country as a functioning land-economic unit.

Congress, weary of the Missouri land issue and greatly impressed with the meticulous detail and overall fairness of the board's work and perhaps also pushed to a decision by the imminent war with Britain, acted quickly on the board's recommendations. On June 13, 1812, barely one month after it received the report, Congress passed a sweeping act that confirmed not only those 1,340 claims for which the board had issued certificates of approval, but also most of the claims in the various classes that the board recommended for confirmation but for which it had withheld issuing certificates for various reasons. These included the village lots, outlots, common fields, and commons of eleven French towns and villages in the territory, approved only as whole villages with enclosing boundaries.[65] Also included was land that was occupied before December 20, 1803, even if not cultivated as farms, such as tracts used for pine timber, animal grazing, and sugar making. Further included was land that met all provisions of the law except permission to settle. By its action, Congress confirmed virtually all of the obviously valid land claims.

But this was only the beginning of congressional action. Two years lat-

64. *ASP-PL* 2:377–79. The categories are clearly described and grouped into five general categories by Pelzer, "Spanish Land Grants," 21–23.
65. *Stats. at Large of USA* 2 (1812): 748–52. The villages are Ste. Genevieve, New Bourbon, St. Louis, Carondelet, St. Ferdinand, Village à Robert, St. Charles, Portage des Sioux, New Madrid, Little Prairie, and Arkansas.

er, in response to concerns raised about concessions made after December 20, 1803, Congress passed another act, on April 12, 1814, the second one titled an Act for the Final Adjustment of Land Titles in the State of Louisiana and Territory of Missouri, which confirmed virtually all Spanish concessions made in the territory of Missouri before March 10, 1804, including those before December 20, 1803, and not inhabited.[66] The move to a cutoff date three months later for land claims in Upper Louisiana should have taken care of all settlers in the territory up to the date that the Spanish actually withdrew from St. Louis, but it also took in the many properties sold for douceurs by lame-duck lieutenant governor Delassus during the winter months that he sat in St. Louis patiently waiting for Stoddard to cross the Mississippi. Thus, Congress, though trying to be fair to ordinary settlers, confirmed many claims for land that many thought unethically obtained and that covered valuable speculative tracts. However, the act of 1814 specifically excluded from confirmation any claim that the board had already judged backdated or otherwise fraudulent, of which there were many, and it also excluded all lead-bearing lands.

Because the board in its 1812 report had raised new questions about claims it could not act on and the suspicion that many settlers had still not filed claims, Congress, in its sweeping act of confirmation of June 13, 1812, also authorized Frederick Bates, recorder of land titles and former member of the just dissolved board, to continue to work with the large number of pending land claims in the territory and to seek out any that might have been overlooked.[67] The deadline for filing claims was extended to December 1, 1812. Already familiar with most of the claims and claimants from his previous six years on the board, Bates was given, in general, the same power and authority that the board had had.

For four years Bates gathered more evidence on outstanding claims and new ones and issued his report on February 2, 1816. He examined 2,555 claims of which he approved 1,746 for congressional confirmation (335 in the Ste. Genevieve District), rejected 801 (102 in the Ste. Genevieve District), and conditionally confirmed the other 8. Reasons for rejection included duplicate and triplicate filings by one person, testimony disproved, no evidence of any inhabitation, presence of lead, and the claimants having moved from the district or territory, resulting in a lack of evidence. Ste. Genevieve District claims confirmed by 1816 totaled

---

66. *Stats. at Large of USA* 3 (1814): 121–23.
67. *Stats. at Large of USA* 2 (1812): 748–52.

504, or 429 if village and common-field lots are excluded. As it did with the board's report, Congress moved quickly by approving Bates's report by the act of April 29, 1816.[68]

While Bates was in the midst of working on the two thousand claims, Congress passed an act on February 17, 1815, for the relief of landowners who had suffered loss of property from the New Madrid earthquakes of the winter of 1811–1812.[69] Injured New Madrid landowners received certificates to take out compensating amounts of land anywhere in the "surveyed public domain," which did not yet exist. At the time, however, Congress thought that the public land survey of Missouri and the sale of public land were imminent. The survey lacking, certificate holders nevertheless forged ahead and claimed land in the unsurveyed public domain. These "New Madrid claims," as they came to be called, came into direct competition with claims to land by holders of unconfirmed Spanish concessions, settlement rights, and those expecting to obtain their land through preemption. All of these parties competed for the best, highest-priced land then available, and consequently the New Madrid claims complicated Bates's work.

The majority of certificates found their way into the hands of St. Louis speculators who openly bought and sold them and advertised them in St. Louis newspapers. In effect, they became the American equivalent of the Spanish floating concessions. They circulated freely from hand to hand because they were the only way an immigrant in the years between 1804 and 1818 could get title to land other than buying someone else's land. As it turned out, no one used any of the 516 issued certificates to claim land in the Ste. Genevieve District, although a few attempted. Major reasons were that the valuable mineral land was specifically excluded from claim by New Madrid certificates and that little valuable agricultural land was left for occupancy. Also, in the post-1814 years, land elsewhere in Missouri, especially in the Boonslick and St. Louis regions, had become relatively more valuable than land in the Ste. Genevieve District. Perhaps, also, land titles in the mining district around Mine à Breton–Potosi were so confused and civil order so fragile that wary land speculators chose to deploy their paper titles in safer regions. Thus, in a negative way, the

68. *ASP-PL* 3:274–330. Totals for the various actions by Bates are from Shoemaker, *Missouri and Missourians*, 1:201. *Stats. at Large of USA* 3 (1816): 328–29.

69. *Stats. at Large of USA* 3 (1815): 211–12. No comprehensive account of the New Madrid claims episode in Missouri history has been made, especially of their locations and effect on settlement. One detailed study of the location of New Madrid claims is Gloria Saalberg, "The New Madrid Claims in Howard County, Missouri."

tribulations of land confirmation and lead-land reservation diverted some of the most egregious land speculation from the district.[70]

The claims that were left unconfirmed after the three congressional actions of June 13, 1812; April 12, 1814; and April 29, 1816, included mineral lands and claims to huge acreages, among other less notable ones. Wealthy and leading public figures in politics and business, both French and American, were responsible for most of these pending claims, and they, or their heirs and assignees, were not about to let their claims lapse. The situation attracted a large coterie of lawyers to St. Louis. Furthermore, the progressive liberalization of land-confirmation decisions, which was clearly apparent to those with outstanding land claims, and the repeated extensions of filing deadlines encouraged these influential persons and their lawyers to continue pressing their claims and to increase the number and size of their claims by searching the nuances of Spanish land-granting regulations for any justification.[71]

Congress found itself confronted with what now promised to become an unending and increasingly litigious continuation of the Missouri land problem, aggravated by a wave of land-seeking settlers after the War of 1812. Congress was mindful that these settlers' search for land was hindered by the absence of the public land survey in the territory, the implementation of which was awaiting resolution of private land claims. Therefore, in 1816, Congress asked William H. Crawford, secretary of the treasury, under whose office federal-land affairs were located, to investigate whether still-pending claims had any merit. On December 7, 1818, Crawford reported that as a result of the continuous "relaxation in favor of land claimants" during the previous decade, "it is extremely improbable that injustice has been done by the rejection of claims which ought to have been confirmed."[72]

Despite that conclusion, Crawford proposed an act by which claimants would have yet another opportunity to file any still-unfiled claims, now fourteen years after the transfer of government. Congress, however, took no action. Thus, with congressional inaction and the diversion of energy in the territory to the movement for statehood, the land-confirmation is-

---

70. *Missouri Gazette,* January 16, 1818; *Congressional Debates,* 19th Cong., 2d sess., Senate, 18 December 1826, 12–13; Treat, *National Land System,* 305–6. Of the 516 certificates issued, 382 were eventually allowed, and only 20 for original claimants. One person held 33 certificates. The federal agent sent to investigate the problems of locating and certifying New Madrid lands concluded that "the New Madrid law . . . has given rise to more fraud and more downright villany" than any other land legislation (*ASP-PL* 4:47).

71. Banner, *Legal Systems in Conflict,* 101–21; Houck, *History of Missouri,* 3:53.

72. *ASP-PL* 3:348–49.

sue tended to quiet down for a few years. Then, relentlessly pressed by wealthy land claimants, Congress on May 26, 1824, expressly permitted land claimants to pursue their unresolved cases through the courts, which several did. Some claimants chose not to, because a negative decision in the courts would bar their claims forever; they chose to keep pressure on Congress through memorials and petitions. Therefore, owners of rejected and new claims did not surrender them, even in cases of proved fraud. If they retained physical possession of the land—and most had—they resolutely refused to give it up. Deep frustration in the first decade of Missouri state government over these unconfirmed lands of some of its leading citizens on some of its most valuable acreages was articulated in an 1831 memorial from the Missouri legislature to Congress that called the situation a manifest evil and requested a new board to settle land claims.[73]

Thus, the unconfirmed claims continued to fester as a political issue and eventually resurfaced by congressional creation of a second Board of Land Commissioners, serving from 1832 to 1835, and, incredibly, a third board in 1860.[74] The second board reviewed thoroughly the actions and recommendations of the first board twenty years earlier, then reexplored even more thoroughly the same arguments. Near the end of his extremely lengthy written opinion on one abstruse point of Spanish law, Judge James Peck wrote, "I am now fatigued, and fear my reader will be equally so," and begged the bureaucrat receiving his opinion to make a copy of it for him, because he had no energy left to write further.[75]

The second board decided to confirm virtually all claims still pending, regardless of size or number. The board took the extremely liberal view that all private property of record at the time of transfer had to be respected, and that it was the government's responsibility to prove fraud rather than the claimant's responsibility to show that there had been none. So much time had elapsed since the transfer, so many documents unlocatable, so many claimants and witnesses deceased that proving fraud was difficult at best.[76] It was the wealthy and their heirs who had

73. *ASP-PL* 3:348–49; *Stats. at Large of USA* 4 (1824): 52–56; *ASP-PL* 6:300.
74. *Stats. at Large of USA* 4 (1832): 565–67; 12 (1860): 85–88; Donaldson, *Public Domain*, 375–76.
75. *ASP-PL* 8:807. The second board had the help of Joseph M. White's detailed "Spanish and French Ordinances Affecting Land Titles," compiled in 1829 (*ASP-PL* 5:218–315) and separately published in 1839 as *A new collection of laws, charters, and local ordinances of the governments of Great Britain, France, and Spain, relating to the concessions of land in their respective colonies*. . . .
76. An example of an acknowledged case of fraud being reversed and the claim confirmed is that of Walter Fenwick's claim to ten thousand arpents at Mine la Motte.

persisted with their claims this long—the small claimants having given up—and it was they who in the end benefited from the liberalization of the land-confirmation process. One should also note that by the 1830s, the perspective of interpreting the past had changed. The commissioners of the second board admired the early settlers, whether French or American, rich or poor, for the hardships and dangers that they had faced and overcome. Admiration of their forerunners replaced suspicion of fraud and unfair economic gain and administrative responsibility to protect government property, if not also concern for fairness under the law.[77]

Surveying Private Lands

The long confirmation process was a political operation that involved personalities and intrigue and aroused strong emotions. It has captured the fancy of historians and biographers interested in understanding the forces operating to establish a society and government in early Missouri. But for settlement geography, it was only the first step toward the legal acquisition and occupation of land. The second step was the locating and surveying of the properties on the earth's surface, and, until that was done, landownership was not fully invested.

Land that had been surveyed or had been ordered for survey by the lieutenant governors universally was confirmed by the board, if it met other basic stipulations. Surveyed property was so persuasive as evidence that claimants rushed frantically to get their lands surveyed by the accommodating Soulard or his deputies, who profited from this work, until his replacement in 1806. One celebrated incident that illustrates the importance of surveying for the official recognition of landownership concerned the attempt by Soulard's deputy surveyor, Thomas Maddin, to survey Ste. Genevieve merchant Pascal Detchmendy's concession in the

---

It was first determined to be "destitute of merit" due to alteration on the original concession, but on August 26, 1835, the second board rescinded the action, declaring that the alteration was in the petition, not the concession (*ASP-PL* 8:107).

77. Pelzer, "Spanish Land Grants," 31. The success of the wealthy in getting ultimate confirmation of their claims, some of immense size and some obviously fraudulent, while so many of the claims of common people were abandoned or not even filed, supports Gates's argument that the wealthy won the West, not the yeomanry celebrated by Frederick Jackson Turner's famous hypothesis. See, for example, Gates's *Jeffersonian Dream: Studies in the History of American Land Policy and Development*, xiv, 6–39. Also see Treat (*National Land System*, 228) and Gates (*Public Land Law Development*, 96–108) on how the second board resolved issues more liberally than the first. Violette concluded that the second board based decisions on what settlers wanted, not on interpretation of Spanish law (*Spanish Land Claims*, 189).

American-settled Bellevue Valley. Maddin and his crew, with Detchmendy along, went onto the concession early in 1804, before the Spanish had turned over Upper Louisiana, obviously rushing to get the tract surveyed at the last moment. Moses Austin, apparently on the scene, was reported to have told American Thomas Rush, who was living on the land that had been conceded to Detchmendy, that "now was the time for [Rush] to save his land, and if [we] could keep the land from being surveyed," Austin would help Rush acquire it. Met by armed Americans on the site, the surveying party retreated and never returned. The board rejected Detchmendy's concession for want of survey and confirmed Rush's claim.[78]

As the board began rendering decisions on claims in late 1808, claimants with certificates of approval had to get their lands surveyed. The secretary of the treasury refused to issue any land patents unless tracts were physically located on the ground and surveyed and the plat filed in the appropriate governmental office.[79] The government-appointed surveyors of private lands received a fee, usually three dollars per mile, which was good pay, yet they could not keep up with the demand.[80] Impatient American settlers, who had been arriving since the late 1790s in full anticipation of a better life in a new land and having gone through the tedious land-confirmation process, now had to face delays in surveying before land patents were theirs.

Confirmation carried the stipulation that the land not overlap or conflict with existing confirmed lands. Settlers had earlier picked out sites specifically for a house, garden, and fenced field, which totaled in acreage far less than the standard 640 acres confirmed. Therefore, a surveyor often could not find enough vacant space in well-settled neighborhoods for a mile-square parcel centering on the owner's improvements, as required in his instructions. As a result, surveys ended up in highly irregular shapes in order to squeeze in all the necessary acreages. This is conspicuous in the compactly settled Bois Brule bottom, where curious jigsaw-puzzle shapes resulted as surveyors maneuvered to find sufficient

---

78. *ASP-PL* 8:74. Others have used this well-known incident as evidence of the independent spirit of Americans. Here it is used to show the importance of surveys for landownership.

79. *Stats. at Large of USA* 3 (1816): 328–29.

80. Amos Stoddard, as cited in Shoemaker, *Missouri and Missourians,* 1:201; Moses Austin to [Albert Gallatin?], [August 1806?], *Austin Papers,* 2:115. Land surveying was a skill in great demand. Land surveying books were among the most popular vocational books in private libraries in frontier Missouri, according to a survey done of libraries that went through probate (Harold H. Dugger, "Reading Interests of the Book Trade in Frontier Missouri," 139, 265).

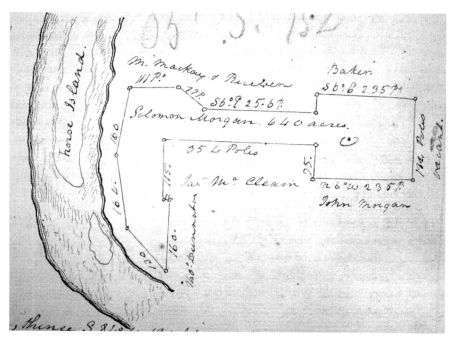

7. *Solomon Morgan Survey, Bois Brule Bottom.* The fifteen-sided polygon is an example of the difficulty of finding one square mile of land in an area of many confirmed claims along the Mississippi River. (U.S. Recorder of Land Titles for Missouri, Record Book of Land Titles, vol. A, p. 523)

land to match all the confirmed acreages. Solomon Morgan's confirmed claim of 640 acres took the contorted shape of a fifteen-sided polygon, and others were as geometrically complicated (fig. 7).[81] Many settlers in the Bois Brule bottom put their physical improvements at closely spaced intervals along the Mississippi River, and surveyors had to run their properties long in depth and narrow in width. The resulting narrow properties of these Americans appear on maps like riverine French long lots. Thus, the actual land a claimant received through the survey may not have corresponded at all to what he thought was his. It was the surveyors, not the settlers, who determined the shapes of many pre-1804 properties and the long-lasting cadastral imprint on the landscape.

81. Private Land Surveys, A523, B78, Missouri State Archives; U.S. GLO Field Notes, Missouri, 209:76 (survey of April 1818). Bernard Pratte's stock ranch in Murphy's Settlement, granted in accordance with Spanish practices as a *sitio* of 1 square league, turned out to be oblong, 720 by 1,308 poles and truncated on the north, when surveyed in 1806 (LPC 16; *Régistre d'Arpentage,* 708).

Some confirmed private lands did not get surveyed until after the U.S. rectangular survey was run. In this case, properties were required to have the public land survey's cardinal directions for their boundaries, regardless of the shape of the property that the owner thought he had. If the confirmed acreage was for recognized units of the rectangular survey (640, 320, 160 acres), the properties had to be one of the standard subdivisions of the rectangular survey. These properties from the Spanish era are utterly indistinguishable from other properties of the American rectangular survey, unless designated with a private survey number or name.

Claims not confirmed until after land went on sale in Missouri bore the risk of having the land sold to others before it could be put in the hands of the original claimants. In this case, claimants received scrip for compensatory land elsewhere in the surveyed public domain, and no private surveys were made for their original confirmed tracts of land.[82] This happened to some of the extremely large claims, not confirmed until the 1830s or later, such as the Maxwell grant of sixteen square leagues on the Black River. Though a confirmed, perfected Spanish grant, it lacked an American survey and does not appear on any cadastral map.

Surveys of villages could not proceed until evidence of the location and extent of individual properties within them was gathered. Though William Rector had begun a survey of Ste. Genevieve village lands in 1814 shortly after village confirmation, he was unable to proceed much. An act of May 26, 1824, required property owners in all villages to present evidence of landownership, and only then, for the first time, could surveyors mark properties in the villages and common fields, their work purportedly reflecting landownership conditions of 1804.[83]

The survey of private lands left a lasting imprint on the landscape of

---

82. *Stats. at Large of USA* 4 (1824): 52–56. A total of 192,701.61 acres of indemnity scrip was issued in Missouri for confirmed and patented private lands that could not be located because of various reasons. Much of this was mineral land in the Ste. Genevieve District. Among those who received indemnity scrip in the historic Ste. Genevieve District were the heirs of land claimants Israel Dodge, Walter Fenwick, and Joseph Gerard, but not until 1855 and 1860 (Donaldson, *Public Domain*, 289–90). Where they located their compensatory land is not known.

Where the size of the confirmed land was less than that of the original claim, the surveyors were required to reduce it to the confirmed size. Later, when the full size of the claim was allowed, usually from a court decision, surveyors had to return to enlarge the property somehow. This would cause the displacement of abutting, unconfirmed claims or squatters with improvements (*ASP-PL* 5:97).

83. William Rector to Josiah Meigs, August 14, 1814, *TP* 14:783; *Stats. at Large of USA* 4 (1824): 65–66. St. Louis had been earlier surveyed and properties recorded in the *Livres Terriens*.

properties of various shapes and property and field lines in all compass orientations that contrasts sharply with the imprint of standard shapes and cardinal directions of lines of the U.S. public land survey. The regularity, predictability, and infinite extension of the public land survey are its hallmarks. The American system, directly superimposed on the natural landscape, gives no indication of the topographic irregularity of the land surface and its varying suitability for human use. In contrast, the private land surveys appear haphazard and geometrically disorderly. Each piece of property seems unique for its place in the topography. On closer inspection, however, the pre-American properties are actually more standardized in shape and orientation than generally understood.[84]

Pre-American properties had three basic shapes. The long lot, defined as a tract whose length is more than twice as great as its width, was introduced into the middle Mississippi Valley with the beginning of French settlement and was virtually the only shape for rural properties until the 1780s.[85] It divided the common fields into private properties and partitioned other river-border lands and other agricultural properties in the interior as they fronted on natural features. The early mining grants also had long-lot shapes, such as Renaut's 1723 grants that measured one and one-half leagues *de front* by six leagues *de profondeur*. Squares occurred in the first French villages as the basic shape of lots but did not appear in rural areas until the Spanish began awarding land for ranches and lead mining. Squares were the required shape for properties surveyed by Americans for settlement-rights claims from the Spanish period. The third shape, irregular polygons, which make up a minor portion of all private properties, occurred only where surveyors could not find enough land for squares or where rivers interrupted the pattern.

Some of the irregular orientations of properties are easily accounted

84. An excellent examination of cadastral patterns is by E. T. Price, *Dividing the Land*. For patterns in Louisiana, which are more complex than those in Missouri, see Hall, "Louisiana Survey Systems"; and Carolyn O. French, "Cadastral Patterns in Louisiana: A Colonial Legacy." Sam B. Hilliard explores the issue of settlement rights and surveying in "Headright Grants and Surveying in Northeastern Georgia." Richard Colebrook Harris describes the long-lot pattern in Canada in his seminal work, *The Seigneurial System in Early Canada: A Geographical Study*, 121–27. Ekberg explores the origin of long lots in the middle Mississippi Valley in *French Roots*, 1–30.

85. For long lots and French cadastral patterns, see French, "Cadastral Patterns in Louisiana," 90; Terry G. Jordan, "Antecedents of the Long-Lot in Texas"; Sam B. Hilliard, "An Introduction to Land Survey Systems in the Southeast"; E. T. Price, *Dividing the Land*, 289–300; Serge Courville, "Contribution à l'étude de l'origine du rang au Québec: La Politique Spatiale des cent-associés"; Pierre Deffontaines, "Le Rang, type de peuplement rural du Canada français"; Max Derruau, "À l'origine du 'rang' Canadien."

for. For example, the process by which the compass orientation of the bluff-top properties overlooking the Pointe Basse was established was described in detail by five *arbitres,* all prominent members of the community appointed by the commandant. On November 15, 1794, the five, accompanied by interested property owners, marched to the bluff where they took several compass readings and, with everyone in agreement, arrived at a reading of twenty-five degrees south of west. This reading fairly well reproduces the direction of property lines in the common field. All other properties along the bluff used the same compass direction.[86]

In many cases, property orientation away from north was clearly a mistake by the surveyor. Written concession documents and plats indicate that the grantor intended certain properties to be laid out according to the cardinal directions of true north and south, east and west, but they ended up at orientations from six to ten degrees east of true north. For example, Soulard's 1798 plat for Israel Dodge had cardinal directions of "Est, Sud, Ouest, Nord," and its accompanying written concession also used those cardinal directions. The plat in Figure 8 includes an arrow showing an east magnetic variation ("Von Est," or *variation est*) of an unspecified number of degrees. When Dodge had his tract marked off on the ground, the surveyor did not correct for magnetic declination of the compass, and consequently the tract, as well as many other properties explicitly intended to have true north-south and east-west boundaries, was surveyed on the ground at an angle of from six to eight degrees to the east of true north. In another case, Soulard admitted that he had surveyed Madame Villars's square league "without regard to the variation of the needle," which was approximately 7°30′ east, and that he had used "north by the needle." This surveying practice resulted in a decidedly offset pattern on the map and in the landscape that is conspicuous around Farmington and in the Bellevue Valley. Had surveyors corrected for magnetic declination when running lines, some one hundred tracts presumably would have had the same directions as the section, range, and township lines of the American rectangular system. They would have appeared to us today as geometrically orderly as survey lines of the American sys-

86. LCS 112. The properties today are aligned at eighteen degrees south of true west, which agrees with the 1794 reading of twenty-five degrees south of magnetic west. Eighteenth-century surveyors laid out lines according to compass directions in the field, uncorrected to true north.

The *arbitres* used the eighteenth-century expression *air de vent* for compass direction, which others have concluded to mean that the *arbitres* found a wind direction to use for laying out the lines, which was associated with the laying out of property lines by the wind under the Spanish Law of the Indies (Williamsen, "Colonial Town Planning," 444–47).

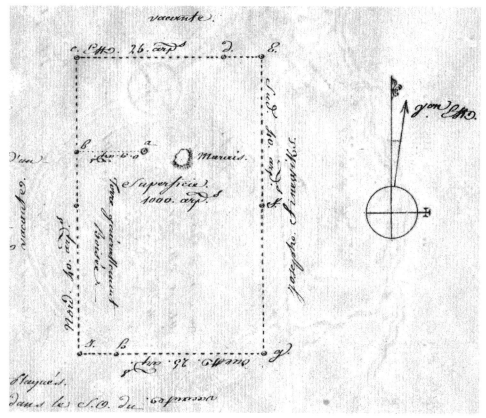

8. *Israel Dodge Survey, February 27, 1798.* As was the case for many Spanish concessions, Dodge was awarded land with boundaries according to true north, and this plat was drawn with boundaries according to true north. However, when surveyors marked off boundaries on the land surface, they did not correct for magnetic declination, clearly shown on the right (the arrow marked "Von Est," or *variation est*), and the properties in the landscape were "tilted" six to eight degrees east of true north. (*Régistre d'Arpentage,* 186)

tem and would have divulged a Spanish-French administrative mind as geometrically oriented as Jefferson's and others who created the American rectangular survey.[87]

Thus, the haphazard-appearing orientations of Spanish and other pre-American properties in the Ste. Genevieve District were not at all random or irrational. These properties had reasons for their orientations and disclose an "order" in the minds of surveyors and owners. What appear as disorderly patterns on maps and in the landscape today and what might be initially interpreted as a lack of coordination or oversight by administrators, especially in contrast to the splendid geometric regularity of the American rectangular land survey, actually show a consistent set of procedures in the surveying of land.[88]

The Landless

The sizable cohort of settlers who began populating the territory after 1804 and did not own land were of no concern to the board. They were

87. *Régistre d'Arpentage*, 186; *ASP-PL* 5:741, 758. "North by the needle" is used in surveys reported in LRC 281, 290. Surveying by the needle was the custom adopted in the Province of Louisiana, according to Soulard (Guibourd Papers). Surveying by magnetic declination rather than by true north was also practiced by American surveyors at the time. In 1798, Rufus Putnam, first surveyor-general of the United States, urged that lines be run by the "magnetic meridians" rather than by true meridians because of the necessity of taking frequent accurate observations in the latter (Treat, *National Land System*, 195). Perhaps Soulard and his deputies thought as Putnam did, that convenience was more important than absolute geodetic control. According to William D. Pattison, French and American surveyors used both true or magnetic north in the late 1700s. Though public land instructions clearly specified true north for late-eighteenth-century surveys in Ohio, some surveyors continued to use magnetic north for convenience (*Beginnings of the American Rectangular Land Survey System, 1784–1800*, 214–16). A general discussion of the issue is in C. Albert White, *A History of the Rectangular Survey System*, 18–29.

Magnetic declination varies over time. It is different today in the Ste. Genevieve District than in the eighteenth century, and the rate of change varies over time. The magnetic variation of 1800 cannot be reconstructed simply by knowing what it is today and projecting backward in time. In the Ste. Genevieve District, magnetic declination varies significantly within short distances due to iron-rich bedrock, and this might account for magnetic declinations varying from five to ten degrees east in Soulard's surveys.

88. An analysis of orientation of pre-American properties in the Ste. Genevieve District is in W. A. Schroeder, "Opening the Ozarks," 188–92. Hall showed order in the pre-American land-division systems in Louisiana by his categorization of properties ("Louisiana Survey Systems"), and Hilliard also determined order existed in the irregularly oriented headright grants in Georgia because the system had common consent and was held intact by trust and faith ("Headright Grants," 428–29). Also see the comments of French on the historical imprint of various notions of cadastral order ("Cadastral Patterns in Louisiana," 1–2).

squatters. The term in its broadest sense refers to the landless, all persons without title to land and who lived on another's land or on the public domain but who made a living from it as farmers, woodsmen, and the like. Squatters fell into two basic groups according to their intentions toward landownership.[89] The first were those who intended to remain on the land they occupied and obtain title to it. They comprised the preemptors who participated in community life and eventually became bona fide owners of the lands they had cleared and improved. For example, John A. Jones built several cabins and cleared about sixteen acres, known to be public land, on the south fork of the Gabouri only three miles west of Ste. Genevieve village. He sold his improvement to Henry Kiel and Edmund Roberts, merchants in Ste. Genevieve, for one hundred dollars on August 11, 1816, noting in the deed that he was a preemptor and entitled to preemption benefits. He actually sold his right of preemption to the land to Kiel and Roberts, as well as any improvements on it.[90] This group of squatters also includes tenants, hired workers, and lessees for privately owned mined land, cattle ranches, and other tracts owned in absentia by Ste. Genevieve and St. Louis residents. An example is John Townsend's Spanish concession at the headwaters of Bois Brule Creek. Townsend sold the land in 1805 to speculator Ezekiel Able, who leased it to widow Margaret McBee, who likely could not otherwise acquire land. The widow's lease was for fifteen years and permitted her, in turn, to take in as many subtenants as she wished.[91] All of these squatters more or less contributed to community building.

The other group of squatters consisted of those who never intended to go through the process of acquiring legal ownership of land. They were backwoodsmen who did not regularly participate in society, chronic movers without any attachment to a place, drifters, those totally unconcerned with landownership and who had no intention of exercising preemption rights. In the Spanish period, they included voyageurs and *en-*

---

89. The term *squatter* was first used in congressional debates in 1806, according to French ("Cadastral Patterns in Louisiana," 31). Clendenen recognized three classes of early settlers in the Ozarks: the squatter-hunter, the hunter-farmer, and the agricultural settler ("Settlement Morphology," 82–85). Throughout the published collection of his letters about the White River country of the Ozarks, S. C. Turnbo categorized as early hunters, gatherers, and stockmen, who were first on the land; and farmers, second on the land. Both categories occupied land without obtaining title to it (James F. Keefe and Lynn Morrow, eds., *The White River Chronicles of S. C. Turnbo: Man and Wildlife on the Ozarks Frontier*). An excellent portrait of the hunters and stockmen who lived outside the market economy in Georgia is in Grady McWhiney, *Cracker Culture: Celtic Ways in the Old South.*

90. LRB 472.

91. LCC 1; LRA 78.

gagés of the fur trade and itinerant traders and hunters who showed up occasionally in the village.[92] Among the Americans, they were chiefly hunters and woodsmen, with or without families, who harvested wild game as much as planted crops. They cleared land by tree deadening and fire, made small clearings, and put up crude cabins, all of which increased the value of the land. As soon as farming settlement approached, commented Frederick Bates, "they sell out their good plantations with vacant cabins & orchards for 50¢ an acre and leave." During his search for souls in the rough hills of the Ste. Genevieve District, the Reverend John Mason Peck observed how an arriving family traded its wagon and yoke of steers to a backwoods family for their cabin and "crap." The backwoods family was departing for the Black River where, they had heard, bear hunters had begun a new settlement.[93]

Needless to say, their neighbors welcomed squatters of the preemption kind as fellow residents helping to transform the wilderness into productive land, including ridding it of fevers, and thus essential to the construction of a vigorous local economy and the development of social and political institutions. On the other hand, the preemptors caused local administrators minor problems and were frowned upon by an eastern establishment that expected frontier settlement to proceed lawfully by orderly land acquisition. Squatters of the backwoods type did not interact much with others and were not usually considered or welcomed as part of a growing community. Political and moral leaders criticized them, as a group, as having gone wild, having lost their civilization.[94]

How many persons were in the two landless groups? What percentage of the population did they constitute? It is not easy to find answers, because of the footloose nature of these people and the fact that Spanish and American population counts may not have included them. Landless, they may not have been taxed, and, if itinerant, they may not have been available for militias, and these were the two chief reasons for keeping

92. Morris Arnold points out that there were two classes of persons at the colonial French post of Arkansas. One small group consisted of merchants and habitants who adhered to European legal traditions and modes of dispute settlement. The other group consisted of hunters, the bulk of the population, who claimed to owe allegiance to no state and steadfastly refused to take part in an ordered, agrarian community (*Colonial Arkansas,* 170).

93. To John Michie, Goochland Court House, Virginia, August 22, 1810, *Bates Papers,* 2:155; John Mason Peck, *Forty Years of Pioneer Life: Memoir of John Mason Peck, DD, Edited from His Journals and Correspondence,* 103.

94. Peck's sharp comments about cultural degradation on the Missouri frontier are echoed by others about other frontiers, such as Frederick Jackson Turner, *The United States, 1830–1850: The Nation and Its Sections,* 289–92; and even by the Spanish, Don Pedro Piernas to Gov. O'Reilly, in *Spanish Regime,* ed. Houck, 1:71.

track of adult males in the population. In colonial times, the transient population around Fort de Chartres may have composed 30 percent of the population. When the British arrived in the Illinois Country in the 1760s and took stock of the population, they concluded that it was difficult to count inhabitants, "as there is always many of them at New Orleans, trading with the Indians, or Hunting, which they go to as regularly as the Savages."[95]

The number of landless increased sharply in the American period, but perhaps not their percentage of the population, and preemptors were certainly more numerous than backwoodsmen. An analysis of Washington County for 1814, the year of its erection as a separate county, gives an indication of the size of the landless cohort.[96] The total number of owned properties was 191. Washington County's adult white-male population by the special census of 1814 was 1,010, which, if doubled for adult females and slaves, comes to 2,020 adults.[97] This is likely too few, because the census missed persons in the remoter mining camps. The result is an average of 10.6 persons per tract. Assuming 6 persons per household, the estimated total population (children included) for 191 tracts is only 1,146, or only 57 percent of the estimated count of adults from 1814. Well over 50 percent of the people of Washington County were living on land that was not in private ownership, either confirmed or claim pending. This proportion compares with a 42 percent calculation of squatters in a central Illinois community in 1830, and an estimate that approximately two-thirds of the population of Illinois was squatting upon government lands in 1828.[98] Clearly, the landless form an important element in lands being settled.

---

95. James M. Delehanty, "Livelihood in the Region of Fort de Chartres under the French, 1720–1763," 21. Ekberg notes that one-third of the militia roster of 1779 at Ste. Genevieve were "'voyageurs,' itinerant traders and boatmen" (*Colonial Ste. Genevieve*, 46–47). According to the 1796 census for New Madrid, one-third of the heads of families (53 of 159) had no property, were not engaged in agriculture, and were presumed to be hunters, trappers, boatmen, or day laborers (Jonas Viles, "Population and Extent of Settlement in Missouri before 1804," 194). Stirling to Gage, December 15, 1765, in *New Regime*, ed. Alvord and Carter, 125.

96. W. A. Schroeder, "Opening the Ozarks," 193–95.

97. *Missouri Gazette*, July 2, 1814.

98. Surveyor William Russell estimated that one-third of the population of Washington County in 1814 was living on public land (William Russell to William Rector, April 20, 1814, *TP* 14:754). It was reported in 1828 that "not one-third part of [Missourians] are possessed of lands," that is, more than two-thirds were landless ("Application of Missouri for a Change in the System for Disposing of the Public Lands," January 26, 1829, *ASP-PL* 5:212). As landless as the Missouri population was, the situation was even worse in Iowa, where whole districts were populated before legal

At times, squatters came into conflict with land claimants and landowners. If private lands were unsurveyed and unmarked on the land, those seeking land did not necessarily know that someone had already claimed it. A case is the land claimed by four stonemasons of Ste. Genevieve: Peter Bloom and Michael, Bartholomew, and Sebastian Butcher. In 1802, they had been awarded concessions of four hundred arpents each in Cook's Settlement on the north side of Mine la Motte for wood for their business operations. The masons' tract lay undemarcated and unoccupied when two newly arrived American families unknowingly intruded on it, built cabins, fenced fields, and later defended their enterprise when challenged with expulsion.[99] Such a situation must have been a common land conflict, especially on large tracts of land owned in absentia, unmarked in the landscape.

Squatters also presented problems for those responsible for maintaining the value of the public domain for future sale and were, in official language, trespassers. During the Spanish regime, authorities usually looked aside at trespassers on the royal domain, but the American government undertook to control them more diligently. Congress forbade all trespassing on the public domain in 1804 in agreement with instructions from President Jefferson to prevent unauthorized settlements on the public domain and destroy any fixed improvements already made.[100] Mineral land was especially to be rid of trespassers, because by their timber cutting for homesteading, they were destroying a resource necessary for the future production of lead.

On April 12, 1814, ten years after the cession and after prodding by restless Missourians who considered squatters as fellow community-building citizens, Congress extended preemption rights into the territory of Missouri with the same general provisions made one year earlier for the territory of Illinois. That should have resolved the problem. However, conditions of land occupance and landownership were more complex in Missouri. Speculators were using New Madrid certificates to claim some of the best land that preemptors had labored to improve, and plenty of private land claims that converged with preemptors' lands were still

---

ownership could be established (Dick, *Lure of the Land,* 261). John Mack Faragher, *Sugar Creek: Life on the Illinois Prairie,* 54–55; Roy M. Robbins, *Our Landed Heritage: The Public Domain, 1776–1970,* 49.

99. LCS 117; U.S. Senate, "Petition of Sands Stuart and John Davis," 26th Cong., 1st sess., S. Doc. 303, 20 March 1840, 1–3. The case was still being settled in 1840, more than twenty years after the squatters arrived as preemptors. In the petition, Bartholomew Butcher is identified by the name Boston Butcher; Michael was deceased.

100. *Stats. at Large of USA* 2 (1804): 283–89; President to the Secretary of War, November 1, 1804, *TP* 13:54.

unresolved. Who had priority where more than one party claimed the same tract under different governmental provisions? In the meantime, on December 12, 1815, President Madison, responding to voices in Congress, issued another futile general proclamation reordering the removal "by force" of all persons unlawfully settled on public land, if they were still on it on March 10, 1816. The Missouri territorial legislature immediately appealed to Congress on January 22, 1816, to allow squatters to stay on land in the territory with full preemption rights. Congress responded favorably with a more inclusive act on March 25, 1816, that provided that all settlers on public lands who were resident on them before February 1, 1816, could apply for their land, up to 320 acres per applicant. The act was in force for one year, then extended for another. The register of the St. Louis land office, Alexander McNair, opened his office on June 13, 1816, to receive applications.[101] Thus, preemptors did get priority to their lands before they went on public sale.

Squatting on the public domain persisted long after land went on sale, though, because backwoodsmen continued their practice of living on the public domain or on someone else's unoccupied land. Their unconcern with legal ownership of land characterized the more sparsely settled parts of the eastern Ozarks well into the twentieth century.

Preemption is another example of the discrepancy between a deeply rooted settlement practice and official policy and law. Governmental attempts to keep settlers off the public domain were unsuccessful, and in most cases no enforcing authority made an attempt at eviction beyond threats. As public-land historian Roy M. Robbins has pointed out, laws allowing preemption for squatters were expedients "to make established law and order conform with the lawless and uncontrollable spirit of the American frontier."[102] Preemption laws made illegal acts legal.

Attempts to Limit Settlement

The free flow of Americans into Upper Louisiana raised the question of whether their settlements should be restricted to a limited space.

---

101. *Stats. at Large of USA* 3 (1814): 121–23; Roy M. Robbins, "Preëmption: A Frontier Triumph," 335. The first federal preemption act was on March 2, 1799, followed by sixteen separate federal preemption acts between 1804 and 1830. For a review of preemption policy and laws, see Hibbard, *Public Land Policies,* 144–70; Donaldson, *Public Domain,* 214–16, 678–95, 1239–47; and Gates, *Public Land Law Development,* 219–47. *Missouri Gazette,* January 27, 1816; "Resolutions of the Territorial Assembly," January 22, 1816, *TP* 15:108–9; *Stats. at Large of USA* 3 (1816): 260–61; (1817): 393; A. H. Rose, "Extension of the Land System," 112–14. The Jackson land office did not receive preemption requests until after 1820.

102. Robbins, "Preëmption: A Frontier Triumph," 332. Also see Meinig, *Shaping of America,* 1:410–11.

Should the federal government exert control of the settlement process by disallowing people to go beyond a certain line?

The federal government had good reasons to limit the geographic extent of settlement in its frontier regions. A state has a responsibility to defend its settlers against indigenous peoples into whose territory the settlers are intruding and against other competitors for the same space. In military terms, defense is easier if advancing settlements are concentrated within a given space and are clustered close to forts. In the history of the American colonial and republican frontiers, defending settlers against Indians or competing Europeans was an expected governmental function. It is also easier to service the needs of settlers if they are not widely dispersed. Religious services and sacraments, provisions and supplies, administration of justice, tax collections, the recording of births and deaths, and just knowing who is where and what is going on—a basic function of a state—are all facilitated by population concentration and hindered by dispersal. Furthermore, the government feared that settlers, if they removed too far from the benefits of civilization, would lose those benefits and live like Indians. Material standards would be degraded, religious values would be destroyed, speech would deteriorate, and culture in general would decay. Though this fear was rarely articulated when limits to settlement were officially declared, it was commonly held by those persons who protected the cultural standards of the country, and it helped form societal attitudes about the wild frontier.[103] The fear supported a policy to keep Indian and white populations geographically separate and establish an "Indian line" as a limit of settlement. Finally, governments wished to keep people within a restricted space in order to build up populations of sufficient density to permit economic, social, and political institutions to develop. Government officials and planners argued that the institutions of a progressive state, especially those of an economic nature, necessitated restrictions on the free dispersal of people into new lands.[104]

The first American attempt to limit settlement geographically in Up-

---

103. Roderick Nash, *Wilderness and the American Mind*, 30; Meinig, *Shaping of America*, 1:284–88, 295–96, 408–10, 413–14.

104. Probably the best-known example of attempts to limit settlement and the one that had major influence in American settlement history was the proclamation line of 1763 (Clarence Walworth Alvord, *The Mississippi Valley in British Politics*, 1:202–3; Meinig, *Shaping of America*, 2:78–79). Full text of the proclamation is in Adam Shortt and Arthur S. Doughty, *Documents Relating to the Constitutional History of Canada, 1759–1791*, 119. In another paper, Alvord argues that the British knew full well at the time that the line was drawn that it would be only temporary as a settlement limit until more Indian removals could be effected and land taken from them ("Genesis of the Proclamation of 1763"). Although the Spanish did not draw lines limiting settlement, they officially required white settlements to be contiguous and instructed settlers not to locate away from the villages.

per Louisiana was Jefferson's proposal to use the broad Mississippi River as the western limit of white settlement. The president proposed moving harassed eastern Indians across the Mississippi into Upper Louisiana and thereby freeing the original United States for Americans. He further proposed to remove white settlers on the western side of the river from the region below Cape Girardeau, because New Madrid and other feeble, scattered settlements were already losing population, and, as the great extent of the wetlands below Cape Girardeau became more fully realized, this region was deemed unsuitable for American settlements. Existing white settlements above Cape Girardeau, including Ste. Genevieve, were, on the other hand, to be strengthened by making them more compact and defensible.[105]

This proposal to rearrange human geography by using the Mississippi River as a limit of settlement, devised in the nation's capital and discordant with conditions on the frontier, was totally unacceptable in the new territory. Residents in the territory simply would not be so geographically manipulated, and they spoke out against Jefferson's plan. Moses Austin, who had staked a fortune in his successful lead operation at Mine à Breton, well inland from the river, voiced his disapproval. If Jefferson's plan to use the Mississippi River as a settlement divide were enforced, Upper Louisiana, he predicted, "would become a Nest of Robbers.... [T]he Navigation of the Mississippi would become more dangerous than to Traverse the wilds of Arabia." Austin was seconded by territorial judge Rufus Easton, who argued in a letter to the president that the west bank of the Mississippi must be well settled in order to maintain control of that valuable artery in times of Indian "disturbance." Easton countered Jefferson's proposal by suggesting that Congress "mark a certain line of longitude including all the settlements west—and say to the emigrant or to the citizen—*this far mayest thou go but no farther.*"[106] Easton did not specify a meridian as a settlement-limiting line.

In 1806, Moses Austin pointed out that "to protect so extended a Frontier... will require more Troops, than the United States have in Service—and a sum of Money which would embarrass the Treasury." Restless men "who are never satisfy'd long in any place... will extend the Frontiers beyond the controul of Government."[107] Despite his belief in Americans'

---

105. Annie Heloise Abel, *The History of Events Resulting in Indian Consolidation West of the Mississippi,* 241–49; Governor Wilkinson to the Secretary of State, August 24, 1805, *TP* 13:189–90; Governor Wilkinson to the President, November 6, 1805, *TP* 13:266; Foley, *Genesis of Missouri,* 177–78; Meinig, *Shaping of America,* 2:78–82.

106. Moses Austin to [Albert Gallatin?], [August 1806?], *Austin Papers,* 2:122; Rufus Easton to the President, January 17, 1805, *TP* 13:83. Emphasis in the original.

107. Moses Austin to [Albert Gallatin?], [August 1806?], *Austin Papers,* 2:121.

restlessness, Austin proposed his own demarcation line, based on his good geographic knowledge of the territory of Louisiana (fig. 9). The line, which remained only a suggestion, would have been no more than sixty miles west of the Mississippi River.

One of the first acts that territorial governor Meriwether Lewis performed upon his tardy arrival at St. Louis was to issue a proclamation on April 20, 1808, establishing a line to limit settlement (fig. 10).[108] He gave as reasons the "extremely inconvenient and almost impracticable" execution of laws by public functionaries to a detached and scattered population, and, of course, difficulties arising from "our precarious standing" with Indians. His proclamation prohibited persons from establishing dwellings or cultivating fields beyond the line, and required the removal of those already beyond it from their lands. North of the Missouri River it was the Indian line of the Sac and Fox treaty, but south of the river it was merely one drawn for the single purpose of limiting settlement.

Yet another line to limit settlement, but not an Indian line, was proposed by Amos Stoddard. Though Stoddard was no longer associated with the administration of the territory when he proposed his idea, he was among the most knowledgeable in the settlement geography of the territory, especially from a military man's point of view. Stoddard believed that a population large and dense enough for Upper Louisiana to defend itself, whether against Indians or a British invasion from the Great Lakes, could most effectively be accomplished by restricting settlers to prescribed limits. "If they be suffered to spread over a great extent of territory, their strength cannot easily be concentrated. Their divided and detached situations will serve to invite hostilities, and probably enable the Indians to destroy them in detail."[109] The line Stoddard proposed was not much different from Lewis's line; it basically delimited all the existing white settlements (fig. 11). Nothing came of Stoddard's proposal. By the time it was published, Indian treaties had been signed and settlement was proceeding far up the Missouri River into the Boonslick region, well beyond the mouth of the Gasconade.

The final attempt to prescribe limits to settlement was through the rec-

108. "A Proclamation, [n.d.]," *Bates Papers*, 1:337–40. Lewis's line north of the Missouri River was the boundary set by the Sac and Fox Treaty and thus constituted the "Indian line," or limit of white settlement, in 1804. One of the first cartographic depictions of the Sac-Fox boundary in Missouri is on the Lewis and Clark Map of 1814 (Moulton, *Journals of Lewis and Clark Expedition*, map 126). It is shown as a north-south line, which may have been intended, from the mouth of the Gasconade to a point on the first river (the "Jeffreon") on the map north of the Salt River. Later maps show it as a more northeasterly line.

109. Stoddard, *Sketches of Louisiana*, 263–64.

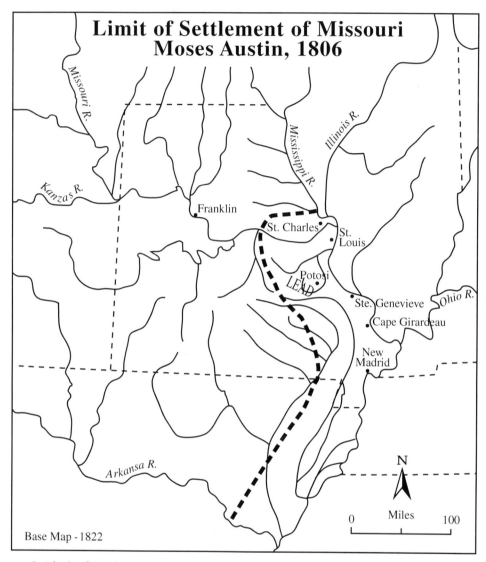

9. *Limit of Settlement of Missouri, Moses Austin, 1806.* Moses Austin proposed to limit settlement at a line approximately sixty miles west of the Mississippi River.

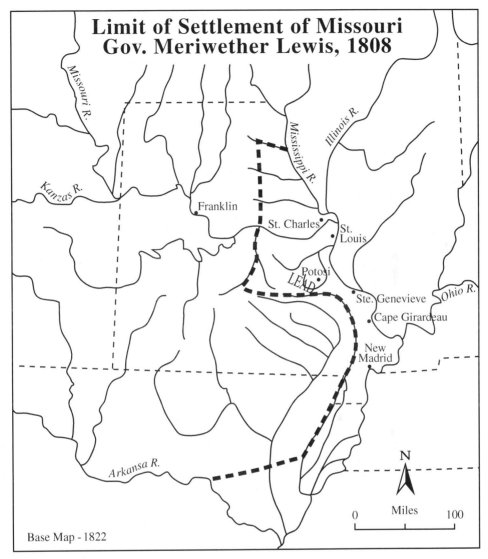

10. *Limit of Settlement of Missouri, Gov. Meriwether Lewis, 1808.* Governor Lewis issued a proclamation that limited settlement to already occupied lands and up to the Sac-Fox Treaty line.

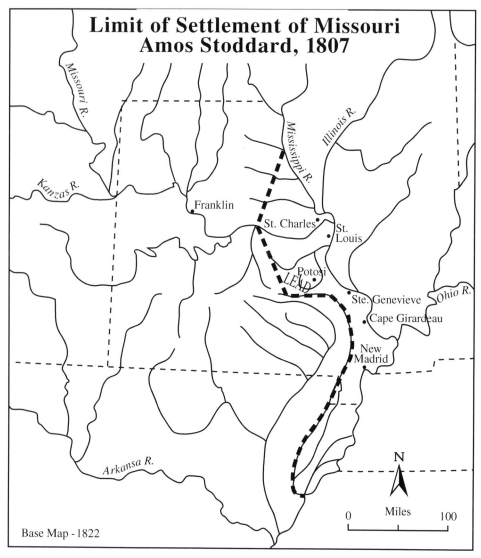

11. *Limit of Settlement of Missouri, Amos Stoddard, 1807.* Captain Stoddard recommended settlement be limited to the lands already occupied by white settlers. His recommendation was not published until 1812.

tangular land survey by the General Land Office. Because surveyed lands were legally open to settlement and purchase and unsurveyed lands were closed, the forward line of surveyed land constituted a definition of a limit to settlement at any given time, up to the extension of preemption rights to the territory of Missouri in 1814, which implicitly allowed settlers to move onto unsurveyed lands.

William Russell, who had been a deputy surveyor under Soulard and then a land speculator, described in 1814, before the government survey began, which lands in the territory were already well occupied and which were under pressure for new settlement. He proposed that the government immediately survey those lands. The outer line of this proposed contiguous tract, which would mark the beginning of the land survey in Missouri, constituted a limit to the lands that settlers could legally occupy (fig. 12). Russell's outer line of settlement was a north-south line, a meridian, from the Missouri River, "70 to 80 miles" west of the Mississippi. The lands south of the Ohio River junction, however, did not merit immediate attention, as they were depopulating due to the recent great earthquakes. On the north side of the Missouri River, the line should be "at least 120 miles" west of the Mississippi and include the Boonslick region.[110]

Between 1805 and 1815, five different lines had been proposed to limit settlement west of the Mississippi River. Whatever form the proposed lines took and for whatever reasons, the effect was the same. Administrative injunctions to contain settlers were futile. Administrators perceived land and people as commodities to be manipulated to achieve various administrative goals, but settlers would not be so manipulated. Settlers in the territory pushed inexorably forward, unrestrained by pronouncements to limit the geographic extent of their settlement.[111]

Discordance between the actual process of settlers occupying the land and government attempts to direct them continued in the Ste. Genevieve

110. William Russell to William Rector, April 20, 1814, *TP* 14:757–58. It is impossible to draw the line between the Meramec and Gasconade Rivers, even on maps then in existence. Perhaps Russell meant a line due *north* to the mouth of the Gasconade or alternatively meant *down* the Meramec to a line, which, "drawn due west . . . ," although this latter is an impossibility on current maps. Even though Russell knew the geography of the territory quite well, this cartographic exercise illustrates how lines conceived to exist in minds sometimes could not exist in reality.

111. At the conclusion of his comprehensive history of American land policies, public-land historian Benjamin Hibbard poses the question, "Might settlement have been controlled?" He answers negatively, except that after the Homestead Act had achieved its purpose, the federal government could have better controlled settlement into remaining semiarid regions, wetlands, and timberlands (*Public Land Policies,* 554–56).

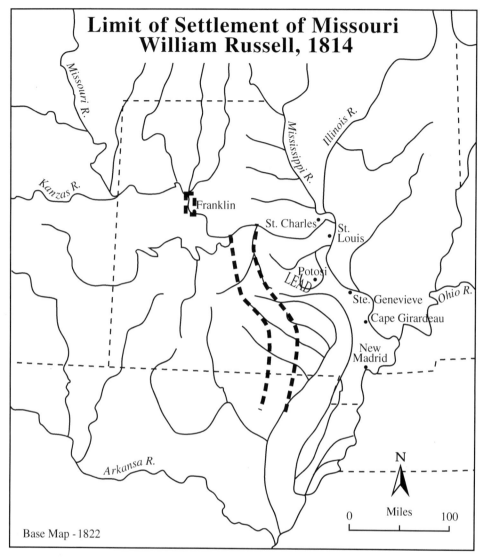

12. *Limit of Settlement of Missouri, William Russell, 1814.* Surveyor William Russell recommended settlement be limited to a zone along the Mississippi River not more than seventy or eighty miles wide, but including the Boonslick, which was then being invaded by white settlers.

District from 1804 to 1818. The change from a loose land-management approach by the Spanish to a more structured American approach replete with laws, policies, and functionaries was confusing enough. But the numerous American laws and their interpretations kept changing, and the long-awaited survey and sale of public lands, which would have brought more order, were repeatedly postponed. While administrators in Washington and St. Louis wrestled with policy and specific provisions of the many land acts, settlers kept coming. Though forbidden by law until 1814 from occupying public domain, they did. Though forbidden by law from mining lead without a lease, they did. Though forbidden by law from occupying Indian land, they did. And though public leaders iterated that a western limit for settlement was desirable and necessary, no line had any effect. American settlers simply went where they saw opportunities.

6

American Period, 1818–1830

The year 1818 defines a sharp break in the settlement of the Ste. Genevieve District and of Missouri as a whole. The single act of putting public land on sale in August 1818, anticipated for well over a decade, created a new state of mind among settlers and land speculators, almost as strong as the change from a Spanish to an American administration in 1804. At last, families could obtain title to lands they occupied, buy and sell land with more confidence, and speculate in lands with less risk. Some unconfirmed and litigated private claims from the Spanish period continued to fester, but they were now individual legal battles, and their effect on incoming settlers was mooted by the lands being incorporated into the American rectangular survey and put on sale as part of the public domain. And there remained disposition of leased lead lands and final resolution of the Indian tract, two issues that muddled land titles and discouraged settlement.

U.S. Rectangular Survey and Land Sales

Hopes had been high in the first years of the American regime that land confirmation would proceed rapidly and that the American rectangular land survey would shortly commence in the territory. The two processes were mutually dependent. On one hand, the land survey could not be run until valid pre-American private lands were located on the land and subtracted from the public domain. On the other hand, to survey confirmed private lands required that they be tied into a comprehensive survey system—otherwise, each private tract would float in space as an unconnected plat. Surveyors of private lands complained repeatedly to Congress of the necessity for a comprehensive grid. Congress responded to this dilemma by putting the land survey, which had been authorized for Missouri as early as June 13, 1812, on temporary hold until more, then still more, of the outstanding claims could be resolved.[1]

When the board reported in 1812, it offered opinions on such a small percentage of the cases that it was questionable whether the U.S. survey could soon begin. Even Bates's subsequent report of 1814, which more than doubled the number of confirmed private claims, left numerous claims unresolved. Nevertheless, before he left office in 1813, Silas Bent, U.S. deputy surveyor in St. Louis, urged the survey of the base line and principal meridian specifically in order to tie in the private lands being surveyed. Then again, in the face of dragging problems of confirmation, William Rector, who replaced Bent, also urged that Congress begin the public land survey.[2] Finally, the secretary of the treasury, Albert Gallatin, trusting Bates's judgment that virtually all of the claims of merit had been confirmed, concluded that the land-confirmation process had gone as far as it could with normal procedures, and upon Gallatin's recommendation, Congress in 1815 ordered the public land survey to begin in the territory of Missouri.[3] The survey was to include all privately claimed land for which title had not been confirmed, even if the claims were in litigation in the courts, with the important exclusion of reserved mineral land claims. This meant that thousands of acres of land, much of it claimed by prominent persons, would be surveyed and prepared for sale as part of the public domain. These claims were to be included in the survey not only because the board and Bates had not accepted their validity, but also because their exclusion would have revived hopes of the claimants.[4]

Surveying began with William Rector's running of the Fifth Principal Meridian, which he began at the junction of the Arkansas and Mississippi Rivers on October 27, 1815, and proceeded northward, reaching the

---

1. *Stats. at Large of USA* 2 (1812): 748–52. It is reasonable to hypothesize that the rectangular land survey may also have been delayed because it would have cost thousands of dollars in expenses paid up front (to be recouped later as land was sold), and the federal government was already in debt and entering a war with Britain. To continue the land-confirmation process, on the other hand, required not much more in expenses than the salaries of a few officials, expected for one or two years at the most.

2. Silas Bent to Josiah Meigs, June 22, 1813, *TP* 14:683; C. A. White, *Rectangular Survey System*, 61; William Rector to Josiah Meigs, July 24, 1813, *TP* 14:688.

3. Edward Tiffin to Josiah Meigs, August 29, 1815, *TP* 15:79; Josiah Meigs to William Rector, September 27, 1815, *TP* 15:83. The 1815 order was for township and range lines only in order to provide a grid skeleton in which to place the numerous private claims being surveyed. A later order in 1816 directed the survey of the interiors of townships to begin in preparation for land sales (Josiah Meigs to Edward Tiffin, March 6, 1816, *TP* 15:124).

4. Josiah Meigs to the Secretary of the Treasury, May 11, 1818, *TP* 15:391.

Missouri River on December 28, 1815. The baseline for Missouri Territory was a true east-west line, a parallel that passed through the mouth of the St. Francis River. Once the coordinate system was in place, surveyors ran the township and range lines and then subdivided the townships into sections of one square mile each. Township subdivision, the most expensive part of the survey because it involved many more miles and points, was not done for an area unless it appeared that settlers wanted to buy land there. William Rector exerted great influence in the selection of which lands to subdivide, and the survey's irregular progress across the landscape in its early years reflects Rector's understanding of where settlers already were and where pressure for land was greatest.

Conducted out of St. Louis, the survey went extraordinarily fast in Missouri. By 1818, several hundred townships containing more than ten thousand square miles were surveyed, including all of the historic Ste. Genevieve District. By 1819, the survey had reached the Kansas River on the west and the Des Moines River to the north. After these initial few years during which surveyors caught up with lands already settled and awaiting preemption claims, no one could claim that the survey held up settlement in Missouri. Although some backwoodsmen were always in advance of the survey, in general land was surveyed before it was entered by settlers. William Rector's influence in getting contracts for himself, his five brothers, and others allegedly was behind the swiftness of the Missouri survey in its first decade. Surveying contracts were profitable enterprises.[5]

However, the public survey in the Ste. Genevieve District encountered problems. Slowness in surveying the numerous confirmed private lands held up the public survey in many local areas, especially in and around the villages where confirmed and public lands intermixed in intricate mosaics. The act of 1824 for ascertaining landownership in the villages (which added Mine à Breton to the original list of eleven villages in the territory) allowed the survey of the three French villages of the district

---

5. Josiah Meigs to the Secretary of the Treasury, November 5, 1818, *TP* 15:456–57, including map between pp. 456 and 457; Josiah Meigs to the Secretary of the Treasury, August 1, 1821, *TP* 15:741. Rector was just too successful. His surveys were good and on time, but he used too much of the land office's budget. He was responsible for 98,299 miles of survey between 1818 and 1822 at a cost of more than $450,000. Despite investigation into his many surveying contracts, nothing irregular ever turned up. However, President Monroe reproved him for nepotism (Malcolm J. Rohrbough, *The Land Office Business: The Settlement and Administration of American Public Lands, 1789–1837*, 168, 187–91). Also see comments on Rector in James W. Goodrich and Lynn Wolf Gentzler, eds., "'I Well Remember': David Holmes Conrad's Recollections of St. Louis, 1819–1823," 21 n.

and their common fields promptly in 1824 and 1825.⁶ These were the first detailed maps of the long-existing villages.

By 1818, enough land had been surveyed in Missouri, appointments made and other essentials in place so that President Monroe declared that land would go on sale at St. Louis the first Monday of August 1818, seven years after the St. Louis land office was authorized.⁷ Not all the surveyed land within a land district went on sale at the same time. The land office announced that sales for specific congressional townships or groups of townships would begin on a designated date and continue for a specific time thereafter (fig. 13).⁸ Staggering the dates for initial sale by township spread out the work at the land office. But in the Ste. Genevieve District, different initial sale dates were necessary because so many local areas continued to have unsurveyed confirmed private properties. For example, townships in the River Aux Vases, Saline Creek, the Barrens, and Bois Brule neighborhoods did not go on sale until February 15, 1824, although surveyed six years earlier. In contrast, townships in the rugged hills west of Potosi, where there were no unsurveyed private properties, went on sale on December 25, 1820, although they attracted much less interest for purchase. Townships in the Brazeau settlement and at St. Michael's–Fredericktown, incredibly, did not go on sale until 1838, a full twenty years after completion of the public survey.

6. *Stats. at Large of USA* 4 (1824): 65–66. Often the same surveyors held contracts for both the private and the public land surveys. For example, Joseph C. Brown surveyed the village of Ste. Genevieve and its common field, held several public land survey contracts, and surveyed the Osage Cession boundary (1816), Missouri-Indian Territory (Kansas) and Missouri-Arkansas Territory boundaries (1823–1824), and the Missouri-Iowa boundary (1837).

7. *ASP-PL* 3:439; *Missouri Gazette,* June 5, 1818; Gary W. Beahan, "Missouri's Public Domain: United States Land Sales, 1818–1922," 16. The St. Louis land office had been authorized by act of Congress seven years earlier on March 3, 1811 (*Stats. at Large of USA* 2 (1811): 662–66). It had opened in June 1816 specifically for preemptors to file claims ("Notice of Opening of St. Louis Register's Office, June 1816," *TP* 15:149–50).

8. Dates when townships would begin to be sold were announced in local newspapers that held contracts with the federal government to print such information. It is important to keep separate the date of authorization of a land office, the date that it was actually opened, and the date when sales actually began for any given township, which may be different by many years. Dates when townships went on sale are in U.S. House, "Letter from the Secretary of the Interior, transmitting A report showing the time the public lands may have been in market . . . ," 34th Cong., 1st sess., H. Doc. 13, 5 February 1856, 90–137, which is the source for the map of Figure 13. The best general reference for information on land office business is Donaldson, *Public Domain.*

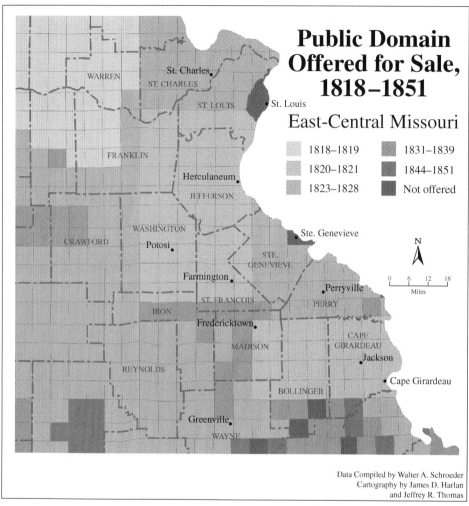

13. *Public Domain Offered for Sale, 1818–1851.* In many parts of the Ste. Genevieve District, land did not go on sale until many years—even decades—after sale was authorized and land was settled. Difficulties with the confirmation of private land and the reservation of lead-bearing lands from sale, as well as surveying problems, caused the delay. (34th Cong., 1st sess., H. Doc. 13, 5 February 1856, 90–137)

Sale was by public auction.[9] The register at the land office would call out a township, then proceed through it from section one to section thirty-six. A person would offer a bid as his section was announced. If more than one person spoke up, bidding proceeded and the highest bid was accepted. For the years after 1820, a minimum price of $1.25 per acre was set by law for all land, regardless of quality, which made $100 the minimum price for a purchase of eighty acres. In places of highly desired land, such as the Boonslick, bidding by land speculators drove the price up to $5 and $10 per acre.[10] If no one bid on a tract, it remained unsold and continued in the public domain.

Accessibility and soil quality were the chief factors that set prices for unimproved land in the Ste. Genevieve District. Mineral land, the most valuable land, was reserved from sale. The discrepancy in land prices, which already existed on the land market before the public domain went on sale, sorted out settlers according to wealth. Those with more cash could bid up the price of good-soil, accessible land until they acquired it, whereas cash-short settlers had to be satisfied with land for which no one else would bid against them. Thus, poorer people tended to get the poorer or less accessible lands and wealthier people the better or more accessible lands. This sorting-out process applied more to regions already partially settled, such as the Ste. Genevieve District, where differentials in land values had been established earlier through the market process, than to regions where the sale of land occurred concurrently with its initial settlement. The process also applied more to regions with significant differences in natural land quality, such as the Ste. Genevieve District, than to regions of more homogeneous land quality.

9. Rohrbough, *Land Office Business*, 75–78. The alienation of the public domain, or the conversion of it to private property, is a major topic in the settlement history of the trans-Appalachian United States. Among the numerous studies are James W. Oberly, *Sixty Million Acres: American Veterans and the Public Lands before the Civil War*, which has a useful bibliography on the topic; Donaldson, *Public Domain;* Hibbard, *Public Land Policies;* Treat, *National Land System;* Rohrbough, *Land Office Business;* Gates, *Public Land Law Development;* Paul W. Gates, *Fifty Million Acres;* Gates, *Jeffersonian Dream;* and Vernon Carstensen, ed., *The Public Lands: Studies in the History of the Public Domain*. Pertinent statistics and information on public land sales in Missouri are in Shoemaker, *Missouri and Missourians*, 1:204–14; and Beahan, "Missouri's Public Domain."

10. Shoemaker, *Missouri and Missourians*, 1:207; Treat, *National Land System*, 395–411. It may be misleading to compare annual land sales among land offices because annual reporting methods varied. The amount paid by a purchaser was usually not credited to the year the land was sold, but to the year it was paid. Sometimes returns for two years were combined when an office reported sales. Furthermore, land sales may not be indicative of actual settlement, as many land entries were not proved up for a patent.

In 1818, payment was by credit with installments paid over a period of years. Problems with this arrangement were legion in the American West after the panic of 1819, and purchase by cash replaced credit in 1820.[11] Almost all the public land sold in the Ste. Genevieve District fell under the 1820 provisions.

Figure 14 shows all of the land entries from the public domain by township in the historic Ste. Genevieve District for the thirteen-year period 1818–1830.[12] The map indicates that people were much more interested in purchasing land in some areas than others. The mineral area was responsible for half of all land entries in the district, but the purchases lay outside the townships reserved as mineral land. Purchasers made most of these before 1825; after then, interest in the lead country waned. Two smaller areas of concentrated land purchases were the limestone basin of good soils around Farmington and the smaller basin around Fredericktown. Both of these new towns were county seats and probably attracted a modicum of land speculation. A fourth area was the karst plain of limestone soils of Perry County, including the Indian tract, which was officially opened to white settlement during this period. In contrast, conspicuous for its sparsity of purchases was the large belt of rough hills. Although settlement was inching deeper into the Ozarks in the 1820s, land entries per township diminished rapidly to the west and south of the mineral area, because land suitable for farming was available only in widely scattered creek valleys, and many cash-poor settlers shunted to these farthest reaches opted for squatting instead of spending $100 for the few uncontested acres that they would put into crops and build their cabins on.

When land went on sale in Missouri Territory, government agents expected preemptors, speculators, and new settlers to be waiting in great numbers to file their entries at the land offices. This was certainly the case at the Franklin and Palmyra offices in central and northeastern Missouri, where initial land sales almost overwhelmed the land registers, but filers did not show up in such numbers at the Jackson and St. Louis offices for land in the Ste. Genevieve District. The Jackson land office reg-

---

11. Beahan, "Missouri's Public Domain," 8–9. "Land entry" was the act of purchasing or otherwise taking out the land, as by military bounty warrant. After all terms of the sale had been met, the government issued a "land patent," a certificate certifying landownership. The patent may have been issued a few years after the land entry. Some who entered land did not receive patents. For an analysis of local land-settlement history using land-entry data, see Walter A. Schroeder, "Spread of Settlement in Howard County, Missouri, 1810–1859."

12. Abstracts of U.S. Land Sales, vols. 1–2, in the Missouri State Archives, Jefferson City, for St. Louis and Jackson land offices.

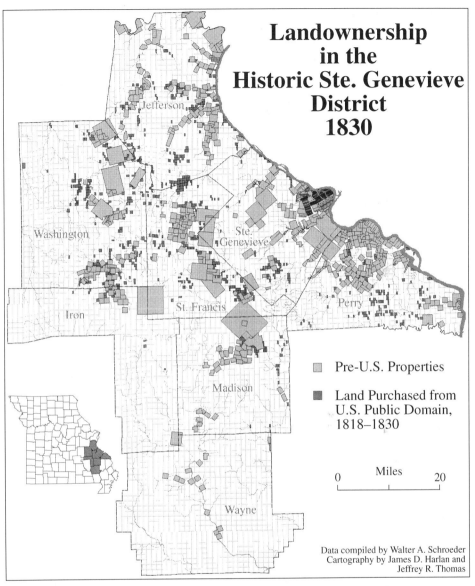

14. *Landownership in the Historic Ste. Genevieve District, 1830.* The map shows all land in private ownership by 1830 divided into two groups: land claimed by 1804 and land purchased from the U.S. public domain 1818–1830. Although many more settlers arrived in the district between 1804 and 1830 than before 1804, substantially less land went into private ownership during the American period than during the Spanish period. The Maxwell grant on the Black River is not shown.

ister, anticipating a rush, had advertised in May 1821 that preemptors should come forward right away to present evidence for their claims before the land went on the market. The disappointing number of preemption filings at his office reflects, in large part, the relatively slow sale of lands in general in the southern half of the Ste. Genevieve District. The register at Jackson reported in 1828 that only 67,199 acres had been sold to anyone, whether preemptors or not, out of a total 4,818,725 acres surveyed and proclaimed for sale in his land district.[13]

Several conditions account for the few land sales in the district. First, new settlers could acquire land by purchasing parts of pre-American concessions and settlement rights that were far larger than necessary for a single family. In compact communities of large families and strong kinship bonds such as the Barrens, the Bellevue, and Murphy's, demand for land was met by subdividing large concessions of eight hundred and a thousand arpents. Private lands covered much of these communities and left little public land to buy. Second, in those townships where sales were delayed for years, preemptors lost their concern over landownership with the passage of time. The slowing up of new arrivals and decreasing competition for land allayed preemptors' fears that someone would seize their improvements. The abject subsistence level of other arriving settlers, aggravated by the economic depression from 1818 to 1822, when prices rose dramatically, may have deterred them from paying even the minimum price for land.[14] If no one else showed interest in their land, and if laws against squatting were not enforced, why bother to file for ownership and pay the fees? In still other cases, those who had come into the district for lead mining found those lands reserved from sale until the end of the 1820s. By then, the boom was off lead mining in the district and both speculators and miners were withdrawing. Some miners had little motive to become resident farmers, and they likely moved on. Others may have been prejudiced against the reputed unhealthfulness of mineral lands for farms.

---

13. The historic Ste. Genevieve District was split between the St. Louis and Jackson land-office districts. The boundary was the north line of Township 37 North, which ran from Ste. Genevieve village westward through Potosi. *Independent Patriot,* May 26, 1821; *ASP-PL* 5:146. Using total acreage figures to represent volume of sales may give a misleading impression of the total number of land sales because tract size varied greatly in this region of preexisting properties. The numerous private tracts at odd angles with the public survey left hundreds of fractional quarters and half-quarters constituting extremely small acreages that adjoining property owners purchased. Thus, a significant number of entries were for tracts of fewer than twenty acres, and they had irregular shapes.

14. Hattie M. Anderson, "Frontier Economic Problems in Missouri, 1815–1828."

When better economic conditions returned in 1824 and continued for the rest of the decade, economic growth in the Ste. Genevieve District was weaker than in most other parts of the state. The great regional magnets of the 1820s, such as the Boonslick and Salt River country, attracted migrants from Kentucky, Virginia, and other eastern states, but also migrants from within Missouri, and this included a share of the residents of the older settled Ste. Genevieve District. For all these reasons, land sales from 1818 through 1830 were much slower than anyone would have predicted when land was surveyed.

The public land survey brought substantial benefits to the settlement process. Above all, it brought spatial order by providing a complete survey of all land. The system was so simple that anyone could understand its essentials: boundaries were straight lines, ran in cardinal directions, and intersected at right angles. All properties were marked on the ground in squares and rectangles, save fractional sections as those along rivers and those abutting pre-American lands. It was simple to lay off equal-size tracts for ownership.

The survey also brought documentary order. Legal descriptions of properties were simplified, and plats were easy to construct. Surveys and descriptions filed in government offices provided protection even if corners and lines in the landscape became obliterated or forgotten by owners. Subdivision of properties and property transfers were simplified. "Stakery" was eliminated, as the survey lines and points were immutable; surveyors could reconstruct them from survey field notes. Because the survey produced so little controversy over where properties were, landowners had confidence that their titles were secure.

One other benefit was psychological. The survey represented the forward thrust of civilization and manifestly displayed a society's and a government's control of the settlement process by casting a net of grid lines over the landscape. In many areas, the survey marked the first tangible evidence of transforming the wilderness into ordered space, part of the dream of settlers entering new land. It said, "We are here to stay."

Lead Lands

A serious problem with the federal policy of leasing mineral lands that made it difficult to administer was the determination of exactly what land should be reserved. After the land survey was run in the Ste. Genevieve District, systematically gathered information for the first time permitted a more precise identification of lead-bearing land ten years after the federal policy was enacted. Surveyors noted the presence of minerals as they walked every foot of survey lines set one mile apart. They recorded in their field notes the presence of "iron ore, tiff, and blossom," "sine [sign] of mineral," and "much mineral blossom." For example, on the east line of section 1, T38N, R1E, the surveyor recorded "mineral, tiff etc. passed through several places where holes have been made to get led ore."[15] In effect, surveyors were serving as prospectors for the public.

In 1824, the commissioner of the General Land Office in Washington, George Graham, compiled a list of townships recommended to be reserved from sale on the basis of the presence of lead recorded in the surveyors' field notes, which Congress approved (fig. 15).[16] The reserved tract included virtually all of the populated parts of Washington County. Not included in the published list were mines in the big bend of the Big River and at Mine la Motte, because they were on private claims and not part of the public land survey.[17]

The tract of reserved land totaled fourteen townships, or 322,560 acres (504 square miles), more than double the 150,000 acres reported in 1814 for reservation before the public land survey was run. In his written report to the secretary of war on March 30, 1824, Commissioner Graham identified the lead country generally as a land bounded by the Mississippi River, the headwaters of the St. Francis River, the Fourche à Courtois, and the Meramec River.[18] Within this general tract, Graham said that the land "more properly regarded as the mineral country of Missouri" embraced "only" 1,512 square miles (almost 1 million acres), which was still three times as large as the tract that Congress reserved from sale.

Because the land survey identified the presence of lead only on section lines and not in the unsurveyed interiors of square-mile sections, the General Land Office continued to add more land to the reserved list as information became available about section interiors. By 1826, the total had risen to 400,000 acres of land for which there was "no reasonable doubt of its containing lead mineral."[19] In 1828, the registers for the St. Louis and Jackson land districts reported a total of 613,308.46 acres (958

15. U.S. GLO Field Notes, Missouri, 197:61, 148, 225.
16. *ASP-PL* 3:575–76. In the 1820s, the first state maps of Missouri based on the rectangular land survey—which was to remain the basis of all detailed maps of the state until the end of the century—were constructed, and many showed the location of lead mines as identified from surveyors' field notes. An example is the Map of the States of Missouri and Illinois and Territory of Arkansas, by E. Browne and E. Barcroft.
17. An independent reconstruction of lead lands from the same field notes produces a different distribution from Graham's. This raises the question of how objective Graham was and whether he was influenced by others in identifying which lands were to be reserved. See the comparison of the two maps in W. A. Schroeder, "Opening the Ozarks," 224–25, 636–37.
18. *ASP-PL* 4:377; *ASP-PL* 3:578.
19. Abramoske, "Federal Leasing System," 35–36; *ASP-PL* 4:373, 550.

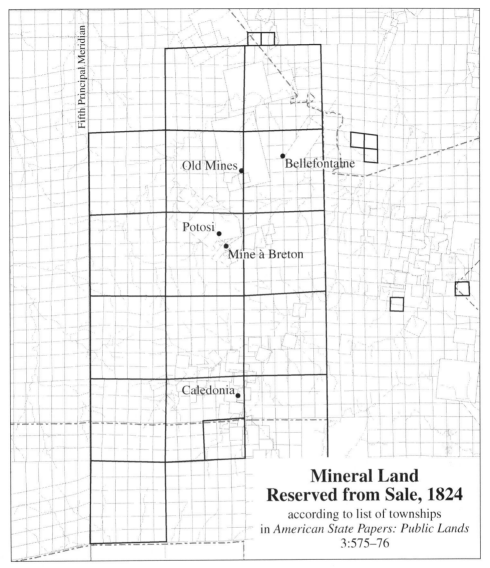

15. *Mineral Land Reserved from Sale, 1824.* The U.S. government defined lead-bearing land reserved from sale on the basis of sections and townships of the U.S. rectangular land survey. (*ASP-PL* 3:575–76)

square miles) of land "withdrawn from market, under belief that it contained mineral." Both Missouri senators, David Barton and Thomas Hart Benton, used an inflated figure of 660,000 acres in speeches before the Senate, and Benton suggested that even more of the state lay in the great lead belt that was believed to stretch from Arkansas to Lake Michigan.[20]

Designation of nearly 1,000 square miles of land as lead bearing and subject to reservation from sale and the associated delayed confirmation of private lead lands vexed residents of the Ste. Genevieve District to no end. People were unsure what were officially designated lead lands or where they were. Should speculators invest? Could settlers buy land with confidence of title? Because of the confused situation, not a single private land claim, not even Moses Austin's perfected Spanish grant, was confirmed for at least twenty years. Missouri's senators exaggerated the amount of mineral land in order to convince Congress of the harm the act was doing to their state: because even more lead-bearing lands had been discovered in central and southwestern Missouri, would not Congress end up reserving most of the state from sale?[21] Although the senators spread their arguments against leasing to include virtually the whole state, clearly harm to settlement was concentrated in the already populated Ste. Genevieve District, where most of the active lead mining was located.

Fortunately, problems with the ownership and leasing of mineral lands did not prevent most mines from operating, although those problems may have kept lead production from rising even faster than it did. Production in the district rose from 425 tons in 1804, to 762 tons in 1811, and to an average 2,000 tons per year between 1818 and 1824, although the rise was certainly not continuous over that score of years.[22] One hundred fifty men worked at the district mines in 1804, and 350 in 1811. In 1819, an estimated 1,130 were employed, including miners, smelters, teamsters, blacksmiths, and slaves, rising to 2,000 in 1824–1825.[23] All of these figures include part-time workers, those who had to take care of their farms and other activities during part of the year. The increase in this population cohort came generally from individual actions as opposed to large-scale, organized recruiting and employment. For example, one man employing two other men had cleared four hundred dollars in

20. *ASP-PL* 5:170–71; *Congressional Debates,* 19th Cong., 2d sess., Senate, 10 January 1827, 52; *Missouri Gazette,* May 24, 1817.
21. *ASP-PL* 4:596.
22. *ASP-PL* 1:191; Brackenridge, *Views of Louisiana,* 154; *ASP-PL* 4:376, 550.
23. *ASP-PL* 1:191; Brackenridge, *Views of Louisiana,* 154; Schoolcraft, *View of the Lead Mines,* 126–28; *ASP-PL* 3:579–80, 4:377, 550.

six weeks in 1821. It was reported that people, "seduced by this prospect of extraordinary profit," deserted farms in Kentucky and settled on the public domain in the mining country, using their profits to buy farmland.[24]

Exhaustion of one mine (numerous "diggings," or pits, in close proximity composed one mine) was compensated by opening of others with or without federal lease. The number of producing mines in the Ste. Genevieve District increased from ten in 1804, to thirteen in 1811, thirty-three in 1816, forty-five in 1819, and fifty in the early 1820s.[25] New finds in the district occurred in the Big River neighborhood; farther west in the Courtois drainage; at Richwoods northwest of Old Mines; and in the drainages of the Joachim, Plattin, and Sandy Creeks of Jefferson County. Some older mines—such as Mine à Breton, Mine la Motte, Old Mines, and Shibboleth—kept producing, although their share of the total production declined considerably.

Of course, factors other than repercussions from the mineral-land reservation and leasing program influenced lead mining in the district. For example, production increased from the rise in prices after the cession of 1804 and again after the War of 1812 and from a rise in the import duty on lead in 1824. Production decreased during the blockage of American ports for export during the War of 1812 and the economic depression of the early 1820s, but the greatest influence was competition from increasing production from the upper-Mississippi lead mines in northwestern Illinois and southwestern Wisconsin in the 1820s.[26]

Although production was expanding and new mines were opening, the number of leases declined. By 1821, the leasing program was virtually extinct, and by 1822, all mine leases had expired.[27] Determined to retain the program, Congress on November 29, 1821, transferred responsibility for managing the lead mines from the General Land Office in the Treasury Department to the War Department. Lt. Martin Thomas of the United States Army, appointed agent for granting leases on public mineral lands on August 18, 1824, was instructed to report new mineral lands to the land office, which would then have to reserve them from sale.[28] Thomas's several reports during the next five years provide a per-

24. George F. Strother to Josiah Meigs, April 11, 1821, *TP* 15:720.
25. *ASP-PL* 1:188–91, 3:576–77, 610; to John Smith T, Cannon Mines, April 29, 1811, *Bates Papers*, 2:173–74; Flint, *Condensed Geography*, 2:85–86.
26. Winslow, *Lead and Zinc Deposits*, 6:277; Gracy, *Moses Austin: His Life*, 123–24, 139–40, 145–46; *ASP-PL* 4:379, 5:5, 76–77.
27. *ASP-PL* 3:564. The list of all lead leases from 1807 to 1824 is in *ASP-PL* 4:370, 373–74.
28. *ASP-PL* 4:370–71, 373.

spective from a person reasonably detached from the various political and economic schemings in the territory. He set about to make the leasing system work with the drive of a military officer. He moved his office from St. Louis to Potosi to be in the center of action and better supervise the reserved lead lands. In order to increase mining and government revenue from leases, he recommended that the government encourage production by building a road from Potosi to the Mississippi River, which he said could be paid for in two years from mine-lease revenue.[29] He argued that the federal government as landowner should build road access to its own property for its development. The people in the mining district, he pointed out, were too poor to engage in such an undertaking and, in any respect, would not build a road to benefit land they did not own.

Thomas also expanded the amount of land reserved from sale, but some of the new land he reserved had been sold after the public domain went on sale. On those sold lands for which the government had already issued patents, nothing could be done. But on those lands already sold but for which the purchasers were still awaiting their land patents to arrive from Washington, Congress intervened and withheld the patents solely on the basis of Thomas's recommendations. The federal government found itself in the awkward position of having to return money to land purchasers and put their lands back into the category of public domain reserved from sale. In his first year alone, ending in September 1825, Thomas issued thirty-four new leases.[30] Clearly, the lieutenant was attending to his charge earnestly and vigorously.

Despite Thomas's determined efforts, the problems with the system were too fundamental to be overcome. Illegal mining by trespassers on the public domain increased. Thomas reported lead production on public lands of approximately 2,100 tons in 1825, but the one-tenth royalty that the government collected in its storehouse at Potosi from leasing amounted to only 135 tons over a three-year period, or a rate of 45 tons per year.[31] Thus, nearly four-fifths of the lead that should have been moving through the leasing system was not.

Everyone in the territory knew the system was not working and scoffed at it. Land claimants powerful enough to work their mines independent of the government openly defied the leasing law, and hardscrabble, independent miners nonchalantly ignored it. The register of the Jackson

---

29. *ASP-PL* 4:377; *Congressional Debates,* 19th Cong., 1st sess., Senate, 6 April 1826, 2:408–9.

30. *ASP-PL* 4:555–58, 370, 374–79. Thomas's model lease is in *ASP-PL* 4:371–72.

31. *Congressional Debates,* 19th Cong., 1st sess., House, 20 December 1825, 2:829.

land office grumbled, "When a new discovery is made of a valuable appearance, a large proportion of the miners, consisting of several hundred, flock to it, and exhaust it in their manner of digging."[32] People stole rock for furnaces, stole timber, stole the ore itself, and bought ore stolen from others. Many individual miners, not troubling themselves for a lease, continued to use the long-standing custom of prescriptive mining from Spanish years.

Local law enforcement officials did not eject trespassers on claimed or government land, or, if they did, their success was only momentary and miners returned when the law left the vicinity. The law inflicted no penalties for illegal mining, and miners were left in possession of the best mines. "Speculative mischiefs" by trespassers, as Frederick Bates warned, "should be treated at origin; otherwise they acquire strength."[33] Bates was correct: intrusions on public land and flaunting of the law grew each year.

It was physically impossible to supervise the mines adequately because they occurred over a thousand square miles, and any one pit was usually worked ephemerally. Public land was interspersed with private land, some of it agricultural and confirmed; other tracts were claims to mineral land that the claimant continued to occupy and use as though he had confirmed title to it. This mingling of different kinds of landownership in a reasonably well-populated area caused Lieutenant Thomas, who incidentally was not a popular person in the mineral area, no end of anguish trying to figure out which lands were his to supervise.

Yet another difficulty with the leasing program was the short-term lease of a maximum of three years. Most were for only one year; lessees' efforts to increase the length were unsuccessful.[34] With no guarantee of renewal, three years was too short for entrepreneurs to invest the necessary capital to make significant improvements. To dig a shaft to get lead deeper than fifteen feet—everyone knew lead was there—required a longer time period to amortize costs. To build a Scotch hearth furnace to replace the grossly inefficient log furnaces took more money than anticipated from the profits from a three-year lease, especially as the richness of a discovery was undetermined when negotiating the lease. Thus, while the leasing system was in place, not a single shaft was dug to get to ore deeper than sixty feet (like Austin's back in Spanish days), and only one

32. As quoted by Thomas in Report of Lt. Martin Thomas to Col. George Bomford, January 1826, in "Lead Mines and Salt Springs," February 8, 1826, *ASP-PL* 4:375.
33. Ibid.
34. Jno. Rice Jones to Frederick Bates, November 20, 1816, in "Salt Springs and Lead and Copper Mines," March 30, 1824, *ASP-PL* 3:605.

Scotch hearth furnace was constructed in the entire mineral district.[35] It is worth noting that these two simple technological advances, which would have notably increased the yield of lead and the stability of mining, became common in the 1830s immediately after abandonment of the leasing system.

Local efforts to end the unwanted leasing system gathered momentum after Missouri statehood when Missourians' voices in Washington strengthened with two senators and a representative. Senator Thomas Hart Benton almost from the day of his arrival in Washington argued the case in each session of Congress.[36] Missouri's other senator, David Barton, was slower to acknowledge the leasing system as a failure, and his unqualified and vigorous support to end it did not come until somewhat later. The state's sole representative, John Scott, was from Ste. Genevieve County and had major interests in lead-land claims. Scott did not want the public land to go on sale until the private claims in the mineral district had been confirmed, lest his and his colleagues' pending claims also be sold as public land. Except for the contingent of those like Scott with as yet unconfirmed claims, the sentiment in Missouri was an extremely unyielding opposition to the leasing program and in favor of sale of the lands on the same basis as other lands of the public domain. Even Lieutenant Thomas, who tried so valiantly to administer federal regulations and be a good custodian of government property, eventually capitulated and recommended the law be repealed and the lands sold.[37]

Congress debated the lead-leasing system throughout the 1820s. The chief argument in Congress favoring repeal was economic: the mines were unprofitable for the federal government. Miners were paying their lease rent in the form of lead that was warehoused at Potosi without much prospect of being sold. The general outlook for the future of the mines of the mineral district was dim, because the lead mines on the Upper Mississippi were booming and miners were leaving Missouri for them.[38]

---

35. *Congressional Debates,* 19th Cong., 1st sess., House, 20 December 1825, 2:829.

36. March, *The History of Missouri,* 1:254. March has a summary of the politics behind the struggle to end the leasing system (250–57).

37. ASP-PL 4:596; March, *The History of Missouri,* 1:254–56; *Congressional Debates,* 19th Cong., 1st sess., Senate, 16 May 1826, 2:738–40, 747–48; John H. Weber for Amable Partenay, to Bates, Mine à Breton, September 28, 1814, *Bates Papers,* 2:278–79. John Scott, with others, knocked Amable Partenay off his horse and beat him when Partenay tried to claim one of the mines that he had leased from the government. Partenay said he could not get justice because of Scott's political prominence (March, *The History of Missouri,* 1:254–55). ASP-PL 5:346–49.

38. Congressional debate throughout the 1820s on leasing of Missouri lead lands, including the effect of import duties on lead and other issues of national scope, is

Opponents further charged that the federal government was creating a class of tenants, not the yeomanry of landowners that was the grand design for settlement of the West. These tenants, a poor and transient lot, did little to develop the country. They were, in Lieutenant Thomas's words, "the most worthless and abandoned part of the community, and are equally disposed to plunder the private as public mines." The entrepreneurs who hired the miners had no incentive to improve the mines, the smelting, and the transportation facilities because of the short leases, compounded by the uncertainty of titles on private land. Property improvements were of the barest necessities. Housing was not built to last; the small huts put up by miners were good for only a year or two. Leaseholders also had little incentive to help develop the neighborhood economically and socially, because their lack of investment in landownership engendered little concern for the future of the communities. The mines put money into the pockets of many people, but "not one dollar [was] invested in permanent works," complained Lieutenant Thomas in 1826.[39]

The numerous intruders on the unleased public mineral lands cut down trees flagrantly, just as the lessees did from the leased lands around the mines. With the wholesale cutting of timber, the mineral country of Missouri was being kept in a state of barrenness. When these lands would eventually go on sale, opponents argued, they would be worth little and bring in little revenue and perhaps remain unsold as "reject land."[40] Benton forcefully pitched his arguments toward lost revenue, knowing that eastern Congressmen regarded the public domain foremost as a revenue-producing federal asset.

The eastern Congressmen also pitched their arguments against sale of the lands in economic terms.[41] The premise for leasing rather than selling certain public lands in the first place was that those lands were in-

---

summarized in Mary C. Rabbitt, *Minerals, Lands, and Geology for the Common Defense and General Welfare,* 31–36; and Daniel Di Piazza, "A History of Federal Policy toward the Public Mineral Lands, 1785–1866," 20–45, 53–62. *Congressional Debates,* 20th Cong., 2d sess., Senate, 18 December 1828, 5:8–9; *ASP-PL* 4:378, 550–51.

39. Report of Lt. Martin Thomas to Col. George Bomford, January 1826, in "Lead Mines and Salt Springs," February 8, 1826, *ASP-PL* 4:375; on housing: *ASP-PL* 3:580; Report of Thomas to Bomford, January 1826, in "Lead Mines and Salt Springs," February 8, 1826, *ASP-PL* 4:376.

40. *Congressional Debates,* 20th Cong., 2d sess., Senate, 18 December 1828, 5:8–9. For a twentieth-century environmentalist's perspective of how early French and American lead-mining practices wasted resources, see Sellars, "Promotion and Development," 146–53.

41. *Congressional Debates,* 20th Cong., 2d sess., Senate, 18 December 1828, 5:8–9; Hibbard, *Public Land Policies,* 495.

herently more valuable than others and more revenue should be derived from them. Base metals occurred only in restricted places, and those lands could theoretically be identified. Therefore, so the argument went, the lead lands needed to be reserved to prevent monopolies and speculation and then sold later at a higher price when the land around them had become populated and economically developed. The eastern Congressmen who opposed sale while the mines were still operating were trying to protect a federal asset and get as much revenue from it as possible.

A bill providing for the sale of lead lands in Missouri at the same rate as other public lands at $1.25 an acre finally passed Congress on March 2, 1829, officially repealing the act of 1807, and Congress thereby abandoned the ill-begotten strategy to lease mineral lands in Missouri. Missouri lead lands first went on sale in October 1830, selling more cheaply than expected, and many mineral tracts even went unsold.[42] They had lost much of their value. Because a significant proportion of the lands in question was still in private claims dating from the Spanish period, those claims could now be adjudicated, a full quarter century after the cession. For example, the 1795 concession to Pierre-Charles Delassus de Luzières, one-time commandant of New Bourbon, of a square league of land at Mine à Gerbore (or Gaboury) on the headwaters of the St. Francis was finally confirmed in 1835 to his heir and son Charles, former lieutenant governor, through a United States Supreme Court decision written by Chief Justice John Marshall. It is noteworthy that the chief justice called attention to the fact that the Louisiana Purchase treaty guaranteed the new citizens protection of the free enjoyment of their property and were afforded all the rights of citizens of the United States. "The perfect inviolability and security of property is among these rights," Marshall wrote. This provision of the treaty, according to the justice, took precedence over

42. *Stats. at Large of USA* 4 (1829): 364; Donaldson, *Public Domain,* 307. The act required a substantial waiting period for their advertisement to be widely circulated in order to prevent immediate sale to local speculators. The sale of reserved salt springs in Missouri was approved by a separate act the next day, March 3, 1829 (*Stats. at Large of USA* 4 [1829]: 364). The federal mineral leasing program continued until 1846 in other states, despite its failure in Missouri (Robert W. Swenson, "Legal Aspects of Mineral Resources Exploitation," 702–8). The problems of the leasing system in Missouri were repeated later in the lead region of southwestern Wisconsin (Joseph Schafer, *The Wisconsin Lead Region,* 114–20), although with much less confusion because the leasing program was set in place before miners came flocking into the region (James E. Wright, *The Galena Lead District: Federal Policy and Practice, 1824–1847,* 1–33; Dick, *Lure of the Land,* 87). On sale of 1830: Rabbitt, *Minerals, Lands, and Geology,* 39.

the acts of 1807 and 1814.⁴³ In other words, withholding all valid Spanish concessions to mines from confirmation had been constitutionally illegal. Had that judicial opinion on "perfect inviolability" of property been issued while the first Board of Land Commissioners was in session, from 1807 to 1811, settlement history and economic development of the mineral district and the entire Ste. Genevieve District would have proceeded quite differently.

Indian Lands

The other lingering federal-land question that needed resolution was the fate of the Spanish governor's grant to the Shawnee and Delaware on the district's southern edge. Most of the Delaware had left by 1815, and by the early 1820s, most of the Shawnee were gone. Actually, pulses of Indians moved episodically in and out of the tract so that it functioned to some degree as a "pass-through" tract for westering Indians.⁴⁴

Negotiations for the official removal of the Indians from their Spanish land grant involved federal administrators, the Missouri delegation to Congress, and local authorities, both public and private. For example, territorial delegate John Scott had discussions with local Indian trader Pierre Menard, who carried propositions for the exchange of territory on behalf of the Indians. The Shawnee knew that the Americans coveted the good soils of their tract and that the land proposed for exchange west of the proposed western boundary of Missouri (now in Kansas) was worse "by a great difference." Negotiations included compensation for moving expenses, for opening new farms, and for their substantial improvements they would leave behind on Apple Creek. Even though Scott was quite familiar with the tract as he lived at nearby Ste. Genevieve, trader Menard had the confidence of the Shawnee to be their intermediary.⁴⁵

43. "Opinion, Supreme Court, in the Cases of C. D. Delassus, Aug. Chouteau, and Others," 24th Cong., 1st sess., House of Rep. Doc 148, 1 March 1836, 1–19. Earlier, May 4, 1828, the U.S. attorney general had issued an opinion withholding a certificate of confirmation and patent, stating that section 1 of the act of April 12, 1814, did not allow confirmation of land containing lead mines, but that the claimant was left to seek redress before Congress or the judiciary, which the claimant did (*General Public Acts of Congress* 2:2 [1838]: 173–74). See Donaldson, *Public Domain,* 367, for other court decisions on Missouri private land claims involving interpretations of "property" and "the free enjoyment" of it, relating to the Treaty of Cession of 1803.
44. The "pass-through" tract for Indians is similar to ethnic ghettoes in American cities for European immigrants of the late nineteenth century, wherein departing persons of a given ethnicity were replaced by arriving persons of the same ethnicity, thus preserving the cultural character of the neighborhood.
45. Delegate Scott to the Secretary of War, September 21, 1820, *TP* 15:645–46.

The formal treaty by which the Shawnee accepted their fate and ceded the land to the United States was not signed until November 7, 1825, well after the Indians had gone, but not until their negotiations had produced assurances from the United States of receiving another tract of similar worth and compensation for their property. Later, on September 24, 1829, the Delaware formally ceded their nation's separate claim to the same tract. In return, the Shawnee and Delaware on October 26, 1832, relinquished all claims against the United States for loss of property and improvements in the Spanish grant. Before the completion of any of these treaties, the United States had confirmed the land claims of Americans within the tract and had surveyed and begun disposal of its lands to Americans. Incoming Americans appropriated for their own gain the Indians' cleared land, cabins, and other tangible property quickly after the Indians left.[46] Thus, the American settlers gained considerably from the Indians' improvements, but they were not responsible for any of the compensation to the Indians. The property exchange worked out by the U.S. government with both Indian nations presumably stood in lieu of direct cash purchase of the Indian tract, which was the usual route taken by the federal government at this time when extinguishing Indian title.

Shawnee and other Indians also left their villages on the Meramec and the upper St. Francis Rivers so that by 1830 probably only a few dozen Indians and Métis remained as residents in the historic Ste. Genevieve District. The aggressiveness of American settlers in occupying land, abetted by a government quite willing to enter into treaties with Indians, secure their lands, and open them officially to whites, was the key to removal and expulsion of Indians from the region. It should be noted that the federal government employed the same settlement policy equally with both so-labeled uncivilized or savage Indians, such as the Osages, and civilized Indians, such as the Shawnee, who had adopted many of the ways of Americans.

Recapitulation

The process of settling new land was a complex interplay between the efforts of people who first assessed the environment for resource opportunities, then occupied and used land according to their practices, and the efforts of governments that sought to direct and supervise the process.

---

46. Kappler, *Indian Affairs*, 262–64, 304–5, 370–72; *Stats. at Large of USA* 7 (1832): 397–99. These several Indian treaties were complicated by the presence of several independent subgroups of Shawnee with their own leaders and living in various geographic locations as they moved west.

Governments attended to the initial allocation of land, which created one level of order to the settlement process, but individuals controlled the process of development of the allotted land, a different kind of order to the settlement process that operated at the family level.[47]

Government settlement policies differed from vernacular settlement practices in large part because they were based on the goals of a central government both geographically distant from the frontier and psychologically removed from the realities of frontier life. The Spanish goal to administer Upper Louisiana as a region to be defended against British and American intrusion, but not Indian intrusion, resulted in forts and immigration restricted to friends of the Spanish crown in distant Europe. Spain expected immigration and settlement to proceed along the model of European peasantry of farmers, in villages, all Catholic and submissive to the authority of the monarch's representatives. In return for their labors and allegiance, the peasantry would be protected against enemies of the crown, provided with priests and churches, and succored with public works. But the Spanish governmental intentions went unfulfilled or at best were inadequate. In the end, European Catholic peasants did not venture into Upper Louisiana, and for anyone who did, the specific regulations for allotting and occupying land were inappropriate for the physical and economic conditions in Upper Louisiana and thus largely ignored. The fact that poorly populated Upper Louisiana lay at the farthest end of the empire put the region near the bottom of priorities in Spanish administration of the Americas.

Meanwhile, French Creoles in the valley created their own settlement system according to their custom and the constraints of physical geography. The tripartite village and prescriptive rights to mining and other resource-extractive activities followed ancient European practices. But in 1795, change came. The Spanish government reversed its immigration policy and let in settlers of a different culture. These Americans brought a contrasting style of land occupation, one of dispersal, and Spanish authorities were neither able to stop the flow nor able to assimilate the new residents. Thus, the vernacular settlement system of Americans won out over Spanish governmental regulations.

The new American regime installed in 1804 introduced strikingly different laws and regulations for settlement. Religious and military concerns were not part of its land policies. The American land distribution system attempted to be fair to everyone (adult white males and their fam-

---

47. See French's comments on this interplay ("Cadastral Patterns in Louisiana," 181). Meinig, *Shaping of America,* 1:411–12.

ilies) and prevent land monopolization and speculation by government favorites or the wealthy. The Americans introduced an areal land-division system with an open, public sale that covered every square mile and discouraged villages and nucleated settlements. The Americans also introduced a policy that strictly separated their space from Indian space. In a more general way, the Americans introduced a new method of acquiring property through cleverness, opportunism, and purchase, which replaced the previous method of acquiring property through prestige, social and political connections, and simple request.

The new American settlement policies lay in abeyance for many years, while the unexpectedly complicated process of confirming private lands from the Spanish regime played out its necessary drama. Although the delay from 1804 to 1818 frustrated both the federal government and its local representatives who had to try to administer the land laws, the delay frustrated even more the few thousand persons migrating into the new territory. Neither could new settlers buy new land, nor could earlier settlers with much confidence sell private land not yet confirmed and surveyed. The degree to which the tribulations of the land-confirmation process affected settlement in terms of numbers of people—those who did not come as well as those who did—and in terms of where people went and how successful they were, remains an area of speculation, but the effect was profound. Undoubtedly, it was greatest in the neighborhood of villages, in mining areas, and on the extremely large concessions of wealthy persons who kept pressing their claims. Indeed, as time went on, Congress kept liberalizing the requirements for land confirmation until various acts of Congress confirmed virtually all claims still remaining, ensuring that the wealthy eventually got title to their lands, so greatly appreciated in value. In this respect, one could agree with Amos Stoddard's prophecy in 1804 that the new American government would be liberal to the citizens of Upper Louisiana in matters of land claims and also with Paul Gates's conclusion that those with money and political power, not Frederick Jackson Turner's yeoman farmer, developed the frontier and achieved ascendancy.[48]

Yet, during the long years of confirmation and until land sales were offered throughout the district, hundreds of families, unaffected by land-

---

48. "Capt. Stoddard's Address to the People of Upper Louisiana," in *Glimpses of the Past,* 2:91. A collection of some of Gates's major essays on public-land law carries out the theme of frontier development by those with money and political power (*Jeffersonian Dream*). One of these essays, "California's Embattled Settlers" (56–83), relates how settlers' problems with private claims in California were quite similar to the Missouri experience half a century earlier.

confirmation struggles, came into the Ste. Genevieve District and occupied land as closely as they could to the same manner that they had in Kentucky, Tennessee, and the Appalachian Carolinas and Virginia. Hardly any governmental land regulation influenced their choice of land and what they did with it. Settlers had a propensity to settle wherever they pleased and dare authorities to oust them, an attitude that helped create disrespect and defiance for law among the immigrants who were rapidly approaching a majority of the population.[49]

Beginning in 1818 and largely completed by the end of the 1820s, more order characterized the settlement process. The federal government adjusted its policies and regulations to meet the conditions and practices of residents of the new state. The settlers themselves became more sensitive to and respectful of "law." Indians left or were expelled from the district, and there was no longer an issue of whites intruding into their space. The public land survey had been run, and land went on sale. Squatters received expected preemption rights. Lingering private land claims from the Spanish period either were dropped or were being settled through court litigation and through a second Board of Land Commissioners tardily convened in the 1830s. The federal government repealed the unpopular lead- and saline-land leasing act, but not until mining was losing its luster in the district. Local developers busily platted new towns, and the district became organized spatially into counties built around early cores of settlement. The process of how to acquire land from the public domain was well known and well accepted; it proved highly workable. In the contest between the two divergent processes of occupying land—by government direction or by custom and practice of the people—neither had won out. They were converging into a single process.

---

49. March, *The History of Missouri*, 1:252–53; Dick, *Lure of the Land*, 17–18.

Part III
Building Communities

7

Settlement Forms

Visitors to the Ste. Genevieve District discover an attractive landscape of mingled architectural styles, sizes of properties, field patterns, spacing of houses, street patterns, and open spaces. Indeed, the visible landscape provides an enduring, tangible record of the district's settlement history.

At its most elemental, the settlement landscape may be reduced to plane geometry, the location of points in space and their designation as dispersed, random, or clustered, and analyzable through spatial statistics.[1] Because geometry only describes and does not explain landscape patterns, we need to invoke historical processes to understand how these patterns came to be. We soon learn that every neighborhood is unique and requires its separate historical treatment. Among the primary, well-recognized historical factors are the cultural practices of the immigrating settlers, physical geography, administrative regulations of governing authorities, and economic activities.

Settlement patterns relate to the identification of "community" and the strength of its bonds. Settlers in a highly dispersed neighborhood related to each other as a community, no less than residents of villages. Some of the dispersed agricultural settlements in the Ste. Genevieve District developed exceptionally cohesive communities. Thus, in the district, the social terms *community* and *neighborhood* are both quite appropriate for the various geographic localities, both clustered and dispersed.[2] In-

---

1. For a quantitative analysis of dispersed, random, and clustered patterns, see Peter J. Taylor, *Quantitative Methods in Geography,* 139–41; for an example of their application to settlement geography, see John C. Hudson, "A Location Theory for Rural Settlement." For an example of expressing the character of settlement patterns in mathematical terms, see Leslie J. King, "A Quantitative Expression of the Pattern of Urban Settlements in Selected Areas of the United States." In this interesting study, "urban settlement" refers to any named, populated place, including hamlets.

2. Carl O. Sauer emphasizes the role of community in dispersed settlements, as opposed to the popular notion that dispersion meant social isolation, in "Homestead

dians who lived in the district were migrants into it the same as others were, and the way they established their settlements may be analyzed in the same way as others.

Although each neighborhood was unique in history and geography, the people of the historic Ste. Genevieve District of the eastern Ozarks were living in seven basic settlement types in the early nineteenth century. Figure 16 displays models of each. Three were types of clustering, three were types of dispersal, and one combined both. Table 3 classifies each of the neighborhoods of the district into one of these seven types, Table 4 gives the estimated 1804 population of the neighborhoods, and Table 5 gives the population of villages and towns in the district from 1769 to 1830.

Planned French Village

Every planned French village in Upper Louisiana was a tripartite village consisting of three geographically separate components.[3] Carl Ekberg's comprehensive and carefully researched study of the tripartite village in the Illinois Country provides insight into the origin and functioning of this type of village. He emphasizes that the tripartite village in the Illinois Country was unique in North America because it required communal actions of the villagers to keep it functioning. Communal decisions both required and produced a *mentalité* of cooperation in these Illinois Country villages.[4]

---

and Community on the Middle Border." Faragher uses strength of community in dispersed settlements as a major theme in *Sugar Creek* and contrasts that theme with Frederick Jackson Turner's frontier individualism in "Americans, Mexicans, Métis." In his studies on colonial New England villages, Joseph S. Wood defines village more on the basis of a social and economic community and notes that the village settlement form may range from dispersed to nucleated and discredits the notion of social isolation in these dispersed villages ("Village and Community in Early Colonial New England," 341; and *The New England Village*). Within the village defined as community, women felt more geographically isolated than men who had more opportunities to travel and associate with others. See Ellen Churchill Semple, "The Anglo-Saxons of the Kentucky Mountains: A Study in Anthropo-geography"; and Adolf E. Schroeder and Carla Schulz-Geisberg, eds., *Hold Dear, As Always: Jette, a German Immigrant Life in Letters*.

3. An important distinction must be made between the common use of *village* for the built part of the community—the collection of structures—and the less common use of *village* for all the lands used by the community, including croplands, pastures, and even woodlands under the control of the village. Context makes clear which use is intended.

4. Among the studies of French villages in the middle Mississippi Valley that include the settlement form of the village, the most prominent are two by Ekberg, *French Roots* and *Colonial Ste. Genevieve*. Other village studies with varying attention

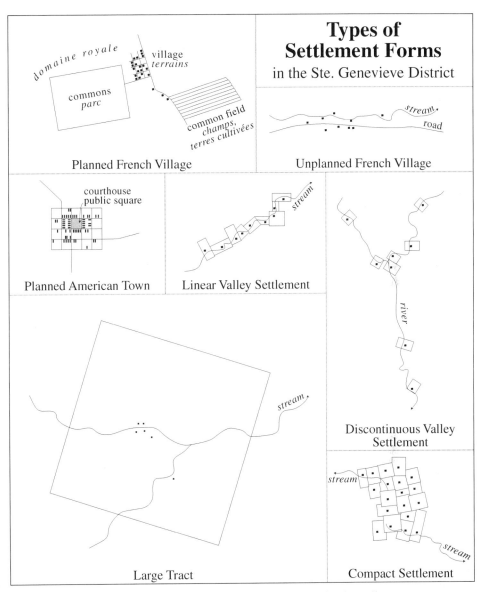

16. *Types of Settlement Forms in the Ste. Genevieve District.* All seven types are drawn to a common scale. The solid square or rectangle represents a residence.

Table 3

CLASSIFICATION OF SETTLEMENT TYPES BY NEIGHBORHOODS HISTORIC STE. GENEVIEVE DISTRICT, EARLY NINETEENTH CENTURY

	French planned villages	French unplanned villages	American town	Linear, continuous	Linear, discontinuous	Compact	Large grant
Old Ste. Genevieve	♦						
New Ste. Genevieve	♦						
New Bourbon	♦						
La Saline		♦					
Establishment River			♦				
River Aux Vases			♦				
Saline River			♦				
Mine la Motte							♦
St. Michel	♦						
Fredericktown			♦				
Owsley's Settlement						♦	
Mine à Breton		♦					
Austin Grant							♦
Potosi			♦				
Old Mines		♦					
Richwoods		♦					
Herculaneum			♦				
Big Bend of Big River						♦	
Murphy's Settlement						♦	
Cook's Settlement						♦	
Farmington			♦				
Plattin River			♦				
Joachim River			♦				
Big River (Jefferson Co.)			♦				
Courtois River				♦			
Isle du Bois Creek and Mississippi River ravines				♦			
St. Francis River (Wayne Co.)				♦			
Greenville			♦				
Bellevue						♦	
Caledonia			♦				
Maxwell Grant (Black River) Bois Brule							♦
Shawnee-Delaware Indian Tract			♦				
Barrens						♦	
Perryville			♦				
Abernathy Settlement			♦				
Brazeau				♦			

Table 4

POPULATION BY NEIGHBORHOODS
STE. GENEVIEVE DISTRICT, 1804
(Calculated from testimony before the Board of Land Commissioners and other documents.)

Neighborhoods[a]	Total Numbered Tracts[a]	Unoccupied Tracts[b]	Occupied Tracts	Estimated Number of People per Tract[c]	Estimated Population	Subtotals
Ste. Genevieve vicinity	44	39	5	5.9	30	
New Bourbon vicinity	68	62	6	5.9	35	
Establishment River	16	8	8	5.9	47	
River Aux Vases	23	12	11	5.9	65	
Saline River	27	15	12	4.92	59	
Bois Brule, Barrens	117	4	113	4.92	556	
Brazeau	20	0	20	4.4	88	
Mine la Motte and vicinity	33	8	25	6.00	150	
St. Francis River	21	1	20	6.00	120	
Bellevue	44	1	43	5.32	229	
Murphy's	11	2	9	5.9	53	
Big Bend of the Big River	38	7	31	5.9	183	
Mine à Breton vicinity	33	10	23	5.9	136	
Isle du Bois	3	1	2	5.9	12	
Plattin River	23	0	23	4.2	97	
Joachim River	32	4	28	5.9	165	
Big River (Jefferson County)	21	0	21	5.9	124	
Neighborhoods Total/Average	**574**	**174**	**400**	**5.354 (avg.)**	**2,149**	
Villages						
Ste. Genevieve[d]					958	
New Bourbon[e]					200	
Mine à Breton[f]					88	
Old Mines[g]					139	
St. Michel[h]					66	
La Saline[i]					10	
Villages Total					**1,461**	**3,610**
Indians[j]						
Ste. Genevieve village, not included in village population					40	
Delawares and Shawnees on Apple Creek[k]					1,500	
Indians Total					**1,540**	**1,540**
Total Population						**5,150**

a. Neighborhoods and tracts (land claims) are shown in Figure 5.
b. Unoccupied tracts are those for which there is no evidence of resident population. These include sugar camps, cattle ranches, and other tracts claimed by Ste. Genevieve, New Bourbon, and St. Louis residents.
c. See text for determination of this multiplier.
d. Estimated by taking 1796 census (773) and adding 3% per year, or total of 24%. Double checked by taking Stoddard's 1804 estimate of 180 houses and multiplying by 5.354, the average multiplier per tract for the entire district.
e. Estimated by combining the 1796 census (270) with Stoddard's 1804 estimate of 35 houses multiplied by 5.354, the average multiplier per tract for the entire district.
f. Estimated by multiplying 22 reported buildings by 4 persons. A smaller multiplier is used because not all buildings may be houses and the mining village probably had more single males in its population. There were 44 lots in the village in 1804.
g. Estimated by multiplying the number of families signing the concession request who were resident in Old Mines (25) by 5.354, the average multiplier per tract for the entire district.
h. Sixty-six persons were specified in the concession request of 1800.
i. Estimated number of laborers and dependents at La Saline who were not counted elsewhere.
j. Indian population includes only those living in or in close proximity to white settlements and excludes visitors and transients. It excludes Indians and part Indians who were part of village households and are counted in those populations.
k. Only part (but a major part?) of this population was in the Ste. Genevieve District. All were officially under the administrative control of Commandant Lorimier of the Cape Girardeau District, but they had other relationships in the Ste. Genevieve District.

Table 5

VILLAGE AND TOWN POPULATION
STE. GENEVIEVE DISTRICT, 1769–1830

	1769–1770	1800	1804	1807	1815–1816	1818–1820	1830
Ste. Genevieve District							
Ste. Genevieve	600	950	1,080	1,200	1,500–2,000	1,500	<1,500
New Bourbon		270	210	(100)	(40)	(40)	(40)
Old Mines			90	(90)	(90)	100	200
Mine à Breton/Potosi		48	156	240	(400)	450–600	(450)
St. Michel/Fredericktown			65	(65)	(100)	300	(250)
Herculaneum					200	180–240	(100)
Greenville						60–75	20
Caledonia						20	(20)
Perryville							(100)
Farmington							(50)
District Total	**600**	**1,268**	**1,601**	**1,695**	**2,330–2,830**	**2,650–2,875**	**2,730**
Approximate Percentage of District Population in Villages and Towns	<100	71	56	≅ 45	≅ 30	21	12
Indian Villages		1,500	1,500	(1,800)	2,200	1,100	0
Jackson						300	(350)
New Madrid		806	(500)	(300)	180	120	(100)
St. Louis	500	1,100	1,039	1,800	2,000	4,000	6,694

Notes: Figures in parentheses are interpolations between estimates.
Old Mines estimates are for only the chief cluster of the dispersed community.

Sources: Spanish censuses; various observers (Flint, Schultz, Beck, Schoolcraft, and Billon); extrapolations from number of houses and buildings; inferences from the U.S. Census of 1840.

The core component of the village was a grid of more or less square blocks usually quartered into four lots of approximately 1 square arpent each.[5] Each lot, called a *terrain* or *emplacement,* typically contained a residence, a household garden and a small orchard, chickens and other *basse cour* (barnyard animals), a barn *(grange)* for grain, cabin *(cabane)* or shed *(appenti)* for corn, and perhaps a slave cabin and a horse and milk cow stable. A high wood-stake fence or stone wall enclosed the lot. Thus, a *terrain* provided adequate space for the daily functions of a late-eighteenth-century agricultural family. The Catholic church occupied a central location in the village with an adjacent open *place* for village assemblies.

The second component was a tract of fertile land devoted to field crops, the *champ,* adjacent to the residential village. It comprised many straight, long, narrow strips (*champs en long* [long lots]) running the depth of the tract. Each village family typically had at least one strip, its *terre,* on which to raise spring wheat, corn, oats, and other field crops. A *terre* averaged a few arpents wide and 40 arpents long, or a total of from 80 to 160 arpents (68 to 136 acres). After crops were harvested, the *champ* became a pasture for cattle and horses, their manure sustaining the productivity of the *champ* as in the open-field system of France. A single fence of upright wood stakes enclosed the entire village *champ* in order to keep domestic and wild animals from the crops during the growing season. Everyone used the same gates. Because the fortunes of the villagers depended greatly on the success of their grain harvest in this single enclosure, village regulations addressed the maintenance of the fence, the timing of planting and harvesting, and the *champ* in general as if it were a single piece of community property. A single enclosing fence to a compact tract of contiguous farm properties prompted English

---

to village form include Balesi, *Time of the French;* Briggs, "Le Pays des Illinois"; Belting, *Kaskaskia under the French;* David Keene, "Fort de Chartres: Archaeology in the Illinois Country"; Bonnie L. Gums et al., "The French Colonial Villages of Cahokia and Prairie du Pont, Illinois"; F. Terry Norris, "Ste. Genevieve: A French Colonial Village in the Illinois Country"; and Denman, "French Peasant Society." Similar studies include Franke, *French Peoria.* For an early-nineteenth-century description of a French village, see Brackenridge, *Views of Louisiana,* 119–20. Reps puts the French villages of the middle Mississippi Valley in a larger context in *Town Planning,* 58–65.

5. Accounts of the square block subdivided into square quarters are in Williamsen, "Colonial Town Planning," 13–16; Flader, "Settlement History," 18–25; F. T. Norris, "Ste. Genevieve," 137–41; Briggs, "The Forgotten Colony: *Le Pays des Illinois,*" 74–75; and Ekberg, *French Roots,* 90, although Ekberg emphasizes "approximately" square. For comparison, size of village lots in seventeenth-century New England villages ranged from one-half to five acres (Glenn T. Trewartha, "Types of Rural Settlement in Colonial America," 569).

speakers to call the *champ* a "common field"; *common field* was definitely not a term used by the French in colonial time. Actually, each strip was individually owned and could be sold by its owner even to persons outside the village without the village's permission. Like village lots, *terres* in the *champ* were part of a capitalist system of private ownership. The English term *common field* may convey the false idea of property held in common.[6]

The third component was a large tract of land less naturally endowed for field crops and usually more distant from the village core. This land belonged to the village as a whole, and in this sense it was a true commons or *commune,* sometimes called a *parc* (or *parque*). In Upper Louisiana, it was either an open woodland or a natural grassland with groves of trees, and it provided herbage for cattle during the crop-growing season; timber for building; wood for fuel and fencing; and special resources such as wild strawberries, plums, and grapes. In some villages, such as St. Louis and Carondelet, the commons was a specific tract with boundaries, but at new Ste. Genevieve and New Bourbon, the surrounding unoccupied land, the *domaine royale,* functioned as the village commons. Where the landscape offered plenty of nonarable land that no one would cultivate there was no need to designate a specific tract for grazing. Whereas the common field was fenced to keep animals out, the commons usually was not.[7]

Thus, the tripartite village efficiently and logically separated the functions of residence and daily activities in the built village, seasonal work with the field crops in the nearby common field, and occasional work

---

6. Among the abundant studies on long lots in agricultural villages, Harald Uhlig offers definitions of terms, compares various traditions, and comments on the advantages of long lots ("Fields and Field Systems"). For the middle Mississippi Valley, Ekberg's discussion of long lots in common fields in *French Roots,* 1–110, is unmatched. In fact, the central theme of his book is to understand French settlement in the middle Mississippi Valley as tripartite agricultural villages that created a strong sense of community among their inhabitants. Other descriptions of how a common field functioned are Henry Marie Brackenridge's account in the *St. Louis Missouri Gazette,* March 21, 1811; and Dorrance, *Survival of French,* 23–25.

7. Specially set-aside commons (for grazing) were not common in agricultural villages in those parts of Europe where extensive woodland grazing was available. Designated commons developed where all the land was arable, and some of it, presumably, had to be reserved for animal grazing, however limited (Uhlig, "Fields and Field Systems," 108). Ekberg treats commons within his larger discussion of common fields and thereby emphasizes their interdependency (*French Roots,* 1–110).

Fencing is an integral part of the land partitioning of common fields and commons. Evolution of fencing laws in Missouri, including colonial French practices, is traced in John H. Calvert, "Fencing Laws in Missouri: Restraining Animals."

with animals and special resources in the distant commons.[8] However, incoming Americans viewed the open-field agricultural village system as quaint, associated with European peasantry, and not in harmony with an independent American society and its closed-field practices. Americans incorporated this difference into their vocabulary and used *village* for French nucleated settlements regardless of size, often derisively, and reserved *town* for their own nucleated settlements regardless of size, attaching to it a positive connotation.

Implicit in the French agricultural village system was a maximum population size. In a region such as the Ozarks, where only a minor fraction of the land is arable, any village could have within daily walking distance only a certain amount of land suitable for field crops. When all that land was allotted, the village could not grow in population except by innovation in agricultural methods or by developing another economic base. Although a *champ* laid out in alluvial bottoms could maintain its productivity for a long time, those elsewhere could decline in yields and cause a decline in the agricultural product of the village, limited in space by its commuting radius. Thus, at least theoretically, the French village could grow only to a certain size before daughter settlements had to be planted elsewhere. Ste. Genevieve was likely born as a daughter settlement of Kaskaskia when it was reaching its limits of accessible agricultural lands. This process appears to have just been coming into play again at Ste. Genevieve when the tripartite village system in Upper Louisiana, under other pressures, began to collapse.[9]

The question arises as to how the idea of the tripartite village reached the middle Mississippi Valley, as that village type occurred neither in Laurentian Canada nor in Lower Louisiana. The seigneurial system in Quebec imposed a linear settlement pattern along the banks of the St. Lawrence wherein each family of habitants lived on its own agricultural *terre*. This unclustered pattern was possible because it was unnecessary to provide

---

8. Braudel noted how the spatial arrangement of fields and commons around a village was efficient for the villagers' time. He also traces the spatial arrangement of agricultural functions far back in time to the notions of *ager* (cultivated fields) and *saltus* (woodland commons) (*The Identity of France,* 140–44).

9. Ekberg suggests that the filling up of the Cahokia common field was responsible for the founding of the new adjacent village of Prairie du Pont and that the filling up of the Kaskaskia common field prompted the founding of Ste. Genevieve (*French Roots,* 61–64). Similarly, Reps states that New England agricultural villages with common fields had a built-in maximum population size (*Town Planning,* 106). Kenneth A. Lockridge's study of Dedham, Mass., links the establishment of daughter villages with the complete occupation of Dedham's finite acreage (*A New England Town: The First Hundred Years,* 139).

for defense against the English or Indians.[10] However, when these Canadians moved into the Illinois Country, they faced a different situation. Authority in the form of landholding seigneurs was not present to allot land and direct settlement. More important, Indian incursions were frequent, and the French grouped themselves into villages for mutual protection. The earliest Canadians, fur traders with Indians around 1700, may have been dispersed, but they were living in small but compact villages within a score of years. The learned John B. C. Lucas, born in France and resident of the United States since 1784 and of Upper Louisiana since 1793, said frankly, "The French settled in villages out of necessity . . . because of Indian attacks."[11] The clustering tendency was reinforced by governors who required that settlers of the Illinois Country live in compact villages for better defense against Indians and the English.

Defense is certainly not the only possible interpretation for a compact village form. Trade in furs, skins, lead, and salt required fixed sites for reconnoitering, producing, warehousing, and shipping. St. Louis is the best example in the mid–Mississippi Valley of a village founded as a trading post, but other villages had these functions in their initial years. Even at trading posts it was necessary to grow food and raise animals, and the founders of St. Louis provided for individually held parcels of cropland grouped into common fields. Ste. Genevieve's central role in lead activities fostered a village settlement, even if that role came after its founding.[12]

It is quite possible that neither defense nor trade had to be invoked for the presence of the unusual tripartite village form in the Illinois Country. David Denman, Susan Flader, Winstanley Briggs, and Carl J. Ekberg have independently examined the initial years of several of the French villages in the valley, including meticulous reconstruction of the arrangement of village lots and common-field properties, and have concluded that villages were initially formed by habitants without supervision by colonial administrators. Freed from the officialdom that imposed settlement systems in Quebec, the habitants reestablished a social community in the French-village tradition. Denman, in his 1980 exploration into the ori-

10. R. C. Harris, *Seigneurial System*.
11. Moogk attributes self-rule, which includes making local decisions on how to allot land, in both the Illinois Country and Acadia to Canadians' geographic isolation from colonial authorities at Quebec (*Nouvelle France*, 73–74). Lucas to Secretary of Treasury, January 29, 1807, *TP* 14:89.
12. Balesi believes that clustered villages in the Illinois Country grew up initially because of trading, as agriculture was undeveloped, although he noted that security remained a primary concern of the commandants of the Illinois Country (*Time of the French*, 199). Ekberg also makes the case for trading (*French Roots*).

gins of village communal activity at Ste. Genevieve, argues convincingly that the "overt explanation for the return . . . to traditional Old World custom rests on the dynamics of cultural conservatism." He notes that traditional peasant culture in both eighteenth-century France and the Illinois Country revolved around agriculture, and "[w]ith the reinstitution of traditional patterns of land use and settlement in the Illinois country and Ste. Genevieve came the close cooperation of individuals in collective action," that is, compact villages. Flader calls the action "a reintegration of traditional communal identity and activity on an Old World model." Briggs notes that people wanted to leave a "stifling seigneurial system" in Canada. Ekberg similarly argues that the entering Canadians "carried with them to the middle Mississippi area group memories of what they or their ancestors had known in France." Ekberg, who has conducted the most sustained research into the nature and origin of the nucleated French villages of the Illinois Country, concluded that although the tripartite village system that evolved "was a system that undoubtedly originated and evolved partly in response to local needs and local geography . . . [the settlers] possessed a mental template for a system that was part of the cultural baggage they had carried with them from the Old World."[13] The Creoles' desire to return to a traditional French-village form, which they presumably remembered after residence in Canada, does not gainsay the roles of defense and trade in supporting clustered settlements.

Stresses on the system appeared early. When old Ste. Genevieve was

---

13. Denman, "French Peasant Society," 39–42. The reinstitution of a settlement form from France, even after the experience of living in Quebec, fitted nicely into the basic cultural conservatism of French colonists in Canada, which is a unifying theme in Moogk's cultural and social history of Canada before 1760 (*Nouvelle France,* xvi). Flader, "Settlement History," 17; Briggs, "Le Pays des Illinois," 31, 35–38; Ekberg, *French Roots,* 108–9, 240.

According to Cole Harris, the association of the nucleated village and its collective agriculture with a "collective tradition" and development of a "community spirit" is being discredited. The alternative view holds that collective agriculture (common fields) and villages seem to have more to do with economic circumstances. Poor, vulnerable people needed the security of being with others, and when people got wealthy they struck out on their own ("The Simplification of Europe Overseas," 471).

Before considering antecedents in France, one should note Braudel's observations that "there is no such thing as a typical French village" and that "there were thousands and thousands of villages in France, no two ever exactly alike" (*The Identity of France,* 74).

A useful work for rural settlements in France is Albert Demangeon, "L'Habitat rural: Villages, hameaux, et fermes." An illuminating essay on the difference between a hamlet *(hameau)* and a village in France and their English equivalents is Pierre Flatrès, "Hamlet and Village."

being evacuated in the 1780s, some families struck out on their own and spurned living in a compact village. After 1795, the American system of dispersed farms began to take over the landscape, as farmer after farmer showed the practicality of living on his land instead of daily commuting to it. Although Ste. Genevieve continued its tradition of villagers going out to their fields in the *champs* and the village continued to enforce community regulations for the private properties well into the nineteenth century, the village population of farmers did not grow. As more land was converted into farms in the neighborhood and population grew, it was done by dispersed farmers and not by farmers living in villages.[14]

The creation of new French villages stopped shortly after the turn of the century. St. Michel, occupied in 1800, was the last in the Ste. Genevieve District, and no new tripartite villages were established elsewhere in the middle Mississippi Valley after 1803.[15]

Unplanned French Village

When the French ventured into the interior for lead, they again grouped themselves into villages of usually no more than ten cabins or fifty persons. At their simplest, these were seasonally occupied and by different persons as the years went on, but some settlements became permanent. Miners, slaves, paid workers, tenants, and others built rudimentary cabins or huts that stood for a few years at most. Some were hardly more than shelters; a few took on the appearance and comfort of a house *(maison)*. As the collection of cabins persisted and became inhabited year-round, property could be improved. Residents enclosed their gardens and small corn plots to keep animals out. In time, the cabin or house with

---

14. Ekberg traces the demise of the open-field system in the French villages of the Illinois Country and notes that it endured longest at Ste. Genevieve. Collapse of the open-field system was, of course, concurrent with the dispersal of villagers onto their properties (*French Roots*, 239–56). Lockridge explains how the collapse of the open-field system at Dedham, Mass., similarly resulted in dispersal of farmers onto their properties and a concurrent decline in the social community (*New England Town*, 91–118). See also C. Harris on the triumph of dispersed settlement over common fields and commons in new lands settled by Europeans ("Simplification of Europe Overseas," 474). The collapse of common-field systems at St. Louis and Carondelet was reported by Lieutenant Governor Trudeau in 1798 ("Trudeau's Report," in *Spanish Regime*, ed. Houck, 2:249) and also described later in the *St. Louis Missouri Gazette*, October 11, 1810. According to Trudeau, the village of Carondelet had abandoned its enclosure even earlier, in 1796.

15. The last were Marais des Liards (Village à Robert) and St. André, subsequently washed away by the Missouri River, both in the St. Louis District (Finiels, *Account of Upper Louisiana*, 70–71; *Régistre d'Arpentage*, 1–61A).

its enclosures could be sold as an improvement, even though no concession to the land had been awarded.[16]

These villages grew up amorphously without any spatial coordination except what common sense would require. Cabins could be oriented in all directions and spaced at different distances, although alignment along creeks was common, as at Mine à Breton and Old Mines. Lacking concessions, these improvements had no surveyed boundaries until forced later by the American government. If tied to lead, the village was vulnerable to collapse and disappearance with depletion of the resource. Such became the fate of several mining camps on the Mine la Motte tract and in the area around Mine à Breton.

Though not French, emigrant Indian villages fit into this category of unplanned villages. The Delaware and Shawnee lived in villages of several hundred residents along Apple Creek and the Meramec, and the Peoria had a small village adjacent to new Ste. Genevieve. Visitors referred to streets in these villages, but because no one described square blocks, standard sizes of lots, or other evidence of planning, we can assume that Indian villages were unplanned and the result of periodic, adventitious immigrations into them and emigrations out. Substantial log cabins in the American style, enclosed gardens and fields, and domesticated animals, however, gave them the appearance of permanency. Little is known of the morphology of Indian villages.[17]

---

16. A collection of cabins, no matter how small, was called a *village* or *petit village*. Some writers have taken *cabanage*, which occurs in descriptions of the mining area, to refer to a group of cabins or huts (Houck, *History of Missouri*, 1:281), but *cabanage* was defined in 1727 as the "ground on which to do our cooking and to sleep," with no mention of cabins, in Drake (quoting Du Poisson) (*Systematic treatise*, 112, 114). The uncoordinated clustering of miners' cabins has largely gone unnoticed as a distinctive settlement form. See Glenn T. Trewartha, "The Unincorporated Hamlet" and "Types of Rural Settlement." Trimble et al. note how settlements tied to seasonally exploited resources are different from those based on agricultural pursuits. They further note that, in general, where extraction of minerals is concerned, there are no ethnic differences in settlement pattern; the pattern is one determined by function ("Frontier Colonization," 187). This assertion needs further examination. Carl Ekberg et al. proposed a model of a mining camp in Upper Louisiana in the 1730s that consisted of a blockhouse for protection from hostile Indians, which served also as the residence of the director of the mines and was surrounded by a scattering of miners' huts. Population, both free and slave, was seasonal and totaled approximately forty-four persons (*A Cultural, Geographical, and Historical Study of the Pine Ford Lake Project Area: Washington, Jefferson, Franklin, and St. Francis Counties, Missouri*, 70–71).

17. Zitomersky explores the form and function of dual French and Indian settlements in early colonial Louisiana (*French Americans–Native Americans*, 361–87), but his discussion of Indian settlement form and function does not apply to the later-arriving Shawnee, Delaware, and other immigrant Indians.

Planned American Town

All the nucleated American settlements in the Ste. Genevieve District were planned in the formal sense that someone imposed a preconceived pattern on the landscape and the pattern had spatial regularity. Like the French planned village, the American town was a grid of parallel and perpendicular streets, blocks of standard sizes subdivided into lots of standard sizes.[18] Figures 17 and 18 show the location and year of establishment of all American towns and French villages in the district. Figure 19 shows plans of district towns and villages, all drawn to a common scale and oriented to north for easy visual comparison.

But the French village and the American town differed sharply in function, which influenced details of form. The French village was populated primarily by habitants, farmers of an eighteenth-century peasantry. American farmers, however, lived on their individual properties away from the town, and merchants, craftsmen, and laborers populated the American towns. Not needing to function as farm properties, American town lots were significantly smaller and much narrower so that more properties could have frontage on streets.

More towns were planned and platted than those that actually materialized, which is proof that American towns did not arise as much from the economic needs of the neighborhood as from the hopes of land developers. Four towns in the Ste. Genevieve District had forced births as county seats, well before the local areas could adequately support them as economic and service centers. All towns provided for public space as French villages did, but planners reserved no space for churches. Because they were more than agricultural centers, American towns had no inherent limits to population growth and their founders expected them to

---

18. A standard source for form of early American towns is Reps, *Town Planning*, wherein Reps notes the great effect of Philadelphia's form on the laying out of towns to the west (141–54). For early planned towns in Missouri, see Russell E. Smith, "The Towns of the Missouri Valley: An Element of the Historical Geography of Missouri"; and Stuart F. Voss, "Town Growth in Central Missouri, 1815–1860." An interpretive study on the generation of towns as a frontier settlement process is Charles J. Farmer, *In the Absence of Towns: Settlement and Country Trade in Southside Virginia, 1730–1800*.

In his study on landscape changes in the mining region of the eastern Ozarks, Hugh N. Johnson makes the remarkable observation that the American town "with its haphazard layout and heterogenous [sic] architecture offered a striking contrast to the mathematical precision and attractive architecture of the older French villages" ("Sequent Occupance of the St. Francois Mining Region," 134). Ekberg et al. noted Johnson's observation and rejected it (*Cultural, Geographical, and Historical Study*, 63). There is absolutely no evidence that planned American towns in the Ste. Genevieve District were "haphazard" in layout.

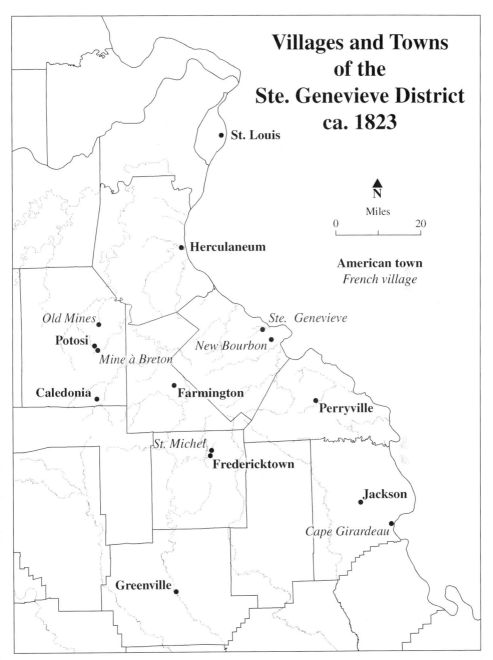

17. *Villages and Towns of the Ste. Genevieve District, ca. 1823.* By 1823, five French villages and seven American towns were in the district among many dispersed rural neighborhoods.

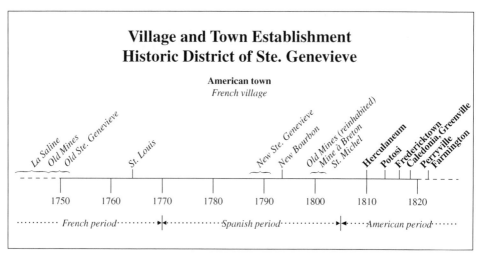

18. *Village and Town Establishment, Historic District of Ste. Genevieve.* French villages ceased being established around 1800, and the first American town was established in 1810.

grow beyond the bounds of the original plat. Anticipation of growth encouraged speculation, absent in French villages. Investors bought lots in expectation of future resale, with the result that the grid of platted streets often comprised an "urban shell" within which many lots remained empty. Though they had smaller lots, American towns had an "open landscape" that contrasted with the "closed village-scape" of French villages in which virtually all lots were occupied and built on.[19]

Linear Valley Settlement

Americans introduced their dispersed settlement to the Ste. Genevieve District in the 1780s at the same time that the first French began to disperse from their open-field villages. Nevertheless, the overwhelming pro-

---

19. Frontier towns as speculative ventures is a common topic in frontier literature. See, for example, Morris Birkbeck, *Notes on a journey in America, from the coast of Virginia to the territory of Illinois*, 103–5; Reps, *Town Planning*, 187–91; and John C. Hudson, *Plains Country Towns*. Meinig contrasts the town as a place to live in a civilized way with the town as a speculative, commercial expression (*Shaping of America*, 2:369–74). Meinig emphasizes that land platting and lot sales were separate actions from construction of buildings and occupancy of lots. Meinig attributes the term *urban shell* to Michael P. Conzen (370).

A quantitative comparison of the geometries of French villages and American towns in the Ste. Genevieve District is in W. A. Schroeder, "Opening the Ozarks," 440–45.

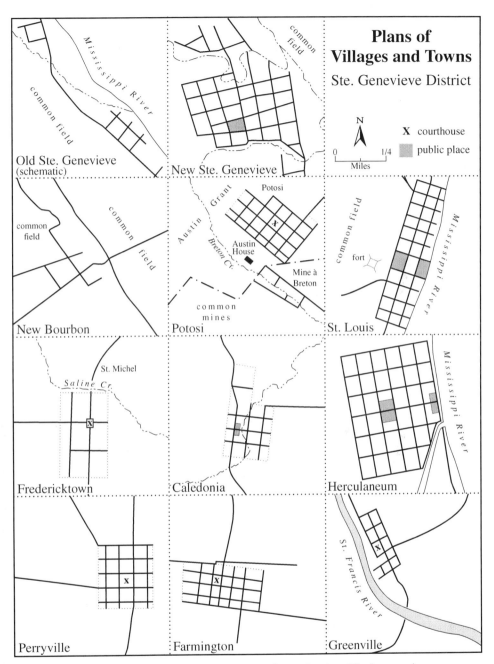

19. *Plans of Villages and Towns, Ste. Genevieve District.* All eleven plans are drawn to the same scale and oriented in the same direction. Shaded areas are designated public places.

portion of people in dispersed locations was American, and most of the French-culture population remained in villages through the 1820s.[20]

The American farmers in the Ste. Genevieve District came mostly from Tennessee, Kentucky, and Appalachian Carolina and Virginia and were of mixed ethnicity, but chiefly English, Scotch-Irish, and German. These ethnicities were submerged into a single, general land economy based on open-range cattle and hogs, a modicum of field crops, and hunting, and required much more land per family than one based primarily on crops. Cattle and hogs roamed open-range woodlands for some distance from farmsteads. Contemporary observers noted the Americans' need for space and freedom for wild pasturage and the ease of raising animals in the open, grassy woodlands of the Ozarks. The Ozarks, they correctly prophesied, would never be settled by full-time cultivators.[21]

The dispersed, linear valley form became the most geographically widespread settlement form in the Ste. Genevieve District. It was associated with the innumerable valleys entrenched one hundred feet or more below the adjacent upland. One could postulate that settlement began

20. Geographer Cole Harris maintains that dispersed settlement of Europeans in their colonies is due to the circumstances of the overseas lands (land availability and absence of markets) and is independent of European national group; all groups formed dispersed settlements ("Simplification of Europe Overseas"). Anthropologist Conrad M. Arensberg, however, argues that the American "open-country neighborhood," or *Einzelhofsiedlung* in settlement nomenclature, may be traced to European antecedents ("American Communities"). James T. Lemon notes that even Mennonites, Moravians, and others dispersed despite William Penn's instructions for settlers to live "in close proximity and in villages" ("Colonial America in the Eighteenth Century," 130). Trewartha concluded that where properties are compact and several hundred acres in size, village residence is virtually impossible and dispersion is required ("Types of Rural Settlement," 584). James E. Vance Jr. considered American preference for isolated farms rather than villages as "the first utopia" in American life ("Democratic Utopia and the American Landscape," 210–16). Ekberg et al. proposed a model of dispersed settlement for the Big River region of the Ste. Genevieve District but supposedly typical of the overall southeastern Missouri landscape of about 1820. Settlers, who were mostly Americans, preferred bottomland and prairie margins (*Cultural, Geographical, and Historical Study*, 70–74).

21. John Solomon Otto, "Migration of the Southern Plain Folk: An Interdisciplinary Synthesis"; Sauer, *Ozark Highland*, 101–2, 112–23; Brackenridge, *Views of Louisiana*, 117; Schoolcraft, *View of the Lead Mines*, 34–35.

Brackenridge presented a vignette of upland southern settlement with Tennessean John Smith T's residence at Shibboleth, ten miles north of Mine à Breton. The residence sat on a sterile dry ridge, with no other improvements nearby: "a space of five or six miles is considered quite near enough for a next door neighbor." Brackenridge was surprised to see that this important person lived so simply in a double log cabin, one side for his family, the other for visitors (*Recollections of Persons and Places*, 216–17).

with a few scattered improvements in a valley, probably kinship related, which then spread up and down the valley until it was full and then stopped, except for subsequent property subdivision. During the period up into the 1820s, the intervening ridges remained unoccupied, though used for grazing, timber, roads, and hunting.[22]

Linear valley settlements created landscapes of vertical zonation. The valley bottom contained crops, pens, and buildings, whereas the bounding steep hillsides remained wooded and the flinty ridge tops, where the woodlands were more naturally open and grassy, were grazed. The compact shapes of properties as depicted on cadastral maps gave no indication of this elevational land-use pattern. Landownership centered on the farmstead in the valley bottoms and extended onto the ridges as much as necessary to incorporate a certain acreage.

In the larger and longer-settled valleys in the district, properties abutted each other without intervening unclaimed spaces. Farms were usually spaced from one-quarter mile to one mile apart yet close enough that farmers could still hear gunshots and see smoke from their neighbors' cabins.[23]

Linear settlement did not promote development of central places because it had no natural focus to the distribution of people except at junctions of valleys. But all of the streams of the river counties lead directly to the Mississippi and have no junctions of consequence until they reach the Mississippi. Low population densities in general for the rough hills also worked against the development of clustered settlements.

Discontinuous Valley Settlement

Where physical geography did not permit continuous settlement for the length of a valley, even after initial stages had passed and the valley was as full as it would likely become, settlement was sparse and discontinuous. American settlers discovered the few directly farmable sites and strung out their improvements like beads on a rosary. In places, miniclusters arose, but in general, the farms were from two to four, even six miles apart. This

---

22. Clendenen presents four reasons for various early settlement forms in the Ozarks: defense, homestead system, habitat and economy, and historical antecedents. He uses the last two, which combine both physiography and culture, to account for dispersed settlements in the Ozarks ("Settlement Morphology").

23. Schoolcraft reported six farms in twelve miles along the Meramec in 1825 and four in eight miles along the White River in 1818 (*Travels in the Mississippi Valley*, 237; *Journal of a Tour into the Interior of Missouri and Arkansaw, from Potosi, or Mine a Burton, in Missouri Territory, in a South-West Direction, toward the Rocky Mountains; performed in the years 1818–1819*, 39- 40). Clendenen reports farms every one and one-half miles along the Current River ("Settlement Morphology," 142).

settlement form most approached true isolation, but at the same time it represented the rather complete occupation of land niggardly endowed for crop farming and not much better for animal herding. It was the settlement form for the lands worst endowed for human occupance.

The irregularly spaced igneous-rock knobs of the rough hills and mining country present obstacles that streams must find ways through. At places, streams flow across the igneous outcrops by way of resistant-rock defiles, called shut-ins, where bedrock roughens the channel and forms bold cliffs that rise abruptly on both sides of the rock channel. These shut-ins effectively block movement up and down a valley, by use of either the channel bed or its alluvial floor. As if to compensate, broad valley bottoms formed on sedimentary rocks occur along the streams just upstream and downstream from a shut-in. Thus, this knob-and-basin topography does not allow continuous settlement in valleys with contiguous landownership tracts, but only tiny pockets of settlement that cannot easily communicate with each other. It also severely hinders the development of social institutions and central places, as well as all economic development dependent on transportation.[24]

Discontinuous valley settlement was characterized by kinship-related neighborhoods no less than other settlement forms. Though families may have been separated from each other by miles of unoccupied space, they were usually related to their neighbors, and long stretches of valleys bore names of family clans. Blood relationships were by default the basis of any sense of community due to the dearth of religious and economic institutions. No central places emerged in areas of discontinuous settlement except at Greenville, when the creation of Wayne County forced the location of an administrative center.

Discontinuous linear settlement occurred on the fringes of the district. It dominated the igneous-knob terrain of the St. Francis River valley below the Fredericktown basin, the Black River drainage to the southwest, and the Meramec tributaries west of Potosi.

Compact Settlement

Where smooth topography presented no constraints, dispersed settlement of Americans spread over the entire surface and took on a compact

24. For accounts of isolation of early shut-in communities, see James Lee Murphy, "A History of the Southeastern Ozark Region of Missouri"; and Clendenen, "Settlement Morphology." Jordan and Kaups identified "loose clustering" among isolated farmsteads in backwoods settlements. They concluded that about 300 to 350 acres per capita was the minimal amount of land needed in the backwoods sytem, or approximately 2,000 acres per family (*American Backwoods Frontier*, 123).

pattern with properties abutting each other. This was possible on the karst plain and in the larger sedimentary basins of the interior.

As elsewhere in the district, Americans who came to these better-soil neighborhoods were upland southerners and their slaves. However, located on more productive soils and forming larger communities with larger markets, they early concentrated on commercial agriculture, both crops and animals, and prospered economically. Farmhouses had an average spacing of somewhat less than a mile, but this declined to a half or quarter mile as owners subdivided properties and population densities increased.

A strong sense of community emerged in compactly settled areas, founded first on blood ties, then strengthened through larger populations and the development of vigorous institutional life built around churches and administrative functions. Aiding significantly the sense of local self-identification was a sharp physiographic delimitation of each neighborhood.[25]

Compact settlements formed in the Barrens, the Bellevue Valley, the Fredericktown basin, and in the three neighborhoods of the Farmington Plain: the Big River Bend, Murphy's Settlement, and Cook's Settlement.

Large Tracts

The awarding of large tracts of land to individuals and companies with the responsibility of developing them with settlers was a colonizing strategy of European imperialists throughout the Americas. The United States and its constituent states continued the practice and tried variations of the large grant to land developers before the instigation of the rectangular land subdivision and public land sales in the Northwest Territory.[26]

The French in Upper Louisiana were familiar with the seigneuries of Canada, but after an early attempt at Cahokia, settlement by seigneury

---

25. Peter S. Onuf argues that the American rectangular land survey with periodic sales through blocks of townships was designed to promote compactly settled neighborhoods of dispersed farmsteads (*Statehood and Union: A History of the Northwest Ordinance*, 38–39). This goal could not be achieved, however, where physical geography prevented the establishment of family farms on every quarter section of land, as in most of the Ozarks.

26. Two accounts of such enterprise are historian Alan Taylor's *William Cooper's Town: Power and Persuasion on the Frontier of the Early American Republic* and geographer William Wyckoff's *The Developer's Frontier: The Making of the Western New York Landscape*. Warren R. Hofstra discusses how English colonial land grants were used to encourage settlement of Shenandoah Valley for geopolitical reasons. Large land grants to the moneyed class encouraged land speculation, but settlers preferred to buy directly from the Virginia government without going through the large landowners ("Land Policy and Settlement in the Northern Shenandoah Valley").

had no place in the middle Mississippi Valley.[27] The Spanish administrators of Louisiana also were knowledgeable of their king's huge land grants to favored persons in the Americas, and several were awarded in Lower Louisiana. The Spanish entertained American George Morgan's proposal to establish a colony at New Madrid but did not allow him to carry through with his plan. Indeed, Governor Miró, cognizant of a long history of difficulties with such colonization schemes, admitted to the royal officer in Madrid that "great concessions to one single individual [in Louisiana] have never produced the desired effect of populating it."[28] After 1795, the Spanish did award concessions and grants in Upper Louisiana that were much larger than necessary for family farms, but they were for extractive resources and stock ranches, not for colonization.[29] Among these were the awards to Pierre-Charles Delassus de Luzi-

27. Priests of the Seminary of Foreign Missions of Quebec, who had arrived at Cahokia in 1699, received a formal concession to their lands *en franc alleu* in 1722. Although identified as a seigneury and the father superior identified himself as seigneur, apparently no one paid dues or otherwise manifested the elements of the Canadian seigneurial system. See Ekberg's account of land tenure at Cahokia (*French Roots,* 41–42, 54–61). Denman surmised that French Canadian habitants entering the middle Mississippi Valley might have been reluctant to settle on land called a seigneury and thus would have bypassed Cahokia for other places ("French Peasant Society," 23). Balesi also noted that the Canadian seigneury was not applied to the Illinois Country (*Time of the French,* 199). Ekberg comments on the role of the royal edict of 1716 against large grants, which were deemed unproductive for colonization at that early time, in preference for the division of land into small parcels of long-lot shape (*French Roots,* 22). One is left wondering whether the edict, clearly intended for Lower Louisiana, was applicable only to agricultural properties and colonization attempts, because the crown continued to award large properties for mining purposes. After 1770, Spanish administrators in Upper Louisiana seemed unconcerned with provisions or even the spirit of the French edict of 1716.

28. To His Excellency, Bayllo Fray Don Antonio Valdes from Esteban Miró, May 20, 1789, in *Spanish Regime,* ed. Houck, 1:277. The large land grants in Lower Louisiana include two made in 1795 by Carondelet (who was also responsible for the only large grant for colonization in Upper Louisiana), one to the baron de Bastrop for twelve square leagues on which he was to settle five hundred families and another for similar purpose in 1795 to the marquis de Maison Rouge for thirty square leagues. Because of Governor Carondelet's largesse, which was not deemed beneficial to the province, the king revoked the governor's authority to grant land in 1798 and transferred it to the office of intendant (Burns, "Spanish Land Laws," 566–69, 580–81). The United States Supreme Court rejected both land claims (Donaldson, *Public Domain,* 373).

29. Most usually a request for mineral land was couched in terms of a stock ranch. Many large concessions were "floating," or unlocated, as in the 1796 concession (actually a *merced*) to Jacques Ceran Delassus de St. Vrain to take out land anywhere he pleased for whatever purposes (LRB 202, 202a). Many large concessions were still unconfirmed after the second board's decisions in 1835. Nevertheless, they did affect settlement, as they were known to exist locally and incoming settlers avoided them.

ères and Moses Austin, both of whom promised to bring others with them to Upper Louisiana, but those others would have to seek their own concessions separate from the grantees. The Spanish made only one enormous land grant in Upper Louisiana explicitly for colonization on the grant: the Maxwell grant of 16 square leagues (150 square miles) on the Black River.[30]

Identification of large tracts as a settlement form, whether for colonization or other purposes, is based simply on the size of the owned property, because the huge size of these tracts by itself directly affected how people spread themselves across the landscape. Large tracts are arbitrarily defined as those properties 1,000 arpents in size (851 acres, or 1.33 square miles) or larger, a size well beyond the maximum size any farm family could effectively use and larger than the maximum 800 arpents and 640 acres permitted by Spanish and American regulations for a family farm.

Population on large tracts was sparse, whether they lay in the rough hills or on the best arable land. Hired workers, tenants, and the owner's slaves occupied the properties on a seasonal or temporary basis in tiny unplanned villages, but although these people may have been temporary in the villages, the settlement may have been permanent. For example, Jean-Baptiste Vallé had different tenants on his River Establishment pinery continuously for ten years, from 1788 to 1798, and he made only occasional trips to it, but the property was permanently occupied during those years.[31] Unlike the French, American owners of large properties, such as Moses Austin and Joseph Fenwick, lived on theirs.

The exploitative and extensive land use on these large properties retarded population growth. This might be understandable on poor-soil tracts, but it was equally the case on good lands. Those who were allowed to occupy the land considered themselves only temporary, unable to purchase land from the owner, and consequently without a future there. No kinship-based communities formed, no social institutions emerged, and no central places of any consequence developed before 1823, except Potosi on Austin's square league.

In general, these large tracts remained in the hands of wealthy owners, usually absentee, who were under little pressure to sell them in parcels or otherwise develop them. Owners succeeded in keeping others

---

30. Of course, before the Spanish regime came in 1770, the French granted land for colonization. Renaut, for example, was to colonize the Illinois Country by attracting settlers to his large land grants as well as to mine lead (Giraud, *History of French Louisiana*, 5:442).

31. *Régistre d'Arpentage*, 87; LCC 176.

off them during the long period when claims were being confirmed. To a remarkable degree, they stayed in family hands as long-term investments and not for speculation or settlement.[32] Thus, through the 1820s, most of these large tracts continue to show up as near voids on the settlement map. After 1804, there was, of course, no longer the possibility of acquiring large tracts of land in a single block from the American government.

Who created these seven basic settlement types and when, how they were created, and their subsequent fate through the 1820s are questions best answered through the diverse accounts of individual neighborhoods. The accounts begin with the initial settlement at the original Ste. Genevieve on the banks of the Mississippi, then proceed through its disintegration and the establishment of new Ste. Genevieve, New Bourbon, and their vicinities. The accounts next move into the mining country at Mine la Motte, St. Michel, and Fredericktown, then to the mining country at Mine à Breton, Potosi, Old Mines, Herculaneum, the big bend of the Big River, and Murphy's and Cook's Settlements. The accounts then leave the mining country and move to the dispersed American settlements of Jefferson and Wayne Counties, Father Maxwell's large colonization grant, and the compact American settlement in the Bellevue Valley and Caledonia. The neighborhood accounts end with Perry County—which presents the most complex variety of settlement types, including dispersed, compact, and Indian—and the planned town of Perryville.

---

32. Moses Austin was one exception to retention of ownership; debts forced him to subdivide and sell his large grant. The Vallé family also sold its square-league pinery on the Establishment River in 1818 to American speculators living in Baltimore (LRC 141).

8

The French Villages

Old Ste. Genevieve

If Indian settlements and the early lead-mining and salt-making camps are disregarded, permanent occupation of the land on the western side of the Mississippi River in the Illinois Country dates from the late 1740s in the Pointe Basse. By that time, buildings were probably on the outer side of the great alluvial bottom at a site now referred to as old Ste. Genevieve, which was as geographically close to Kaskaskia as possible. "Ste. Genevieve of Cascaskias" was an early name. In the 1740s, Kaskaskians, apparently needing more alluvial land for an expanding population than their original common field provided, had begun to acquire additional bottomland to the north of the village.[1] The spread across the river was probably part of this village expansion onto lands beyond the daily commuting range and necessitated a new daughter village. Accustomed to navigating the Mississippi, habitants did not regard the river as a barrier to movement. There being no specific act of founding of the new village, Kaskaskians individually erected houses and began cultivating the bottomland in the years from 1749 to 1752. A 1752 census of the Illinois Country listed eight households with thirty-four arpents of land on the west bank, but another document that year specified twenty-seven inhabitants with ninety-three arpents of frontage. Probably some of the landowners still resided in Kaskaskia.[2]

Other sources corroborate a mid-eighteenth-century date for the founding of Ste. Genevieve. On the Bellin map of 1755 is written "Ste. Geneviève

---

1. Ekberg, *Colonial Ste. Genevieve*, 35; Belting, *Kaskaskia under the French*, 56. Using Macarty's observation that there was no more land to sow or to assign on the Kaskaskia side, Flader concludes that a scarcity of agricultural land induced Kaskaskians to seek land across the river in the 1740s ("Settlement History," 6–8).

2. Accounts of the founding of Ste. Genevieve are in David Denman, "French Peasant Society"; Flader, "Settlement History"; Ekberg, *Colonial Ste. Genevieve*, 2–25, and *French Roots*, 88–96.

village François établi depuis 3 ans," and an account written probably in 1756 relates that at one league from the Saline, on the same side, approximately fifteen families had been established on the riverbank for several years *(depuis quelques années)* and that the place was named Ste. Genevieve.[3]

By 1752, the form of the settlement was in place as an incipient tripartite village. *Terres* were being laid off in long, narrow strips, a direct transferral of the Kaskaskian pattern of long lots. Apparently, a major portion of the Pointe Basse was already so partitioned by 1752, and virtually all was in private ownership by 1766. Jean-Jacques Macarty, the new commandant of the Illinois Country, had arrived in Kaskaskia in fall 1751 and, with engineer François Saucier, immediately began the regularization of land already occupied in the Pointe Basse and made new concessions in an orderly fashion. He gave *terres* the same orientation, which was that of perpendicularity to the riverbank at its outermost bend—or so it is assumed because that was the practice elsewhere along the river—and that made the *terres* also perpendicular to the bluff line at their other ends. The compass orientation is basically the same as the orientation of the long lots systematized in 1720 across the river at Kaskaskia, which were also apparently laid out perpendicular to the same river bend. The long lots in the Pointe Basse ran from the river to the bluffs without specifying the customary forty-arpent depth for properties. Their width was also apparently standardized at three arpents, although a few departed from this. A single fence enclosed the *terres* to create a common field. Susan Flader concludes from her reconstruction of properties in these early years of old Ste. Genevieve that the relative equality in landholdings suggests that the land early put into ownership in the Pointe Basse was needed more for its food production for an expanding population and less as a source of wealth and status.[4]

A small grazing commons lay between the fence surrounding the

---

3. Ekberg traces the identification of Ste. Genevieve on three Bellin maps, 1750–1755 (*Colonial Ste. Genevieve,* 6–18). A strikingly similar map is the one attributed to Villiers du Terrage of 1755 in Alvord, *Illinois Country* (pl. between pp. 154–55), and in Tucker, *Atlas,* pl. xxiv.

The French text reads: "A une lieu de la Salinne, du même bord, il s'y est établis depuis quelques années sur le bord du fleuve, une quinzaine d'habitans. . . . Cet endroit se nome Ste- Genevienne" (Cruchet, "La Vie en Louisiane," 67).

4. This paragraph is based upon the reconstruction of early Ste. Genevieve by Denman and Flader as reported in Flader, "Ste. Genevieve," E1–8; and "Settlement History," 1–18. Linear villages on the riverbank were also made at Prairie du Chien, Wisconsin (Ekberg, *French Roots,* 102–8); at Peoria (Franke, *French Peoria,* 36, 49, 52–53); and by the Métis in the Red River Valley of Assiniboia (Faragher, "Americans, Mexicans, Métis," 101), and elsewhere.

croplands and the river, a position analogous to the grazing commons at Kaskaskia. Unallotted lands at the extreme southern and northern ends of the common field also probably served as grazing lands, as well as adjacent bluff lands of the royal domain.[5] The total area of the Pointe Basse in the mid-eighteenth century was approximately eighty-five hundred arpents, or more than eleven square miles.[6]

Unlike compact Kaskaskia and other east-bank villages, some habitants in the initial years up to 1752 built their houses at the river ends of the long lots, resulting in an exceptionally long village.[7] This suggests that Ste. Genevieve houses were accumulating on the west bank only as an extension of Kaskaskia, and that the new settlement functioned as an agricultural adjunct—each house on its *terre cultivée*—and, in its initial conception, the community group functions of church, commerce, and civil and military administration would continue in Kaskaskia.

The initial linear settlement pattern does not support interpretation that Ste. Genevieve arose as a fur-trading or lead-shipping point on the west bank. A site where the Mississippi washed the bluffs, where the channel and its banks would be stable, and at the mouth of a tributary such as the Saline or Establishment Rivers would have made more sense, if Ste. Genevieve were to have had its beginnings as a river shipping port. It is also illogical that lead-laden animals and carts from the interior hills would roll over another one or two miles of soft alluvial ground and marshy swales to get to the river, when other places were available where the Mississippi banks were next to firmer ground. Ste. Genevieve was not established as a lead-shipping port.

In association with his work of regularizing the *terres* of Ste. Genevieve, Commandant Macarty had François Saucier lay out streets and 1-arpent *terrains*, four to a square block, and provided for a church building and a location for it. Thus, the new settlement, within a few years of its found-

---

5. Ekberg, *French Roots*, 92. Ekberg notes that Ste. Genevieve did not have a formally designated commons as the east bank and St. Louis–area French villages had and that the Ste. Genevieve commons developed ad hoc as any land not within the enclosed common field or the village.

6. *Missouri Gazette*, March 21, 1811. Measurements and estimates of the size of the fenced common field varied because of different locations of the fence and river erosion (PC 208b-427). The total area of the Pointe Basse as a landform was, of course, significantly larger than the fenced common-field portion of it, Le Grand Champ or Big Field. For example, Edmund Dana gave a figure of ten thousand acres for the entire bottom in 1819 (*Geographical sketches on the western country, designed for emigrants and settlers . . . including a particular description of all the unsold public lands, collected from a variety of authentic sources. Also, a list of the principal roads*, 296).

7. Susan L. Flader, "Final Narrative Report, Ste. Genevieve: An Interdisciplinary Community Study," 4.

ing, acquired its own identity as a planned, compact village with the most essential community function: a church. However, David Denman's meticulous reconstruction of properties in old Ste. Genevieve shows clearly that rather than a single village, several nodes of embryonic grids developed along the river, which makes it more accurate to describe the form as a hybrid of both a compact element with square blocks and *terrains* and a linear element of houses along the river. (The nodes are shown schematically in Figure 20.) Consequently, descriptions of the village up to its demise in the 1780s can appear to contradict its existence in blocks and village lots. The incoming Spanish lieutenant governor, Pedro Piernas, described Ste. Genevieve in 1769 as a settlement of six hundred persons where houses were "separated and scattered." English captain Philip Pittman said that the seventy families of Ste. Genevieve were dispersed along the river for almost one mile. Lieutenant Governor Cruzat noted in 1782 that houses were too far apart from each other to make it possible to defend the village.[8] To an observer from the river, the village's linearity aspect would have been more visually prominent than the small nodes, which may have been partially hidden.

The nodes comprised square blocks of approximately 380 feet by 380 feet and divided into four lots, each approximately 1 arpent square. Terry Norris confirmed this standard size and shape in an archaeological study in the 1980s that identified house sites at an average distance of 180 feet, or virtually 1 arpent. These nodes grew by the arrival of new residents who were not farmers and by habitants who cultivated their long lots at some distance in the common field, and this represented a shift from living on one's own long-lot farm to living on a square lot in a compact village. Perhaps some moved into the nodes due to bank caving of the river ends of their long lots, others by a desire to live closer to relatives. Some habitants, however, continued to live at the ends of their long lots, Laurentian style, but their numbers probably did not grow.[9]

8. Reconstruction by Denman and Flader as reported in Flader, "Settlement History," 18–25, app. C, E. Except where noted, the following paragraphs on old Ste. Genevieve are based on the work of Denman and Flader. Don Pedro Piernas to Gov. O'Reilly, in *Spanish Regime,* ed. Houck, 1:70; Pittman, *Present State,* 50, 59. If seventy houses stretched for 1 mile, they would be an average of 75 feet apart, or less than one-half arpent. However, Pittman also noted that houses were 130 feet apart, which leads to the conclusion that some must have been in nodes in their 1-mile stretch along the river and on lots of sizes more like the 1-arpent village lot. "Council of War Held at St. Louis, July 9, 1782," in *Spain in the Mississippi Valley,* ed. Kinnaird, 3:39. Similarly, the German settlements in Lower Louisiana persisted in linear form despite repeated Indian attacks (Blume, *German Coast,* 125–29).

9. F. T. Norris, "Ste. Genevieve," 139–41; F. T. Norris, "Where Did the Villages Go? Steamboats, Deforestation, and Archaeological Loss in the Mississippi Valley," 88–89.

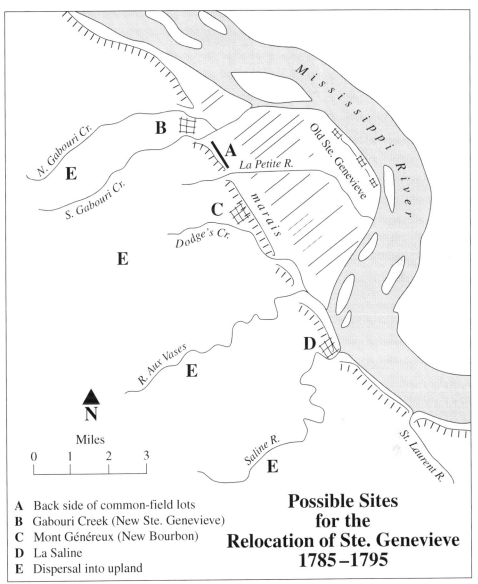

A Back side of common-field lots
B Gabouri Creek (New Ste. Genevieve)
C Mont Généreux (New Bourbon)
D La Saline
E Dispersal into upland

Possible Sites for the Relocation of Ste. Genevieve 1785–1795

20. *Possible Sites for the Relocation of Ste. Genevieve, 1785–1795.* Villagers leaving old Ste. Genevieve could move to five different locations. The three bluff locations (B, C, and D) were promoted by three different community leaders, but most villagers chose site B between the forks of the Gabouri Creek, which merged with site A to become the new village of Ste. Genevieve.

Of the several reconstructed village nodes, the one in the north, which coalesced with the next two on its south, became the largest and most important. It had the church, cemetery, and the commandant's house (Ste. Genevieve had a commandant after it became Spanish), all powerful forces for settlement concentration. Houses occupied *terrains* on both sides of a main street *(grande rue),* cross streets, and backstreets. The grid was interrupted by a small coulee, probably a wet swale of a former Mississippi channel. Another important node lay somewhat to the south along the channel of the Petite Rivière, interpreted as the lower course of present Mill Creek, possibly joined by the waters of the Gabouri River before later changes diverted its mouth to the north. Indeed, the Petite Rivière served as the village's safe, quiet harborage for shallow-draft riverboats and most certainly was the reason for a node at that location. The *terrains* in this node were not located as close to the Mississippi River bank as in the northern nodes, if the point of boat loading was up in the mouth of the Petite Rivière.[10]

Because the site of Ste. Genevieve was initially chosen for its agricultural potential, it was not necessarily suitable for nonagricultural functions, and some residents and travelers began to describe the site as unhealthful and generally "poor." Farmers needed to live on or close to their farmland, but others did not. Farmers could tolerate a certain number and severity of floods, but others would not, if options existed. And no one could long tolerate Mississippi bank caving, except by persistent, periodic removal of structures away from the slumping banks. Those who expected Ste. Genevieve to be something more than an agricultural village, such as a port for river traffic, could well have become restless staying at the village's riverbank location.

Consequently, it was only a matter of time before the Ste. Genevieve community would spread beyond the riverbank and the sharp topographical confines of the productive Pointe Basse. Although the great flood of 1785 is the most cited cause for the removal of the buildings of the village from the alluvial bottoms into the hills (it is convenient to seize upon specific events for reasons to take action), and most likely was the single most galvanizing event that affected everyone at Ste. Genevieve, other conditioning forces had been at work.

First, as the population grew over a generation of time, the Pointe Basse was running out of land to allot to families, and land that was al-

---

Norris estimates that approximately two-thirds of the remains of the original Ste. Genevieve was destroyed by lateral channel shifting of the Mississippi River during the nineteenth century. Flader, "Settlement History," 18–25.

10. Flader, "Settlement History," 18–25.

lotted was being lost to chronic bank caving. In any one year, the lost land was manageable, but the cumulative effect was not, especially at the upper end. Land lost to the river needed to be replaced, but where? Land was being accreted to the lower end of the bottoms, but it appeared as sterile sandbars colonized by willows and would not become arable land by natural processes for a few decades. The bluff lands offered the only recourse. North enders who lost land by bank caving were the first to seek replacement *terres* in the hills and actively cultivate them. They began to farm in the hills behind their disappearing properties in the late 1770s, but this replacement land lay three miles from the village houses, a distance that stretched the time spent in a daily commute and suggested new residences closer to the bluff land being farmed.[11]

After a generation of occupance, the limited timber and wood supply of the Pointe Basse was depleted, including trees that helped to keep the riverbanks stable, and this required increased usage of the wood in the bluff lands. Resources tend to draw people closer to them, the more frequently they are exploited. One would also like to think that the French would sooner or later recognize the relative healthfulness of the uplands in contrast to the malaria-scourged marshes and wet swales of the Pointe Basse, but little direct evidence supports this line of reasoning for removal, however logical in retrospect. Finally, fear of Indians, which had been strong in the 1770s when the Osages repeatedly drove the French and their slaves from mines in the interior, was abating in the 1780s.

People had been gathering experience living and working in the immediate bluff lands and creek valleys by pasturing stock there, trapping and hunting, cutting cedar and other choice timber, annually tapping maples for sugar, hauling to and from the mines, and eventually cultivating some small fields of corn and wheat.[12] These various accumulated experiences provided a knowledge base that led the villagers to place permanent settlement inward from the Pointe Basse. Formal requests for upland locations in the 1780s state that the petitioners had been using the land for many years.[13] The initial moves to settle in the hills, there-

11. Ekberg says that residents considered leaving the old village as early as 1775, a decade before the 1785 flood. One landowner, Caron, said in 1771 that his land was being washed away and that he needed a new *terrain* (*Colonial Ste. Genevieve,* 418–19, 421).

12. For example, when Louis Lasource requested lands in the bluff lands because of the floods, he wanted it next to the *sucrerie* (sugar mill) that the Lasource family had been using for a long time (LCS $58\frac{1}{2}$).

13. LCS 62, 77, 211. The move to the uplands involved some dispute over land. For example, an advance party of six established themselves there and sowed jointly $11\frac{1}{2}$ arpents on the royal domain and presumably fenced it in. Their action was objected

fore, were by individuals or family groups, without coordination, done under pressure for more land and resources, and preceded the great flood of 1785. The move into the bluff lands for permanent habitation represented a significant break in the French settlement *mentalité*, because the habitants went into a new environment of upland soils, removed themselves from a compact village community, and occupied land in other than long, riverine lots.

Nevertheless, it was the series of floods and accelerated bank cavings in the 1780s, culminating in the exceptionally high flood of 1785, that forced a decision by many to remove their structures to higher ground. Beginning in 1780, the community experienced six consecutive years of bad or lost harvests.[14] The great flood of 1785 began in April, and by May the villagers had abandoned their homes and withdrawn to Vallé's mill tract and other refuges on the bluffs overlooking their drowned fields.[15] This time, the entire bottom was under water; all houses and buildings, including those of the commandant and soldiers, were deeply submerged. In such deep water, wooden structures and fences became buoyant, floated, and washed away. During that exceptional flood, which lasted all summer, families living temporarily in the bluff lands probably broke some land for survival food crops such as corn and root crops, which could be planted on virgin soil and at a later date in the year. If habitants had been considering removal, this event would have made up their minds. By September, convinced they would have to relocate, some requested concessions in the karst lands, and by October a new town with a church was being proposed at Vallé's mill in the bluffs.[16]

For those who remained, another flood in 1788 accompanied by "illness" and still another severe flood in 1794, which again covered most of the Pointe Basse with up to five and six feet of water and washed away fences and destroyed the wheat, made continued living in the old village

---

to by those who used the uplands for grazing. The lieutenant governor decided that the six could continue cultivating fields broken before the 1785 flood and had the promise of possibility of title to the land (Petition for land by Bte Morriseaux et al., January 30, 1787; Petition by J Bte Vallé et al., January 31, 1787; Response by Francisco Cruzat, February 9, 1787, LCS 8). According to Julien LaBriere's testimony in 1825, three habitants of the old village had already moved to the uplands before the flood of 1785 (*Hunt's Minutes,* October 22, 1825, reproduced in E. M. McCormick, "The Coleman Family History," 34).

14. Cruzat to Miró, August 23, 1785, PC 11-581.
15. Peyroux to [the governor?], May 1, 1788, PC 201-948.
16. Cruzat to de Oro, September 9, 1785, PC 117b-247; Cruzat to de Oro, October 20, 1785, PC 117b-257.

insupportable.[17] The church, which villagers had rebuilt in 1788 after flood damage, was in 1792, at best, a "propped-up" church. By then, the sacred relics had been removed because so many villagers had departed that only eight to ten families remained in the village and the church was deemed unsafe from robbers on the river.[18] So few *terrains* remained occupied in 1792 that the fence that separated the village from the *terres* was removed and all cross-valley fences were extended straight through the evacuated village to the river, eliminating many linear arpents of fence between the *terres* and the village.[19] Thus, the *terrains* of the village were physically incorporated into the common field, and any structures that remained on them were likely cannibalized for parts in the building of the new town. Pigs and cattle could roam the former village streets and abandoned properties. Bank caving eventually ate into the derelict village and its cemetery. Coffins protruding from the riverbank were an unforgettable, macabre sight to young Anne Lucas when, as an eight year old in 1805, she traveled upriver on a keelboat poled and pulled along the bank at the site of the old village.[20]

Numerous requests for land in the bluffs specify that floods had forced the petitioners to seek new land. However, villagers had no intentions of surrendering their valuable Pointe Basse croplands, which they continued to use uninterruptedly for crops. They were abandoning and relocating the built part of the village only, although their requests for land in the hills often included additional land for farming. Thus, many habitants enlarged their total landholdings when they received concessions for land in the bluffs and lots in a new village.

The process of relocating the community took more than a decade. In addition to residents of old Ste. Genevieve, it involved Kaskaskians who had been just as severely flooded and other newly arriving persons, some of them quite wealthy. The process also involved local commandants and supervising Spanish authorities who were well aware of the tragedy in-

17. Peyroux to [the governor?], November 20, 1788, PC 201-959; de Hault Delassus to the Baron de Carondelet, April 6, 1794, PC 208a-480; Luzieres to [?], April 15, 1794, PC 209-643.
18. Zenon Trudeau to the governor, September 1, 1794, PC 197-715; Paul de St. Pierre to [?], October 1, 1792, PC 206-736.
19. François Vallé to Manuel Perez, April 11, 1792, letter number thirteen in "Letterbook of François Vallé" concerning governing the district, 1791–1793, PC 207b-335.
20. Anne Lucas Hunt, "Early Recollections," 44. Because the cemetery stood behind the old village according to Hunt, this part of the old linear village between the cemetery and the river had to have been completely lost to the river before 1805.

volving physical destruction and potential community dissolution, but their slowness of action, if not absence of it, allowed local people considerable latitude to make their own decisions.

There were several possibilities for relocation (see fig. 20). Those with *terres* in the Pointe Basse could go to the colluvial slopes at the opposite, inner end of the common field. Others could go to where the two forks of the Gabouri form a wide break in the bluffs, the *petites côtes*. Another possibility was the bluff land immediately behind the Pointe Basse, the *grandes côtes*. A fourth possibility was La Saline, the saltworks where the Pointe Basse pinched out at its southern end. Finally, habitants could disperse onto individual properties in the karst upland and in the several creek valleys of the hills. Villagers tried all of these possibilities during the 1780s and early 1790s.

Colluvial Slopes

It would seem logical that residents with *terres* in the Pointe Basse would consider relocating to the far end of their own properties that extended onto the colluvial slopes just beyond the common-field fence. In the northern half of the Pointe Basse, the alluvial land merged gradually into colluvium washed down from the hill slopes, and these colluvial slopes provided elevations higher than all but the highest floods. In addition, small ravines formed breaks in the bluff line, suitable for house sites and room for barnyard activities of a *terrain,* but these sites were irregularly spaced and did not occur at the back ends of all common-field properties. In the southern half of the bottom, however, the *marais* (marsh) near the base of the bluff forced the *chemin des petites côtes aux salines* (the road on the back side of the common field) to rise into the bluffs in its southern half.

A line of houses did develop on the colluvial slopes and in the ravine breaks in the bluff line south of Gabouri Creek, which is along the present St. Mary's Road, but it is unclear whether owners built them on the back ends of their properties or on new properties (A on fig. 20). If these houses were built on the back sides of common-field lots, then no new concessions were required and documentary evidence would be lacking. Later, true village lots *(terrains)* were conceded to individuals in this location, but these may have been awarded well after their occupation. The development of a linear village on the colluvial slopes in the northern end of the Pointe Basse makes sense, because land in the northern end of the old village was lost to the river first and in gravest danger of being further eroded. Removal to colluvial slopes may have reflected more concern with bank caving than with flooding, and may have begun before the great flood.

The row of houses appeared in contemporary accounts of Ste. Genevieve in the 1790s. The priest wrote in 1792 that a chain of houses joined new Ste. Genevieve and Mont Généreux. On a 1798 plat, "Village de Ste. Genevieve" is written along the colluvial slopes at the base of the hills below Vallé's mill tract, even after other residents had established new Ste. Genevieve in grid form between the Gabouris.[21] There is no indication that this linear string of houses was a coordinated settlement process, although the 1825 surveyor who reconstructed conditions of 1804 put cross streets between some lots. This location accommodated only a few of those who left old Ste. Genevieve.

New Ste. Genevieve

"New" Ste. Genevieve, the Ste. Genevieve of today, was the primary beneficiary of the exodus from the alluvial plain. The site is in a major break in the bluffs between the two Gabouri Creeks and just behind the bluff line projected across the break (B on fig. 20). The double Gabouri Valley offers easy, low-gradient access up into the karst plain and had long been in use for a route to the interior. The Gabouri between new Ste. Genevieve and the Mississippi had no current and held backwater from the Mississippi the depth of which varied with the elevation of the Mississippi. The new site was still close enough to wetland on the floodplain to be malarial and in this regard was not any better than the old. Most important, the site was adjacent to the Pointe Basse *terres*, albeit on their northern end, and habitants could continue to commute daily to their fields.

New Ste. Genevieve, as its form crystallized in the 1790s, was at its center an imperfect, but obvious, grid of square blocks approximately 350 to 400 feet on a side, divided mostly into four square *terrains* of generally 180 to 190 feet on a side (fig. 21).[22] Streets were subparallel and met cross streets at right angles or nearly so. When Americans surveyed the village in 1825, the first done for the village, the surveyor attached a note

---

21. Paul de Saint Pierre to [?], October 1, 1792, PC 206-736. The depiction of new Ste. Genevieve on the 1797 Finiels map shows a row of houses along the road at the base of the bluffs (*Account of Upper Louisiana*, 22). Similarly, the April 4, 1798, survey of François Vallé's mill tract contains the words *village de Ste. Geneviève* along the colluvial slopes at the base of the bluffs below his property (*Régistre d'Arpentage*, 62).

22. Dimensions from property deeds. The unsubdivided properties still fit their original dimensions today.

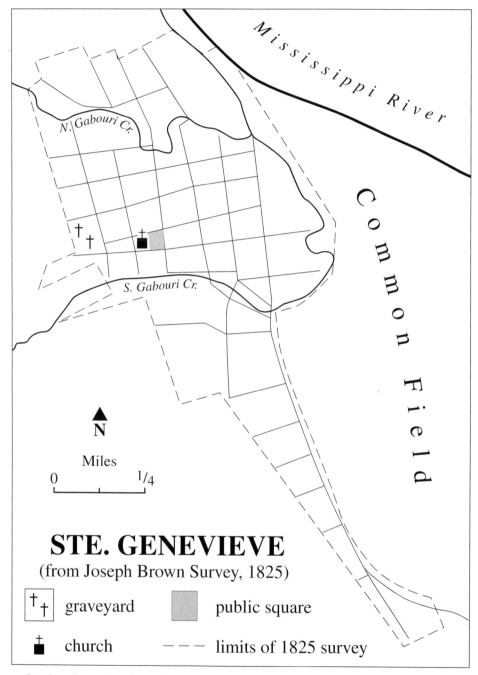

21. *Ste. Genevieve.* American surveyor Joseph Brown produced the first survey of the village that connected the various lots, streets, and public places as a way to confirm ownership of individual parcels.

to the connected plat that the true meridian was "laid down on this copy according to the course of Fifth Street, which comparatively runs N 79°40′ E or S 79°40′ W and is the only straight street in this Town, following one single course of the compass from one end to the other."[23] Any attempt the American surveyor may have made to regularize the imperfect grid into true squares with right angles was not possible because long-established property lines and buildings jutted into the rights-of-way. That the grid is not geometrically perfect may be explained by the original staking out of properties, maybe in haste, without a professional surveyor and the later unintended minor encroachments into public rights-of-way due to ordinary land use. The professional survey of 1825 had to respect fences, gardens, and buildings that extended beyond the intended lot lines. The grid form is most nearly perfect in the village center, where the early lots were laid out, but where the smooth terrace falls off on both sides to the Gabouri Creeks, the blocks and lots are less regular. The grid is oriented at approximately N 80° E and N 10° W, which conforms to the shape of the terrace and the orientation of the two flanking creeks.[24]

The village layout contains a *place* near its center. Reservation of open space had to be done by mutual consent before too many lots were staked off. A chapel was erected in new Ste. Genevieve, presumably by 1787, and later, in 1793, the collapsing church at old Ste. Genevieve was disassembled and moved to the *place* and enlarged when rebuilt.[25] Thus, the village arising on the Gabouri inherited the basic religious institutional func-

---

23. U.S. Surveyor General, "Plat of the Village of Ste. Genevieve, 1825," Sundry Surveys, by J. C. Brown, Missouri State Archives. This is a schematic plan (not a map drawn to scale) that shows the blocks of the central portion of the village in perfect squares and all streets perfectly parallel and perpendicular. It was constructed before the actual survey of village properties to help the surveyor understand the relative arrangement of properties that he was to survey. Fifth Street had to be the point of departure for the survey, as it was the only reasonably straight street he could find. Ekberg notes that the U.S. surveyors at Vincennes, Indiana, experienced similar difficulties on squaring blocks and lots in 1792 (*French Roots*, 50).

24. The grid of St. Louis, both street width and alignment and lot size and shape, is much more regular and precise than that of Ste. Genevieve, a contrast that is attributed to the presence of a professional surveyor at St. Louis at the time of laying out streets and lots (Peterson, *Colonial St. Louis,* 6–11; Ekberg, *French Roots,* 90, 99–100).

25. On chapel erected by 1787: De Hault Delassus to Carondelet, May 1, 1793, PC 214-4; on collapsing church: Zenon Trudeau to the governor, September 1, 1794, PC 197-715; Zenon Trudeau to Henrique Peiroux [sic], August 9, 1793, PC 207a-524; Father de Saint Pierre to [?], May 29, 1793, PC 208a-487. Commandant Vallé's concession in the center of the village was for a lot of "from four to six arpents," much larger than the lots of others, and it occupied one full square block (LCC 174, LCC 174a).

tion of the old village, which, as much as any act, determined that the new village would be the primary legatee of old Ste. Genevieve.

The grid form of the new village has provoked speculation as to whether it was planned and, if so, by whom and under what Spanish town-planning regulations. Clearly, the new village had forethought to its layout. Straight streets bounding approximately equal and square blocks quartered into square lots imply that someone, with the cooperation of the group, had to have marked and staked out properties. The *place* had to be identified and reserved. The simplest explanation is that the acknowledged community leader, François Vallé, took charge. The district commandant, Henri Peyroux, was concurrently trying to establish a village on the Saline as post of the district. Peyroux later gave Vallé and the village priest, Paul de St. Pierre, credit for establishing the new village. Members of the community held Vallé in respect and would have looked to him in their time of leadership need. The tight-knit community of interrelated families was described by Lieutenant Governor Trudeau: "Nearly all of them being related, blood binds them, to maintain the fast friendship and harmony which has always existed among them." Their leader, Trudeau continued, is "esteemed by those inhabitants . . . [and] is their true friend and protector."[26]

Vallé, a longtime resident of the Illinois Country, and all the other villagers were familiar with the grid pattern for a village. Kaskaskia, where many families had roots, had a grid form regularized in the 1720s. Cahokia also had a grid form, and old Ste. Genevieve had square blocks and lots nearly from its beginning. Even the dimensions of blocks and lots and widths of streets at new Ste. Genevieve are quite similar to those of preexisting villages in the valley. Vallé and other villagers were also aware of the grid plans of St. Louis, New Orleans, and Montreal, all French towns. There is absolutely no question that the grid form was part and parcel of French-village tradition in the Mississippi Valley for much more than a half century.

Others, however, have reached conclusions that new Ste. Genevieve was influenced, if not determined, by Spanish town-planning ideas from the Law of the Indies. Urban historian John Reps contends that the "regularity of Ste. Genevieve's streets and its imposing central square may well have resulted from the application of the Laws of the Indies." Landscape historian John R. Stilgoe believes that the Spanish planned new Ste. Genevieve: "Spanish officials insisted that it be layed out in a compact

---

26. Henri Peyroux to [?], November 14, 1793, PC 207b-373; Trudeau to Governor, St. Louis, January 15, 1798, in *Before Lewis and Clark*, ed. Nasatir, 2:535.

grid, about a plaza, and divided into blocks roughly 350 x 400." He gives no documentary evidence for the Spanish insistence. As for St. Louis, architectural historian Charles E. Peterson does not directly credit Spanish town-planning ideas for its grid plan, but simply notes that St. Louis is like New Orleans, which "may have inspired it."[27]

The Spanish Law of the Indies likely did not directly influence the plan of new Ste. Genevieve, even though the village was laid out during the Spanish administration of Upper Louisiana. Tora Williamsen, who both searched for documentary evidence and compared the law's provisions with Ste. Genevieve's layout, concludes that there is no reason to believe that Spanish ideas had any direct or significant role. In reality, she says, it is hard to separate French and Spanish town-planning ideas.[28] The grid plan, with a public square, was common to both. Therefore, the fact that the grid form of new Ste. Genevieve—and of St. Michel, Carondelet, St. Charles, Portage des Sioux, Florissant, and New Madrid, all of which were laid out during the Spanish administration—reflects stipulations of the Law of the Indies does not prove that those who directed the planning did it pursuant to those laws, because those same elements were already present in traditional French-village plans. It is, of course, reasonable to believe that community leader Vallé would have done nothing to incur disapproval of his Spanish superiors in St. Louis and New Orleans and would have done whatever he could to keep in their favor.[29]

27. Reps, *Town Planning*, 62. Also see Reps's discussion of the Law of the Indies in *The Forgotten Frontier: Urban Planning in the American West before 1890*, 5–29. John R. Stilgoe, *Common Landscape of America, 1580–1845*, 215–16; Peterson, *Colonial St. Louis*, 6. Peterson points out that, although we do not know whether founder Pierre Laclède was aware that St. Louis was on Spanish soil, "it is interesting to note how closely Laclède conformed" to the Law of the Indies (4).

28. Williamsen, "Colonial Town Planning," 112. Williamsen observes that the initial linear settlement along the colluvial slopes of St. Mary's Road may be French, but the town center with its grid may be Spanish; also, the alignment of the grid with the Gabouri Creek seems French, but the grid offset to deflect winds conforms to Spanish rules (117–19). Faragher remarks that the Spanish-Mexican settlement pattern had as much to do with Hispanic folk tradition as with imperial intentions, thereby discounting the significance of the Law of the Indies ("Americans, Mexicans, Métis," 100–101).

29. The one village that was unquestionably planned by Spanish officials in Upper Louisiana, but never laid out, was the village to adjoin Fort San Carlos on the lower Missouri River, a few miles above its mouth. Directions issued by Governor Ulloa in 1767 for its construction were for lot sizes and blocks utterly dissimilar to those of St. Louis and Ste. Genevieve, but also nonconforming to the Law of the Indies ("The Spanish Forts at the Mouth of the Missouri River," in *Spanish Regime*, ed. Houck, 1:9–10, 17). Judge James H. Peck thoroughly reviewed Spanish law on the laying out of villages and the size of village lots in a lengthy document he prepared for the second Board of Land Commissioners (*ASP-PL* 8:810–47).

Furthermore, it was the practice of the Spanish lieutenant governors at St. Louis not to interfere with local activities. They did, of course, formalize actions with their signatures, and thus the lieutenant governor did give his approval for the removal of the church and district commandant post to new Ste. Genevieve. Finally, Carlos III, king of Spain, had instructed the first Spanish governor of Louisiana that "no law or custom prevailing in the Indies will be applied" in the French-populated province.[30] If later governors adhered to this order, then they would not have used the town-planning ordinances of the Law of the Indies for Ste. Genevieve.

The best argument against Spanish law and ideas in the planning of new Ste. Genevieve is the sheer absence of documentary evidence. In all the rich archival record of the Papeles de Cuba that treats the move from old Ste. Genevieve, there is absolutely nothing about Spanish involvement in site selection or in planning the town, its lots, the *place,* the grid orientation, or anything. We are left with nothing but the speculation that François Vallé might have been aware of Spanish ideas and, with his community friends, did nothing that would seriously violate them. New Ste. Genevieve village was laid out according to French custom as directed by its community leaders.[31]

A new village of only *terrains* would not be sufficient to relocate the community. The Pointe Basse, with a finite amount of land, was insufficient for a growing population. Additional *terres* had to be identified. On November 16, 1797, after the new village had been well established, Lieutenant Governor Trudeau awarded the villagers of new Ste. Genevieve a large new common field of 2,520 arpents in the karst upland, from one to three miles or approximately a twenty- to forty-minute walk, south of the new village.[32] It was no farther away than the common field in the Pointe Basse. As with the latter, the new one was "common" only by virtue of a projected enclosing fence. There is no evidence that it was ever enclosed by a single fence, as the old common field was, and its nonenclosure severely compromised any function as cropland unless individual plots were fenced. To the habitants it was simply *le parc* to distinguish it from *le grand champ* in the Pointe Basse. Villagers later referred to it in English as the Ste. Genevieve Park or the Grand Park common field.[33]

---

30. As quoted in Weber, *Spanish Frontier,* 200.
31. The Papeles de Cuba are similarly without any reference to Spanish ideas in laying out other villages in the district: New Bourbon, St. Michel, or the proposed village at La Saline.
32. *ASP-PL* 2:499–502.
33. LCS 67. The term *parc* indicates that it was not plowed. In contrast, the Pointe Basse common field was not called a *parc.*

The *parc* lay in the grove-studded prairie in the karst upland of broad, shallow solution depressions dominated by the Grand Marais. It was actually composed of two sections, D and E on Figure 22, each separately conceded. Section D, the first, had the traditional 40 arpents width, even though it did not front on anything, but the other section was only 30 arpents wide.[34] Long lots were added later to the southern end, farthest from the village, in rougher land much less suitable for cultivation. The *parc* and its property lines were oriented N 65° E, continuing the direction that had been established a few years earlier for bluff-land properties at New Bourbon. Virtually all properties in the bluffs and karst plain between new Ste. Genevieve and New Bourbon adhere to this compass direction, which indicates an effort to bring geometric order to cadastral boundaries, perhaps so that they would fit together without leaving unconceded space between them.

Most of the individual strips in the *parc* were 2 arpents wide, but a few were wider for the more prominent families. The correlation between wealthier persons and strips closer to the village, better soils, and wider strips is too obvious to be random. Clearly, some families were favored in the way land was apportioned in the *parc,* and this practice was a departure from the earlier division of the Pointe Basse common field where a more uniform width prevailed among habitants. As in the Pointe Basse common field forty years earlier, villagers individually requested and received concessions for strips. Many were requested by April 1797, actually before the *parc* as a whole unit was conceded later that year. Those who requested land first, who were also the more prominent families, were awarded it beginning on the end closest to the new village. Those who obtained land in the new *parc* retained their *terres,* if any, in the Pointe Basse, so that many families had landholdings in both common fields.[35] The new common field turned out to be less an area of expansion for new families and more an expansion of lands for established families.

Although the new common field was intended for agriculture, there is no strong evidence that habitants ever used it as *terres cultivées*. The *parc* instead supplied the town with firewood, and animals grazed its prairie

34. The standard 40-arpent depth of concessions was distinctly a French custom that the Spanish administration of Louisiana accepted early (Hilliard, "Land Survey Systems in the Southeast," 9).

35. LCS 98. Although the *parc* was set aside as a single unit, families requested individually their strips in the common field and received separate concessions and American confirmations for them. Of fifty-four landowners in the *parc* in the uplands, twenty-seven continued to own land and presumably farm in the older Pointe Basse common field. The other landowners consisted of their married children and newcomers to the community.

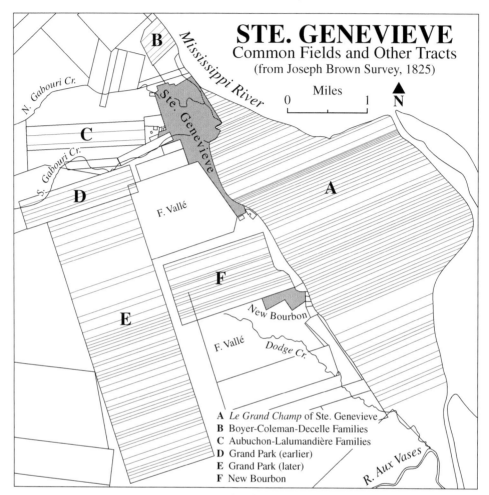

22. *Ste. Genevieve, Common Fields and Other Tracts.* American surveyor Joseph Brown produced the first survey that connected the individual-owned common-field strips and other properties around the village. In addition to the original Grand Champ (Big Field) in the Pointe Basse, at least five common fields were laid out in the bluff lands around new Ste. Genevieve. All may not have had a single enclosing fence to create a "common" space.

grasses, usages that correspond to the villagers' word *parc* for it.[36] In other words, the so-called common field actually functioned as a commons, and it was so referred to in testimony before the Board of Land Commissioners. It made sense that as long as a family retained sufficient land in the richly productive alluvial Pointe Basse, it had no reason to break out land in the new one. Land claims in the *parc* may have been made for prestige, as a land bank for the future, or as originally intended for the removal of crops from the Pointe Basse to the uplands. But the persistence of agriculture in the *grand champ* common field and its lack of development in the new one are more evidence that flooding of agricultural lands was not the real reason for village relocation, but rather the effect of flooding and bank caving on village structures.

Families new to the community, especially those arriving from the east bank, received concessions for lands elsewhere in the bluff lands and the karst plain, especially in the community of New Bourbon that was being developed at the same time. Thus, the failure of the new common field to replace or supplement the Pointe Basse common field must be interpreted within the general process of the scattering of habitants across the landscape and the early stages of the disintegration of the tripartite village and its open-field system.

Despite the slow demise of the traditional farming system in the French village, the cadastral element of the tripartite village persisted in the landscape. The individual properties of the common fields continued long into the nineteenth century, slowly undergoing reconfiguration into more manageable shapes. The built part of the village grew in population in the first decade of the American period as it acquired more mercantile and river-shipping activities. Six stores were in the village in 1811 with sales of $140,000 a year, while the Delaware still camped on its outskirts. The principal employment of villagers remained farming, but few men had not tried their hand at some aspect of lead mining or boating on the big river. A clear sign of the presence of Americans in the village was the organization of the first masonic lodge in Missouri in 1807, but even at that, Ste. Genevieve remained a thoroughly French village through the 1820s, with only a "few Americans."[37]

36. *ASP-PL* 2:499–502, 560.
37. Paul Wilhelm, duke of Württemberg, *Travels in North America, 1822–1824*, 175; *Missouri Gazette*, March 21, 1811; *Jefferson City Daily Tribune*, July 17, 1926. In his treatise on farming systems in the French villages of the Illinois Country, Ekberg concludes that Ste. Genevieve had the most enduring system of open-field agriculture and that it persisted longest into the nineteenth century (*French Roots*, 249).
Giovanni E. Schiavo, a promoter of the contributions of Italians in Missouri,

New Bourbon

The *grandes côtes* immediately west of the Pointe Basse was the floodproof land closest to the majority of the residents of old Ste. Genevieve. Several ravines rend the sixty-foot-high bluff face, so people, animals, and carts had little trouble ascending or descending it. The Vallé family, community leaders, had managed to come into ownership of most of the bluff-top land before the flood of 1785 and by concessions when habitants were scrambling for land in 1787.[38] François Vallé II used the land to make deals with important people who entered the district, land deals in the form not of traditional long lots for habitants, but of larger, compact shapes intended for other purposes. He sold a large tract on April 4, 1791, to John Dodge, an American immigrant from Kaskaskia, whose family soon became prominent in the Ste. Genevieve community. Another piece of the bluff land went to immigrant nobleman Pierre-Charles Delassus de Luzières in 1793. Vallé and Luzières also saw to it that Joseph Fenwick and his doctor son received land on the bluffs. Although ordinary habitants did not acquire title to land on the *grandes côtes*, it certainly was not closed to their use. Vallé built a water mill on a small spring-fed creek, a site that was, in fact, just as close or closer to the wheat crops of the Pointe Basse as the old horse-powered mill in the village of old Ste. Genevieve, which presumably did not function after the 1785 flood. Another community mill in the bluff lands was on Israel Dodge's land, on St. Cloud Creek before it tumbled down to the marsh at the base of the bluff. Habitants cut wood freely on Vallé's bluff-top lands to such a degree that he allowed them to continue only if they stayed away from creek sides.[39]

---

claimed that "many" "unmistakably Italian" surnames in early Missouri were rendered as Spanish, French, or Canadian. Among the "Italians" in colonial Ste. Genevieve were Michael Lami (Lamy), and elsewhere in Upper Louisiana were Emilian Yosti, Bartholomew Berthold, and Antoine Vincent Bouis, but the Spanish records seem not to indicate any Italian presence in Ste. Genevieve (*The Italians in Missouri*, 31–33).

38. François Vallé's first concession on the bluffs was for forty arpents' frontage on the edge of the bluffs *(sur les bords des dites côtes)* and extending back indefinitely to the rear of the lands of the post. When it was surveyed in 1798, its depth was reduced to the standard forty arpents of all bluff-facing properties (LCS 28; *Régistre d'Arpentage*, 62, 185; Amoureux-Bolduc Collection, Missouri Historical Society, St. Louis). Although Vallé did not have concessions for all of the bluff land between Ste. Genevieve and New Bourbon and beyond, he had effective control of it and its allocation after becoming commandant of Ste. Genevieve.

39. On Dodge: LR 304, 304a, 304b, 304c; and Dehault Delassus to Carondelet, April 30, 1793, PC 214-1; on Luzières: LR 304c; on Fenwicks: *Régistre d'Arpentage*, 188; and LR 112. Joseph Fenwick also received one of the widest strips in the New

The French Villages 245

By the end of the summer of 1785, enough people had moved to the bluffs near Vallé's mill, many as refugees from that year's long-lasting flood, that the site comprised a village (C on fig. 20). Bazile Vallé and François Leclerc had built residences there, without official concessions, and a commons was projected, apparently in anticipation of others. There was even talk of building a church near Vallé's mill. Five years later, in 1790, commandant Henri Peyroux described "a new village just formed on a very high bluff between the Saline and the village of the Rivière à Gabourit, named Calvaire by its people." The water was good and the view beautiful overlooking the Pointe Basse fields. Farmers were harvesting good wheat on the friable loessial soils without the rank growth common on alluvial lands. And two years later, in 1792, the parish priest reported excellent farm harvests at "Mont Généreux," yet another name for the *grandes côtes,* a site that he identified as one of the locations that his congregation had fled to from the flood seven years earlier. A 1793 map shows twenty houses for Mont Généreux (not labeled New Bourbon and therefore predating establishment of that village), although their orderly arrangement in a regular grid pattern may be merely a cartographic convention and not a representation of reality.[40] Thus, refugees and others without concessions were occupying the large tracts owned by Vallé and others on the *grandes côtes.*

This strong interest by habitants in the *grandes côtes* could well have led to its selection for the primary village to replace old Ste. Genevieve, had Vallé chosen to offer his tracts for a relocated Ste. Genevieve. However, the chief problem with its location was its two- to three-mile hauling distance from the Mississippi River. No tributary allowed harborage for boats.

Thus, this nearest upland to old Ste. Genevieve was partially occupied when Pierre-Charles Delassus de Luzières arrived in the Ste. Genevieve

---

Bourbon common field and the one most conveniently located next to the village (*Régistre d'Arpentage,* 206–7). On Vallé's mill: LRB 1, LRB 550a; on mill on Dodge's land: LCS 28, LR 304c, Pittman, *Present State,* 50; on woodcutting on Vallé's land: LCS 28.

40. On Vallé and Leclerc: Cruzat to de Oro, September 9, 1785, PC 117b-247; on church: Cruzat to de Oro, October 20, 1785, PC 117b-257; Peyroux de la Coudreniere to the governor, September 2, 1790, PC 203-402. The site was also known as Mont Généreux and La Barrera (PC 2363-331). The road that traversed the Pointe Basse common field and reached the bluffs at the foot of La Barrera is labeled on the AGI 1773 map as "chemin de la barrière." It had gates through the common-field enclosure at both ends. Paul de Saint Pierre to [?], October 1, 1792, PC 206-736; map drawn by Luzières in 1793 in the Archivo General de Indias, Seville, and reproduced in Ekberg, *French Roots,* 94.

District in 1793. This refugee of the French Revolution and Governor Carondelet were both Flemings from Hainault, a French-culture region of the Spanish Netherlands, and their families were well acquainted. While in New Orleans negotiating his friend's planned settlement in Upper Louisiana, Carondelet awarded Luzières a large grant with promises of more. Upon arrival in Upper Louisiana, Luzières was warmly greeted by Lieutenant Governor Trudeau, who recognized his potential for promoting settlement in the slow-growing Spanish Illinois. By his financial resources, his connections with other refugees, his figurehead leadership as a nobleman, and his knowledge, Luzières could further the economic and demographic development of the territory. Because such an important person as Luzières should have some administrative responsibilities, Trudeau, perhaps influenced by Carondelet, awarded him the position of commandant of a new district, created in 1793 especially for the nobleman, but with fewer responsibilities than other district commandants. The District of New Bourbon (Nueva Borbón or Nouvelle Bourbon), dedicated to the memory of martyred, beheaded Louis XVI, was carved from the southern part of the Ste. Genevieve District.[41] The boundary between the two districts split the Pointe Basse common field and passed between existing properties in the bluffs and karst upland, but as it moved westward into poorly settled lands, it became indistinct and can be determined only by noting which commandant signed land petitions. Luzières held 710 arpents on the *grandes côtes,* part of it coming from François Vallé's lands, and there he built his house overlooking the Pointe Basse common field, then already in cultivation for a half century and presenting the most completely Europeanized landscape in all of Spanish Illinois. It was a supreme view from *la maison du commandant* for *l'homme de l'ancien régime* to survey his domain.[42]

41. Archibald informs about the Luzières family in "Honor and Family." See also Houck, *History of Missouri,* 1:363–65; 2:136. Albert Tison and Frémon Delaurière testified in 1832 before the second Board of Land Commissioners that Luzières was a personal friend and "allied by blood to the Baron de Carondelet" (LPC 18m). On Trudeau: LCS 247; on Louis XVI: PC 2363-331 and Koehler, "French New Bourbon," 14. Koehler's thesis is the best treatment of New Bourbon. Also see Houck, *Spanish Regime,* 1:374.

Place-names in the New World that commemorated the Old Country, such as New Bourbon, had a counterpart in the Old World. When settlers pushed into new lands being cleared in the Loire Valley of France in the eighteenth century, their little groups of cabins were known by names such as "Canada," "Mississippi," and "Le Nouveau Monde" (Braudel, *The History of France,* 149).

42. On district boundary: PC 218-374, 214-870. Archaeologists from the Wickliffe Mounds Research Center in Kentucky and cooperating institutions were studying the Luzières house and site in the late 1990s.

A commandant needs a *poste,* a village center for his district. The site chosen for a village in 1793 by Luzières and his associate Barthélémi Tardiveau, with Vallé's collaboration, was the bluff top one mile northwest of Luzières's house and one mile southeast of Vallé's mill. The bluff top was already occupied by the collection of houses and fields of habitants of Mont Généreux. These existing improvements were probably incorporated into the new planned village, which would explain their absence from later records and nonappearance in land claims before the American board that confirmed land claims. As was the custom, the new village and the district had the same name, New Bourbon.[43]

The nobleman's decision to put the village at some distance from his house suggests an intention to put social distance between him and the habitants, a class separation of *la noblesse* from *les paysans,* which was a class and spatial separation that did not exist between Creole François Vallé and the habitants of Ste. Genevieve. Vallé lived in the middle of his village. Others of high social rank who came to live at New Bourbon also chose not to live among the habitants in the new village. René Guiho de Kerlegand, another émigré from the revolution, bought property well away from the village during his short, contentious tenure in the community. Similarly, Dr. Walter Fenwick, the American physician for Ste. Genevieve and New Bourbon, had his residence on an 828-arpent tract of land one-quarter mile west of the village, which does not seem far, but it was a great psychological distance from the 1-arpent-square village *terrains.*[44]

Luzières's initial plan was to populate New Bourbon with refugees from the revolution. A primary source of settlers was the collapsing colony of refugees at Gallipolis on the Ohio River, and Luzières and his associates made strenuous efforts, virtually all in vain, to lure them to New Bourbon. The connection between Gallipolis and New Bourbon was based largely on associate Barthélémi Tardiveau's ownership of land at New Bourbon and the prospect of profits from lead-mining there. Some did

---

43. PC 207a-532, 207b-371. New Bourbon must have been laid out or platted before October 21, 1794, when a 1-arpent-square lot was sold (LR 100). Koehler gives 1793 as the year New Bourbon came into existence ("French New Bourbon," 1).

44. Walter Fenwick acquired the land on August 18, 1798, from the extensive concessions awarded to his father, Joseph Fenwick (LR 112).

In a somewhat analogous situation to Luzières and New Bourbon, Alan Taylor points out the symbolism of William Cooper's residence in the center of the village that he created and thoroughly dominated. Then, as the village grew and Cooper grew more elitist, he built a new, imposing residence well away from the village center and surrounded it by a picket fence to clearly demarcate his domain from the town. He subsequently lost his authority over village affairs (*William Cooper's Town,* 259–62).

come to Spanish Louisiana, but they went elsewhere than to New Bourbon. As a receiver of refugees from the revolution, whether from Gallipolis or elsewhere, New Bourbon was a failure.[45]

After 1795, when it was clear that French from Gallipolis and other émigrés were not going to come and when the province was opened to American immigration, Luzières began his bold strategy of advertising in newspapers in Kentucky for settlers. Of the many Americans who responded, most went to individual farms elsewhere and only a few to the village of New Bourbon.[46] Most successful for populating the village was a flow of French Creoles from Kaskaskia, still reeling from assaults on their social and political systems by Virginia troops and a general lack of civil order. Some had tried new Ste. Genevieve first, in the late 1780s, before resettling in New Bourbon. Finally, there was a sizable cohort of Creoles from depopulating old Ste. Genevieve who preferred to move to New Bourbon instead of to the new village forming on the Gabouri, for whatever reasons.[47]

The village, as it emerged in the 1790s, was a collection of different groups of people who formed interrelated clans, such as the large Caillot dit Lachance and Deguire dit Larose families. They had not yet established an attachment to any place in the district and certainly had no association with their new commandant fresh from France or any of his business associates. It was a community gravely split between the habitants, diverse and not yet tied by any tradition to the place, and their leader. This situation contrasted sharply with the social cohesion of Ste. Genevieve, a community of persons with a generation or more of deep roots to the land they lived on and great respect for their leader.

If some habitants were already living at Mont Généreux in the 1780s and their lands were incorporated into the new village of 1793, then the village probably did not have a preconceived plan of square blocks and *terrains* that could be carried out, as in the case of new Ste. Genevieve. The only cadastral map of the village was the attempt of the American surveyor in the 1820s to reconstruct landownership as of 1804 (fig. 23). Before 1804, however, many properties had been transferred and the village had begun to disintegrate, and therefore the reconstruction, no mat-

---

45. Koehler, "French New Bourbon," 9–20; Howard Crosby Rice, *Barthélémi Tardiveau, a French Trader in the West*, 40–47; PC 203-520; Houck, *History of Missouri*, 1:362–64.

46. LR 216; Koehler, "French New Bourbon," 93–97.

47. Koehler, "French New Bourbon," 26–33. Koehler gives a full account of the large Deguire dit Larose and Caillot dit Lachance families at New Bourbon (42–55).

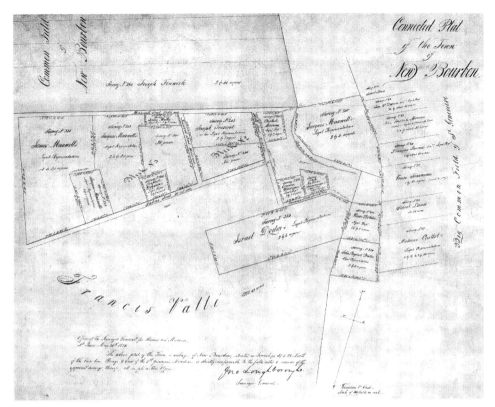

23. *Connected Plat of New Bourbon.* Laid out in 1793, only a few years after new Ste. Geneviève took form, New Bourbon was to be a separate post village for a separate administrative district. The connected plat, probably made in the 1820s to show landownership as of 1804, indicates that the village had not grown very much. (Miscellaneous Plats, Missouri Department of Natural Resources, Division of Geology and Land Survey)

ter how accurately done by the surveyor, probably did not represent the greatest development of the village of the late 1790s. Consequently, the map is quite inadequate for a depiction of the village. Other methods to discover the village properties of New Bourbon are similarly unsatisfactory, as individual land concessions were apparently not made for most of the town lots or else have been subsequently lost. Buying and selling of town lots were recorded, but the surviving documents cannot account for the large number of lots needed to account for the thirty-eight families listed as resident in the village in the census of 1797. Perhaps some *terrains* had more than one house on them, or perhaps residents had houses on land that continued in the ownership of important people. According to the 1797 census, the village of New Bourbon had a total population of 270, including 83 black slaves.[48]

The Finiels map of 1797–1798 shows two parallel rows of houses on either side of what might be interpreted as a *place;* at least it greatly resembles the depiction of the *place* at new Ste. Genevieve on the same map. New Bourbon probably did have a *place* where Luzières intended to erect a post for posting official notices, according to an expense account he submitted to Governor Gayoso.[49] The *place,* possibly for the location of the intended church, indicates a degree of coordination in planning. Certainly, with so much lead time for planning for the arrival of the Gallipolis French, someone could have marked off streets and lots on the ground. Perhaps Vallé, who donated the land for the village, had a hand in laying it out, especially if he were also involved in the layout of streets and lots in new Ste. Genevieve. Yet, there is no documentary evidence for "planning," whether by Vallé, Luzières, Tardiveau, or Spanish authorities.

Property descriptions also provide insight into the form of New Bourbon. At least sixteen properties in the village changed owners between 1795 and 1804, and at least twenty-one properties changed hands during the American period until 1819.[50] Virtually all lots for which dimensions are given are the standard one arpent square. Many list two and three bounding properties, but never more, which suggests four lots to the block, also standard. The documents mention the *chemin des côtes,* the road north to Ste. Genevieve, which followed the base of the bluffs and lay along the Pointe Basse common-field fence; the *chemin à la Sa-*

48. Miscellaneous plats, Missouri Department of Natural Resources, Division of Geology and Land Surveys; "Census of New Bourbon and Its Dependencies, 1797"; PC 2365-345. A good start at a much better plat map of the village could be made from three sources. One is the private land surveys in the Missouri State Archives, Jefferson City. A second is the numerous descriptions of properties sold that are in the record of property sales in the Ste. Genevieve Archives (before 1804) and in Ste. Genevieve County Deed Book A (after 1804). These descriptions include dimensions, shapes, sizes, frontages, streets, bounding owners, and buildings. The third source is Antoine Soulard's detailed surveys in September–October 1802 of René Guilho de Kerlegand's several disputed properties in the village and common field that are in the Papeles de Cuba, folder 291, items 426, 431, 436–38.

49. Delassus de Luzières to Gayoso de Lemos, June 6, 1798, PC 215b-629. Gayoso refused to pay the expenses for the post (and other submitted expenses) and told Luzières that he could post notices on the side of his house (Gayoso de Lemos to Delassus de Luzières, September 7, 1798, PC 215b-622).

50. LCS 47, 53; LRA 3, 26–27, 32, 42, 47, 49, 58, 61, 68, 71, 82, 86, 106, 108, 160, 223–24, 231, 236b; LRB 2, 40–42, 47, 97, 100–102, 129, 131–32, 158, 244, 270, 407, 434, 442, 460, 481, 498, 503, 529–30, 541, 561, 592; LRC 31, 47–48, 158, 162, 267, 274, 285.

line, the road south to the Saline; and the *chemin à la Pointe Basse,* the road that descended the bluff and went through the common field to the Mississippi River. Property descriptions include *rues,* but without any identification of a main street or cross streets. Curiously, no mention is made of a public *place* in any of these thirty-seven property descriptions.

Some *terrains* lay at the base of the bluffs along the *chemin des côtes* and bounded the Pointe Basse common-field fence. In fact, one New Bourbon house was on the bluff end of a *terre* in the common field. Only two of these lots are shown on the American survey map of Figure 23. Therefore, the built village not only was on the bluff top as traditionally described, but also descended down the bluff to include houses and lots at its base.[51]

A traditional French village required associated *terres cultivées,* or a common field. The New Bourbon common field, which went by the name of *le grand parc des grandes côtes,* was officially laid out in 1798 on the karst plain to the north of the village on land donated for that purpose by the Fenwicks, who had earlier acquired it from François Vallé (F on fig. 22). However, habitants had much earlier broken that land and cultivated it, because the village of New Bourbon had been in existence for several years before 1798.[52] The delayed official act probably indicates an attempt at regularizing the sizes and shapes of existing properties to establish a single common field. As in all other common fields, each villager was to have a *terre* and own it outright. Several *terrains* backed up to the *parc,* as shown on the American survey map. Individual long lots were similar in size (most of them the standard 2-by-40 arpents) to those of the adjacent new Ste. Genevieve common field. The total area of the New Bourbon common field was much smaller, only 715 arpents, or about one-fourth the size of the new Ste. Genevieve common field. Like the Pointe Basse and new Ste. Genevieve common fields, the New Bourbon common field grew by adding new properties to its western end, aligned with existing ones. Thus, Alexis Griffard requested a strip 5-by-40 arpents at the end of the common field, and later Jérome Matis (or Matisse) requested a *terre* $2\frac{1}{2}$-by-40 arpents at the end, next to Griffard. Although villagers had been using their long lots for several years, they were not surveyed until landholders so requested in 1798. Landholders

51. The thirty-seven property transfers include descriptions of nineteen houses. They specify *poteaux en terre* (vertical posts in the ground) construction five times and *maison sur sol* (vertical posts on a sill) only once. Several slave *cabanes* are identified. Lots held barns, corn sheds, chicken houses, gardens, and orchards. Five properties are described as enclosed, one with oak posts.

52. LRA 223; LR 203.

asked that the *terres* take the orientation of N 65° E established by the *arbitres* in 1794 when Vallé's mill tract was surveyed.[53]

The New Bourbon common field was conceded specifically for cultivated crops, and a single fence enclosed the entire tract. Several persons who were allotted land in the New Bourbon common field also had land elsewhere, especially in the Pointe Basse common field, and likely they preferred its continued cultivation to the breaking of new land. The New Bourbon portion of the Pointe Basse field had its own fence and gates and was traversed by its own road across it to the Mississippi River. Several residents of New Bourbon village also had *terres* in the new Ste. Genevieve common field. New immigrants from Kaskaskia and elsewhere got land in the old Pointe Basse field, as long as it was available. Lieutenant Governor Trudeau had earlier, in 1794, reserved all remaining land in the lower end of the Pointe Basse for those coming from Gallipolis, and their nonarrival left this space available to other New Bourbon villagers. For many, however, the New Bourbon common field on the bluffs constituted their sole *terres cultivées*. When village lots were sold, accompanying *terres* in the common field went with the sale, continuing the French custom of pairing the two disjoint pieces of property.[54] New Bourbon, like new Ste. Genevieve, had no specific tract of land committed as a grazing commons.

The village, as a settlement scheme, was not successful. Already in 1804, Amos Stoddard, in his carefully written inventory of Upper Louisiana, paid no attention to the village and utterly ignored the existence of a New Bourbon District with its own commandant. He assumed both to be part of Ste. Genevieve. New Bourbon had declined to an "inconsiderable village" in an 1811 account in the *Missouri Gazette,* published in St. Louis. By the 1820s, it had fallen into utter dereliction, although the toponym has remained on maps and is still marked in the landscape at the beginning of the twenty-first century.[55] Except for a sharp bend in a county road that respects original property lines, its site passes today virtually unnoticed.

Why did New Bourbon fail as a settlement with such prominent backing and great promise? First, except for the civil commandant's house, it

53. Though an archaeological survey of the site of New Bourbon would be helpful in the understanding of the colonial and territorial history of the Ste. Genevieve area, none has been done. On Griffard: LCS 34; on Matis: LCS 67; on 1798 request: LCS 225; on orientation: LCS 112.

54. On specific use of concession: LR 49; on New Bourbon common field: PC 218-374; on Trudeau: PC 207a-542; on sale of *terres:* PC 218-501, 218-506.

55. *Missouri Gazette,* March 21, 1811; H. S. Tanner, map, Illinois and Missouri, State Historical Society of Missouri, Columbia.

had no institutions. In his negotiations with Luzières and Tardiveau, Governor Carondelet had promised a church for New Bourbon as part of the process of relocating the church from old Ste. Genevieve to new Ste. Genevieve, but the New Bourbon church never materialized.[56] Villagers had to walk to new Ste. Genevieve for mass, a distance of two and one-half miles, or forty-five minutes. The two villages were too close to each other to be separate parishes, and the decision had been made to place the church at new Ste. Genevieve.

The village was to get two mills, according to Lieutenant Governor Trudeau, one of them to be built by Luzières as part of his deal with Carondelet, but ended up with none as the nearby Vallé mill fell into disrepair. New Bourbon developed no commercial activities, no fur exchange, no Indian trading. The 1797 census identifies a few craftsmen, but no merchants.[57] Too far from the Mississippi River, the village could not serve as a lead depot and had no lead-marketing families except that of Luzières, a subordinate partner of Vallé, who, of course, used his own town of Ste. Genevieve as a lead depot.

Luzières's residence away from the village and the diversity of its habitants did not help matters. In addition, one might surmise that this older nobleman—his life and most certainly his wife's shattered by the bloody revolution and their strenuous flight from it, and then overcome with chronic high fever (yellow fever?) that made him bedfast and energy sapped—may not have had the drive to be a leader and promoter in a new land so strange to him. His visitors commented on his languor and unhappiness. Having fallen heavily into debt, he lacked servants, leaving his own family to work the land and do menial housework. His excellent ideas for economic and demographic development received polite responses but no financial support from the Spanish. Except for the youngest, Luzières's grown children established lives away from New Bourbon. Charles de Hault Delassus was commandant at New Madrid in 1796, then promoted to lieutenant governor at St. Louis, and expended considerable effort sustaining his father's good name in Upper Louisiana. Luzières's son Jacques St. Vrain went to St. Louis and became prominent in the fur trade. His daughter and son-in-law, Pierre-Augustin Derbigny, moved to New Orleans. Only the youngest child, Camille, stayed in New Bourbon. Luzières's fellow aristocratic refugee in the district, the sieur de Kerlegand, did not offer the moral or public support to Luzières that he could have, and the Kerlegand family and slaves moved to New

56. PC 207a-532, 214-4.
57. On Trudeau: PC 2363-320, 329, 207a-532; on 1797 census: PC 2365-345.

Orleans in 1802. Contrasted to Vallé, Luzières comes across as an ineffective leader.[58]

Military commandant Henri Peyroux had reservations about the site for New Bourbon immediately upon hearing of its choice for a village. It was, he said, badly situated and would never hold one hundred families. There was no good water on the bluff tops between the two mill sites, which were not surrendered by Vallé and Dodge to the proposed village; streams ran dry for six months. Peyroux pointed out that the location was a few miles from the Mississippi and the road across the alluvial plain was bad, across fields and marshes and closed at both ends by gates of the common field. But the commandant was intentionally accentuating the negative about New Bourbon, because he was trying his best to make every site look bad in comparison with La Saline, which he was concurrently promoting for district post. Nevertheless, the negative geographical characteristics that Peyroux attributed to the site were true, even if the positive went unmentioned.[59]

In the 1790s, lead mining held a powerful attraction for the young men of New Bourbon and other villages. Settlers at New Bourbon village, such as the large Lachance family and its extensions by marriage, already psychologically uprooted by the move to New Bourbon from Kaskaskia, had not put their roots deeply into New Bourbon soil before they opted to pick up once again and move. For example, eight New Bourbon families left to establish the lead-mining village of St. Michel in 1790–1800, and this drain on the village was not replaced. Curiously, though, a relationship between St. Michel and New Bourbon continued through the 1810s, as residents of the two French villages exchanged properties.[60] Americans

---

58. On Luzières's condition: PC 208a-446; on family situation: Zenon Trudeau to Baron de Carondelet, April 11, 1794, PC 209-638; on excellent ideas: PC 214-4 and Koehler, "French New Bourbon," 77–83; on family: Archibald, "Honor and Family"; on Kerlegand: Koehler, "French New Bourbon," 26. Koehler concluded that aristocrat Luzières failed to develop a social and working relationship with the habitants of New Bourbon (7, 98).

59. On Peyroux's reservations: PC 207b-3717. Koehler listed four reasons for the demise of New Bourbon: the lure of St. Michel; a lack of merchants, church, and meetinghouse; the presence of American settlers who broke up the French-based community; and a lack of a relationship between the commandant and the habitants ("French New Bourbon," 87–98). Elsewhere, Koehler offered a fifth reason: the Spanish government did nothing to aid the project (77–83).

60. For example, in 1812, Robert and Elizabeth Franserge Crawford exchanged their town lot and common-field strip at St. Michel with a New Bourbon town lot and common-field strip in the Pointe Basse owned by Joseph and Elizabeth Levrard Tesserot (LRB 270). Finiels noted in 1797 that some residents of New Bourbon were lead miners, which was before their move to St. Michel (*Account of Upper Louisiana,* 119).

from the St. Michel community, the Kellys and the Crawfords, bought New Bourbon lots, and other Americans moved into New Bourbon. For example, James Moore from New York bought four *terres* in the *parc* and several *terrains* in the village before abruptly leaving for Lawrence County. Thus, the fragile social cohesion of the community was further rent by the entrance and exit of persons of a different culture.

Finally, the few pillars of the community died and were not replaced. Luzières died in 1806, surviving his unhappy wife by only a few months. The pair were buried in Ste. Geneviève because New Bourbon never achieved a church and cemetery. Father James Maxwell, priest of Ste. Geneviève who lived at New Bourbon, died unexpectedly in 1814, and his fine house and eight-arpent lot were sold to an American in 1818. Israel Dodge, a major figure in the Illinois Country whose residence was on the bluff top near Luzières's, died in 1806. His house and league-square property were sold in 1819.[61]

Thus, the village with a location "fresh and healthy and a bit more cheerful" and a commanding view of the Mississippi and the lush Pointe Basse declined as the nineteenth century unfolded. "A beautiful view is not sufficient for a successful settlement," wrote Peyroux prophetically of the site in 1793. New Bourbon falls into the group of settlement strategies of privileged aristocrats who found their undertakings in America, however well intentioned, overwhelmed by the availability of land, alternate opportunities, a strangely different society, a population that would not be commanded, and, in general, circumstances beyond their control.[62]

La Saline

In several respects, La Saline, the salt-producing community in the Saline River Valley, was like lead-mining villages (D on fig. 20). Its site was determined by a local extractive resource, not by arable land. La Saline was always small, with a temporary and diverse population comprising

---

61. Archibald, "Honor and Family," 40; Rothensteiner, "Father James Maxwell," 154; Louis Pelzer, *Henry Dodge*, 13.

62. First quote: Finiels, *Account of Upper Louisiana*, 45; Peyroux quote: PC 207b-371. The general availability of land and alternate opportunities destroyed many colonization schemes in the New World. Meinig so explained Lord Calvert's lack of success in Catholic Maryland (*Shaping of America*, 1:151). Meinig also noted that it was early observed that a transplanted aristocracy could not survive in the "American forests" (303). Sigmund Diamond explained the failure of seigneuries in colonial French Canada in terms of availability of other activities that attracted farmers away from the seigneurs' lands ("An Experiment in Feudalism: French Canada in the Seventeenth Century").

more single men than families. It seemed to be in continuous litigation over ownership and leases during both Spanish and American regimes and over prescriptive rights to the resource. It also had the same irregular landscape form of unplanned, semipermanent, lead-mining camps.

Several salt springs or seeps, the surface resurgence of groundwater from the extensive karst plain to the west and south, occurred along the lower Saline Valley. The settlement of La Saline formed around the lowermost: La Grande Saline. Salt was also extracted from other springs a few miles up the valley, called La Petite Saline (Les Petites Salines), although more erratically.[63]

Indians occupied the site from approximately A.D. 700 to 1500, and the French learned about the salt springs from Indians during their first encounters in the seventeenth century. A Euramerican settlement at the site appears on a map dated 1686. A settlement of French was reportedly at the site in 1700, and the "Salinas River" is marked on a map of 1700. The saline was conveniently located right where the Saline River emptied into the Mississippi and therefore wonderfully accessible to all who moved on the big river. It may have been a factor in the Kaskaskia Indians' choice of a site for their village just across the Mississippi River and, in turn, for the French to locate their mission settlement of Kaskaskia next to the Indian village. La Saline was providing the necessary salt for the valley in the 1750s.[64] Although La Saline never became a population center of any consequence, it was a site of regional economic significance as well as a place for individuals of different economic circumstances to turn a handsome profit by the exploitation of an essential local resource.

63. Trimble et al., "Frontier Colonization"; Denman, "History of 'La Saline.'" The Petite Saline is identified in MN 30 and on the Finiels map.

64. On 1686 map: Trimble et al., "Frontier Colonization," 167; Richard O. Keslin, *Archaeological Implications on the Role of Salt as an Element of Cultural Diffusion.* Keslin traces Indian occupation of the Saline River Valley back to approximately A.D. 700 (146). On 1700 settlement report: "Relation de Pénicaut," in *Mémoires et documents pour servir à l'histoire des origines français des pays d'outre-mer; découvertes et établissements des français dans l'ouest et dans le sud de l'Amérique septentrionale (1683–1724),* by Pierre Margry, 407. Pénicaut's attributed words in 1700 were "la petite rivière de la Saline, nommée ainsy parce qu'il y a deux sources d'eau salée. . . . Il y a présentement en cet endroit un establissement de François." On 1700 map: Jacques Bureau, map of 1700, pl. xii (also see Guillaume Delisle, map of 1703, pl. xiii), in *Atlas,* by Tucker; on location: Cruchet, "La Vie en Louisiane," 66; MN 25; Trimble et al., "Frontier Colonization," 169–72. Belting noted that Kaskaskians made extensive use of La Saline and that the road to it was well traveled already in the 1730s (*Kaskaskia under the French,* 22). After 1763, the English in Kaskaskia were jealous that the valley's salt supply lay on the Spanish side (Denman, "History of 'La Saline,'" 313).

La Saline also served as a lead-shipping point. Lead from Mine la Motte, opened in the 1720s, came by animal or cart over ridge roads and then down the Saline River Valley to its mouth to be loaded on Mississippi River boats. La Saline was the closest landing on the Mississippi to Mine la Motte and did not require hauling lead over the wet alluvial lands of the Pointe Basse to reach old Ste. Genevieve.[65]

The Saline was early conceded to Antoine de Gruy in 1744 during the French regime, well before any riverine land was conceded at old Ste. Genevieve. Despite this, and especially after de Gruy's departure, individuals came to the Saline to make salt, some with leases from the owners, some without. The making of salt was officially viewed similarly to the mining of lead through prescriptive rights, although it is not clear whether such salt making was allowed only for one's own supply or also as commercial ventures. Thus, a curious complex of rights developed to the land, to the brackish water, and to the adjacent timber for the nearly perpetual fires. The assertion and enforcement of these rights shifted according to personalities and situations.[66]

The concession passed through several confusing ownerships—was the land being bought or only the rights to make salt, and did the rights to make salt include rights to cut timber?—until Henri Peyroux, military commandant of the Ste. Genevieve District, purchased it in 1787. The concession at that time was forty arpents' frontage on the Mississippi, extending one league into the interior, typical French property dimensions and shape. Peyroux asserted strong control over the land. He noted that because people had been building furnaces since before de Gruy's concession, he would allow those already there "to keep that right," which could also pass to others by sale or inheritance. Thus, an American salt maker named Dugan sold his *fourneau* (the fire pit to boil the saltwater) to John Duval, who then sold it to David Rhorer, all transactions in the "village of La Saline" apparently recognized by landowner Peyroux and others. However, Peyroux objected to additional persons moving onto his land.[67]

---

65. Part of this route is the "Three-Notch Road" (Kedro, "Three-Notch Road Frontier"); MN 22ao.

66. On de Gruy: MN 26c, 26f. Flader links de Gruy's 1744 concession to general expansion at Kaskaskia in the middle 1740s, including Kaskaskian interest in agricultural settlement at Ste. Genevieve ("Settlement History," 5–6). On complex rights: MN 25; MN 25a; MN 25c; MN 26f. Ekberg also noted the difficulty of determining landholding patterns at La Saline (*Colonial Ste. Genevieve*, 158–61).

67. On Peyroux: MN 26b, 26c, 26f. Trimble et al. conclude that Peyroux was never able to exert effective control over his land ("Frontier Colonization," 187). On furnaces: MN 25, 25a, 26f. As with lead *fourneaux,* the word *furnace,* a cognate of *fourneau,* may be an inappropriate translation for salt makers' uncovered *fourneaux.*

In late 1787, shortly after buying the old concession of forty arpents' frontage, Peyroux obtained a new concession that enlarged the property to eighty arpents' frontage on the Mississippi, which doubled the size of his property to one league square. Peyroux purportedly wanted to secure a larger supply of wood for the salt furnaces and to establish a *vacherie*. However, it was impossible to survey the square league without encroaching upon other concessions recently made in the Saline River Valley, some of them also for wood supplies for salt making. Lacking a single map of connected concessions, the granting officers created trouble by awarding land where not enough unconceded land remained. The land problem also arose in part because of an incorrect mental map of the tributary St. Laurent River, which granting officers thought ran perpendicularly to the Mississippi River, when in fact it ran more parallel to it; land descriptions used the St. Laurent as a central, perpendicular axis for land fronting on the Mississippi, which was impossible on the land surface. Thereupon commenced a series of confrontations among commandant Peyroux and many others.[68] Although the chief issue raised was the wood supply, figuring in also was Peyroux's attempt to move old Ste. Genevieve, then disintegrating, to his property on the Saline River. This could have been an unstated motive to double the size of his property in the years after the flood. Peyroux's chief rival in both the Saline land dispute and the relocation of Ste. Genevieve was none other than François Vallé, at that time leading the move to the Gabouri. Neither Peyroux nor Vallé lived at La Saline. Ironically, Peyroux's continued litigation over the Saline River land may have deterred some people who were leaving old Ste. Genevieve from moving there. People wanted land for themselves, not land in dispute or leased from Peyroux.

François Vallé had received a concession with a salt spring and the necessary timber supply in September 1796 for 7,056 arpents (1 league square) adjacent to Peyroux's and just upvalley on the Saline River. Upon surveying the next September, there was insufficient land for the full league, and Vallé received only 6,056 arpents. He then obtained another

---

Perhaps *fireplaces* or *fire pits* would be better translations. On Dugan et al.: MN 27; on Peyroux's objections: MN 26.

68. On Peyroux's new concession: LCS 237, PC 2365-523. The incorrect mental map of the orientation of the Mississippi River as hardly deviating from a north-south line along longitude ninety-one to ninety-two degrees west and of the orientation of its tributaries as eastward flowing is depicted on the Soulard 1795 map (W. Raymond Wood, "The Missouri River Basin on the 1795 Soulard Map: A Cartographic Landmark," 185) and several other contemporary maps, although by 1810, the orientation had been corrected and shows St. Louis at a longitude farther west than Cape Girardeau (Moulton, *Journals of Lewis and Clark Expedition,* maps 2, 6, 123, 124, 125).

concession in 1799 for the remaining 1,000 arpents, this time specifically for woodcutting for salt making. Vallé and Peyroux disputed their league concessions. Vallé, like Peyroux, had others living on his land. Hapless Ephraim Carpenter, an American trying to establish a simple farm in the Saline Valley, was driven off his land first by Indians, then again by Vallé when the latter enlarged his property by the additional 1,000 arpents.[69] Other American farmers in the valley received the same treatment from Vallé. Thus, the Saline Valley came to be dominated by large landholdings whose owners made it uninviting, if not impossible, for others to live there independently. The Gabouri site did not have such trouble, as the timber on it was not thick enough to claim in league-size tracts as on the Saline.

The form of the settlement at La Saline is known only sketchily. Judging from its long history as a place of mineral exploitation and the lack of references to streets or a pattern, La Saline was an unplanned collection of dispersed houses, cabins, *fourneaux,* and pertinent salt-making landscape features (fig. 24). English-speaking Thomas Hutchins called La Saline a "hamlet" in the early 1770s. Finiels described it as a "little village of little consequence . . . of 5 or 6 families" who "cultivate wheat, maize, mountain rice, vegetables, & raise cattle; work at extracting salt." In a Mississippi river man's guide of 1808, the settlement was a "village." *Fourneaux* and houses were being sold by 1780, but also being sold were village lots of 1 and 3 arpents size. According to the property deeds, the lots had houses, some of stone; fenced-in yards and gardens; and sheds and other outbuildings. In general, the lots were similar to a village *terrain*. Residents grew garden vegetables but tilled no large fields, and they raised barnyard animals. The village had no *terres cultivées,* no *parc.*[70]

Archaeologists working at La Saline in the 1980s discovered evidence of a house of the French *poteaux en terre* construction with dimensions of approximately twenty by twenty-six feet. They assembled evidence of up to twelve *fourneaux* for salt making, and one built with stone walls and dimensions of twenty by twenty-five feet, possibly for sugar making. Plenty of English ceramics suggest that the village had residents other than French and French Creoles. They concluded that the community at La Grande Saline was a nucleated village, but the community up the valley at La Petite Saline was dispersed and much smaller.[71]

69. On September 1796 concession: LCS 106; LPC3a; *Régistre d'Arpentage,* 194, 243A; on size increase: *ASP-PL* 2:407.
70. Hutchins, *Louisiana and West Florida,* 110; LR 243; Finiels, *Account of Upper Louisiana,* 44–45; river man quote: Z. Cramer, *Navigator,* 87; on sales: LR 243, 282; on lots: MN 25, LR 282.
71. Trimble et al., "Frontier Colonization," 172–86.

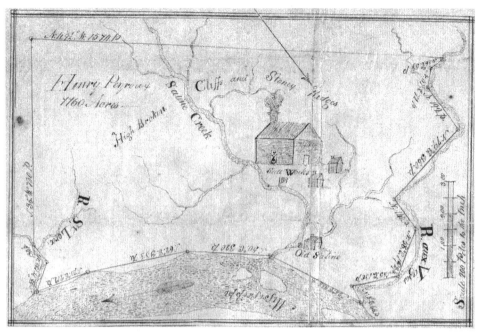

24. *Henri Peyroux Concession at La Saline, February 22, 1806.* The large Henri Peyroux concession of December 24, 1787, which included the saltworks at La Saline, was surveyed by Americans in 1806 for purpose of land confirmation. This plat shows four buildings, the large one with what is believed to be a female figure in the doorway and smoke from the chimney, and two boxed-in springs (documents of the *American State Papers, Public Lands [for Missouri]*, Decisions of the Second Board of Land Commissioners, 1st Class, Decision No. 166; in Missouri State Archives). Another version of the same survey plat also shows four buildings, all of which have smoke coming from chimneys. On that plat, the building at the salt spring closest to the Mississippi River has a double chimney and an arched doorway (Record Book of Land Titles, vol. C, p. 258; in Missouri State Archives).

The detailed 1797 census of the New Bourbon District presents an interesting profile of the Salines and their environs and captures the settlement during its highest development. It listed forty-one households for a total population of 166, including fourteen single-man households. It listed 17 slaves in only three households: Henri Peyroux, the French-born, now former district commandant; Mr. James (DeGimes), an American farmer; and Mr. Samuel, an Irish-born salt maker. Males constituted 70 percent of the total white population and 77 percent of the working-age population, which was fitting for a mining village. The census revealed a quite ethnically diverse population for so small a settlement. Although Americans dominated La Saline with twenty-eight households, five other "nationalities" were represented: French, French Canadian, Creole, Irish, and Scottish. Four different religions were identified: Anglican (twenty-three households), Catholic (eleven), Presbyterian (five), and Anabaptist (two). In addition to 10 farmers, who probably lived along the branches of the Saline River away from the village, no less than ten other occupations were listed. Subtracting the 10 farmers would produce a resident population of the village proper of from 50 to 80. However, a list of those who contributed to the defense effort in 1799 and gave La Saline as their residence includes many names that are not on the census.[72] La Saline was a village of an ethnically diverse and transient population.

La Saline was a primary place in the district for single Americans to establish the required residency of three years before requesting land in Spanish Illinois. For instance, James Farrell worked at the salt furnaces for six years (from 1793 to 1799), established residency, and then acquired his own concession up the Saline. In this way, La Saline served as a staging area for Americans, especially bachelors, before and after the liberalization of immigration rules.[73]

As the Mississippi River water spread across the Pointe Basse again in 1788 and banks slumped into the channel, district commandant Peyroux recommended to Governor Carondelet in New Orleans that La Saline be the new post *(chef-lieu)* for the district. He noted that Louis Lorimier and Hypolite Bolon, both of whom held keys to good relations with the Shawnee who then figured in Spanish defense strategies for Upper Louisiana, had taken up residence at La Saline. Carondelet replied with a consent to proceed with the plan for La Saline—nothing yet was organized elsewhere to relocate old Ste. Genevieve and shift the district post—but there would be no money to support the project. A buoyant Peyroux believed he had full support to build a village at La Saline and wrote back of his intentions to move the district's troops there as soon as lodging could be built for them. He noted that the only problem of movement between La Saline and the rest of the populated area would be the crossing of the River Aux Vases, but a bridge, he said, would be easy to build. As the flood year drew to a close, Peyroux, who had now seized the official initiative on relocation, was upbeat about his post-to-be. Although he was still paying on the purchase of the large property, he reported that houses were going up on lots and a blacksmith and gunsmith were taking up res-

---

72. "Census of New Bourbon and Its Dependencies, 1797," PC 2365-345; "Patriotic Donations and Loans Made by the Residents of Upper Louisiana to Aid Spain in the War—1799," in *Spanish Regime,* ed. Houck, 2:292-96.

73. LCS 101, 146, 163, 168, 186, 191, 224.

idence. He informed the governor of his plans to build a boatyard and church at La Saline.⁷⁴

These early efforts produced little tangible result. During the next few years, those evacuating old Ste. Genevieve followed François Vallé and the village priest to the Gabouri. Peyroux, who, after all, was commandant of the district, refused to cooperate with that flow and stubbornly remained in his house in old Ste. Genevieve. Lieutenant Governor Trudeau, apparently with the governor's support, bluntly told Peyroux not to interfere with the move to new Ste. Genevieve and that as military commandant, he should move from his residence in the old village to new Ste. Genevieve because the arms and the men he commanded had already moved there.⁷⁵ Yet, Peyroux continued to hold out for the post village on his property at La Saline.

By this time, 1793, the high-profile New Bourbon settlement project of Luzières and French émigrés was unfolding, and Commandant Peyroux saw a new opportunity to gain control of the relocation process from Vallé and his group. Referring back to the 1788 permission from Carondelet to build the district post at La Saline, Peyroux acknowledged that new Ste. Genevieve on the Gabouri was indeed a fait accompli, but argued to Governor Carondelet that the proposed New Bourbon should become the administrative post for a new and greater New Bourbon District whose boundaries, Peyroux proposed, would extend from the mouth of the Ohio north to the Joachim River and thus absorb the Ste. Genevieve District. Its commandant would be directly accountable to the governor at New Orleans and bypass the lieutenant governor at St. Louis. The New Bourbon commandant, Peyroux further proposed, would have the authority to concede land, as the lieutenant governor at St. Louis had, and the proposed new village of New Bourbon would not be built on the bluffs but at Peyroux's new village at La Saline. Clearly, Peyroux was energized by a combination of his own grandiose thinking with everyone's high hopes for the immigration of large numbers of French to New Bourbon.⁷⁶

Peyroux laid out a carefully constructed argument to the governor why the New Bourbon project should be located on high ground at La Saline rather than on the *grandes côtes*. Peyroux's site was on the low bluffs lying along the Mississippi River between the mouths of the Saline and the Aux Vases, which were "suitable and commodious ports for pirogues and boats." The location lay directly opposite the mouth of the Kaskaskia Riv-

---

74. On Peyroux's recommendations: PC 201-941; on Carondelet's response: PC 201-943, 201-945; on Peyroux's reaction: PC 201-954, 201-957, 201-1001.
75. PC 207a-524, 526, 532.
76. Henri Peyroux to the governor, November 14, 1793, PC 207b-375.

er around which were well-populated American settlements. The site was fertile and had plenty of fish. Stone for construction and millstones were on the site. The two streams had sites for "more than one hundred water mills" capable of milling all year long. The adjacent Bois Brule bottom of the Mississippi River, just below La Saline, was five leagues long and "covered with woods of all kinds." Behind this bottom was an upland prairie (the barrens) "covered in hay and beautiful trees which contains around one hundred square leagues," every arpent of which, according to Peyroux, was arable.[77]

Peyroux also suggested what the governor could do to ensure the success of a post village at La Saline. La Saline was far enough from new Ste. Genevieve to have its own church and priest. The government should build flour and sawmills and should maintain a company of riflemen at the post. Peyroux also recommended changes in dealing with the local Indians.[78]

All of Peyroux's ambitious plans, all of Peyroux's letter campaigns to the governor, all of Peyroux's efforts to change the settlement geography came to naught. The relocation process out of old Ste. Genevieve continued to flow primarily to the Gabouri, and others arriving at the time also went there or to New Bourbon on the bluffs, but not to La Saline. The failure may be explained in part by the rivalry between François Vallé and Peyroux. Peyroux, a political appointee, and Vallé, the valley Creole, held quite different opinions on how to manage local affairs, and these differences came to a head during the process of relocation of the Ste. Genevieve community. At one point, Peyroux had the Creole leader under house arrest for twenty-four hours for insubordination. Little wonder that habitants chose to follow their fellow Creole over the imperious Peyroux.[79]

But the relocation also needs explaining in terms of availability of land. The huge private tracts in the Saline Valley were not attractive to habitants for relocation or to new immigrants. Peyroux did not offer to put tracts of his large concession into private ownership. In addition, the site contained no tract of arable land large enough for a common field. Most important, the site was too distant from the middle and upper ends

---

77. Ibid.; Peyroux to Miró, March 8, 1788, in *Spain in the Mississippi Valley*, ed. Kinnaird, 3:245–46. In retrospect, Peyroux's arguments did show a good understanding of the potential productivity of the land. Perry County, the land tributary to Peyroux's proposed village, eventually developed into a much richer agricultural region than Ste. Genevieve County, although the Mississippi River abandoned the village site during an avulsive channel change in the late 1800s.
78. PC 207b-371, 375.
79. PC 207b-373.

of the Pointe Basse common field, which continued the primary *terres cultivées* for the district and required daily commuting.

The population of La Saline village may never have exceeded one hundred and was more often much less, although the number of dispersed farms in the several branches of the Saline River increased with American immigration. Salt production peaked during the decade centering on 1800. After the United States Saline on the Ohio River in southern Illinois came into production in the early years of the nineteenth century, salt making at La Saline rapidly declined, serving only local markets. In 1822, seventeen workers still used 100–150 kettles, but by 1825, all production had ceased.[80] As a village without any other economic base and with Peyroux's efforts to make it a district post having failed, La Saline depopulated and was no longer a central place in the Ste. Genevieve District. Its location passes unnoticed in the twenty-first century.

Dispersed Settlements

Although interpretation of French settlement in the middle Mississippi Valley has always relied on the village form, a small cohort constituting only a few percent of the total French population did live separately on land away from the villages. The number of French who lived outside the villages and mining camps increased with time and especially after the floods of the 1780s. The psychological break with "place" brought about by the floods encouraged those who wished for more freedom of movement to exercise that option. These newly freed former villagers went onto the karst plain and up creek valleys in the rough hills (E on fig. 20). When they finally did request concessions in the 1790s under the pressure of arriving Americans, they reported having been on their lands for ten years or longer.

The French moved onto the lower Establishment River in the 1780s. François Coleman's 1788 concession was for land that he had cleared and enclosed some years before and on which he had already built a house. In 1797, Nicolas Plante and Vincent Lafoix received a concession for farming in the cramped valley of the Fourche à Duclos, ten miles northwest of Ste. Genevieve. It had the customary forty-arpent depth, even though it did not front on anything. The Aux Vases and Saline Rivers received French settlers during and after the flood, some to long-used family sugar camps *(sucreries)*. Jean-Marie LeGrand took his family from their flooded house and fields in the Pointe Basse and moved onto the

---

80. *Missouri Gazette,* July 12, 1812; Denman, "History of 'La Saline,'" 317–19.

Saline River, accompanied by several former residents of the old village who moved onto the Aux Vases and Saline Rivers.[81]

Some habitants who chose not to accompany their fellow villagers to new Ste. Genevieve selected land farther up the twin Gabouris, west of the developing village. Jean-Baptiste Hubardeau went into hill land northwest of the village without a concession and enclosed twenty-five to thirty arpents. No one raised any objection. Other former old-village residents occupied several tracts of land along the Gabouri. Four groups of long lots, which look like miniature common fields on cadastral maps, appeared in the 1780s and 1790s. Two of them, B and C, are shown on Figure 22. One group of thirteen had *terres* of three arpents by forty in depth, the others two by sixteen, six by forty, and three by twenty. Lands taken out extended westward into the Prairie Gauthier grasslands in the karst plain, where the Gabouris headed.[82]

The French still dispersing from old Ste. Genevieve in the 1790s were accompanied by arriving Americans. Some Americans found vacant land on the karst plain near new Ste. Genevieve and New Bourbon, which, according to Finiels, "had been poorly thought of" by the French and "responded to effort and produced handsome wholesome grain, as bountiful as in the Pointe Basse." Finiels ascribed the slow spread of the French into the interior to a lack of oxen for hauling, as opposed to oxen for cultivation, whereas the Americans had plenty of the former. Trudeau commented in 1798 on the proved success of the new American farmers in the New Bourbon community, and contrasted it with the French who regularly lost two out of five of their spring-wheat harvests in their Pointe Basse fields.[83] Such visible successes within their midst must have impressed the French who then took renewed interest in the uplands for wheat.

Dispersal onto individual properties in the interior even received official encouragement. Carondelet asked the commandants of Upper

---

81. On Coleman's concession: Houck, *History of Missouri,* 1:386; on Plante and Lafoix's concession: LCS 232; on Aux Vases and Saline Rivers: LCS 3a, LCC 15, 126, LR 72; on the LeGrands: LCS 217, PC 2365-523, LCS 236, *ASP-PL* 2:488.

82. Letterbook, letter 65, from [?] to Zenon, May 26, 1794, PC 209-598; Houck, *History of Missouri,* 1:380; Petition for land by Antoine Aubuchon, May 3, 1787, LCS 2; Petition for land by Antoine Aubuchon, May 15, 1789, LCS 1. Lockridge says the dispersal of farmers away from Dedham, Mass., spared them the daily trek from house to fields and thereby formed new "precincts" in the countryside (*New England Town,* 94).

83. Finiels, *Account of Upper Louisiana,* 47–48, 132–33; Trudeau to Governor, St. Louis, January 15, 1798, in *Spanish Regime,* ed. Houck, 2:248; PC 2365-382.

Louisiana "to induce, if possible, all the poorer inhabitants of the villages to settle on the land instead of living in indolence and idleness; to give to each head of a family a quantity of land in proportion to the number of persons in the family especially at Ste. Genevieve." Frémon Delaurière, the Ste. Genevieve *greffier* (district clerk), acting upon orders of his commandant, drew up "a great number of petitions" for concessions for these poorer folk, and the commandant distributed them to all those in need of land. Delaurière reported that these poorer French went to "the banks of the streams and rivers in the vicinity of St. Genevieve."[84]

Thus, the French made their first agricultural penetration of the Ozark hills on individual properties as old Ste. Genevieve was disappearing and new Ste. Genevieve and New Bourbon were appearing. They were joined by Americans who had prior experience living on their individual properties in narrow, wooded valleys of hill country and whose lessons were not lost on the dispersing French.

This dispersal most certainly did not mean that the French lost community cohesion and began social "isolation." Their new *établissements* were less than eight miles from the villages, and more commonly less than five, and trips to and from the villages were easily made in one day. Dispersed families continue to show up in village and church records as if they participated in the community's life no less than their relatives and friends living in the nucleated villages.

---

84. Testimony of Frémon Delaurière, November 18, 1833, in "Land Claims in Missouri," January 21, 1834, *ASP-PL* 5:709.

9

A Traditional French Village in the Hills

St. Michel

Even though the uprooting of habitants from old Ste. Genevieve provided the impetus for the French to leave their villages and live on individual farms dispersed in the hills, one group did establish a tripartite village far into the interior in the final year of the eighteenth century. St. Michel was a French island of centuries-old European tradition in an American sea of dispersed farms. It represents one of the last attempts to reproduce a traditional open-field French settlement form in the middle Mississippi Valley, and its account sheds light on why the form could no longer be successful.[1]

The story of St. Michel begins almost a century earlier with the first lead mines at Mine la Motte, because the French who established St. Michel were attracted there by lead mining. Mine la Motte was one of the first French explorations in the hills west of the Mississippi, the site taking its name from Antoine de Lamothe Cadillac, who visited it in 1713. Philippe Renaut received a grant to mine lead there in 1723, and his slave-assisted enterprise apparently produced significant amounts of lead. Renaut returned to France in 1744, and, according to terms of the grant, after three years of abandonment, the mines and land reverted to the crown, although much later under the American regime, Renaut heirs in France tried in vain to push their claim to ownership.[2]

---

1. In *French Roots,* Ekberg does not mention St. Michel in his otherwise comprehensive treatment of open-field agricultural systems in French villages in the Illinois Country.
2. *ASP-PL* 3:669. The early history of lead mining at Mine la Motte is told in Rothensteiner's "Earliest History" and *Chronicles of an Old Missouri Parish. Historical Sketches of St. Michael's Church, Fredericktown, Madison County, Missouri. A Souvenir of the Centenary of Its Erection into a Parish, 1827–1927.* Also see Giraud, *History of*

In 1757, François Vallé I, or more likely two of his slaves, found ore along the Rivière au Castor, which skirted the older mines to the south, and worked the Mine au Castor with slaves profitably for at least six years.[3] In the 1750s, a French visitor to the province reported that the lead mines provided all the lead necessary for the whole colony of Louisiana, from cannon balls to shot for hunting. By 1773, other residents of Kaskaskia and Ste. Genevieve were also involved in mining, separately or in partnership with Vallé. Apparently, these entrepreneurs requested no concessions, and none were issued. Their mining activities were based on traditional prescriptive rights to mines and so acknowledged by the Spanish government. Nevertheless, controversy swirled around who owned which of these shallow surface pits and over the selling and buying of rights to mine. In addition to slaves working for the wealthy, ordinary habitants also mined the land, either independently or as wage employees for others. In the 1770s, some of these miners lived there year-round in cabins scattered across the land, although most of the miners worked seasonally. One miner's contract to work at Mine la Motte in 1773 specified fifteen days reserved for harvest of the miner's wheat, apparently back in his home village.[4] Ore was crudely smelted on site, then hauled by horse to La Saline on the Mississippi River.

In 1774, Osage Indians forcefully asserted their dominion of the eastern Ozarks. They attacked the white intruders at Mine la Motte and Mine au Castor and killed seven mine workers, including twenty-year-old Joseph Vallé, son of Ste. Genevieve leader François Vallé I.[5] The French

---

French Louisiana, 5:440–45; Donaldson, *Public Domain,* 373; Hanley, "Lead Mining in the Colonial Period," 30–37, 50–85; Willms, "Lead Mining, 1700–1811," 1–35; and Winslow, *Lead and Zinc Deposits,* 7:646–59. On Renaut's heirs' claims on the mines, see *ASP-PL* 2:522.

3. MN 19, 22ap, 27. This Castor River is a short tributary of the Little St. Francis River in northern Madison County and should not be confused with the much larger Castor River in eastern Madison County that flows southward into the swamplands. The mines on the Castor River may have been as productive as the original la Motte mines, but the term *Mine la Motte* has long since applied to the general area, including the Castor (Beaver) mines and others, probably because the large land grant that encompassed all of them has always been known as the Mine la Motte grant.

4. On French visitor: Cruchet, "La Vie en Louisiane," 66–67; on other residents: MN 18a, 19, 21, 22ap; on controversy: MN 21, 22ao, 22at, 23, 23a, 23d; work contract of Joseph Trudel, June 3, 1773, MN 22t.

5. Houck, *History of Missouri,* 1:378. Houck describes the encounter as "undoubtedly the bloodiest massacre in upper Louisiana during the Spanish regime." The seven killed included the Creole Vallé, three Canadians, a Frenchman, an Englishman, and a black slave.

immediately retired from the mines, and some years passed before they returned, at which time claims to mines became more confused than ever. The wealthy continued to assert ownership, but numerous others worked the mines, defying expulsion.

When the richer Mine à Breton was discovered about 1774, miners turned their energies to it, but Mine la Motte did not fade away. Its promise was sufficient to keep luring persons who sought the opportunity of mines and to retain the attention of the wealthier for investment. In 1800, after the arrival of Moses Austin and other Americans in the district and amid rumors of the possible change of regimes, four wealthy claimants to the mines—Jean-Baptiste Vallé, Jean-Baptiste St. Gêmme Beauvais, François Vallé II (now the district commandant), and Jean-Baptiste Pratte—requested a huge 2-league-square ($37\frac{1}{2}$-square-mile) concession to the land around the mines.[6] The petitioners' request was explicitly for a timber supply to work the mines, noting that smelting was "very far removed from the said mine because of the consumption of it which was made by the former exploiters." The request, which had to be sent to the intendant at New Orleans because Lieutenant Governor Delassus thought it was too large for him to approve, effectively gave the four control over the mines proper by controlling the timber supply and superseded all previous grants and claims, except for continuing prescriptive rights for pits being worked by others. The 1797 census of the New Bourbon District did not list anyone as resident at Mine la Motte, indicating that it had no permanent residents.

At the same time that the four wealthy residents of Ste. Genevieve were obtaining their enormous concession to the mines and timber, a group of habitants working at the mines decided to establish permanent residence there. They chose land just beyond the southern periphery of the concession, but still a mere three miles from the mines, a walking distance similar to that covered daily by villagers commuting to their croplands in common fields. Unlike other settlers who decided to live at mines in the district, this well-organized group of thirteen families, mostly from New Bourbon, formally requested a single concession for the entire group. The request, made in early 1799, specified the establishment of a

---

6. MN 15. The tract was clearly intended to be a perfect square with the corners pointing in the cardinal directions and the sides intercepting meridians at angles of forty-five degrees, which, incidentally, is both the form and the orientation of Spanish *sitios* or *vacheries*. However, when surveyed, the tract had a slice of land added to the square's southeast side in order to compensate for land taken from the southern corner for an earlier confirmed claim. Thus, the east point of the tract lies south of the west point.

village to be named St. Michel.[7] The name St. Michel was purportedly chosen to honor the New Bourbon commandant, Pierre-Charles Delassus de Luzières, who was a knight of the Grand Cross of the Order of St. Michel. If so, the action evokes an image of French Creoles trying to preserve the old order and a traditional way of doing things. Moreover, Luzières had a contract with Vallé to supply lead to New Orleans. Did Luzières therefore have an interest in having miners live at the mines to increase lead production? He could have encouraged or sponsored the founding of St. Michel for this reason.[8]

The village was designed to be a traditional French tripartite village. It had the three components of a compact village of square blocks quartered into one-arpent *terrains,* an adjacent common field divided into linear *terres* and enclosed with a single fence, and a grazing commons. Only one concession was awarded for all the lands of the village. Thus, the thirteen families were treated as a single social and economic community — a kind of covenant community — but with privately held lands. Within their community, they decided themselves who would get which village lot and which common-field strip. Individuals did have the right to buy and sell property to persons either within or outside the community. In all these land matters, St. Michel was a deliberate attempt to copy basic elements of the centuries-old European village form introduced a few years earlier at new Ste. Genevieve and New Bourbon.

The group had reasons for the founding of St. Michel other than proximity to the mines they worked. In their 1799 petition to the lieutenant governor, which contains no mention of lead mining, the concessionaires complained that their lands at New Bourbon and Ste. Genevieve were too eroded and no longer supplied sufficient pasturage for their cattle or enough grain for their families' subsistence. Had the virgin loessial soils of the bluff lands become too severely eroded from only ten or fifteen years of farming? Had the families grown too large to be supported by their fields in those villages, regardless of erosion? Did parents require more land for their numerous children of marrying age, land no longer available around New Bourbon?

---

7. On establishing permanent residence: Henry Clay Thompson II, *A History of Madison County, Missouri,* 12; on 1799 request: Rothensteiner, *Chronicles,* 9–10. Original is in the archives of the Missouri Historical Society, St. Louis.

8. Floyd C. Shoemaker, "Madison County," 2–3; *ASP-PL* 6:809–12; PC 209-666. H. C. Thompson II identifies Nicholas Caillot dit Lachance *père,* one of the older founders of St. Michel and probably its leader, as a French royalist and a knight of the Grand Cross of the Order of St. Michel (*History of Madison County,* 10). This could be a different source for the village name although it is unlikely that this resident of the valley was a knight.

The Spanish did not award concessions to ordinary folk for mining purposes. If the St. Michelins intended to be miners, their request for land would have to be couched in terms of farming. In fact, they would have to do some subsistence farming, as they were too far from the agricultural lands of Ste. Genevieve and New Bourbon.

There is evidence for yet another reason for the group's move. The habitants were unhappy with the social environment of New Bourbon and Ste. Genevieve. A Tesreau (or Tesserot) family history alludes to a deterioration of religious life under the Irish-born priest of Ste. Genevieve, James Maxwell, at this time. Confessions were heard only once a month. Gambling was excessive. A resident of Ste. Genevieve, Henry Marie Brackenridge, alleged that Father Maxwell "took more pleasure in his dog and gun than in the celebration of the mass and the spiritual concerns of his flock." Anglo Catholics, who were entering the district at the time, also complained publicly about the priest's excessive absences from the parish and even asked the bishop to remove him. The fact that New Bourbon never got its own church and priest (although Maxwell resided in New Bourbon, not Ste. Genevieve) may have displeased the devout at New Bourbon, several of whom had just moved from Kaskaskia, which had a resident priest. Complaints led one of the founders of St. Michel, Joseph Tesserot, to conclude bluntly, "I don't want to raise my children at Ste. Genevieve."[9] The habitants, by all indications, were a socially conservative group who wished to retreat to a more isolated place where they could re-create their version of a traditional French Creole community.

Finally, the Creoles may have felt pressured by incoming Americans to request concessions for the land near the mines that they had already been using for some time. American farmers had begun to enter land on the south side of the mines in the closing years of the century. Jonathan Owsley's petition for land to build a grist- and sawmill on the St. Francis a few miles downstream from St. Michel, dated November 25, 1799, mentions that numerous families were already in the neighborhood, called Owsley's Settlement.[10] The mill, he claimed, would prevent families, including nearby Creoles, from having to go twelve miles away to make

---

9. Mary Ellen C. Tesreau, *Search and Find: Tesreau Genealogy,* 22–23; Brackenridge, *Recollections of Persons and Places,* 204; "Two Maxwell Letters"; Tesreau, *Search and Find,* 22. Even the official historian of the St. Louis Archdiocese, John Rothensteiner, who would not be expected to expose priestly shortcomings, alludes to loss of piety during the tenure of Father James Maxwell when he notes, "Ste. Genevieve owes to Father Pratte [who replaced Maxwell] the renewal of piety" (*History of the Archdiocese of St. Louis,* 1:362).

10. LCC 133, 230; LCS 74; LRA 77; Houck, *History of Missouri,* 3:239; H. C. Thompson II, *History of Madison County,* 166.

flour. It is noteworthy that apparently none of the newly arriving American families at Owsley's Settlement came to the area to get involved with the French-owned Mine la Motte mines. They were farmers, not miners.

The founders of St. Michel were a cohesive band of thirteen families, French Creoles of the Illinois Country with roots in Canada and a few native French who had lived for some time in the valley. About half had moved with their slaves from Kaskaskia to the New Bourbon area between 1787 and 1790.[11] Three were single, ten had wives, and among them were twenty-two children and twenty-one slaves, for a total of sixty-six persons.

Ten of the thirteen families were closely related. Six were Caillots dit Lachance from New Bourbon, all brothers, the sons of Nicholas Caillot dit Lachance, himself a New Bourbon resident who died in 1799, the year of the petition request. (Their father's death, incidentally, may have figured in the decision to leave New Bourbon.) Three were Deguire brothers from Ste. Genevieve, and one of them was married to a sister of the six Lachance brothers. Also, a Deguire sister was married to one of the six Lachance brothers. These two large families were interrelated in other ways through their parents. A tenth concessionaire, Pierre Chevallier, was married first to another Deguire sister, then to another Lachance sister. The remaining three concessionaires were apparently not directly related to the others. One, Jérome Matis, had occupied the lot adjacent to one of the Lachances in New Bourbon and also owned land in the New Bourbon common field. Pierre Viriat and Gabriel Nicolle were miners with interests in the Big River mines and Mine la Motte. Viriat had bought property rights in 1796 at Mine la Motte. These three miners probably had been associated with the Lachances and Deguires in lead-mining activities.[12]

A seventh Lachance brother, Benjamin, was just shy of majority at the time of the concession request and perhaps omitted because of his age. The following year, he obtained his own concession of 400 arpents (the same amount that each St. Michel concessionaire had requested in the group petition) adjoining the St. Michel concession across Saline Creek to the south.[13]

The picture that emerges is one of a tightly knit, multiply interrelated community of Creole and French habitants, most with young and growing families, who covenanted to establish a settlement in the form they

---

11. Alvord, *Kaskaskia Records, 1778–1790*, 5:50, 414, 450; Ekberg, *Colonial Ste. Genevieve*, 431–32. Caillot dit Lachance probably had mining interests at Mine à Breton and Mine la Motte before moving across the river to New Bourbon.
12. W. A. Schroeder, "Opening the Ozarks," app. H; LRA 178, 179; MN 24.
13. LCC 93. The age of majority was twenty-five, unless married.

were familiar with from Kaskaskia, Ste. Genevieve, and New Bourbon. As regards land division and control, all families were equal.

The size of the land request was determined by asking for 400 arpents for each of the thirteen petitioners, or a total for the group of 5,200 arpents. The 400-arpent figure is interesting, for at that time, the Spanish government was using the Gayoso formula for awarding land according to the size and composition of individual households, including slaves, which would have varied widely the amount of land per family. Requesting the same size for everyone may have been a conscious attempt at equalizing the landholdings of the constituents, regardless of family size, age, or material circumstances.

Spanish policy for new towns and accompanying agricultural lands required a minimum population size of thirty families (single men were apparently not considered for residency) but allowed for as few as twelve. Another important element of Spanish policy was permanency, which is why new villages had to have a self-supporting agricultural base.[14] Although these basic stipulations fitted St. Michel, they are far from constituting evidence that the village was Spanish planned or Spanish influenced because they also fitted a traditional French village. Although the concession was granted by a Spanish administrator, nothing suggests any Spanish influence in the establishment and laying out of the village.

Lieutenant Governor Trudeau immediately granted the concession of 5,200 arpents (6.9 square miles) on May 12, 1799 (fig. 25). Soulard made a first survey of only the village and its common field for 990 arpents on February 19, 1804, during that hectic month before Spain relinquished Upper Louisiana to the Americans. Later, on April 27, 1805, Soulard's deputy Thomas Maddin made a second survey of 4,008 arpents that extended the earlier survey to include the commons. Both of these surveys, however, demarcated only the boundaries of the concession and left all internal land subdividing to the families. The Board of Land Commissioners initially confirmed only 3,820 arpents of the claim on June 20, 1806, rejecting Trudeau's concession ratio of 400 arpents per claimant and instead strictly applying the Gayoso land-allotment formula. The board reasoned that although Trudeau had not followed his own Spanish land-granting authorities, it would. However, the same board later (on September 28, 1810) confirmed the concession for the full 5,200 arpents, without comment, and directed a survey of it.[15] It was when this Amer-

14. *ASP-PL* 5:730–31; Martin, *The History of Louisiana*, 276.
15. On Trudeau concession: Minutes, 2:25–26; on surveys: *Régistre d'Arpentage*, 576, 576A, 576B. The plat for Soulard's first survey has not been located, but it is referred to in Minutes, 2:25–26. On 1806 concession: Minutes, 2:25–26; *ASP-PL* 2:594.

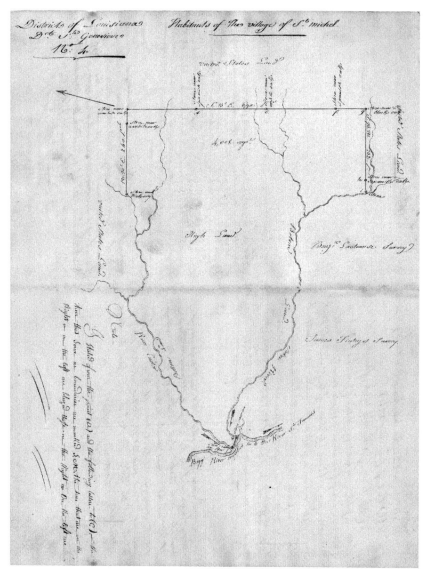

25. *St. Michel Survey by Thomas Maddin, April 27, 1805.* Because the Spanish concession of May 12, 1799, was made to the village as a single community, the 1805 land survey by Thomas Maddin was for the village as a single tract of land without individual private properties. The American government did not make separate surveys of individual properties within St. Michel as it did for Ste. Genevieve, New Bourbon, Mine à Breton, and other French villages. North is to the left on this plat. The Little St. Francis River (labeled Bigg River) flows from left to right at the bottom of the map. (*Régistre d'Arpentage*, 576)

ican survey was made that an additional 1,192 arpents were added to the 4,008 arpents surveyed earlier by Maddin; this comprised an oblong tract on the east end (shown east of the dashed line on Figure 26), the land farthest from the village.

Three of the claimants (Paul DeGuire, François Lachance, and Jérome Matis) built houses in the village in November 1800, and the others moved in with their families the following year. Property deeds record land sales in the village as early as February 1803. In 1804, the Osages raided the American farms to the south on the St. Francis River, and the Americans retreated into more defensible St. Michel with its closely set buildings. By 1808, the village included Americans living on its town lots among the French Creoles.[16]

The St. Michel concession has the shape of a westward-flying arrowhead, the chipped edges of which are Saline Creek to the south and Castor Creek (renamed Village Creek by the early 1820s) to the north, which also flowed through the Mine au Castor (fig. 26). The arrowhead point is formed by the junction of these two streams with the Little St. Francis River (which was invariably and confusingly referred to on plats and in documents as simply the St. Francis River). Straight lines from the Castor to the Saline formed the butt end of the arrowhead.

The concession lay in a sedimentary basin, now called the Fredericktown basin, of better soils and visibly lower relief than the circumjacent region of igneous knobs. Major portions of the basin were open woodland. None of the streams of the basin were navigable, not even the main St. Francis because of several gorgelike shut-ins. Smaller tributaries with more manageable flows, such as the Rivière aux Cannes (Cane Creek, now Mill Creek), provided excellent mill sites. The concession lies thirty-two miles southwest of Ste. Genevieve, most of that distance through rough hills.[17]

Because the concession request specified the names of the families, the board could identify the claimants and issue individual certificates of confirmation to them. They could not do this at other French villages, except Old Mines, because individual families were not identified at the time of settlement. On 1810 concession: Minutes, 2:25–26.

16. On built houses in 1800: PC 217b-216. Some of the concessionaires were already living in the vicinity because they were actively mining at Mine la Motte. On families moving: Minutes, 2:25–26. Those who came to St. Michel by the end of 1805 were also chiefly from New Bourbon as deduced from signers of the "Memorial to the President by Citizens of the Territory," December 27, 1805, TP 13:341. On property deeds: LRB 420; on the Osages: ASP-PL 8:108; on 1808 village makeup: LRB 485.

17. For the distinctive topography of the Fredericktown basin, see Walter A. Schroeder, "Landforms of Missouri: A Geography of the Configuration of the Land Surface of Missouri." Owsley's Mill (or Callaway's Mill) was on Mill Creek and likely is the reason for the name change from Cane Creek (LCS 17, 133; LRC 72).

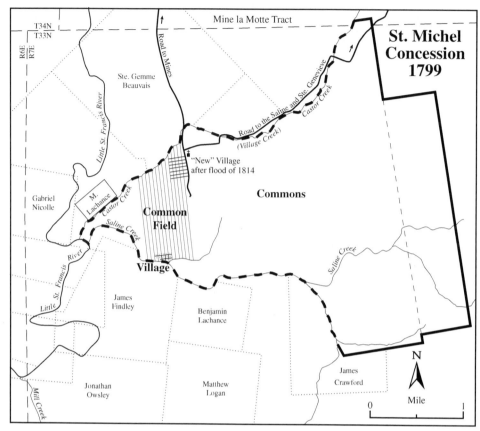

26. *St. Michel Concession, 1799.* The original French village of St. Michel consisted of a small grid of streets and blocks quartered into lots on the north side of Saline Creek. The common field, enclosed by a single fence, extended from the village northward to Castor Creek. After the flood of 1814, a new village cluster arose on high ground along Castor Creek where the road left the concession for Mine la Motte, the villagers' workplace. The grazing commons of the tripartite village occupied the tract east of the common field. This map also shows the rectangular tract of land on the east end of the concession, added to the original survey in order to bring the total land area to the full 5,200 arpents awarded by the American Board of Land Commissioners.

A three-mile road connected St. Michel village with the lead mines, which were populated only by tenant miners, then continued northward to Cook's, Murphy's, and the settlements at the big bend of the Big River.[18] Another road went up Castor Creek Valley to the northeast, then joined on its left the lead-hauling road from Mine la Motte, later called the Three-Notch Road, which went northeastward to meet the Mississippi at La Saline, and from there along the low bluffs up to New Bourbon and Ste. Genevieve.

The Americans who lived on the south and west sides of St. Michel held concessions and settlement rights that were later surveyed with variable orientations to take in as much good land as possible along the St. Francis River as it twisted its way among the igneous knobs. By 1810, the total number of Americans living downvalley from St. Michel was much greater than the population of the village. A road led south from St. Michel to Owsley's Mill, the focus of the Owsley Settlement, and then farther south as a trail for the woodsmen along the St. Francis in present Wayne County. During the American period, this trail became the Red River Road, the primary overland route from St. Louis into Arkansas, Lower Louisiana, and Texas. Therefore, St. Michel, which lay on the farthest edge of settled country when founded, developed into a node in the regional road network by 1819 and a jumping-off point to the Red River country, Natchitoches, and Texas.

Evidence suggests that the St. Michel Creoles had close and friendly relations with local Indians. Gabriel Nicolle had land on the Little St. Francis just west of the common field that he had purchased from two Indians who occupied it until shortly after 1803. Maria Tomar, an Indian woman, claimed 850 arpents next to Gabriel Nicolle, but Frederick Bates later rejected her claim, and she did not pursue it. The multiracial nature of the community is further shown by concessionaire Pierre Viriat's marriage to a mulatto whom he had purchased as a slave and manumitted two days before their wedding.[19]

Numerous property deeds from 1803 to 1818, when Madison County was created, allow a partial reconstruction of the tripartite village.[20] Lots

18. *Missouri Gazette,* March 21, 1811.
19. On Nicolle: *ASP-PL* 3:318. Nicolle had another tract confirmed to him (*ASP-PL* 5:799–800); on Tomar: *ASP-PL* 3:318–25; on Viriat: Ekberg, *Colonial Ste. Genevieve,* 210–33.
20. Reconstruction of the village in the following paragraphs is based on numerous deeds from 1803 to 1818 that describe properties and improvements on them, bounding lines, and bounding owners: LRA 225–27, 234; LRB 21–22, 29, 66, 81–82, 99, 132, 223–24, 270, 345, 397, 413–14, 423, 425–26, 450, 456, 483, 483a, 483b, 484–85, 487–89, 489a, 490–92, 499, 509, 542, 545, 554, 558–59, 565, 565a, 569–

and blocks that are completely described by precise compass directions are square, and a planned, regular grid pattern is evident. Lacking a surveyor, landowners could not put their buildings and fences in perfect alignment and orientation, so the resulting layout contained slight irregularities in the compass directions of lots and blocks, like the new village of Ste. Genevieve. Property deeds do not divulge the exact location of the village grid along Saline Creek. The location shown on Figure 26 is an approximation based on oral and published histories and the presence of old graves.[21] The grid could be shifted a few hundred feet eastward or westward along the creek and perhaps slightly north or south. The village had to have been on ground low enough to have been flooded in 1814.

Property deeds describe two streets parallel to the creek and at least two, possibly three, cross streets. Some *terrains* backed up to the common field and some to the creek. All these distributional requirements from the land records are met in the postulated village street pattern on the map.

The deeds also provide a glimpse of the built environment. None of the houses described were French-style *poteaux sur sol* or *poteaux en terre* (vertical logs set on a sill or in the ground). All houses for which construction type is given were described as *pièces sur pièces* (horizontal logs), except one, a stone house built in 1816. As a general rule, Creoles in the middle Mississippi Valley considered American-style *pièces sur pièces* construction inferior to their vertical log construction. That French Creoles in St. Michel built in the American style suggests either that St. Michel housed a lower economic class or that Americans nearby aided the Creoles in house raising and the Creoles accepted the American style of log construction.

Stretching north from the village was the *champ* (common field) subdivided into privately owned *terres*. Some of these long lots stretched the entire length between Saline and Castor Creeks; others went from village lots to Castor Creek. Though the total number of strips is unknown, it is likely that each of the thirteen families had at least one strip, if the custom of matching *terre* with *terrain* was followed. Several deeds describe a

---

70, 576, 578; LRC 15–17, 30, 35, 38, 51, 56–57, 68, 73–74, 81–82, 95, 107, 112–13, 126, 128, 155, 160, 166, 166a, 167, 197, 223. More details of the village as reconstructed from these deeds are in W. A. Schroeder, "Opening the Ozarks," 309–13.

21. Brief, general accounts of the village of St. Michel are in Rothensteiner, *Chronicles;* H. C. Thompson II, *History of Madison County; Fredericktown Bee,* May 20, 1869; Shoemaker, "Madison County," 2–3; and Paul Skaggs, interview by author, Fredericktown, August 15, 1985. During the celebration of Fredericktown's bicentennial, the local *Democrat News* published an article about the founding of "St. Michael's Village," in which the original village was described as near the old railroad depot, which would place it slightly north of the site determined from the deeds (July 7, 1999).

village lot with the common-field strip "that goes with it," or vice versa, as though the two were paired and inseparable. The map shows a schematic arrangement of thirteen strips, each two arpents wide, as postulated to exist at the time of original layout of the common field (see fig. 26). The strips differed somewhat in length because of the varying distances between the two creeks, but they averaged about thirty arpents in length. A single fence with a "big gate" enclosed the common field, at least partially, as late as 1816. The community had rules for the maintenance of the fence, and land purchasers had to agree to uphold them.

The common-field properties were laid out nineteen degrees west of magnetic north, which converts to a compass reading of about thirteen degrees west of true north. This uncommon orientation fits perfectly with the trend of a pronounced linear valley in which the common field lay, and the orientation was in all likelihood determined by that valley's lineament. This orientation is several degrees off the village grid at the south end of the common field, and the lack of correlation of the two, which would have been expected, indicates separate topographic control in this hilly landscape.

The west side of the common field occupied the lowest part of the linear valley. On that side, the valley is only about ten feet above the two creeks at their ends, and, because the valley soils are poorly drained fine silts formed from alluvium, the valley must be interpreted as a former valley of either Castor (Village) or Saline Creek.[22] It lies low enough to be flooded by those streams. The east side of the common field rises gradually to about fifty or sixty feet above the west side. The entire common field would have been an excellent agricultural area of productive alluvial loams, probably the largest such tract for many miles around the mines. The St. Michelins, familiar with the alluvial soils at New Bourbon and Ste. Genevieve, recognized the agricultural potential of this unusual linear valley and specifically chose it to reproduce their tripartite village. No site within a dozen miles of the Mine la Motte mines could have provided a better setting for a common field of a traditional French village.

The deduced location of the common field is proved by the field notes of the U.S. rectangular land survey. When the section line between sections 8 and 9, T33N, R7E, was surveyed in 1818, the surveyor noted that "this mile passes through the St. Michael's fields."[23] (Because the U.S. survey lines were run before the St. Michel concession was confirmed, they were run as if the Spanish concession did not exist.) The original common-

22. W. A. Schroeder, "Landforms of Missouri."
23. U.S. General Land Office, Field Notes, Missouri, 216:179.

field survey lines still show up in subdivision, street, and lot lines in north Fredericktown two centuries later, even though the Creole village no longer exists. Catherine Mine Road, Buford Boulevard, and Villcar Street retain the compass direction of the property lines of the old common field.

The village and the common field account for about one-fifth of the total area of the concession. The remaining four-fifths of the concession served as a commons for animal grazing and woodcutting. Because the Board of Land Commissioners confirmed the entire concession as a single entity on the ratio of 400 arpents per family, the commons eventually had to be subdivided into individual landholdings so that individuals could sell all or part of their share of the concession. Land sales in tracts of 200 arpents and larger began in 1806. Just as for a *terre* in the common field, a tract in the commons was sold along with a *terrain* in the village with a comment to the effect that a "house in the village of St. Michel . . . belongs to the said tract of land" in the commons.[24]

Some of the deeds for land in the commons read "from the Castor to the Saline," and their widths ranged from five to nine arpents. By such clues, it is postulated that individual properties in the commons were strips like those of the common field, but significantly wider, so they contained 200 to 400 arpents each.

Present property and field lines in the former St. Michel commons show the legacy of its early land subdivision. Major north-south property lines in the early twenty-first century are spaced from 1,100 to 1,800 feet apart (5.8 to 9.4 arpents) and continue interruptedly from the Castor to the Saline, even though the ends are in different ownership today. The orientation of present property and field lines is exactly the same as in the time of the St. Michel commons.[25]

After a disastrous flood in 1814 that washed away log houses along Saline Creek, some of the residents of St. Michel moved north to higher ground near Castor Creek, but still within the concession.[26] Unfortunately, property deeds from 1814 to 1818 do not usually distinguish between the two villages. Only three post-1814 deeds specifically refer to the "old village," which indicates the existence of a newer one, but no deeds specifically refer to a "new" village. Only two were definitely in the

24. LRB 501, 502, 421, 422.
25. Measured on Fredericktown air photograph (79–7–2.1 dated March 1, 1984; photography of September 28, 1981) in the Madison County Tax Assessor's Office, Fredericktown.
26. According to H. C. Thompson II, the new village lay "east of the Budenholzer homestead," apparently a place in existence at the time Thompson wrote (*History of Madison County,* 24); Rothensteiner, *Chronicles,* 11–12; Skaggs, interview.

new village, because the main road was described as north-south, and the shape of lots was linear (30 feet by 150 feet), a shape that could not fit into the square blocks of the original village.

Local Fredericktown lore puts the site of the new village at a place known as Stringtown on the Mine la Motte Road near its junction with St. Mary's (Village Creek) Road. Other evidence for this location is its high elevation and the presence of long, narrow lots on both sides of the road, those on the west side backing down to the common field. In addition, at the junction of the two roads is the place where mass was said before a church was built in St. Michel. The map of Figure 26 schematically shows the long lots of the new village at this location according to these pieces of evidence. Lots and houses were simply lined along both sides of the main road leading to the mines with sufficient land for gardens and barnyard animals stretching out behind the houses. The change in village morphology suggests that the Creoles had discarded the compact-village form.

The new village did not replace the original one. The new village had twelve families in 1816, whereas the original village was still recovering from the flood. Henry Rowe Schoolcraft visited in 1819 and recorded fifty houses and several stores in the village along Saline Creek. In that same year, a Catholic log church was built in the original village, not in the new village as earlier proposed, and villagers gradually drifted back to the original location.[27]

The dispersed population of the hilly region around St. Michel, composed completely of Americans and their slaves, grew steadily and faster after 1804 than the population of the Creole village. Creoles and Americans who worked the mines on the Mine la Motte grant were unable to acquire land there from the grant holders and left to settle other places, only to be replaced by other miners. The region reached the minimum population for erecting a county in 1818, in which year the territorial legislature created Madison County. The commission charged with the selection of a site for the county seat ignored Creole St. Michel, the only existing village in the region and located ideally at its center, and selected a site owned by a prominent American and promoter of the new county, Nathaniel Cook. The site was the high ground directly across Saline Creek from St. Michel, and the commissioners gave the town platted on it the name Fredericktown.[28]

27. Rothensteiner, *Chronicles*, 12; Schoolcraft, *View of the Lead Mines*, 48. Fredericktown was in the process of being laid out across the creek when Schoolcraft passed through. Rothensteiner, *Archdiocese of St. Louis*, 1:363.
28. Houck, *History of Missouri*, 3:181.

Fredericktown was laid out on the distinctive "Lancaster square" plan in which the courthouse sits on its own square with four entering streets extending out from the midpoints of the square. It is a plan that was popular in Tennessee and Kentucky, from which the Americans in the neighborhood came.[29] The blocks of the original plat were rectangular and subdivided into narrow lots that required alleys for access to their back sides. The courthouse square was placed atop the rise on the south side of Saline Creek, directly overlooking the French village. The grid was oriented to true north; the cardinal lines of the U.S. land survey had just been run through the settlement a few months earlier. The original platted town was small, but it met the minimum size of fifty acres for a county seat.

Meanwhile, the French village across the creek was being Americanized. "St. Michel" became known as "St. Michael's," and more Americans than French moved into the village. Interestingly, those Americans who bought strips in the common field accepted the authority of the corporate body of villagers in making decisions, that is, they agreed to maintain the fences of the *parc* and conform to other practices of the French community.[30] It is also interesting to note that most deeds of land sales at St. Michel continued to be written in French through 1818 and probably later. These deeds continued to use the old French expression *Poste de St. Michel,* even though the term *poste* was discontinued on documents elsewhere in the district in 1804, immediately after the end of the Spanish administration. Elements of the French tradition at this Ozark outpost persisted.

Mass was said quite early at Mine la Motte, probably in the eighteenth century, and at St. Michel, first in the original village, then at the new village after the 1814 flood. The log church that was built sometime between 1815 and 1822 in the original village was disassembled and moved to Fredericktown in 1827. A Baptist congregation among Americans was organized in 1812 in St. Michel, but built its first church in Fredericktown in 1827. Formal church organization and structures helped to secure permanency and growth for the combined town of St. Michael's–Fredericktown. However, Baptist missionary John Mason Peck did not think much of local religion when he passed through in 1818. He found it a "very wicked place" where French and Baptists intermarried.[31]

29. Marian M. Ohman, *A History of Missouri's Counties, County Seats, and Courthouse Squares,* 31; Marian M. Ohman, "Diffusion of Foursquare Courthouses to the Midwest, 1785–1885," 185.

30. Deed of sale by Gabriel LaChance and Mariane Louviere LaChance to Robert Crawford, April 14, 1809, LRB 29; translation by Walter A. Schroeder.

31. On Catholic church: Rothensteiner, *Chronicles,* 16; and H. C. Thompson II, *History of Madison County,* 13, 25; on Baptist church: H. C. Thompson II, *History of Madison County,* 25; Peck, *Forty Years of Pioneer Life,* 120.

The combined town of St. Michael's–Fredericktown continued to grow in the 1820s. Cook's Settlement on the north side of Mine la Motte failed to develop any community focus, and some of its families followed namesake Nathaniel Cook to Fredericktown. In 1821, Schoolcraft noted that St. Michael's had been in a state of decline, but had rapidly improved since Fredericktown's designation as the seat of justice. St. Michael's was still a toponym on 1861 and 1869 maps of Missouri, although it was by then considered a neighborhood of Fredericktown.[32]

St. Michel was the only attempt to create a traditional French agricultural village in the Ste. Genevieve District away from the Mississippi River, and it may have been one of the last attempts to lay one out in the Mississippi Valley. Residents in the region perceived it as a unique place long after its uniqueness faded. That its New Bourbon founders were successful in creating it attests to the persistence of tradition. But the score or so of French Creole families, unaugmented by further Creole immigration, were no match for the superior numbers of American settlers. The French learned American ways, not vice versa. The French were distracted from agriculture by mining, which has a less permanent attachment to the land. Significantly, the Creoles at St. Michel did not have the capital, the tradition of economic organization, or the political ties with the territorial legislature to embark upon the building of a real town, as Nathaniel Cook and other Americans had in the establishment of Fredericktown.[33]

---

32. Schoolcraft, *Journal of a Tour,* 89; Flint, *Condensed Geography,* 2:100. Of the fifty-one persons who signed the 1818 petition for postal service to Fredericktown–St. Michael's, only one had a French name. Of long lists of jurors and others involved in public life in early Madison County, only the name of Nicholas Lachance from the French community appears (H. C. Thompson II, *History of Madison County,* passim). Lloyd's Official Map of Missouri; George F. Cram, New Sectional Map of the State of Missouri.

33. On uniqueness: H. C. Thompson II, *History of Madison County,* 12. Curiously, in 1838, thirty-four years into the American regime, the laying out of a new village of Fryberg at Mine la Motte sounded an echo of the planned French village tradition. The proposal by French Creole landowners Vallé and Pratte of Ste. Genevieve consisted of two parts: a compact residential part of one-half-acre village lots and an agricultural part of five-acre outlots in a "proximate tract" set aside specifically for cultivation. Each resident would have a village lot paired with an outlot. Fryberg was not built. Its stillbirth had more to do with the declining lead economy than with the proposed return to a spatial separation of village residence and outlying farm (Thomas G. Clemson et al., *Observations on the La Motte Mines and Domain, in the State of Missouri, with some account of the advantages and inducements there promised to capitalists and individuals desirous of engaging in mining, manufacturing, or farming operations*).

10

The Mining Country

Mine à Breton

By the 1820s, the mining district that centered on Mine à Breton and Potosi had become the largest area of settlement in the Ste. Genevieve District, both in numbers of people and in area occupied, and had the district's strongest and most diverse economy. Potosi had been considered for the capital of the new state of Missouri. What happened in the mining country is central to understanding the settlement of the historic Ste. Genevieve District.

Europeans entered the mineral area at Old Mines in the 1720s and worked the surface lead deposits off and on. Continuous, permanent settlement, however, began in the 1770s at Mine à Breton, which in turn caused the reoccupation of Old Mines and the establishment of other mining camps and villages in the general mineral district. Around the turn of the century, farming neighborhoods arose and prospered when lead mining and smelting became full-time occupations and those workers relied on others for their food supply. American-planned towns subsequently developed as administrative and commercial centers for the fast-growing population. A key figure in this growth and spread of settlement was Moses Austin, accompanied by several other American entrepreneurs, such as John Smith T, James Bryan, Andrew Henry, and William Ashley, all of whom lived in the mineral district, and members of the Ste. Genevieve and St. Louis French Creole communities who retained residences in those places. Investment capital flowed into the region from St. Louis and the American East. Thus, the account of the settlement of the mining district integrates the community stories of Mine à Breton, Potosi, Old Mines, Herculaneum, the Big River Bend, and Farmington.

According to tradition, surface lead ore at Mine à Breton was discovered in 1774 by one François Azor dit Breton, who had been mining earlier at Old Mines, and his guide, Pierre Boyer, also of Old Mines. The place came to be referred to as *la mine à Breton,* following the custom of naming mines after their discoverers and using the *dit* (vernacular) names.[1] The unusual richness of the mines soon attracted Creoles, Canadians, and Americans, but French Creoles clearly were in the majority. The potential of the mines also attracted wealthier Ste. Genevieve residents, but they remained in the village and sent their slaves and hired men to work. All of these workers continued the traditional Creole method of mining and smelting as practiced since the 1720s and the traditional respect of mine ownership by prescriptive rights. They built their cabins and cut timber on the *domaine royale* and periodically hauled the smelted ore to Ste. Genevieve for sale and shipment. Apparently, there was little development of community life and little economic cooperation. Little dissension arose between wealthy miners with their slaves and the independent miners in respect to rights to the land. There was enough ore for all to dig. No one person tried to—or could—control the mining area, whether for ore or timber. Besides, the lead dug and smelted by the independent, hardscrabble miners eventually ended up in the possession of the Ste. Genevieve wealthy for marketing. In stark contrast to Mine la Motte, only a few Ste. Genevieve operators bothered to request official concessions during the first twenty years of mining at Mine à Breton.[2]

Neither did residents at the mines request concessions or receive allotments for the lands they built cabins on and enclosed for gardens. Again, common consent prevailed. Mining camps, mere groups of cabins (*cabanes,* to be distinguished from *maisons* [houses]), were occupied all year long sometime before 1791.[3] Enough people resided in the area for Jean-Baptiste Placet to have a mill in operation in 1795 and for Bazile

---

1. The year 1774 is deduced from a note signed in 1775 that indicates he discovered the mine a year earlier (MN 34). Some years later Azor dit Breton (he reportedly lived to be more than one hundred) could not remember the year of discovery (Hanley, "Lead Mining in the Colonial Period," 86–88). For a biographical sketch of the legendary Azor (Azau) dit Breton, see Houck, *History of Missouri,* 1:284–85. Mine à Breton occurs in shortened forms Breton, Briton, Burton, and Barton. Mine à Breton Creek is now Breton Creek. Many Americans used, and still use, Burton in both pronunciation and spelling. Mine à Breton is also rendered as Mine au Breton.

The dit names or nicknames were so popular that some eventually became family surnames. For example, the Caillot dit Lachance family more often occurs in records as simply Lachance. A list of dit names is in Houck, *History of Missouri,* 2:244–46.

2. One of the few was Jean-Baptiste Ste. Gemme Beauvais (in Spanish as Juan Baptista Sanchem), who, upon moving to the Spanish side of the valley in 1788, requested formal ownership of land for sixty feet in circumference around pits and then defended those claims against intruders (LCC 150; MN 3b, 3c, 3d).

3. MN 66; *ASP-PL* 3:611.

Vallé to operate a "business house" in 1792 in a cabin at the head of Mine à Breton Creek, a place called La Fontaine de la Prairie, which was the junction of roads joining Ste. Genevieve, Old Mines, and Mine à Breton.[4] In December 1797, Osage activity caused workers to leave, and the mines lay unoccupied and unworked that winter. Pierre Boyer, one of the original miners in the 1770s and identified as much as anyone with the early fortunes of Mine à Breton, withdrew in 1797 to Terre Bleue Creek halfway back to Ste. Genevieve. Consequently, when the American Moses Austin visited the mines late in 1797, he claims to have found the site uninhabited, the Creole miners "holed up" in Ste. Genevieve.[5]

However, when Austin returned to stay at Mine à Breton in 1798, Creole miners had reinhabited the mining camp. The settlement, now more appropriately referred to as a village, lay between Mine à Breton Creek and the mines and consisted of a double row of cabins and gardens on the south side of the creek. A sketch map drawn by Austin himself depicts the cabins as simple structures, each with access to the creek.[6] If the distances on the map are reasonably proportionate to reality, the cabins of the little village lay quite close to each other. It was populated, according to Austin, by only eight "French" families, except during the height of the mining season, when many black slaves also lived there. Certainly, many other miners still worked on a seasonal basis. The actual number of people living in the village of Mine à Breton probably did not total more than one hundred at any one time before 1800, but perhaps as many others lived within a few miles at other mines.

Moses Austin was the dominant figure in the mining activities of Missouri from his arrival in Louisiana in the fall of 1796 to his death in 1821. Capt. Amos Stoddard depended on him to provide information about the valuable lead resources of Upper Louisiana when the American regime took over, and later, in 1816, the U.S. government requested him to prepare a report on the status of the lead mines. The value of his mining property in the Ste. Genevieve District was reckoned at $190,000 in 1810, an enormous sum for a frontier territory.[7] Year after year, he employed

---

4. Mine à Breton survey, roll 49, p. 115, Record of Surveys, Missouri State Archives; Houck, *History of Missouri*, 1:367.

5. On Osage activity: PC 2365-323; on Boyer: Houck, *History of Missouri*, 1:367; Austin, "'Memorandum of M. Austin's Journey,'" 519.

6. Mapas y Planos no. 198, PC. Reproduced in Gracy, *Moses Austin*, 72. Also on AGI-PC microfilm at the University of Missouri–Columbia.

7. Gracy, *Moses Austin: His Life*, 128. The most complete biography of Moses Austin is Gracy, *Moses Austin: His Life*. Also see Gardner, *Lead King: Moses Austin*. Other Americans showed interest in lead mining in Spanish Upper Louisiana about the same time as Austin. In October 1796, just in advance of Austin, a mining company

the largest number of workers in the mining district, and the value of his lead production was higher than that of any other operator during most of his Missouri years.

Austin introduced major changes in lead mining, lead smelting, and lead marketing. Before he arrived, twenty small "French" log smelters or furnaces were in the vicinity. By 1802, merely five years later, there was only one large furnace, Austin's, and the miners brought their lead to it. This was a reverberatory furnace, vastly more efficient for recovering lead from the ore. He also established a manufactory for shot and sheet lead and sank the first deep shaft in the territory. He invested some $8,000 in his enterprises.[8] Furthermore, Austin effected significant changes in the settlement geography of the lead-mining district. He reintroduced the large land grant specifically for mineral land, not granted since the 1720s. He attracted numerous settlers to Mine à Breton and worked to bring order to the village. Later, he laid out a county-seat town, Potosi, for a county that he was largely responsible for creating, Washington County, and he also laid out a river port, Herculaneum. He built bridges and roads with his own money to help establish Potosi as the commercial hub of southeastern Missouri, if not the capital of the state of Missouri.

Austin immigrated into Spanish Upper Louisiana at a propitious time. A resident of Virginia, where he engaged in lead mining and smelting, he heard about the lead mines in 1796 in the pamphlet that Luzières had circulated in the United States promoting Spanish Louisiana. Austin visited Upper Louisiana and its mineral district in 1796 and found that his financial resources and knowledge of lead mining and smelting dovetailed nicely into Lieutenant Governor Delassus's and Commandant Vallé's contract to furnish lead to Governor Carondelet at New Orleans. In fact, Austin formed a partnership with the two and fellow American John Rice Jones to that end. Much to the French officials' delight, Austin also promised to bring with him, in addition to his own extended family, thirty American families of tradesmen and builders, whose skills and services were sorely needed in Upper Louisiana. District leaders greeted the American with open arms.[9]

---

led by James Twyman requested "reserved" land on Castor Creek near Mine la Motte (LCS 99). Nothing is known of the company's fate.

8. LRA 176.

9. Gracy, *Moses Austin: His Life,* 53–54, 64; on partnership: *Austin Papers,* 2:29–31. The partnership was to last for ten years, and in 1807, it was formally dissolved and Austin and Jones reconstituted a two-man partnership (LRA 176). The original partnership was effectively dead by September 1805 (*Austin Papers,* 2:99). LCS 226,

In January 1797, Austin wrote to Governor Carondelet requesting an enormous 16 square leagues of land (150 square miles) centering on Mine à Breton. Austin intended to control as much of the mines as he could, including the heaps of slag or ashes left over from previous inefficient smelting. He wanted to make sure his resource base was as large as possible, especially a wood supply, in order to safeguard the thousands of dollars he intended to invest. But the American's desire to own as much of the mineral land as possible, whereas local miners "owned" only the mines they actively worked at any one time, struck the Creole miners as excessively greedy and an aggressive attempt to take over their pits and traditional mining rights.[10]

The governor took no action on Austin's land request. Finally, armed with Soulard's survey of the claim, Austin traveled to New Orleans in 1801 to acquire the grant, but even then was unsuccessful. It was not until July 5, 1802, five years after his arrival in Spanish Louisiana, that Intendant Juan Ventura Morales issued the grant. Morales had reduced the grant to 1 square league (9.4 square miles), or only one-sixteenth of the original request, but at least Austin had his perfected title, one issued from New Orleans.[11]

Meanwhile, Austin's agents had arrived in 1798 to start the building of the carefully planned great enterprise, and Austin himself arrived soon after with his family, which included his wife, Maria (Mary) Brown; his son, Stephen; his daughter, Emily; and his brother-in-law Moses Bates, together with slaves and other workers. Professional American builders, who had been sent in advance, constructed the furnace and mill and the

---

227; *Austin Papers*, 2:31, 84; Houck, *History of Missouri*, 1:367–73; Gardner, *Lead King: Moses Austin*, 57.

10. Gracy, *Moses Austin: His Life*, 63–64. Austin lived for a short while in Ste. Genevieve before moving to Mine à Breton (LCS 109; LR 235), and it was the *greffier* (district clerk) at Ste. Genevieve who prepared Austin's petition for his Spanish grant. The *greffier* reportedly was so impressed with Austin, who did not know French or Spanish, that he wrote the petition for an unbelievable 12 leagues square (Houck, *History of Missouri*, 1:370). Austin's attempt to gain control over ashes left behind by log-furnace reduction has an interesting sidelight. When Pedro Vial returned in 1797 to Upper Louisiana from his trip to Santa Fe, Mexico, to enter lead mining at Mine à Breton, he brought back with him Francisco Luna, who had convinced Vial that he knew how to extract silver from lead ashes. Luna failed to be able to do so, and Vial sued him to recover various costs (MN 7, 7a–e). Vial's interest in lead ashes was different from Austin's, but Vial was aware of Austin's intentions to control leftover ashes.

11. *ASP-PL* 3:582–83, 591–92, 607; Gracy, *Moses Austin: His Life*, 89–90. Intendant Morales's disapproval of the enormous size of some grants issued a few years earlier by Carondelet may have been a factor in his decision to reduce the size of Austin's grant to only 1 square league. That Austin was American may have been another reason.

Austin family residence, Durham Hall. This assemblage lay on the north side of Mine à Breton Creek, just downstream from and opposite the Creole mining camp. Durham Hall was in the style of a southern mansion and pretentious for the frontier.[12] The contrast between this sturdy, professionally built, multistory structure with the adjacent small, crudely constructed log cabins of the miners must have been striking to anyone. It was the great house of a lead plantation juxtaposed to the laborers' modest cabins, just as Commandant Luzières's *grande maison* overlooked the habitants' houses at New Bourbon.

Austin's huge grant had two repercussions on landownership. First, it did not sit well with the Creole miners, and this was a problem from the beginning. Austin understood that his concession included the mines up the slope from the village. If it did, then Austin in effect had usurped the Creoles' traditional rights to climb the hill south of their village and dig in their pits and open new ones. Austin commenced his mining operations before the bounds of his square league were surveyed (fig. 27), an act that brought on small acts of vandalism against him. Austin believed his land should be a square with straight sides, whereas miners insisted that Austin's property be bounded so as to exclude both their village and their mines. After great pressure from Commandant and business partner François Vallé, who instructed Austin in the customs of the people, Austin reluctantly agreed to cut out of his square league a 105-arpent trapezoid containing the village and the mines as a public commons for the Creoles to continue working by prescriptive rights. In compensation for this excision, he received a 790-arpent triangle attached to the east.[13] Though the miners thereby won their rights to mine the hillside commons, they still found themselves economically subordinate to the American, because they had to sell their ore to him for smelting. An adversarial relationship developed between Austin and the Creole community that was never overcome. The disagreeable experience with local miners and with local authorities in getting the land surveyed engendered adverse feelings in the wealthy and energetic American who was introducing new ideas and challenging established practices. He seemed to encounter French tradition and inertia and Spanish bureaucracy and delays at every step.

---

12. Cynthia R. Price, "Report of Construction Monitoring at Durham Hall, 23WA77, in the Town of Potosi, Washington County, Missouri"; Gracy, *Moses Austin: His Life*, 73–75, 163–64. A disastrous fire consumed Durham Hall and many adjacent buildings in 1871 (*History of Franklin, Jefferson, Washington, Crawford, and Gasconade Counties, Missouri*, 519).

13. On commencing before survey: Gracy, *Moses Austin: His Life*, 78–83; on excision: *ASP-PL* 3:611 (Austin's account) and PC 217b-193 (Vallé's account).

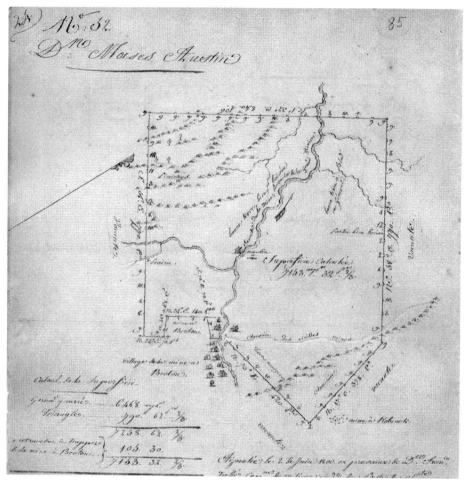

27. *Moses Austin Survey, June 2, 1800.* Moses Austin's grant was intended to be a square. The Mine à Breton mines, long worked by Creoles as common property, were excised from the southern corner of the square as a trapezoid west of the Creole village of Mine à Breton shown by a double row of cabins along the creek. In compensation, Austin was awarded a triangle on the opposite side of the creek. The single building on the north side of the creek is Austin's residence, Durham Hall. The two buildings on the south side opposite Durham Hall and within Austin's grant are his furnaces. The building below the stream junction near the center of the grant is a mill. A pinery is identified in the western portion of the grant. The road to Old Mines *(chemin des vieilles mines),* where other miners lived, crosses the grant in a northward direction. (*Régistre d'Arpentage,* 85)

A second repercussion of Austin's grant was that it prompted others, mainly entrepreneurial St. Louisans and Ste. Genevievians, to obtain their own large concessions for mineral land, an action previously deemed unnecessary. Members of the Luzières, Delassus, Pratte, Labeaume, and other families received concessions, some of them also a square league.[14] These mineral concessions awarded by the lieutenant governor in the closing years of the Spanish regime constitute a substantial percentage of the backdated claims of the district.

Austin's improved smelting operations caused the opening or reopening of other lead mines, the ore from which he bought to satisfy the appetite of his large-capacity furnace. During the five-year period from 1799 to 1803, six productive new mines (each mine a congeries of pits) were opened within a twenty-mile radius of Austin's furnace.[15]

The resident population of Mine à Breton grew from all this activity, not only because of more miners, but also because of more woodcutters, smelters, and haulers. According to Austin, twenty-six American and French families lived at Mine à Breton, which he estimated to comprise 208 persons, excluding slaves. Austin estimated 728 residents in the region around Mine à Breton.[16] Concentration of smelting in one furnace also brought about a geographic centralization of all ancillary activities at Mine à Breton village. Thus, in these few years after Austin's arrival, Mine à Breton became the true central place of the mining district and was no longer merely one mining camp among many.

Austin's frontier lead operation and settlement functioned like the semifeudal iron and lead "plantations" of the eastern states, and Austin, who earlier had a lead plantation in western Virginia, probably planned it that way. These iron and lead operations were typically located in sparsely settled, timbered country, and they required the developer himself to build the works and the houses for workers and provide stores and services for the community, over all of which he exercised paternalistic management. Thus, Austin's Mine à Breton "lead operation" included not only the lead smelter and shot and sheet factories but also his store, blacksmithery, sawmill, and flour mill. Austin was responsible for bringing in tradesmen with specialized skills, such as carpentry and masonry. However, Austin did not control the Creoles or their village. Mine à Breton had no church, and Austin did not plan for one. The lieutenant governor, recognizing the growing population and its distance from Ste. Genevieve, did appoint a *commissaire de police* for the village. Comman-

14. LCS 24, 25, 82, 83, 86, 102, 174.
15. Austin, "Description of Lead Mines," *ASP-PL* 1:188–91.
16. Ibid., 191.

dant Vallé recommended Creoles for the position; no American should be appointed, Vallé said, because the mining village was mostly Creole, despite its American nabob.[17]

The social and personal gulf between Austin and the French Creole community that opened at the time of the land survey never closed. Austin, who refused to learn French, had little to do with the Creoles except to smelt their ore and use their labor as necessary. From 40 to 50 persons worked on Austin's own property, but most of them were Americans. Austin's attitude toward the Creoles (they were always "the French" to him) reached disdain when the Osages attacked in 1799 and again in 1802 and Austin and fellow Americans fought them off, receiving absolutely no help from the Creole community.[18] The two settlement cohorts lived geographically juxtaposed, but socially in two different worlds.

Though lead mining prospered and the population grew after 1804, civil order at Mine à Breton deteriorated. The village descended into a lawless place with the "rudest and most savage society," according to Brackenridge, who visited it in 1809. Sheriffs could not check the violence. Miners stole others' ore and ashes that were left over on the ground from log smelting because they could now be used like raw ore in Austin's furnace. Gamblers and adventurers appeared in the population of the bonanza community. The place was a "constant scene of warfare" where people carried guns and weapons all the time; even Moses Austin's house was attacked.[19] These descriptions likely exaggerated the real situation, but they clearly indicated that the smoothly functioning society of pre-American days was past.

Disorder was rooted in conflicting ownership, rights, and access to mines and to land, exacerbated by newcomers impatient for quick prof-

---

17. François Vallé to Charles de Hault Delassus, July 10, 1802, PC 219-297. A detailed description of an iron plantation is in Arthur Cecil Bining, *Pennsylvania Iron Manufacture in the Eighteenth Century,* 29–48. In *Frontier Iron,* J. D. Norris describes the Maramec Iron Works west of Mine à Breton as an iron plantation. In his seminal work on the history of the exploitation of American forests, Michael Williams describes an iron plantation as a place where the ironmaster exercised a paternalistic, even feudal, oversight of people and the land (*Americans and Their Forests: A Historical Geography,* 147). An account of Austin's earlier lead plantation in Virginia is in David Gracy, *Moses Austin: His Life,* 46–47.

18. On gulf: PC 216b-206 and Gracy, *Moses Austin: His Life,* 76–83; on Austin's attitude: PC 219-330; Houck, *History of Missouri,* 1:369–72; and *Austin Papers,* 2:88.

19. Brackenridge, *Recollection of Persons and Places,* 212. Governor Lewis wrote about a riot and resistance to order by "an armed force," and he ordered the militia to fire on "lawless Banditti," if necessary (Meriwether Lewis to James Austin, November 10, 1808, *Bates Papers,* 2:38–39). *Missouri Gazette,* February 14, 1811; February 7, 1812.

its. Scarcely five years after having fought off Austin's attempt to seize their community mines, the Creoles—with some Americans in their midst—now had to fight a U.S. government that recognized no prescriptive rights to mine or, for that matter, to live and farm, or even to cut timber and fuelwood on the public domain. The Breton Creoles had no cabal of resident, wealthy French to speak up for them in public debate with the new administration, as the habitants at Ste. Genevieve had in the person of François Vallé, nor did they have wealthy French with similar economic interests on their side, as Creole miners at Old Mines did. They had only one way to assert their cause: to continue to mine on public property and defy the new American authority by civil disobedience, and that they did. They held on to their cabins in the village and to their traditional rights to mine the hill to the south, despite its being coveted by so many others and the exhortative pronouncements to move from the U.S. government.

To confuse matters, some French and Americans brought forth floating concessions to claim mineral land in the vicinity. No one was better at this than John Smith T, who used several floating concessions to claim ownership of various mines, including some on Austin's perfected grant. Smith T was supported by tough men and gunslingers—Smith T himself used his gun effectively in duels and otherwise—who, by intimidation or force, usually got what they wanted. The constable was usually not Smith T's equal in these disputes; the arm of the law was not long enough in the territory, and the litigation process in the federal territory over land claims in the mineral area was endless, a situation on which Smith T's strategy counted.[20]

Congress's decision in 1807 to extend its policy of reserving and leasing mineral lands to its new territory west of the Mississippi tended to dampen the wild scramble for ownership of mines and deferred the necessary hard decisions to a later time. Uncertainties in landownership and delays in land-confirmation proceedings elsewhere in the territory discouraged further settlement, but in the mining country and especially around its center at Mine à Breton, where profits could come fast without making time-consuming improvements such as clearing land and building houses and where the distance from administrative centers was greatest and enforcement least effective, those uncertainties and delays actually encouraged and quickened settlement. Confusion over land led to landgrabbing and rapid exploitation before authorities knew about it

---

20. For John Smith T, see Steward, *Frontier Swashbuckler* and "'With the Scepter of a Tyrant': John Smith T and the Mineral Wars"; *ASP-PL* 2:389, 3:599; LRA 156.

and rightful ownership could be enforced. People simply came and occupied land, undaunted by the cumbersome legal procedures for orderly settlement brought by the new republican government.

Mine à Breton and its neighborhood grew rapidly in the first decade of the American regime, the decade of "mineral-mania," as territorial secretary Bates observed. Americans moved into the Creole village. In 1806, there were forty houses in the original village, and a second row of houses went up along a new street on the opposite side of Breton Creek. Stores in competition with Austin's sprang up in the village. Austin built three sturdy bridges to tie the two parts of the village together: one, ten feet high and from sixty to eighty feet long, was over Breton Creek, and two were over town ditches that Austin had dug to convey water from the mines.[21] His paternalistic interest in the welfare of the fast-growing community did not languish.

By 1811, violence was far less prevalent than earlier, although disputes in landownership had hardly changed. Growth and prosperity inflated wages and prices in the mining community. Slaves were hired for eight to fifteen dollars per month in 1810. Miners could make as much as thirty dollars a day for weeks at a time, when top wages for laborers in the American East were ordinarily six dollars a month. The federal land agent reported that miners commonly earned fifty to one hundred dollars in one day.[22]

Growth of the village of Mine à Breton was accompanied by a great expansion of settlement in the neighborhood and in the mineral district in general. Some of this was increased mining activity as miners annually discovered more ore within a twenty-mile radius of Mine à Breton. Much more of the growth, however, was agricultural. Miners raised crops only for subsistence and sometimes not at all. As mining and smelting became full-time, year-round activities, Mine à Breton developed into a market of some significance for farm products; by 1814, it was the largest local market in the Ste. Genevieve District. The food needs of slaves and hired men had to be met by purchase. The cost of hauling food by wagon from Ste. Genevieve was too high. Already by 1801, some miners were turning to farming and animal raising, because it could be just as prof-

---

21. To Albert Gallatin, February 9, 1808, *Bates Papers,* 1:280–83; on houses: Christian Schultz Jr., *Travels on an Inland Voyage,* 2:48–54; on stores and bridges: *ASP-PL* 2:683, 3:591–92.

22. On slave wages: to John Michie, Goochland Court House, Virginia, August 22, 1810, *Bates Papers,* 2:154–56; on miner wages: van Ravenswaay, *St. Louis,* 85; on federal land agent report: Carr to Secretary of Treasury, November 14, 1805, *TP* 13: 274.

itable, was less arduous and hazardous, and had more security.[23] Americans were almost totally responsible for the increased farm output, and they did not neglect their fields when participating in aspects of the mining industry.

Two neighborhoods flourished as food suppliers for Mine à Breton and later Potosi: the Bellevue to the south and the Big River Bend to the east. Both were within fifteen miles of Mine à Breton, a reasonable distance for wagon hauling. Farmers in the Bellevue, especially, also furthered the growth of Mine à Breton by serving in the off-season as lead haulers, woodcutters, and carpenters. The countryside around Mine à Breton itself was considered too heavily mineralized for crops and animals.

Potosi

By 1813, the mining region was populous enough that the territorial legislature created Washington County out of the western portion of Ste. Genevieve County. It was the first county added to the original five districts-counties carried over from the Spanish period and the first county not bordering the Mississippi River. Its creation identified the mining area as the major geographic focus of settlement in Upper Louisiana in the years between the Louisiana Purchase and the War of 1812. Austin and his American associates—obviously excluding any Creoles—chose the site for the county seat, which, of course, would have to be at Austin's residence and mining establishment. To this end, Austin donated thirty acres of his league tract close but not adjacent to the village of Mine à Breton, and business associate John Rice Jones added another ten.[24] They platted a grid-pattern town aligned with straight-flowing Breton Creek, but not along it. This made the town aligned more or less with the two rows of cabins stretched along the creek in the village of Mine à Breton, a mile or so upstream. However, the creek-side streets of Mine à Breton did not directly flow into any of the platted streets of the new county seat. The Americans clearly designed their town to be an entity completely separate from the mostly Creole, unarticulated mining village.

Austin wanted a new and dignified town, befitting American enterprise and set apart from miners' cabins. It was a settlement strategy of spatial

---

23. To Richard Bates, December 17, 1807, *Bates Papers,* 1:238; Gardner, *Lead King: Moses Austin,* 103. Michael Quin, a longtime Mine à Breton miner, turned to farming and stock raising in 1801 and requested a concession for two thousand arpents in the chert hills nine miles west of Mine à Breton (LCC 134).

24. George Showalter, "Potosi, Missouri, Bicentennial, 1763–1963: A Bicentennial Scrapbook Commemorating the 200th Anniversary of Potosi at the Heart of the Historic Mining Country of Missouri," 6.

separatism similar to European colonial towns in Africa and Latin America that were laid out in the geometric grid of rationalists adjoining the vernacular layouts of the common people into whose territory they were intruding. Austin first named his new town St. George, but within a few months he changed it to Potosi, a name earlier used during a Fourth of July celebration of 1811 at Mine à Breton at which he raised a toast to "the lead-mines of Louisiana, may they increase and multiply in wealth, and merit the name of the Potosi of the United States." The name referred to the fabled wealth of the silver mines of Upper Peru (now Bolivia), and had nothing to do with the former Spanish administration that had welcomed Austin fifteen years earlier.[25]

Sale of lots in Potosi was brisk despite the War of 1812 and a downturn in lead sales. They were advertised in the *Missouri Gazette* in St. Louis, and within a three-day period in July 1814, 79 lots of the 147 platted were sold, probably more for speculation than for immediate occupation. The sales showed confidence in Austin and Jones's town venture and continuing investment in the lead-mining region, especially by the founders' numerous St. Louis and Ste. Genevieve business contacts. Sales also identified Potosi as an almost purely American venture: of the 79 lots sold, all but 1 went to Americans. Those investing in the future of Potosi included such contemporary and upcoming notables as William Ashley, a gunpowder manufacturer at Potosi and later fur trader and lieutenant governor of Missouri; Andrew Henry, a lead businessman and fur trader; Daniel Dunklin, a lead businessman, land speculator, chief founder of the academy at Potosi, and later governor of Missouri; Andrew Scott, a participant in territorial government; William Perry, a merchant; Lionel Brown, Ashley's partner in gunpowder manufacturing, first sheriff of Washington County and its representative in the territorial assembly;

25. For a description of the grid plan and layout of Potosi, see W. A. Schroeder, "Opening the Ozarks," 328–29. The colonial town with separate forms and functions for its juxtaposed colonial and indigenous populations is a well-known phenomenon in settlement geography. Probably the best known is Delhi–New Delhi, India. Other examples are Spanish colonial towns adjacent to Indian villages in Latin America. A comprehensive study of one such colonial town is in Ann E. Larimore, *The Alien Town: Patterns of Settlement in Busoga, Uganda*. Zitomersky explored the dual settlement form of French and Indian villages in the Illinois Country before 1770 (*French Americans–Native Americans*, 359–87). On St. George name: *Missouri Gazette*, June 4, 1814; quote: *Missouri Gazette*, August 8, 1811. The association of the toponym *Potosi* with riches was made a generation earlier in North America, around 1765, by residents of New Orleans who said in a memorial that trade with Indian nations "promises treasures more valuable than those of Potosi" (Usner, *Indians, Settlers, and Slaves*, 117 n).

James Austin, Moses Austin's son; and Moses Bates, Austin's brother-in-law and business associate. Only Amable Partenais dit Mason of the French community bought into the new town, and he had major business deals with the Americans, major mine interests in Mine à Breton, and legal skirmishes of his own with the Creole Bretoners.[26] Conspicuously absent from investing in Potosi were the established French families with lead interests, such as the Vallé, Chouteau, Luzières, and Pratte families, who remained comfortably domiciled in Ste. Genevieve and St. Louis.

The American backers were a different type of person for the territorial interior and more resembled the types then entering the commercial and speculative life of St. Louis; they gave the district a cosmopolitan flavor. Many had been born and raised in towns on the East Coast, particularly New England, or in Britain, and thus represented a cultural element on the Ozark frontier far different from the Kentuckian and Tennessean farmers of surrounding neighborhoods. Potosi, the incipient urban center of the booming lead region of southeastern Missouri, was to have a type of civic leadership that planned for a great future that would set the town apart from all others in the eastern Ozarks. Indeed, Potosi appeared on the brink of becoming one of the future great cities of the West.

Platting of Potosi in 1814 occurred before the U.S. land-survey lines were run and before most of the private properties in the vicinity were surveyed. The amorphous Mine à Breton village had never received a concession as a single community, although a few small properties in the village had been conceded. No survey had been made of the 105 acres agreed to be the common mines of the village at the time of the survey of Austin's grant. Mine à Breton, both village and mines, continued to be regarded as some kind of undelimited "commons," open to anyone, even though ten years had passed since the Spanish regime. In general, the status of landownership and the bounds of properties in the village and vicinity left much to be decided.

When Congress confirmed the properties of the traditional French villages of Upper Louisiana in 1812, Mine à Breton, a mere mining village without a common field for crops, had not been included. On February 1, 1817, villagers, suffering under increasing confusion over properties along the banks of the creek, petitioned Congress to be placed "on the same footing" as the other French villages of Upper Louisiana, but Con-

26. *Missouri Gazette,* September 30, 1815, and subsequent issues; Showalter, "Potosi Centennial," 6, 10–13. Firmin Desloge arrived from France in 1823 and operated a store in Potosi. He quickly married into the American community.

gress did not get around to confirming Mine à Breton properties until May 26, 1824.[27]

After that group confirmation, the private properties of the village as of 1804 had to be surveyed so that titles could be conveyed. Unlike the agricultural villages of square lots and common-field strips, Mine à Breton presented a real challenge to draw lines equitably that delimited the inhabited tracts of the disorderly mining camp of providentially located structures. Incredibly, the survey was not done until October 13, 1844; it is hard to believe that any survey could have been done with accuracy to represent conditions forty years earlier.[28]

Better than any other survey in the Ste. Genevieve District, the survey of Mine à Breton village shows the triumph of geometric orderliness of the American concept of towns over the vernacular, but topographically sensitive, clustering of Creole mining villages (fig. 28). The surveyor regularized properties into rectangular shapes. He straightened the two streets and drew property lines perpendicularly to them so that all properties fronted on one of them. Some cabins, however, lay in odd positions that made it awkward to find street frontage and right-angled property lines. For example, the surveyor ended up locating one of his property corners "within Dr. Bryan's office." Another corner had its stone set forty inches from the corner of a "negro house," and yet another cornerstone was set alongside a stable.[29]

If land was vacant or could not be otherwise assigned ownership, it was designated school land, as it was in Ste. Genevieve and other French villages. The surveyor marked public passageways to cross the creek and grandiosely labeled them streets. Mine à Breton (Jefferson) Street, the original street on the south side, was laid out only forty-five feet wide, significantly narrower than main streets of American towns, but the space was simply not there for a wider street. High Street on the north

---

27. On exclusion: *Stats. at Large of USA* 2 (1812): 748–52. However, some individual lots in the Creole village had been conceded by Spanish authorities as early as November 10, 1799 (LCS 81½). On petition: 15:692 n. The petition has been lost. It would be of interest to know whether the petition was signed by Creoles or by Americans, and whether a wealthy person initiated and pushed it, as in the case of Old Mines, another Creole mining village. On confirmations: *Stats. at Large of USA* 4 (1824): 52–56; *Congressional Debates,* 18th Cong., 1st sess., 26 May 1824, app., 3269–70.

28. Mine à Breton survey, roll 49, pp. 1–116, Record of Surveys, Missouri State Archives.

29. Ibid., 4–5, 56. The survey of thirty-nine inlots and nine outlots uses chimneys and other conspicuous building features for precise locations. The sites of Samuel Perry's house and shop and Daniel Dunklin's house and millrace can be located through the field survey notes of Mine à Breton.

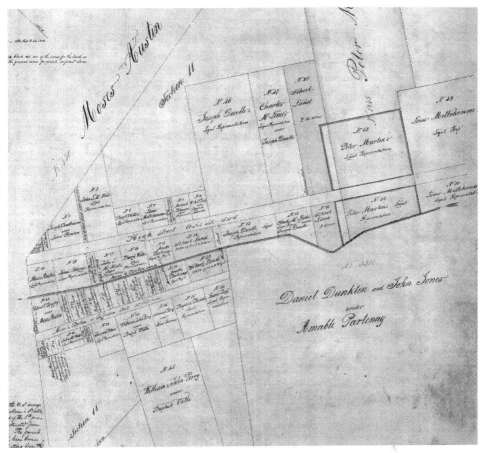

28. *Connected Plat of Mine à Breton.* The surveyor in 1844 transformed the irregular tracts of the mining camp into squares and rectangles of different sizes aligned to two streets on opposite sides of the creek. The street and lots of the Creole south side are narrower than those on the American north side. The village abuts Moses Austin's grant and his platted town of Potosi. (Miscellaneous Plats, Missouri Department of Natural Resources, Division of Geology and Land Survey)

side was not so cramped and was laid out to be the standard sixty feet wide, but even at this width, houses and buildings stuck out into the straightened public right-of-way. Consequently, though the surveyor was instructed to lay out a street sixty feet wide, the same width as the street's continuation onto Austin's grant, he could make it only fifty feet wide because of the irregular setbacks of buildings. To compensate for this narrowing, he added an arbitrary five feet on the back sides of lots, in case the street should be restored to its full width. His supervisor questioned this procedure, to which the surveyor, obviously extremely frustrated from trying to create spatial regularity out of disorder, angrily replied in his official notes, "I said 50 feet. Do you dispute me?"[30]

Austin's plan for a new, orderly town for Americans to sit above and look down upon the old creek-side village may be interpreted as his vision for American Potosi to rise above the superseded French Creole way of doing things. Furthermore, he intended Potosi to become an important city in a new state. Indeed, the lead area was economically vigorous in these years and was attracting considerable investment capital from the American East. In 1820, the territorial legislature considered Potosi for the temporary seat of state government as preparations for the new state of Missouri turned into reality. Though Potosi did not win this designation, it was the location of some sittings of the Missouri Supreme Court. Austin, who had an interest in architecture, had a role in designing the Washington County courthouse as a structure suitable for the first capitol of the state. In original design, it was to be an unusually large building, but the sale of town lots, which brought in $5,080 for the new building, did not provide enough money for the whole structure to be built. Two wings were never added. Even at that, the courthouse with its Greek Revival architecture symbolic of the American republic was pretentious for its time and place on the Ozark frontier.[31] A sketch of "Potosi alias Mine à Burton," originally published in 1819 in Schoolcraft's *View of the Lead Mines of Missouri,* shows the columned courthouse sitting proudly atop a hill neighbored by only a few houses, Austin's columned

---

30. Mine à Breton, Village à Robert, and St. Ferdinand surveys, 1842–1862, Record of Surveys, Missouri State Archives. Some buildings in Potosi (not the original ones) still protrude several feet into the High Street right-of-way.

31. On territorial legislature: *Missouri Gazette,* December 6, 1820; *Missouri Herald,* August 19, 1820; and Shoemaker, *Missouri's Struggle,* 274–76; on Missouri Supreme Court: Marian M. Ohman, *Twenty Towns: Their Histories, Town Plans, and Architecture,* 73; on original design: *History of Franklin, Jefferson . . . Counties,* 484–85; on architecture: Ohman, *Encyclopedia of Missouri Courthouses,* 219. Meinig describes how Greek Revival architecture was symbolic of democracy and the American republic (*Shaping of America,* 2:405–7).

29. *View of Potosi.* View looking north. The columned Washington County courthouse sits on the hilltop overlooking the town of Potosi and the Creole village of Mine à Breton in the valley of Mine à Breton Creek. The furnaces are the buildings farthest to the left. The mines occupy the slopes leading off the left side of the view. (Henry Rowe Schoolcraft, *A View of the Lead Mines of Missouri,* courtesy of the State Historical Society of Missouri)

Durham Hall mansion on the creek below, and a bustling creek-side Mine à Breton village of cabins and stores (fig. 29).

Austin and associates were not successful in promoting the new town of Potosi as a commercial center. Few businesses located there and not a single one around the courthouse square as intended. Instead, old Jefferson Street of Mine à Breton and its creek-side extension onto Austin's grant (not the Potosi plat) became the main street of merchants, and, lined with pack mules and ponies, it served as the major outfitting point for most of the territory west and south all the way to Arkansas, the Red River country, and Spanish Texas. High Street, on the opposite side of Breton Creek, also continued onto Austin's grant quite close to Durham Hall, and it also became a commercial street of businesses and hotels. Schoolcraft noted in 1819 that Potosi and Mine à Breton combined had eighty buildings, a courthouse and jail, a post office, an academy founded in 1817, two distilleries, two flour mills, one sawmill, nine lead furnaces, and shot- and sheet-lead manufactories. Population of the twin towns jumped to one hundred families in 1817, then to approximately five hundred persons in the mid-1820s. People commonly referred to the twin towns as "Burton," which is what Austin himself used, possibly as a holdover in usage before Potosi was created, but much more likely because all the business activity was in "Burton," whereas Potosi had only residences and that imposing courthouse up on the hill. The two towns built toward each other and merged. The two original plats never actually joined, but the two nodes of activity did. The two were formally combined into one entity upon the incorporation of Potosi in 1826.[32]

By the early 1820s, Potosi–Mine à Breton had become a regional commercial center, equaling, if not exceeding in some respects, Ste. Genevieve, fifty miles distant. To this end, Austin helped develop a regional infrastructure. He improved the public road linking Old Mines with Mine à Breton to expedite Old Miners' travel to work at Mine à Breton. He built a road to Herculaneum, a port on the Mississippi, and he tied the Renault mines to his furnace at Mine à Breton with another road in 1808. A road already connected Mine à Breton with the Big River mines and farms to the east and on to Ste. Genevieve. Another road connected Mine à Breton with the Bellevue to the south. Finally, Austin had talked territorial authorities into building a road from Potosi all the way to the Boonslick on the Missouri River in the central part of the proposed state.[33] Acknowledging that most of the thousands of new immigrants to Missouri came down the Ohio and headed for the Boonslick, he was determined to deflect this current overland through Potosi rather than letting them take the natural but longer river route through St. Louis. The Potosi-Boonslick road also had scheduled mail service. Thus, a road network focused on Potosi–Mine à Breton, which brought more people and products into the town. No other town in the Ste. Genevieve District had experienced such a determined, coordinated effort at building roads to develop trade.

---

32. Gardner, *Lead King: Moses Austin,* 105; Schoolcraft, *View of the Lead Mines,* 48; Gardner, *Lead King: Moses Austin,* 105; *Missouri Gazette,* March 30, 1816; May 24, November 22, 1817; Works Progress Administration, *Missouri: A Guide to the "Show Me" State,* 537.

33. On Old Mines–Mine à Breton road: *ASP-PL* 2:683, 3:591; on Herculaneum road: to Moses Austin, February 7, 1808, *Bates Papers,* 1:273–74; and to Albert Gallatin, February 9, 1808, ibid., 281; on Renault mines: *ASP-PL* 2:683, 3:591; on Potosi-Boonslick road: *Missouri Gazette,* March 23, 1816. According to Steward, John Smith T, Austin's nemesis in the mining country, developed farms in Saline County in the Boonslick after 1820 and had the farm goods sold in the mining region. Austin's Potosi-Boonslick road would have been ideal for this movement, if the movement were by land. Most likely, it was by water to Selma on the Mississippi, then wagoned into the mining region (*Frontier Swashbuckler,* 166–67).

After 1815, Potosi's future dimmed as miners began to leave when new mines opened up elsewhere in Missouri and in the Galena District on the upper Mississippi River, and they could not be adequately replaced by higher-cost slaves. Just as bad was the country-wide financial collapse of 1819, which brought about Austin's bankruptcy, sale of his property, and near destruction of his great furnace, manufactory, and other enterprises.[34] His death in 1821, a year after his celebrated trip to Spanish Texas to arrange for a new colony, and the subsequent departure of remaining members of his family and associates to the new grant in Texas, deprived Potosi and its region of leadership and unquenchable enthusiasm. No single person replaced Austin for his promotional skills, and no resource replaced lead for a basic economic activity. Potosi, once expected to be a leading center in the new state, began receding into a future as just another county seat in the eastern Ozarks.

Old Mines

The Creole community of Old Mines, which lay only six miles north of Mine à Breton, was one of the many neighborhoods that developed within the mining district centered on Potosi. However, miners had begun exploiting the lead ore in and around Old Mines a half century earlier than at Mine à Breton. Philippe Renaut, who began working in the area in 1719, received a lead-mining grant in 1723 for one and one-half leagues fronting the *"Petit Merrimac"* and extending up *"la première branche"* for six leagues to include the *Cabanage le Renaudière* (a camping ground), for a total of nine square leagues.[35] The exact location of the

---

34. For the financial collapse of 1819 as it affected Missouri, see Dorothy P. Dorsey, "The Panic of 1819 in Missouri." For Moses Austin's bankruptcy, see Gracy, *Moses Austin: His Life,* 183–91. What was left of Austin's Mine à Breton grant and his lots in Potosi went on sale in 1820 (*Missouri Gazette,* September 30, 1820). Austin's square-league grant at Mine à Breton, one of the few perfected by the governor at New Orleans, was finally upheld by the United States Supreme Court on January 5, 1888 (Gracy, *Moses Austin: His Life,* 220).

35. *ASP-PL* 2:163, 3:590. Renaut's six grants in the Illinois Country are also identified and described in Martin, *The History of Louisiana,* 152. Though now usually spelled Renault as in Fourche à Renault Creek and the Renault Mines near Old Mines, the French man signed his name Renaut, according to Ekberg (*French Roots,* 35) and various documents in *ASP-PL*. Renaut is the spelling used for the person in this work. The name is also spelled Reno, especially in Spanish documents. Renaut's grant on the Meramec would seem to be the location of François Tayon's concession on the Fourche à Renault, which fronts on Big River (*Régistre d'Arpentage,* 120). Giraud noted the confusion that results from two men with similar names engaged in mining at the same time in the Illinois Country: Renaut and Renaudière (*History of French Louisiana,* 5:441–42). An excellent, brief account of the early history of mining in the Old Mines–Meramec area is in Ekberg et al., *Cultural, Geographical, and Historical*

grant is unknown. The Little Meramec is now the Big River, and the first branch could be the present Fourche à Renault (Fourche Arno); that name itself is evidence. Renaudière was one of several French who were in the lead country some years earlier searching for silver but finding lead. A reference to "la Grande Mine" suggests the existence of more than one mine, although the French meaning of *mine* as an ore deposit and not necessarily an exploitation suggests that not all of the places called "mines" had been worked this early. Renaut's grant and mines could have been either at present Ebo on the Fourche à Renault or at present Old Mines on its principal tributary, Old Mines Creek, or on both. According to one strand of local tradition and field evidence, which includes ruins of furnaces, Renaut mined on Old Mines Creek.[36] The name itself, Vieille Mine, especially when used in the midst of so many other early mines, suggests that lead ore at Old Mines was among the first exploited in the Meramec River drainage.

The French abandoned the Renault (or Meramec) mines in the 1730s, but reopened them in the early 1740s when Antoine Valentin de Gruy visited the site in 1743. A village of some sort, perhaps a collection of cabins, existed at or near Old Mines in 1748, when Pierre and Marie Wifvarenne, married at Fort de Chartres, listed "village des mines" as their residence.[37] If so, the village did not continue, but the toponym did in many continuous references to the location as Vieille Mine or Vieilles Mines. Renaut's French grant having lapsed, no other official grant of the land was made.

Discovery of Mine à Breton in the 1770s, a few miles to the south, drained Old Mines of its remaining miners. However, the site may never

---

Study, 8–18. For mining under the names of Renaut (Renault), Renaudière, and Old Mines, also see Hanley, "Lead Mining in the Colonial Period," 1–29, 136–38.

36. Showalter, "Potosi Centennial," 30; McCormick, "The Coleman Family History," 8.

37. Ekberg, "Antoine Valentin de Gruy"; Houck, *History of Missouri*, 1:378. The toponym is variously recorded as La Vieille Mine and Les Vieilles Mines, both names with and without capitalization, and in English as Old Mine and Old Mines. The latter is standard today. Because the French *mine* refers to both a worked mine and the mere presence of ore, the early usage of the term *vieilles mines* could have meant unexploited ores as well as those exploited by Renaut and others. *Mines* could also have meant a general region of ore deposits rather than the location of specific deposits worked or unworked. Finally, *mine* could refer to either the physical ore deposits without any human habitation or to a human settlement that grew up at the mines. The latter is clearly the way the term *Old Mines* is used today (a populated place without any physical diggings of ore), although it is also used in a historic sense to refer to the physical diggings of the ore at Old Mines.

have been utterly depopulated, except during times of sustained Osage activity, because it was close enough to Mine à Breton for miners to continue to live at Old Mines and work at mines a few miles south. Jean-Baptiste Miliet requested a concession at Old Mines in 1792, and this appears to be one of the first formal requests for land at Old Mines since Renaut's grant and evidence of its reinhabitation. Ste. Genevieve Church records make mention of people at Old Mines in 1793.[38]

Enough persons were resident at Old Mines in 1796 and 1797 for them to petition as a group for a concession for agricultural occupancy, as mining could be done by prescriptive rights on royal domain. The request received no action; it was later alleged to have been lost. Both Moses Austin and Pascal Detchmendy (of Ste. Genevieve) applied for mining concessions at Old Mines in 1797, but the commandant told both that the tract had been promised to its inhabitants, which suggests that their request had reached him. The nonresidents' petitions were not granted. Austin's request for more lead mines particularly perturbed the Old Miners who viewed it as an American's attempt to take land away from Creole miners of long residence. During the years from 1796 to 1803, Creole miners and families apparently mined at both Old Mines and Mine à Breton and moved back and forth between the two places, perhaps keeping cabins at both. Osage visits to this part of Osage territory during these years disrupted mining and life in general at Old Mines, which lacked Americans to fight them off.[39] It was a period of instability for building a community at Old Mines.

Nevertheless, mining at Old Mines was definitely going on, with improvements (such as residences and enclosed gardens) in place just before the turn of the century. One concession request of 1799 refers to the "little village of Boyer's family" at Old Mines. But another account reports that Old Mines was "re-inhabited" in 1802 by twelve French families who had been living and working at Mine à Breton. They claimed that Amable Partenais dit Mason had polluted the creek water at Mine à Breton by ore washing and animals upstream, and their grievances to Commandant Vallé had not received justice. In their words, they were "compelled to leave our habitations and remove to the Old Mine which was then a wilderness and destitute of a single inhabitant." Other families, mainly

---

38. Houck, *History of Missouri,* 1:379. According to Houck, Miliet abandoned his concession, and it was reconceded in 1799 to Jacques Guibourd of Ste. Genevieve, who was interested in the mines with Joseph Pratte. On church records: Ida M. Schaaf, "Henri Pratte: Missouri's First Native Born Priest," 137.

39. LPC 1m; *ASP-PL* 5:721–25; PC 2365-323.

more miners from Mine à Breton, joined the group shortly afterward, but some of the earlier group of twelve moved back to Mine à Breton.[40] It is almost impossible to determine who among this fluid group of miners was actually resident at Old Mines or Mine à Breton in any given year. As with the residents of New Bourbon who left to establish St. Michel, the miner settlers may also have left Mine à Breton, which had come under the domination of Austin and the Americans, to establish a more homogeneous and cohesive French community built around a few interrelated families. Whatever the reason, Old Mines was settled permanently before the nineteenth century by French Creoles and a few Canadians and French.

Old Miners again ran into trouble in 1804 with the same Amable Partenais dit Mason, a large-scale lead operator with many slaves, when he also moved to Old Mines and began working ore in the headwaters of Old Mines Creek and polluted its water. Just as Commandant François Vallé dismissed the miners' complaint of 1802 at Mine à Breton, his brother Jean-Baptiste Vallé, the new district commandant, dismissed their 1804 complaint addressed to Capt. Amos Stoddard.[41] The Old Miners had predicted this response from Vallé, because Mason owed debts to Vallé and they expected Vallé to favor Mason in the dispute. The Old Miners had wanted "disinterested Americans" to arbitrate their grievance. Remembering their earlier treatment at Mine à Breton, they distrusted their own French commandant, stating that they "by no means wish to be left to the decision of Frenchmen." For his part, Stoddard had refused to get involved with local affairs of the French population and had passed the grievance down to the district commandant. This episode illustrates that the poorer French recognized inequities of class distinction of the old regime and hoped for better treatment under the new democratic regime.

Meanwhile, as a result of Austin's success at Mine à Breton and the growing market for lead, the mines at Old Mines and its surrounding hills also attracted mining interests from Ste. Genevieve and St. Louis families. They employed hired hands or slaves to dig, smelt, and haul, though they continued to live in Ste. Genevieve and St. Louis. Their pits intermixed with those of the local miners.

When word reached Old Mines of the retrocession of Spanish Illinois

---

40. On 1799 request: LPC 19a; quote: PC 219-328. Moses Austin said fifteen, not twelve, French families moved to Old Mines ("Description of the Lead Mines in Upper Louisiana," *ASP-PL* 1:191). Names of miners at Mine à Breton in 1788–1802 overlap to a large degree with names of inhabitants at Old Mines during the same period (MN 6b; Houck, *History of Missouri*, 1:367–68).

41. MN 40; Willms, "Lead Mining, 1700–1811," 75–76.

to France, those working the mines became concerned about ownership status of the productive mines. The absentee lead operators knew that any change in government would make "the position of our properties . . . precarious," and it would be important to have title to the land. When it was remembered that the original group petition of the Old Mines community had not been awarded (it had probably been drawn up without the St. Louis–Ste. Genevieve interests included as petitioners), a new petition was drawn up, this time including the wealthy absentee miners. Nicholas Boilvin, an agent for François Vallé, was responsible for gathering signatures. Testimony before the second American board for land confirmation in 1832 alleged "many irregularities" in the drawing up of the list of names for the request.[42] Omitted from the list were some of the earliest miner settlers at Old Mines, yet it included absentee miners and other nonresidents. Thus, the list of thirty-one names on the concession request was not an accurate account of the actual settlers in the Old Mines neighborhood at the time.

The petition of May 25, 1803, which was written by Commandant François Vallé himself, explained that although the petitioners had felt secure on their lands under the Spanish and therefore did not seek title, they were now worried because they had heard of retrocession to France. Vallé, as commandant, forwarded the petition to Lieutenant Governor Delassus. Did Vallé write the request because he was interceding on behalf of his wealthy Ste. Genevieve and St. Louis friends with lead-mining interests at Old Mines? Was it a way for him and his friends to get in on a cut of the expanding lead business and exclude Austin from Old Mines? The group petition went forth as a hybrid between resident miners and nonresident entrepreneurs and is misleading as to the true nature of the Old Mines settlement, which was actually quite a homogeneous and cohesive community of interrelated Creole families working as independent miners.

It was necessary for the nonresidents, especially Commandant Vallé, to participate in the endeavor so that, with their connections and knowledge of how to get things done, the petition could proceed expeditiously through the bureaucracy. It was equally necessary for the nonresidents to include the resident Creoles in the petition because they were the real occupants of the land and gave the request its necessary credibility and agricultural justification. The two groups with contrasting ways of life merged their goals to get title to the land and were successful in getting

---

42. Quote: *ASP-PL* 3:593; on new petition: *ASP-PL* 3:593 (French); *ASP-PL* 5:721–25 (English); and LPC 1a; on testimony: *ASP-PL* 5:723–25.

one group concession, whereas they likely would have been unsuccessful if they had instead sought thirty-one separate concessions. Consequently, the mining camp of Old Mines took a different direction and developed a different settlement form than the neighboring mining camp of Mine à Breton, which lacked wealthy nonresidents to lead its Creole miners through quasi-legal procedures in the closing months of Spanish control.[43]

The petition included thirty-one heads of families, thirteen wives, seventy-two children, and eighteen slaves. The figures for wives and children did not include members of the six nonresident Ste. Genevieve families. Twelve "heads of families" without wives and children out of twenty-five resident households suggest a young population.[44]

As miners and laborers, the residents of Old Mines appear to have been poorer in material wealth at the time of the concession request, less interested in the conventions of either French or Creole society, and ignorant of or inattentive to administrative procedures. Though most had lived earlier in Ste. Genevieve village and wealthy merchant Vital Beauvais was godfather to François Coleman dit Racola, the resident miners seemed not to relate much to the established families of Ste. Genevieve. They formed their own community, demographically buttressed by their large families. Evidence to support this perception of them as a community of less means and socially removed from Ste. Genevieve comes from a variety of sources. They held little land elsewhere. Those who had lived in Ste. Genevieve did not have land in the Pointe Basse fields but worked their own lands on the north side of the village away from other villagers. Some just abruptly appear in the district documentary record at Old Mines or Mine à Breton. They typically lived in *cabanes,* not *maisons.* Seventeen of the twenty-three Creole residents could not write their names. Their surnames had various spellings in documents, suggesting that someone else wrote them as they were pronounced: for example, Portel, Potel, Porter, Portais, Porte; Declue, Duclos, Declos; and Roxe, Russ, Ross, Rose. They regularly replaced their given baptismal names with nicknames in important official documents: Charlot for Charles; Perod and Perote for Pierre; Poline for Appolline; and Polite for Hypolite. Certain vital statistics are missing for some, indicating a transient past or that they did not bother to have them recorded. They would not or could not pay

---

43. John B. C. Lucas saw through the hybrid petition and referred to it as an ingenuous "contrivance" in a sweeping criticism of the landgrabbings of the Chouteaus, Cerrés, Clamorgans, and others (*ASP-PL* 3:587).

44. LPC 1a. For a detailed reconstruction of the Old Mines families and their interrelatedness, see W. A. Schroeder, "Opening the Ozarks," pp. 338–39, app. 1.

the small surveyor fees to Thomas Maddin even when sued. John B. C. Lucas described them as "poor mineral diggers [for whom] cultivation was trifling and merely incidental" and *"créoles du pays"* (country folks), as though they had lived in the valley for so long that they had lost the culture of Canada or France and retrogressed to something else. They built their cabins in the American style of horizontal logs, which elsewhere was derided in French communities as construction for outhouses.[45] They held fewer cattle per family, and those they had were let to roam and forage in the woods American style, because the community had no commons.

The miners justified their 1803 request for land as farmers. Even the nonresidents, whose interest lay only in lead, signed the petition for agricultural land. The amount requested was 400 arpents per family, the same as at St. Michel. Families of greater means requested no more land than those of lesser means or single-person households. Thus, the total group of thirty-one asked that a concession be granted to them as a single group "for a quantity corresponding to their population, to be divided in equal portion to each of them." This sounds egalitarian, but it was a clever way for wealthy nonresidents to obtain land that they did not live on and never intended to cultivate. To request the mineral land on the basis of agriculture may have been consistent with the Spanish policy of conceding lands, but it also proved to be wise strategy later, because the American board applied the rule of land cultivation and residence, not mining, for approval of claims. Pascal Detchmendy went to great lengths in his 1833 testimony to state that nonresident claimants (whom he openly admitted had not been residents of Old Mines) had someone else on their lands to inhabit and cultivate them, not to mine lead, as they really were there to do.[46]

Lieutenant Governor Delassus granted the request within ten days on June 4, 1803, during the window of time when he was actively awarding land to his friends and others who paid him for his largesse, and one suspects that the nonresident petitioners of Old Mines also paid him to get title to this mineral land. Delassus awarded 13,400 arpents (17.8 square miles) for the concession, which is exactly 400 arpents for each of the

---

45. Variations in spellings of family names have increased since 1820. See Kent Beaulne, Judith Escoffier, and Patricia Weeks, *Les Noms des vielle famille de la region de la Vielle Mines* [sic] (Old family names of the Old Mines area in Missouri). On vital statistics and surveyor fees: LPC 1m and *ASP-PL* 6:728–29; on Lucas: *ASP-PL* 3:587, 593; on cabin style: Peterson, "Early Ste. Genevieve," 217–18.

46. On single concession request and quote: *ASP-PL* 3:593, 6:729; on Detchmendy: LPC 1e and *ASP-PL* 3:587.

petitioners. Thomas Maddin surveyed only the outer boundaries of the concession on February 3, 1804, during the rush to get properties surveyed before the Americans took over.[47] The American-born surveyor, who had a geometrical puzzle to solve, set the concession width at exactly 88 arpents, an unexplained 4 arpents longer than 1 linear league used in French surveys, and more than twice as long as the traditional 40-arpent *terre* (fig. 30). The length of the concession was determined by how much land was needed for thirty-one 400-arpent tracts, after subtracting for two overlapping claims at the southern end. The linear axis of Old Mines Creek (*"branche de la Vieille Mine"* on the plats) determined the long axis and orientation of the concession and was what the concession "fronted" on, except that it was in the middle of the tract rather than on one side. It was fortunate that the tract, laid out beginning in the south where the mines were, did not have to continue much farther north, because Old Mines Creek curved to the east and would have been lost as a central axis.

The topography of the hilly concession is simple. It is a broad, shallow trough sloping northward with the creek in the middle. The active mines lay at the upper southern end, and any waste from lead washing would flow the entire length of the concession. The two long sides of the concession lay on flanking ridges of shallow, flinty soils, and they were not directly useful for much except the necessary supply of firewood.[48] At the time, however, the whole concession was thought to be potentially productive for lead ore. No one could possibly have selected this tract for its agricultural worth.

The concession could not include the common property of the active mines, just as Austin's grant could not include the common property of the mines at Mine à Breton and the St. Michel concession could not include any unconceded mines on its north side.

On December 20–22, 1805, surveyor Maddin, now employed by the United States, divided the interior of the concession into thirty-one equal divisions along its long axis so that each division had frontage on the creek. Lot numbers begin in the south, at the mines, and end in the north (fig. 31). No land was set aside for a village of streets and square blocks and lots, for which these thirty-one strips would then have served as *terres* in a *champ*. Old Mines would not be a tripartite French village with a clustered village, common field, and commons. In order to arrive at divisions of exactly 400 square arpents for lengths of 88 arpents, divisions

47. On concession: *ASP-PL* 3:593; on Maddin survey: LPC 1b, 1g.
48. U.S. General Land Office, Field Notes, Missouri, 1:270–73; *ASP-PL* 3:611.

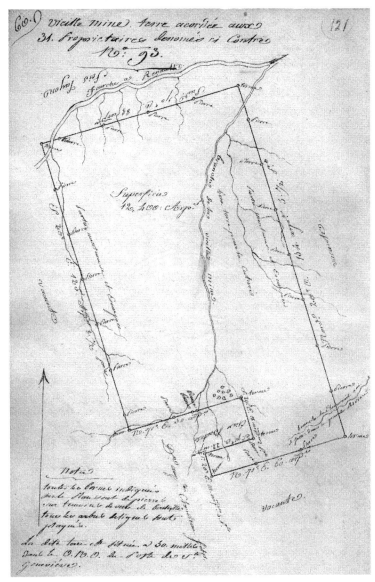

30. *Old Mines Survey, Original Exterior Survey, February 3, 1804.* The rectangular concession lies astride Old Mines Creek with two claims subtracted from its southwestern corner. The mine pits, shown with circles, are in the headwaters of the creek at the southern end of the concession. The size of the concession was determined by allotting 400 arpents for thirty-one families. Surveyor Thomas Maddin notes that the boundary markers are stones over bottle glass. The singular form, Vieille Mine or Old Mine, is used on this survey. (*Régistre d'Arpentage,* 121)

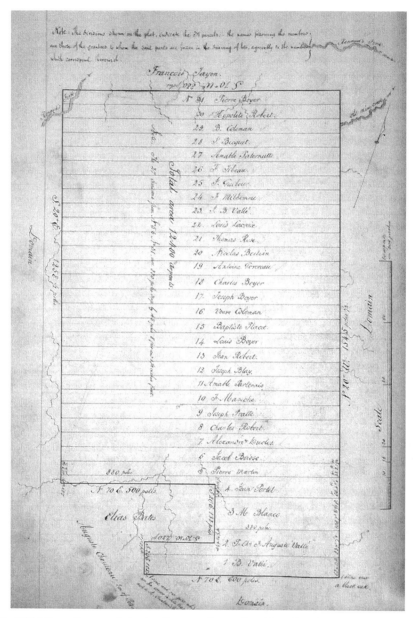

31. *Old Mines Survey, Interior Survey, December 20–22, 1805.* Each of the thirty-one families received four hundred arpents in long strips extending across the concession and including a portion of Old Mines Creek. Assignment of location was made by drawing lots by chance. The village of Old Mines is along the creek on lots 4, 5, and 6. The village of Racola is along the creek on lot 16. (Record of Surveys, Recorder of Land Titles, vol. C, p. 334)

had to have widths of 4.545 arpents, which equaled 818 feet and $1\frac{1}{4}$ inches for measuring in the American period.[49]

As stipulated in the concession approval by Delassus, the long, narrow divisions were numbered and lots drawn by chance to assign owners to specific tracts of land. Length of residence at Old Mines determined the order of drawing. Although François Vallé had assured the lieutenant governor that "no objections will disturb the good harmony that exists between claimants by this allocation" of the concession, residents did object to the procedure because they recognized immediately that it was highly improbable that the lots they received by chance would contain their personal cabins and improvements.[50] How indeed could people living in clusters of houses be assigned to lots in a row for residences? This unusual procedure of awarding specific tracts of land to settlers by chance years after they had built houses and enclosed fields was devised by the absentee owners in order to appear part of the community. Vallé most certainly wrote the provision into the concession request without the knowledge of the Creole residents. Although surveyor Maddin dutifully carried out the plan and assigned lots, settlement customs prevailed, and the Creoles persisted in living where they were and wanted to be, rather than redistributing themselves onto assigned lots and creating a linear village with houses $4\frac{1}{2}$ arpents apart. Which miner would want to live on the distant northern end of the concession, 5 miles away from the mines at the southern end? Who would want to distance himself and his family from his cousins, parents, aunts, and uncles? The Old Miners would never knowingly have agreed to such a procedure, and, in the end,

---

49. On Maddin's survey: Record of Surveys, Recorder of Land Titles, vol. C, p. 334; LPC 1g; *ASP-PL* 2:591. The French measure for lot dimension is 880 poles deep by 45 poles, 8 feet, $1\frac{1}{4}$ inches wide. The amazing $\frac{1}{4}$-inch measure was apparently carried out in field survey (Record of Surveys, Recorder of Land Titles, vol. C, p. 334). The surveyors used an equivalent of 180 feet for 1 linear arpent.

50. On Delassus's stipulations: LPC 1a and *ASP-PL* 5:721–25; on quote: 3:593. Drawing lots for properties has a long precedence in such contrasting land-division systems as those in New England and those governed by the Spanish Laws of the Indies (Reps, *Town Planning*, 29, 101; Faragher, "Americans, Mexicans, Métis," 100). One could interpret the plan to assign lots to individuals in other ways than by chance. For example, there is no other obvious way to divide a rectangle of 12,400 arpents into thirty-one equal divisions with frontage on the creek, unless land were taken out for a tripartite village, as in the case of the group concession at St. Michel, and then divide the remaining space into a common field and a grazing commons. Such a traditional village for the single Old Mines concession would not have made sense for all concessionaires, however, as the group petition included so many persons who would not have lived in a village on the concession. At St. Michel, everyone who was part of the group concession was a resident of the village.

the tradition of family village clusters won out over the official stipulations of the concession.

Because a village had already formed near the southern end of the concession on the land closest to the mines, virtually everyone ended up living on someone else's 400-arpent tract. For example, Joseph Boyer, Alexandre Duclos, Jean Robert, Jean Potel, and Polite Robert and their families had houses and enclosed gardens in the Boyer family village on lots 4, 5, and 6, which were assigned to others, whereas their own assigned lots were scattered throughout the concession.[51]

Maddin stated that "the claimants [understood] that, in the event of an individual who had made an improvement being thrown on a tract where there was no improvement, he was to be paid for his labor." Who was to make the payment? This issue became moot shortly. The residents in the village simply stayed where they were, and within one year most had sold their lottery-assigned, unoccupied 400-arpent strips to larger mining entrepreneurs and speculators. Among them were John Smith T and Amable Partenais dit Mason, against whom the Old Miners had twice filed grievances and who was no friend of the community. Partenais dit Mason, a Creole who had developed a flourishing lead operation in the mining district, was particularly successful in acquiring ownership of eight lots in addition to his assigned lot 11. In September 1804, ownership of lots 7, 8, and 14 had passed to Seth Craig and Company, a Philadelphia firm speculating in lead, even before the lots were surveyed.[52] The price paid for 400-arpent tracts of mineral land was minimal, as low as only twenty or thirty piastres, or approximately twenty or thirty dollars. The lead speculators were acquiring potentially valuable land for a trifle. All land sales, of course, were ultimately contingent upon confirmation of the Old Mines concession by the United States and supposedly under lead land–leasing provisions.

During all this maneuvering over landownership, miners and their families continued to reside in their simple village. It all worked out to everyone's general satisfaction. The miners received money for sale of

51. *ASP-PL* 2:547; LPC 1k.
52. Quote: *ASP-PL* 6:728–29; on Partenais: 3:594. Amable Partenais dit Mason had forty-five slaves, probably all of the men used in mining, and ranked as one of the largest slave owners in the district. On Seth Craig and Company: LRB 8. Some lots were sold in fractions. For example, the company of Easton, Byrne, and Cook bought one-third interests in four lots in 1812 (LRB 311). When John Smith T bought Old Mines land from the local Creoles, he made them responsible if the land did not get confirmed. However, when he bought land from the prominent Delassus family, he assumed responsibility for the transaction, in the event that the land did not get confirmed (LRA 152, 153, 156).

land they never intended to use or occupy, and the speculators acquired title to mineral land, otherwise difficult to acquire. The miners continued to live where they wanted and dig ore for a living, and the wealthy had either a labor supply or a lead supply at Old Mines. There was no friction between these two dissimilar groups, except when John Smith T arrived in their midst. Smith T's residence, first a log cabin, then an impressive three-story brick house, was at Bellefontaine, three miles east of Old Mines village and just outside the eastern boundary of the concession near his Bellefontaine and Shibboleth mines. In addition to buying some lots in the Old Mines concession, Smith T claimed other lots under different pretenses and kept their ownership in question. Those claims were rejected in 1832 when the second board finally took up the issue of confirmation of the Old Mines concession.[53]

In addition to Smith T's mines, several others were opened up on the Old Mines concession or next to it. Much of the area was dug into by pick, and its landscape two centuries later still bears the scars of shallow pits and rocky waste piles gently smoothed out and revegetated over the years. Earlier, miners had built several log furnaces on the concession, but as time went on, they hauled more of their ore to Smith T's or Austin's furnaces, or to others in the vicinity.[54]

When Nicolas Boilvin collected signatures for the 1803 petition, he told residents that a "village would be built," probably upon instructions from François Vallé, his employer.[55] Was he suggesting a planned village of square blocks and square lots as at St. Michel, New Bourbon, and new Ste. Genevieve, with the $4\frac{1}{2}$-arpent-wide lots of the concession serving as long lots of a common field? Or did Boilvin mean a one-street linear village, the length of the concession, a house on each strip and therefore $4\frac{1}{2}$ arpents ($\frac{1}{6}$ of a mile) apart? Who would build the village? Or was the promise only an insincere ploy to secure signatures?

A possible plan for the village layout appears in a twentieth-century historical account of the Coleman family at Old Mines. According to this document, the concession was "laid out as any 15th or 16th century village of Normandy would have been."[56] The village was to consist of one

53. [Natalie Villmer], "250th Anniversary Historical Program Pageant Book of Old Mines, Missouri, 1973," 16; *ASP-PL* 3:599, 6:730–31; Gardner, *Lead King: Moses Austin,* 110–21; Steward, "'With the Scepter of a Tyrant.'"

54. The Old Mines Area Historical Society at Cadet and the Mine au Breton Historical Society at Potosi have knowledge of several historic furnaces within and around the Old Mines concession.

55. LPC 1k.

56. McCormick, "The Coleman Family History," 14. This reference to a Normandy village is unaccompanied by any drawing and lacks any written source; it stands as

long central street parallel to the creek and flanked by farms on both sides, 1 or 2 arpents wide, running back ½ league (40 arpents) to the two opposite concession boundaries for a total length of 1 league. The central street and creek would presumably account for the extra 4 arpents width of the concession. Farmhouses would line the central street at 1- or 2-arpent intervals, fields behind them, with the lead mines, as known at the time, lying at the far southern end of the concession. This pattern would have been, minus the mines, similar to the linear one-street villages of Laurentian Canada, except that the street would be in the center of the league-long property rather than on one end. The reference to Normandy, or to any place in France or even Canada, as a model for settlement may be unique in the Mississippi Valley. The Coleman account remains an uncorroborated explanation for the peculiar long-lot division of the concession and the nature of the village that did emerge, but it could also be nothing more than the surmise of a family descendant vaguely aware of the form of Normandy villages.

The community of Creole miners, acting on their own, produced a settlement form unrelated to the concession shape and quite different from both a one-street "Normandy" village and the tripartite villages at Ste. Genevieve, New Bourbon, and St. Michel. The settlement took the form of an unplanned village of clustered houses, already in place before the concession, but property deeds do not divulge enough information to reconstruct it in any detail. In only one document is spacing mentioned—houses 1 arpent apart—but no lots are mentioned nor compass orientations given. The village seems to have been amorphous. Cabins and a few houses were probably positioned according to minor topographic features such as water sites and grouped in extended family compounds. Each family had a small enclosure that was regularly referred to as a *jardin* (garden) (pointedly not a *terre cultivée*). John B. C. Lucas observed the placement of "their cabins near each other, and ... a small patch behind each." Livestock roamed the encircling woodlands. There was no designated common land. Jean Robert used the term *common land* on his lot 13, but context indicates that this expression referred to the royal domain beyond the two ends of his lot.[57] Paths and cart trails intercon-

---

oral tradition. A Normandy connection was also established by Dorrance, who concluded that the French speech of Old Mines residents was brought "fairly intact" from Canada, which was populated largely by peasants from Normandy and other places in northern, western, and central France (*Survival of French,* 47–48).

57. On form of settlement: *ASP-PL* 6:728–29 and LRA 29; on spacing: LPC 1m; Lucas quote: *ASP-PL* 3:587; on Robert: LRA 29.

nected the cabins. In such an informal setting of relatives, no carefully surveyed property lines were necessary.

The main village was located along Old Mines Creek on lots 4, 5, and 6, in the southern portion of the concession, the end nearest Mine à Breton and the mines of Old Mines. Austin said the village had 15 French families in February 1804 and estimated its population at 120. In 1816, Austin reported French families in a "small village of about 20 houses . . . built of logs." We do not know much from documents about the kinds of buildings in the village, except that they were cabins of horizontal logs, American style. One house, however, is described in the Coleman family history as whitewashed and built of hewn square logs set vertically in the "French provincial type that was found throughout Upper Louisiana." Lot 13 had a slave cabin on it. As time went on, extended families established new clusters on the concession, the chief one of which was at Racola on lot 16, the center of the concession, $1\frac{1}{2}$ miles north of Old Mines village.[58]

The U.S. township line between ranges 2 and 3 east runs along Old Mines Creek. When the surveyor ran it on August 17, 1817, before the Old Mines concession was confirmed, he recorded "a little village or settlement" 25 chains (1,650 feet, or about $\frac{1}{3}$ mile) N 10° W of the southeast corner of section 13. These directions place the village close to present St. Joachim's Church, located near the southern end of the concession and therefore at the original village on lots 4, 5, and 6. This is where the toponym *Old Mines* is applied today. The surveyor identified the cabins of Charles Boyer and Bernard Coleman along the creek at this place. The surveyor recorded a second cluster of cabins, a separate village, at 15 chains (990 feet, or about $\frac{1}{5}$ mile) west of the northeast corner of section 12, or at present Racola.[59] This cluster lay on the widow Coleman's lot

---

58. On 1804 estimate: Austin, "Description of Lead Mines," *ASP-PL* 1:191; Austin quote: *ASP-PL* 3:611; McCormick, "The Coleman Family History," 16; on slave cabin: LRA 104; on Racola: McCormick, "The Coleman Family History," 5.

59. U.S. General Land Office, Field Notes, Missouri, 197:150, 152. The Old Mines concession was surveyed into sections, ranges, and townships in 1817 because the concession had not been confirmed. Property transactions, however, continued to use the lot numbers of the Spanish concession. Confirmation of the concession in 1833 legitimated the use of the lot numbers as subdivisions of Private Survey no. 3039, the new legal identification of the concession. The American rectangular survey subdivision has never been used for the concession for landownership purposes.

The name *Racola* comes from François Coleman II dit Racola, who received the alias from his godfather (McCormick, "The Coleman Family History," 5). Cola is a shortened form of Nicolas.

16, which she retained in her possession. If other clusters of cabins were on the concession by 1817, they were away from the township survey line and unnoticed. Old Miners today, by tradition, refer to the Racola nucleus as the site of the "old mine" *(la vieille mine)*.

The multinucleus community grew up without an institutional focus. A priest from Ste. Genevieve provided religious services, but a log church, St. Joachim, was not built and consecrated until October 29, 1820, later than the church at St. Michel.[60] Its placement in the village at the southern end of the concession ensured that the cluster would permanently bear the name *Old Mines*. St. Joachim also served Catholics who lived at Mine à Breton and Potosi. St. Joachim Church has continued to the present as the strong and undisputed center of community life for the widely dispersed population of northeastern Washington County.

Although Americans bought land in the concession, it was for speculation only, and the villages remained impervious to outsiders as a place to live through the 1820s. Moses Austin made a determined effort to attract the Old Mines residents to Mine à Breton both for their lead ore and for their business by improving the road between the two villages. John Smith T, Austin's equally aggressive rival, also attracted Old Miners' lead and business at his establishment on the east.

The group concession of 1803 was finally confirmed by the second board in 1833 as a single claim, "leaving it to the parties interested to make the subdivision of it according to their several interests, and this relieves [this office] from the responsibility of investigating and deciding upon the separate interests of the grantees." Thus, even the nonresidents got their lots confirmed along with bona fide residents. The United States House committee recommending its confirmation reported that "the lands . . . are not of any value or importance in consequence of the mineral supposed to be in that region [and] the mines in that section of the country would be no consideration in the sale of the land." Whatever surface lead was on the concession had been exhausted, or so the committee thought. In 1867, the Missouri state geologist reported continued diggings, furnaces, cabins, and frame houses on the concession, all of whose locations he identified by lot numbers of the original concession. Surface

---

60. Natalie Villmer, "History of the Old Mines Area, Washington County, Missouri: A Resume," 4; Houck, *History of Missouri*, 2:326. In 1830, John Smith T, who was until that time no friend of the Old Mines villagers, deeded lot 7 of the concession for purposes of building a new church. Steward surmises that the donation was in recognition of the Presbyterian's daughter's conversion to Catholicism but notes that local historians interpret the act as a memorial to Smith T's deceased wife (*Frontier Swashbuckler*, 180).

lead mining also persisted on the western and eastern sides of the concession, but by the 1870s, introduction of deep drilling had revolutionized the mining industry. Drilling deep into bedrock commenced, and a mechanized industry eventually replaced surface diggings of lead ore by pick and shovel.[61]

Another French island in an American sea, Old Mines became a pocket of tradition. Few American farmers ventured onto this poor-soil land to introduce different types of agriculture. No commercial agriculture of consequence developed, unlike at St. Michel where American farmers intermixed with the French. Farming—perhaps *gardening* is a better term—remained subsistent and incidental to other ways of making a living, just as it was when Old Mines was founded.

Despite its residents' ambivalent attitude to landownership and the lack of a formal village with streets and properties, it is at Old Mines, a village of Creole miners of lesser means, that French ways have survived longest in the entire historic Ste. Genevieve District. Twentieth-century investigations into the remarkable preservation of French Creole language and other remnants of French culture in the Old Mines area invariably invoke geographic isolation in the backwoods as a reason. As a matter of fact, Old Mines, located as it was on lead-hauling routes, was most certainly not geographically isolated through the 1830s. Geographic isolation set in much later, in the post–Civil War years, when wagon hauling of lead was replaced by railroads, which bypassed Old Mines. What is more important in explaining the persistence of tradition is that Old Mines had not been organized into a functioning central place—except for the church, which served only the local in-group—when geographic isolation set in. Central-place functions of an economic nature

---

61. *ASP-PL* 5:725, 6:734. Confirmation testimony before the second Board of Land Commissioners is in *ASP-PL* 5:721–25, 6:728–34. Vallé's 1803 request is printed in French, and Lieutenant Governor Delassus's reply is printed in Spanish in *ASP-PL* 3:593. LPC 1a–o; LPC 25a–d; *ASP-PL* 6:386. The American land surveyors described the land of the concession as mostly "2nd rate" and unfit for cultivation (U.S. General Land Office, Field Notes, Missouri, 1:270–73, 197:148–53). On state geologist: Benjamin F. Shumard, *A Geological Report on the Old Mines Property of William Long, Esqu., in Washington County, Missouri*. A similar document showing the status of lead mining at the same time is Benjamin F. Shumard, *Report on the Chouteau League Tract, or Spanish Mineral Land Grant, Known as U.S. Survey No. 2066, in Washington County, Mo., supplemented by a report on the same tract, now the property of the St. Louis Lead Mining Company, by N. W. Bliss*. On deep drilling: Winslow, *Lead and Zinc Deposits,* 6:292, 296. In northeastern Washington County, barite mining, by pick and shovel in hand-dug pits, began to replace lead mining in the 1870s (Villmer, "History of the Old Mines Area," 6; William Arthur Tarr, *The Barite Deposits of Missouri and the Geology of the Barite District*, 1–4, 100–106).

do not easily arise where people live at a near-subsistence level. And without central-place functions, few people with different ideas moved into or visited this neighborhood that became increasingly ingrown. Isolation of Old Mines is less of a geographic nature and more the defensive, psychological isolation that poverty begets.[62]

Herculaneum

The lead-hauling road from Mine à Breton to Ste. Genevieve was a long thirty-five miles over rough hills with steep grades, although Austin reckoned it to be more like forty-five miles. The cost of hauling to the Mississippi made Mine à Breton lead less competitive at New Orleans and East Coast markets than it could have been. It cost seven dollars to float one thousand pounds of lead down the Mississippi from Ste. Genevieve to New Orleans, but this expense was doubled to haul the one thousand pounds from Mine à Breton to Ste. Genevieve.[63] Austin's lead would compete better if only he could reduce the cost of the overland haul. After smelting, hauling was the major cost of the lead business at Mine à Breton.

At the beginning of the American period, the lead-shipping business at Ste. Genevieve was completely in the hands of the French community with whom the independent-minded Austin had difficulty working. In fact, Austin had set up his own lead depot in Ste. Genevieve to circumvent the Ste. Genevieve establishment. Then, within a few years, he found a better route to the river. It lay northeastward through the populated country around Old Mines to the easily forded, gravel-bedded Big River, then across a low divide and down the ridges of the American-populated Joachim River Valley to its junction with the Mississippi, approximately midway between Ste. Genevieve and St. Louis. Plans to use this new route began in 1806, and Austin laid it out in 1808. Though it was scarcely any shorter than the old route to Ste. Genevieve—Austin thought it was shorter by fifteen miles—wagons could negotiate the new route to the Missis-

---

62. Unplatted, unincorporated, amorphous villages still characterize the settlement pattern of the Old Mines concession and vicinity. Some villages are no more than clusters of five or ten houses or trailers in the style of extended-family compounds; *village* is no longer used locally.

Several persons have contributed to the understanding of the survival of French Creole traditions in the Old Mines community. Among them are C. Ray Brassieur, "Expression of French Identity in the Mid-Mississippi Valley"; Dorrance, *Survival of French;* Rosemary Hyde Thomas, *It's Good to Tell You: French Folktales from Missouri;* McMahon, "Ozark Mining Community"; and Gerald L. Gold, "Lead Mining and the Survival and Demise of French in Rural Missouri (les gens qui ont pioché le tuf)."

63. On distance: Gardner, *Lead King: Moses Austin,* 106; on cost: Gracy, *Moses Austin: His Life,* 124; and *ASP-PL* 3:498, 605.

sippi more easily, and the terminus was a place Austin himself could control as a lead transshipment point. Austin and Samuel Hammond formed a partnership to establish a new town and depot.[64]

The Joachim River enters the Mississippi through a narrow opening in high limestone bluffs, forming a natural amphitheater to observers on the river, "an immense door cut open in the cliffs." A small Pleistocene alluvial terrace lay along the Joachim behind the bluff line, high enough to be above known floods. It was a superb site that combined both good loading facilities in the quiet-water mouth of the Joachim and a flood-proof terrace with easy access into the growing interior.[65]

The river terrace had earlier, in 1798, been conceded to Francis Wideman, with the stipulation that he build a bridge over the Joachim River, which was muddy and not readily fordable so near its mouth. Recognizing its potential as a lead-shipping port, Austin had his front man, Judather Kendal, purchase the concession, then Austin purchased it in 1806 from Kendal.[66] He promptly platted a town, laid it out on the ter-

64. David Gracy gives an account of the dissolution of Austin's business partnership with François Vallé and other French (*Moses Austin: His Life,* 77–78). Difficult relationships with the French at Ste. Genevieve may be a reason for Austin's decision to leave Ste. Genevieve and develop a new lead port at Herculaneum, but simple geography may also have played a role. Austin thought Herculaneum was significantly closer than Ste. Genevieve, the road better, and the cost of hauling less. Austin thought the distance from Mine à Breton to Ste. Genevieve was forty-five miles but only thirty miles to Herculaneum (*ASP-PL* 3:611). Thus, the high cost of overland lead hauling should have been reduced by one-third, a significant saving. But Austin underestimated the distance to Herculaneum. Frederick Bates thought the road to Herculaneum was twenty-five miles shorter than any other and was convinced that it would solve the problem of high costs of hauling to the river (to Albert Gallatin, February 9, 1808, *Bates Papers,* 1:281). Lieutenant Thomas said it was thirty-six miles (*ASP-PL* 4:377), which is close to the actual distance. The Bullitt and Quarles report on mineral lands in Missouri gives thirty-five miles as the distance between the two points (*ASP-PL* 3:579). On new route: *ASP-PL* 3:591 and Gracy, *Moses Austin: His Life,* 124–25; on distance: *ASP-PL* 3:611. On Hammond partnership: *Austin Papers,* 2:123–24; and Gracy, *Moses Austin: His Life,* 124–25.

65. Unsigned article, but generally attributed to Henry Marie Brackenridge, in *Missouri Gazette,* March 21, 1811. Herculaneum's amphitheater setting, with its edges of bounding limestone strata worn down to resemble seats, opened onto the river for all travelers to admire and was the inspiration for its name. Roman Herculaneum, south of Naples, sat in a natural amphitheater opening onto the Bay of Naples (G. W. Featherstonhaugh, *Excursion through the slave states, from Washington on the Potomac, to the frontier of Mexico; with sketches of popular manners and geological notices,* 73). A detailed description of the site and its unusual geomorphic origin is in W. A. Schroeder, "Landforms of Missouri."

66. On Wideman: *Régistre d'Arpentage,* 91, on which plat the Joachim is phonetically spelled Shuashan; Houck, *History of Missouri,* 1:379. The stipulated bridge had not yet been built when the tract was brought before the board in 1808, a fact that

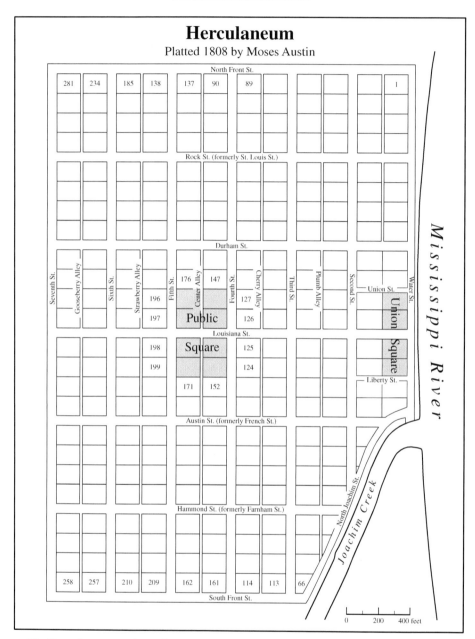

32. *Herculaneum, Platted 1808 by Moses Austin.* Moses Austin laid out Herculaneum as a lead-shipping port on the Mississippi River in the distinctive American form of a rigid grid of rectangular blocks cut by alleys. Austin moved from Potosi to Herculaneum in 1817. (Deed Book G-1, pp. 298–301, Office of the Recorder of Deeds, St. Louis)

race in late 1808, and named it Herculaneum. Herculaneum was thus the first American-planned and -platted town in the Ste. Genevieve District, and the second in Missouri Territory, after Colonel Morgan's aborted planned town at New Madrid.

The original town of 1808 can be reconstructed through the description of its physical layout in the deed of sale of the town as a single piece of property in 1818 (fig. 32). Herculaneum was a rectangular grid oriented to the bank of the Mississippi River, which was generally north-south, with the southeastern corner interrupted by the mouth of the Joachim River. The north end of the grid rose onto the steep, rocky bluff, and the south end was against the Joachim. But most of the grid was a nice topographic fit on the flat alluvial terrace. The grid's geometry was entirely consistent with that of planned American towns of the eastern states and represents the first transferral to the trans-Mississippi West of American town-planning principles. Austin built his own house on the terrace's riverfront with the intention of moving there from the mines, which he did in 1817.[67]

As a river port for lead, Herculaneum had several physiographic advantages over Ste. Genevieve. The Mississippi limned the terrace and town. At Ste. Genevieve, the Mississippi had unstable banks, and it was usually half a mile from the village; the landing in the early 1800s had shifted to Petit Rocher (Little Rock) a mile north of town. But at Herculaneum, boats could either tie up along the Mississippi itself or come up the deep mouth of the Joachim. The one-hundred-foot-high sheer bluffs at Herculaneum lent themselves admirably to making lead shot by dropping the molten lead vertically through sieves in towers built along the bluff face. They were probably the best sites practically accessible in the entire Ste. Genevieve District. John Maclot, a recent arrival directly from France and probably informed of Austin's plan to haul lead to the site, seized the opportunity to erect the first shot tower west of the Alleghenies in 1809 on the south bluff at Herculaneum. Austin erected a second shot tower in 1810 on the north bluff of the amphitheater (fig. 33).[68] Un-

---

provoked discussion about imposing additional conditions for confirming concessions, but the board voted to confirm by a split vote (to Albert Gallatin, December 25, 1808, *Bates Papers*, 2:52–53). On Kendal: *ASP-PL* 2:690; *Austin Papers*, 1:123–24; Kendal and Magt to Austin, January 9, 1809, Deed Book C-1, p. 91, Office of the Recorder of Deeds, St. Louis. *Judather* also occurs as *Jeduther, Jeduthan, Jaduthan,* and *Jonathan*.

67. Austin and Wife, and Hammond and Wife to Trustees for Herculaneum Town, June 8, 1818, Deed Book G-1, pp. 298–301, Office of the Recorder of Deeds, St. Louis. A detailed description of the town plan is in W. A. Schroeder, "Opening the Ozarks," 349–50. On Austin's house: Gracy, *Moses Austin: His Life,* 168.

68. Billon, *St. Louis in Its Territorial Days,* 115, 178.

33. *View of Herculaneum, Sketched by Charles LeSueur, April 9, 1826.* Charles LeSueur, a French visitor, sketched this watercolor view when Herculaneum was already in decline. Austin's shot tower rises on the high bluff above the town, and his house occupies the terrace on the Mississippi River bank. The Joachim River enters through the amphitheater break in the bluffs just to the left, beyond the painting. (Courtesy of the State Historical Society of Missouri and the Museum d'Histoire Naturelle, Le Havre, France)

like Ste. Genevieve, where the low bluffs were not suitable for making shot, Herculaneum could also serve as a manufacturing center for entrepreneurs such as Austin who wanted to do more with lead than ship it out in an unmanufactured form, as the French did.

Herculaneum, promoted vigorously in St. Louis newspapers by news items and advertisements, quickly grew from lead shipments from Austin's Mine à Breton furnace and others. Herculaneum received at least one-half of the lead mined, probably much more in most years. The shot manufactories provided employment. Herculaneum became a major Mississippi River ferry crossing to the Illinois town of Harrisonville directly opposite, itself a river port of some consequence for the eastern side. Herculaneum was a stop for the first steamboats on the Mississippi. By 1811, three years old, it had a population of two hundred, twenty houses, and, in addition to the two shot towers, a store, blacksmith, hatter, taverns, ropewalk, and even an establishment that made barges. Several mills were nearby on streams with a good fall to them. By 1818–1819, Herculaneum had grown larger: from thirty to forty houses, four stores, a post office, a school, a lead warehouse, and, in the vicinity, two sawmills, two gristmills, three distilleries, and a tanyard. Exports shipped from Herculaneum in 1818 included $173,000 of lead and shot; $41,000 of flour to the United States Army at St. Louis; $32,000 of whiskey, pork, bacon, wheat, and pine and oak boards. "Herculaneum will certainly become a place of consequence in a few years," wrote Col. George C. Sibley in 1813 when looking for a place to retire. "It is here that all the lead trade of the country centers to in an increasing amount."[69]

As long as the lead business prospered, Herculaneum prospered. When Jefferson County was erected in 1818, Herculaneum, the only nucleated settlement and platted town in the county, became the county seat as much by default as by promotion by its sponsors. Austin, whose fortune was tottering toward the end of the decade and who was in the process of paying debts by selling Herculaneum, had neither the resources nor the energy to build another courthouse in another county. On March 24, 1820, James Bryan, Austin's son-in-law, and wife, Emily Austin Bryan, donated the east half of lot 129 to Jefferson County for the site of public buildings for the county. However, no courthouse was ever erected in Herculaneum; the court met in various buildings, "wherever circumstances and convenience permitted."[70]

The high cost of hauling lead overland to the river was never reduced. In 1816, the difference between hauling from Mine à Breton to Ste. Genevieve and to Herculaneum was in Herculaneum's favor, eight dollars per one thousand pounds for the former, and seven dollars per one thousand pounds for the latter, but still this cost was too much and made Missouri lead high priced in eastern markets. Shippers kept hoping to rid the Big and Meramec Rivers of shoals so that lead could float by barges the eighty river miles to the Mississippi River, but their hopes were not realized. When the Galena mines of northern Illinois were brought into production after

---

69. *Missouri Gazette,* November 16, 1809; and September 30, 1815; on steamboats: Z. Cramer, *Navigator,* 134 (Z. Cramer spells *Joachim* as *Swashing*); on 1811 population and businesses: *Missouri Gazette,* March 21, 1811; Billon, *St. Louis in Its Territorial Days,* 35; and Brackenridge, *Views of Louisiana,* 181; on 1818–1819 population and businesses: Schoolcraft, *View of the Lead Mines,* 46; on exports: *Missouri Gazette,* June 19, 1818; quote: George C. Sibley Papers, Missouri Historical Society, St. Louis.

70. A rival lead-shipping port, New Hartford, was proposed one and one-half miles north of Herculaneum, but it was never built (W. A. Schroeder, "Opening the Ozarks," 351–52). *History of Franklin, Jefferson . . . Counties,* 387; Ohman, *Encyclopedia of Missouri Courthouses,* 99; *History of Franklin, Jefferson . . . Counties,* 388.

1815, their location close to the Mississippi more than made up for their long river distance above St. Louis. Before long, the lead manufacturers at St. Louis began to use as much Galena lead as Missouri lead.[71]

Increased prices for lead following the War of 1812 brought a welcome boom to the Missouri lead industry, but when normal prices returned and general economic decline set in from the panic of 1819, hard times befell the industry and brought about Austin's financial collapse. He had been selling off tracts within his league grant at Potosi to keep afloat and then, unable to raise enough capital that way, offered the whole grant. He offered his business at Herculaneum for sale in 1818, then lost what remained of his town of Herculaneum in a debt arrangement in 1820. He kept only his riverfront residence, where he preferred to live rather than Durham Hall in Potosi. Austin died on June 10, 1821, in the midst of negotiations to restart his life as an *empresario* in Spain's Texas colony. Then, as Galena and the upper Mississippi mines drew more miners and investment money from Missouri, Austin's Herculaneum stagnated, then withered. "The Town of Herculaneum is all going to Naught. . . . This place isn't the same. . . . [H]ow my heart pines . . . for whatever," lamented Austin's daughter during a visit to Herculaneum in 1828.[72]

Because of the rough terrain both north and south along the Mississippi River, Herculaneum was not located on the most direct land route south from St. Louis to Potosi, which bypassed it a few miles to the west. The land around it was settled by farmers scattered along narrow stream valleys, and Herculaneum could not convert to a farm-service center once the lead economic base faded, as St. Michael's–Fredericktown did. For example, flour and sawmills were built not in Herculaneum but up the Joachim and Plattin Valleys amid the farms. Although the bridge over the Joachim at Herculaneum was built by 1819, Herculaneum was not

---

71. John Rice Jones, Austin's business partner at Potosi, provided the most detailed analysis of the cost of mining, shipping, and marketing Missouri lead to eastern markets and argued for the necessity of tariffs on imported ore (*ASP-PL* 3:604–6). *Nile's Weekly Register* 10 (August 10, 1816): 399.

72. *Missouri Gazette,* June 12, 1818; Austin and Wife, and Hammond and Wife to Trustees for Herculaneum Town, June 8, 1818, Deed Book G-1, pp. 298–301, Office of the Recorder of Deeds, St. Louis; Gracy, *Moses Austin: His Life,* 190. Austin had moved to Herculaneum in 1816–1817 and was living there when Schoolcraft first visited him in 1818–1819, and the two developed a relationship in which Schoolcraft learned about the lead ores and business and Austin expected help from Schoolcraft in disposing of his property (Schoolcraft, *Travels in the Mississippi Valley,* 247; Gracy, *Moses Austin: His Life,* 183–84). Austin also divided up his Mine à Breton tract into lots and advertised them for sale in 1820 in St. Louis, Louisville, and Nashville newspapers (*Missouri Gazette,* September 20, 1820). Emily M. Perry to J. E. B. Austin, November 29, 1828, *Austin Papers,* 2:149.

centrally located in the county, an acknowledged principle for the placement of American county seats, and the county commissioners removed the seat of justice from Herculaneum in the 1840s to a central location in the county, Monticello, now Hillsboro.[73] Thus, Herculaneum, the first American-planned and -platted town in the Ste. Genevieve District, its lead base shriveled and its promoter dead, and neither favorably located to be an agricultural center nor centrally located to be the county seat, virtually disappeared from existence after 1850.

The entire town site was purchased later in the century by the St. Joseph Lead Company for its smelter, and the various lots and streets of Austin's town were amassed into one piece of property. Then, the St. Joseph Lead Company laid out a completely new company town next to its huge smelter, and Herculaneum rose again, phoenixlike, with lead once again its raison d'être, but now brought to the site by railroad.[74] The present town of Herculaneum has no morphological connection whatsoever to Austin's town, except its river-terrace site and name.

Big River Bend

The road between Ste. Genevieve and Mine à Breton passed across a broad limestone plain at the big bend of the Big River. Not until the turn of the century, when surface lead was discovered in significant quantities in the northern portion of the plain and farmers recognized the potential of the plain's fertile soils and grasslands, did settlers, virtually all Americans, begin to occupy the plain. Three distinct neighborhoods emerged: the Big River Bend in the north, Murphy's Settlement in the center, and Cook's Settlement in the south. They formed the populated core of St. Francois County when it was created in 1821 with a new planned American town, Farmington, as its county seat (fig. 34).

The plain was an oak woodland with an abundance of grass. In places, the grasses so dominated that the French applied their term *prairie*, as at Prairie Salle at Cook's Settlement and prairies on the Terre Bleue River.

---

73. The fate of the bridge was a point of discussion in the board's confirmation of the tract because the bridge was stipulated to be built in the original Spanish concession to Francis Wideman, a stipulation Wideman himself proposed, and had not been built by 1808 (to Albert Gallatin, December 25, 1808, *Bates Papers*, 2:52–53). When Schoolcraft visited Austin at Herculaneum in 1819, a "substantial wooden bridge" over the Joachim was in place (*View of the Lead Mines*, 42).

The new county seat for Jefferson County was to be named Monticello for Jefferson's Virginia home, but by the 1840s, Monticello was used for another Missouri county seat, whereupon the county selected Hillsboro, a translation of Monticello.

74. Mary Joan Boyer, *Jefferson County, Missouri, in Story and Pictures*, 104; "St. Joseph Lead Company."

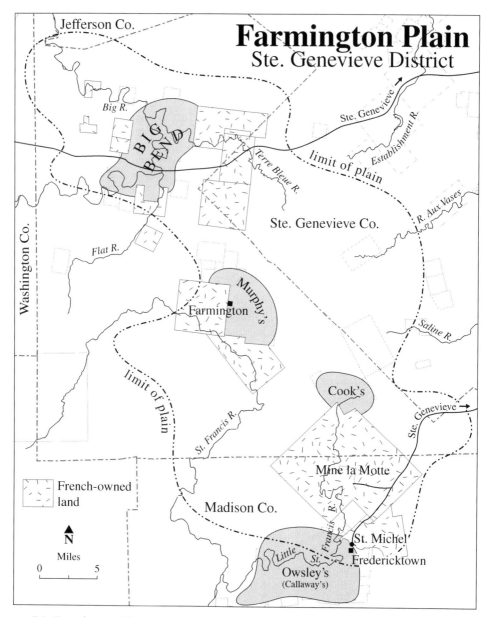

34. *Farmington Plain, Ste. Genevieve District.* Three neighborhoods arose on the limestone plain in the first decade of the nineteenth century, each with large French-owned properties for lead mining and stock raising and smaller American-owned and -occupied farms in compact settlements.

The chert-free limestone soils, among the best in the whole Ste. Genevieve District, were the great attraction to land-seeking Americans. Surveys of pre-American concessions on the plain regularly noted that the land was good for wheat and corn *("bonne terre à bled et à mays"),* but the extensive plain did not interest the French for farming.[75]

The plain occupies a large, well-defined basin surrounded by higher and much rougher country of igneous knobs to the west and cherty hills and valleys on other sides. The Big River cuts across the northern part of the plain in a big, looping horseshoe bend—Austin described it as "nearly three parts of a circle round the mines"—but most of the plain is watered by smaller streams and creeks. The central part of the basin is the flattest, as it straddles the nearly imperceptible drainage divide between the major drainages. The drainage pattern of the basin, being radially outward from the center of the plain, provided no natural geographical focus for settlers who were otherwise space oriented by drainage systems.[76] Therefore, although the plain had natural unity as a single basin surrounded by rough lands, it was fragmented in its settlement pattern, sociological characteristics, and economic activities.

Persons traveling to and from the mines eventually recognized the potential of the big bend of the Big River for agricultural settlement. Among the first to locate there as permanent farmers in 1790, possibly aware of mineral in the area, were Joseph Gerrard (Girrard dit Megar, Garrard, Gérard, Gerau, Gerar, Jaret, or Jarrett); his eighteen-year-old son-in-law, Patrick Fleming (Flamand or Flammand); two Macagné brothers, Laurent and Joel; and presumably their families. Both Gerrard and Fleming were from Ste. Genevieve families. The commandant recommended them for a single concession of one league square on the Big River, $1\frac{1}{2}$ leagues above Shallow River (Flat River), but Lieutenant Governor Perez divided and reduced the request on July 7, 1790, to four adjacent tracts of 5-by-40 arpents each, a traditional long-lot shape and size for French properties. Then Ste. Genevieve District commandant Henri Peyroux curiously changed this on July 17 to four concessions of 7-by-30 arpents each, next to each other, for a total of 840 arpents.[77] These tracts were never sur-

---

75. On *prairie:* LCS 190 and Houck, *History of Missouri,* 1:373; on chert-free soil: Burton L. Brown, *Soil Survey of St. Francois County, Missouri,* 23–24; Schoolcraft, *Journal of a Tour,* 52, 90; and Sauer, *Ozark Highland,* 107; on surveys: *Régistre d'Arpentage,* 239–42.

76. Austin, "Description of Lead Mines," ASP-PL 1:191; W. A. Schroeder, "Landforms of Missouri."

77. ASP-PL 3:593–96. The concession by commandant Vallé identifies Fleming as Gerrard's son-in-law, but Ste. Genevieve family records indicate that he was his step-

veyed; hardly any properties outside the villages were surveyed at that time anywhere in Upper Louisiana. Had they been, they would have appeared on maps as a distinct minicluster of long lots fronting on the Big River, just as clusters of long lots were being laid out along the Mississippi, on the Gabouri, and on the *côtes* at Ste. Genevieve and New Bourbon at the same time, which was the years following the great flood when settlers were expanding outward from the Pointe Basse.

These four and their families, and possibly other families on nearby lands, were driven off their farms during 1793–1794 by Indians. When Gerrard and Fleming returned in 1797 without the Macagnés, they were accompanied by a score of American families from Kentucky and Virginia. Many of these families were kinship related or otherwise became related soon after their arrival. Farmers all, they clearly had come for land. Some of them, such as carpenter Abraham Eads and his wife and seven children, had accompanied Moses Austin to help construct his lead works, but then chose to return to a farmer's life.[78] The lands these American immigrants selected formed a compact neighborhood along the Big River, the Terre Bleue, and creek tributaries, and on the adjacent limestone plain.

The American farmers received Spanish concessions for their lands, even though it had been common knowledge since 1783 that some land in the big, looping bend was lead bearing and that the Mine à Gerbore (Gaboury) on the northernmost headwaters of the St. Francis River (known then as the Rivière à Gaboury) had produced as early as 1745. Farmers made several lead finds in the late 1790s as they got to know their properties better, and lead mining thereafter became a significant factor in the settlement and economy of the Big River bend.[79]

As elsewhere, discoveries of lead ore attracted wealthy Ste. Genevieve

---

son. Gerrard's wife, Jane Donaghoe (Donohue?), had been previously married to Patrick Fleming Sr., who died in 1780 (*ASP-PL* 3:593–96). One of Patrick Fleming Jr.'s daughters married Andrew Henry after he returned in 1818 from the first of his Rocky Mountain fur-trading expeditions (Houck, *History of Missouri*, 3:97).

78. The family surnames are in W. A. Schroeder, "Opening the Ozarks," 356. On settlers: *ASP-PL* 3:593–96; Houck, *History of Missouri*, 1:387; LCC 136, 137, 162; LCS 121, 140, 141, 142, 145, 147, 182, 184, 222, 234. Houck lists names of American settlers on the Big River (*History of Missouri*, 1:373–74, 376), but his geography can be confusing because he uses the geographical locator "Big River" for its entire length and often does not distinguish among its portions in the Bellevue Valley in Washington County, the Big Bend in St. Francois County, and its long course through Jefferson County.

79. Winslow, *Lead and Zinc Deposits*, 7:270–76, 660–75; Austin, "Description of Lead Mines," *ASP-PL* 1:190.

families. They used two strategies to get control of land in the Big River bend, probably uncoordinated among them but known to each other just the same. One was to request unconceded land in a large tract, expecting ore would be found somewhere on it, if not already known. They made such a request on the basis of stock ranches and mentioned mineral only in passing, if at all. Luzières family members made several requests based on the governor's generous promise of unlimited arpentage to the noble family upon its arrival in Upper Louisiana. Prominent Ste. Genevieve families requested several large tracts, some in the name of their underage children, as for Antoine and Athanase Villars and Pierre Auguste Pratte, and had the concessions backdated.[80] The second strategy was blatantly to lay claim to mines being worked by others, using devious procedures that included physically intimidating the workers as they worked and asking that officials backdate their concessions to make it appear that the claimants had owned the land before the lead ore was discovered or before Americans were on the land. Two cases illustrate these strategies.

One case involves Philippe-François-Camille Delassus's concession near the mouth of Flat River and land claimed and occupied by Americans Jacob and Susanna Doggett. In October 1799, Philippe-François-Camille Delassus, known as Camille, the youngest son of the New Bourbon commandant and recently married, asked for a large concession of 2,500 arpents, based on the open-ended land grant of 1792 to his father that Camille had not yet taken advantage of because of his young age. The land he selected was suspected to have lead on it, because it lay just north of his father's concession on which there was an active mine. In fact, Camille said that he could oversee his father's mining operations from his concession. Camille's tract was conceded on October 12, 1799, and immediately surveyed, if we accept without question the dates on these documents.[81]

An American, Samuel Pearceall, discovered lead on the north end of the tract, along the Big River, and worked his find and marked it in accordance with the customary prescriptive rights of mine finders. That summer, the Doggetts moved onto Pearceall's land, and, with the help of other American settlers, built a substantial 18-foot-square cabin on it. Pearceall turned his marked and occupied claim over to Doggett and left, but Doggett had no interest in mining, only "to have a plantation." In late

---

80. LCS 24, 25; LPC 18, 28, 50; *ASP-PL* 6:809–12, 7:786–88, 837–39; *Régistre d'Arpentage*, 97, 234, 548. On Villars and Pratte: LCS 86, 107; and LPC 31.

81. *ASP-PL* 3:596–97, 7:837–39; LCS 24; LPC 28, 48.

1801, young Camille Delassus or his employees accused Doggett of being on Delassus land, which Doggett was totally unaware of, and threatened to take him to "the calaboose" if he should refuse to leave the premises and surrender his improvements and claim. It is probable that Camille Delassus did not actually have title to the property but was then getting interested in it because of Pearceall's mine, and he had his agents or agents of his commandant father or those of his lieutenant-governor brother intimidate the Doggetts into thinking he had title to it. (The Delassus concession and survey were given the early 1799 date at a later time by the lieutenant governor and surveyor at St. Louis.) Doggett, described as "upset," took the threat seriously and surrendered his improvements. He signed a statement to that effect, left the premises, and died later that same year. Widow Susanna, with four children, asked for compensation from Camille Delassus for her husband's cabin and improvements that Delassus had seized, but the young man refused. She then bought an improvement two miles to the west and moved onto it, where she died in 1806 or 1807. Her children continued to live on the new farm. Delassus's claim was confirmed by a United States Supreme Court decision in 1836 when virtually all mineral land claims still being pursued were confirmed. Even assuming that the claim was not backdated and Delassus was fully within his rights to exert control over all his land and it was most unfortunate that the Doggetts had built without that knowledge, this incident illustrates how an open-ended concession was used to acquire income-producing mineral land, not for purposes of residence and at the expense of those who were there to establish farms and build communities. In the long run, such actions, or just the threat of them, held back the settlement of the land by excluding serious-minded farmers from it.[82]

A second case concerned the American Mines, or Mine à Maneto, discovered and worked by Americans but taken over by the Pratte family of Ste. Genevieve by clearly fraudulent means. Facts of the case came out in testimony in August 1806 before the Board of Land Commissioners, which was acting on Jean-Baptiste Pratte's claim to the land. Pratte, one of Ste. Genevieve's leading and wealthiest citizens, who had nine children and forty-five slaves, called nine witnesses, all from the French community, some of them in his employ and none of whom lived in the Big River neighborhood. For its part, the "United States" called six witnesses, all

---

82. Willms presents conclusive documentary evidence of backdating the Delassus claim ("Lead Mining, 1700–1811," 84–85). U.S. House, "Opinion, Supreme Court, in the Cases of C. D. Delassus, Aug. Chouteau, and Others," 24th Cong., 1st sess., H. Doc. 148, 1 March 1836, 1–19; *Delassus v. United States,* 9 Peters 117.

Americans and all actual residents of the vicinity of the land in question. It was a sharply divided hearing.[83]

William Alley, John Baker, and other Americans who had been farming nearby since 1797 discovered the ore early in the summer of 1800 when it became known as the American Mines. They immediately set to work digging, but word of their find spread. Manitto (a black slave), someone named O'Connor, and Jean-Baptiste La Breche, all in Jean-Baptiste Pratte's employ, came to the mine in October of the same year and told the Americans that Pratte had filed a request for the land. The Americans, apparently intimidated and probably still feeling alien in a country under Spanish rule, ceased digging. Word spread to the nearby Mine à la Platte, where other Americans stopped working because of the affair at the American Mines. That same fall and winter, Pratte's men and some of his slaves raised three cabins (a "half camp") and built fences to enclose some fields in which no crops were cultivated until 1802. Cabins, enclosed fields, and cultivation were inconsequential acts in order to establish a Spanish claim, but they constituted important evidence for a claim being heard by the American board. Lieutenant Governor Delassus awarded a concession in April 1801, but he backdated it to September 4, 1799, in order to "prove" that Pratte had the land before the Americans discovered the ore and before the Treaty of San Ildefonso when Spain supposedly surrendered authority in Louisiana. When Pratte's concession was later surveyed (the survey also bore an earlier date of November 5, 1800), it was done in such a manner to also include Abraham Eads's farmstead. Pratte's concession was backdated to be three days earlier than Eads's and therefore superseded it. Eads lost his improvement, but received six hundred arpents to Pratte's east.

At the hearing, Pratte's witnesses were vague on dates and years. They stated that Pratte's workers were already mining in 1799 and early 1800. Charles Boyer went so far as to state rather unconvincingly that he had shown Pratte the land in 1797 and said that the land would be "good to settle." According to Boyer, Pratte replied that he "was about to apply for a concession for it."

The board denied Pratte's claim to the thousand arpents of Mine à Maneto (the name had changed to identify Manitto, who had become supervisor of the mine) on the grounds of a backdated concession, but when Pratte resubmitted it later, after removal of the reservation of mineral land, it was confirmed by the second board, which by that time had

---

83. Austin, "Description of Lead Mines," *ASP-PL* 1:189–90; *ASP-PL* 3:594–95, 603.

thrown up its hands in frustration at so many similar cases relentlessly pursued by the wealthy. Pratte acquired by the same backdating technique an additional adjacent eight hundred arpents of mineral land in the name of his son Jean-Baptiste Pratte, and another concession of five hundred arpents two miles to the west for son Antoine, and still another nearby six hundred arpents of mineral land to underage son Pierre-Augustin Pratte (born in 1786). All were subsequently confirmed by the second American board.[84]

The pattern of landownership that emerged in the Big River Bend community by 1804 was one of a mixture of regular-size American tracts for farming and larger-size French tracts for mining and stock ranches, absentee owned and tenant occupied. Austin thought that the seizure of the lead mines by Ste. Genevieve families had slowed their exploitation, because Americans would have worked them more assiduously than the French. The French were more interested in acquiring them either for speculation or to keep them out of American hands and did not have the labor to exploit them. Some of the finds along the Big River did not amount to much and were exhausted in a few years, but new ones continued to be discovered through the 1810s. The total amount of lead produced in the Big River Bend (all French properties) was small when compared to other mines for the year 1804: 30,000 pounds in the Big River Bend; 366,667 pounds at Mine à Breton; 133,333 pounds at Old Mines and vicinity; and 200,000 pounds at Mine la Motte.[85]

The continuous new finds along the Big River kept interest alive. The absentee French held on to their large tracts, but the American farmers sold their properties, or portions of them, to American lead speculators. By 1818, major lead operators had ownership of twenty-four of the thirty-eight properties in the big bend (fig. 35). James Bryan, Austin's son-in-law, purchased at least thirteen tracts on the Terre Bleue and its vicinity between 1808 and 1818 and developed Bryan's Mines.[86] He moved his residence from Mine à Breton to one of his Terre Bleue properties, and it was there that Moses Austin died in 1821. Rufus Easton and other St. Louis interests also recognized the potential of Big River Bend land and bought several properties, such as Armstrong Diggings and Gerrard and Fleming's Mine à Joe. The irrepressible, omnipresent John Smith T also invested in several tracts in the Big Bend.[87]

---

84. LPC 29, 31; *ASP-PL* 7:788–89, 791–92; *Régistre d'Arpentage*, 107.

85. Austin, "Description of Lead Mines," *ASP-PL* 1:189–91.

86. LRA 179, 196; LRB 52, 113, 115, 116, 231, 446, 588, 589; LRC 7, 8, 115; Gracy, *Moses Austin: His Life*, 144.

87. LRA 22, 23; LRB 57; *ASP-PL* 3:583, 585. On Smith T: LRA 156; LRB 434; *ASP-PL* 2:389, 3:599; Willms, "Lead Mining, 1700–1811," 83–85.

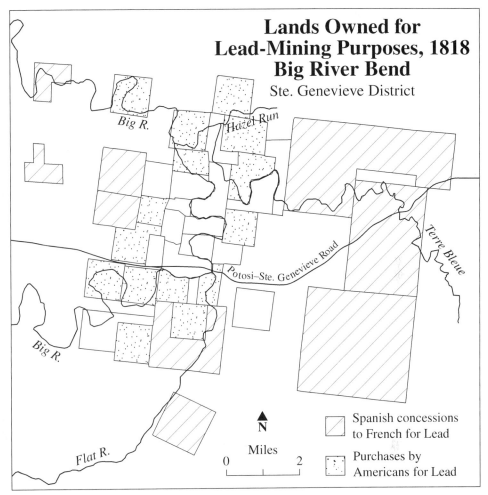

35. *Lands Owned for Lead-Mining Purposes, 1818, Big River Bend, Ste. Genevieve District.* Lead-mining interests dominated landownership in the big bend of the Big River as new discoveries were made after the Louisiana Purchase. The area emerged in the last half of the nineteenth century as the center of the world's largest lead-mining district.

Farming persisted despite the lure of money to be made in lead. Farmers supplemented their income from farm operations by part-time jobs at the lead mines, especially woodcutting and ore hauling in the farming off-season, and by providing meat and grain to the growing market of full-time miners.

The population of the Big River Bend grew steadily all this time. Austin reported only 8 American families, or 60 persons, in 1799, but 30 American families, or 240 persons, in 1804.[88] In the late 1810s, the population had probably increased to approximately 500. Virtually all of it was American; the French had not moved in despite their ownership of land.

No central place emerged among the farms and mines of the river bend. The small, dispersed surface diggings had no concentrating factor. No single furnace had centralized smelting functions, and miners hauled ore to Austin's furnace in Mine à Breton, to Smith T's at Shibboleth, or elsewhere. The population was one of various Americans without the cohesion of kinship, except for the substratum of the first families of 1794–1802, and without any organized religious groups. In fact, itinerant preachers bypassed the neighborhood in the years before 1820. By that time, the lead economy was in considerable decline, although mines still employed 50–100 men in 1824. The fuel demands of mining had gravely reduced the quality and quantity of timber in the neighborhood.[89] Big River Bend settled down to continue its existence as a compact and reasonably prosperous agricultural community for the next forty years.

Farmington Plain

The beginnings of Murphy's and Cook's Settlements, farther south on the limestone plain, were similar. American farmers founded and populated both in the closing years of the Spanish administration. Their namesakes provided both neighborhoods exceptionally strong local leadership, and both areas grew in size and prospered steadily by attention to farming. Both were fringed by huge properties in absentee ownership by members of the Luzières and other French and Creole families.

William Murphy Sr. chose the land that bore his name during a scouting expedition in 1797, and he received a concession for it in 1798 with the promise of more land for other members of his family. Murphy was a well-known Baptist preacher, and the awarding of sizable tracts of land to this group was as clear an example as any of the change of Spanish immigration policy to allow Protestant Americans into Upper Louisiana.

---

88. Austin, "Description of Lead Mines," *ASP-PL* 1:191.
89. *ASP-PL* 3:580.

Murphy died while in Tennessee returning for his family, but during 1801 to 1803, his widow, Sarah, and several of his six sons, their families, and slaves (son William alone brought twelve) arrived and established farms near the center of the fertile limestone plain. Other friends and relatives of the Murphys also came with their slaves just before the cession and in time to receive Spanish concessions or to establish settlement rights. Still others came after 1804 and had to wait to get title to land.[90] The concessions in this settlement area were at the maximum for ordinary settlers—most received 800 arpents—and all seem to have survived confirmation without reduction to the standard maximum of 640 acres. Perhaps the presence of many slaves helped keep the sizes large. As time went on, owners divided these large concessions, which were much too large for one family, and sold the pieces in response to the demand for land by new immigrants and by children reaching adulthood. The extra income was welcome in the initial years, and the process of land subdivision to relatives and friends allowed a multigenerational and homogeneous community to develop. Murphy's Settlement is noteworthy on account of hardly any litigation over properties.[91] There was no lead; there was no land speculation.

In February 1819, when the settlement was less than two decades old, Schoolcraft admired Murphy's as "a large and flourishing neighbourhood of industrious farmers, [which] presents many well-cultivated fields, fenced in a neat and substantial manner. . . . All these things indicate the wealth, the industry, and the intelligence of the inhabitants." By that time, it was the oldest and strongest Methodist community of the territory of Missouri, and the religious focus added to community solidarity.[92]

---

90. LCC 131; *Régistre d'Arpentage,* 240; Houck, *History of Missouri,* 1:375–76. Houck has the Murphys settling first on the Gaboury Creek at Ste. Genevieve and then moving west to the St. Francis River. He is mistaken by not recognizing another Gaboury. The Murphys settled first and only on the Gabourie fork of the St. Francis River. LCS 20, 71; LPC 16; *Régistre d'Arpentage,* 239, 241, 242, 254, 255, 598, 599, 601; LRA 112, 123; LRB 170A, 170B, 369, 384, 402, 403, 543, 599; LRC 96, 103, 237, 251; *ASP-PL* 2:509. Lead speculator John Smith T bought land from John Capeheart (Kephart) adjacent to the Luzières Mine à Gerbore square league in 1806 but produced no lead from it and disposed of the tract in 1814 (LCC 27; LRA 151; LRB 434).

91. As an example of land division, Joseph and Sarah Murphy sold part of their concession to Robert Dumville in 1813, who then gave it to his grandson Robert Sims in 1818 (LRB 384; LRC 103). Apparently, the only dispute over properties before 1821 occurred when the improvements of Joab Line (or Lene) were surveyed as part of David Murphy's property (LRB 170A, 170B; *ASP-PL* 2:425, 3:318).

92. Schoolcraft, *Journal of a Tour,* 90; Houck, *History of Missouri,* 3:239. Although Murphy was said to be Baptist, his widow and others embraced Methodism, beginning in 1804 (*History of Southeast Missouri: Embracing an Historical Account of the*

During the same years, a few miles to the south, a separate, somewhat smaller neighborhood was also emerging on the same limestone plain. Nathaniel Cook had scouted a location at the same time as Murphy and subsequently led a group of slave-owning Kentuckians into Upper Louisiana. They requested and received concessions in 1799–1800 to land on the north side of the huge Mine la Motte concession opposite St. Michel on the south side, but their land was not lead bearing.[93]

Like Murphy's, Cook's Settlement was a Methodist community, and like the people at Murphy's, the families at Cook's in general did not get involved in lead mining, despite the large, historic mines next door. They also prospered on their farms on large concessions, but population did not increase as much from later immigration, although they too subdivided their concessions for children and others. Nathaniel Cook's ambitions took him beyond farming in the local neighborhood. He served as deputy surveyor for the new American government in the region. In 1812, he bought land in the Old Mines concession in partnership with St. Louis lead speculators. He bought land in the fast-growing Owsley's Settlement on the opposite side of Mine la Motte and adjacent to St. Michel, and it was upon his land that Fredericktown was platted in 1819. Cook became a leader in the affairs of Madison County. He also served in 1820 as a representative of that county in the Missouri constitutional convention and was a candidate in Missouri's first senate election.[94]

Surveyor Nicholas Biddle Van Zandt, in an 1818 report on western lands for the U.S. government, reported that the limestone plain was "in a high state of cultivation and improvement by a large number of excellent farmers," and this complimentary description would be echoed by Schoolcraft the following year. The road network on the plain had no focus. The northern portion of the plain lay astride the chief road between Ste. Genevieve and Potosi, and lead-country traffic passed across it. A north-south road developed through the plain, and the junction of it with the Ste. Genevieve–Potosi road provided a kind of focus. This junction was briefly known in the period of 1815–1820 as Hale's or Hale's

---

Counties of Ste. Genevieve, St. Francois, Perry, Cape Girardeau, Bollinger, Madison, New Madrid, Pemiscot, Dunklin, Scott, Mississippi, Stoddard, Butler, Wayne, and Iron, 252, 532).

93. LCC 65, 79; LCS 166; *Régistre d'Arpentage*, 246, 247, 248, 250; Houck, *History of Missouri*, 1:376–77, 3:239. Nathaniel Cook's two brothers were also prominent in public affairs. Daniel Cook was an antislavery leader; Cook County, Illinois, was named for him. John Cook was a judge and otherwise active in Missouri government.

94. LRA 202, 203; LRB 199, 291. On Cook's ambitions: Houck, *History of Missouri*, 1:377; on partnership: LRB 311; on Fredericktown: LRC 15, 86, 124; on Cook as leader: Houck, *History of Missouri*, 3:249, 267.

Crossing, from a "house of private entertainment" (a place that accommodated travelers), but it did not develop any further.[95] The plain's human geography remained unorganized.

Guided by the principle of centrality, commissioners in charge of finding a county seat for the new St. Francois County (erected in 1821) chose in 1822 a fifty-two-acre vacant site on land donated by David Murphy in the dispersed Murphy's Settlement. The site lay adjacent to, but carefully avoided, the huge square-league concession awarded to one-time New Bourbon commandant Pierre-Charles Delassus de Luzières as lead mine cum *vacherie* and still in the hands of absentee owners. The commissioners platted a town and named it Farmington. Its grid was oriented according to the lines of David Murphy's Spanish concession, which was eight degrees east of true north, that is, in alignment with magnetic north, because Murphy's concession was one of the many for which the surveyor disregarded magnetic declination when running supposedly north-south lines. The courthouse square was of the Shelbyville plan, apparently brought to Missouri by the settlers from Tennessee where it had been popular in the preceding two decades. The county erected its courthouse in due order, but it sat for a long time among the trees in the platted but mostly unbuilt town. Farmington, created by administrative fiat and not from economic forces, was slow to grow as a commercial center; it reportedly had one store and a hotel in 1823.[96]

---

95. Nicholas Biddle Van Zandt, *A full description of the soil, water, timber, and prairies of each lot, or quarter section of the military lands between the Mississippi and Illinois rivers*, 136; Schoolcraft, *Journal of a Tour*, 90; Peck, *Forty Years of Pioneer Life*, 100; Schoolcraft, *Journal of a Tour*, 90.

96. On county seat selection: *History of Southeast Missouri: Embracing an Historical Account*, 440; on courthouse: Ohman, *Counties, County Seats, and Courthouse Squares*, 33; and Ohman, "Foursquare Courthouses," 185. The surveyor of Farmington was Henry Poston, a native of North Carolina who came to the lead country of Spanish Louisiana in 1800, attracted by reports of Austin's mines (Robert Sidney Douglass, *History of Southeast Missouri: A Narrative Account of Its Historical Progress, Its People, and Its Principal Interests*, 277). On Farmington's growth: *History of Southeast Missouri: Embracing an Historical Account*, 440.

11

Valley Settlements in the Hills

Plattin Valley

In the rough, rocky hills between Ste. Genevieve and the Meramec River, streams enter the Mississippi through deeply entrenched valleys, and bold cliffs face directly upon the Mississippi channel without any alluvial bottom large enough for a French village. No Creoles settled in this region. The earliest Americans had arrived on the Meramec in the 1770s but receded under Osage pressure. Only a few Americans had picked out advantageous spots for isolated farms in hidden breaks along the Mississippi River bluffs in the 1780s.[1] But when Americans started entering Upper Louisiana in large numbers in 1796, it was the first area to receive them as settlers, because it lay directly on the Mississippi River. These hill people settled the area in the linear-valley patterns that came to characterize much of the Ozarks.

Typical of the beautiful streams of the Ozark rough hills, Plattin Creek carried a sustained, perennial flow in a shallow, gravel-filled channel with numerous bars that formed no impediment to travel. Wagons, horses, and people could ford the creek virtually anywhere and could move up and down the valley directly in the gravel bed of the creek itself. Bottomlands of rich alluvial soils were narrow—usually only three hundred yards wide—but quite adequate for a couple dozen acres of corn. Well-drained soils held no standing water after rains; valley sites were "salubrious." The Plattin Valley provided a welcoming setting familiar to Americans arriving from hill lands in Kentucky, Tennessee, and Appalachia.

Ignored by the Creoles, Plattin Creek was completely unoccupied when Americans entered in 1796, and by 1800, Americans had taken out land in the lower half of the valley by means of Spanish concessions. Set-

1. Houck, *History of Missouri*, 2:73–74; Finiels, *Account of Upper Louisiana*, 52.

tlement then proceeded into the upper (southern) half of the valley during the next three years, but these settlers, who likely came without asking permission, claimed the land by settlement rights. The Board of Land Commissioners readily confirmed all claims and concessions. Thus, within seven years, Americans had occupied the twelve-mile-long Plattin Valley. The valleys of adjacent Joachim, Sandy, and Glaize Creeks experienced the same swift initial-settlement history.

Although all Plattin settlers fell broadly in the category of upland southerners, they came from various traditions and Appalachian regions. Peter McCormack was Irish born, a Methodist of some fame who had served in the Revolutionary War and had led his family through Virginia and Kentucky to Missouri. John Doolin, who had the only Spanish concession for a stock ranch in the valley, came directly from Pennsylvania. Jacob Donner was of German ancestry. The four Horine brothers were sons of Frederick Horine, a German immigrant from Alsace-Lorraine, and the Stricklands last lived in Kaskaskia. There were father-and-son teams (John Sturgus and son; Humphrey Gibson and son), and they represent a common migration phenomenon whereby an unmarried son over twenty-one years of age moves with his parents and other underage brothers and sisters but establishes his own farm upon arrival because his age allowed him to do so and then claims the land by settlement rights after the cession. Several families owned a few black slaves each. Also, there were the usual clusters of related families, such as the Wideman family and its "connections," who were given permission in 1799 "to settle on the frontier, fifteen miles in front of the settlements."[2] All the families built log cabins, had gardens and small orchards, grew corn and oats in small enclosures, and let their cattle and hogs roam the hills. This was a purely American neighborhood without any association with earlier Creole or Spanish influences.

The board, when confirming claims, reduced most of the Spanish concessions to 640 acres and set all settlement-rights confirmations also at 640 acres. Thus, the valley landownership pattern in 1804 is one of

2. On McCormack: Boyer, *Jefferson County, Missouri*, 17, 107, 153; on Doolin: *ASP-PL* 3:306 and Houck, *History of Missouri*, 1:380; on Donner: *ASP-PL* 2:440, 515, 529, 3:297–98; and Houck, *History of Missouri*, 1:380; on Horines: Zoe Booth Rutledge, *Our Jefferson County Heritage*, 22–23; on Stricklands: LCC 163; *Régistre d'Arpentage*, 82–83; *ASP-PL* 2:445, 447, 465, 466, 521, 522, 537; and Houck, *History of Missouri*, 1:379–80; on father-son teams: LR 5; LCC 66; *Régistre d'Arpentage*, 82; *ASP-PL* 2:440, 425, 447, 529, 590; and Houck, *History of Missouri*, 1:379–80. Like many of the Americans who entered Spanish Louisiana, Sturgus was a Revolutionary War veteran. On Widemans: LPC 42b; *ASP-PL* 2:528, 561, 3:289, 306, 307, 323, 324, 7:827–28.

equal-size properties. However, the farmstead improvements lined up and down the valley were so close that square-shaped properties could not be surveyed for everyone, and several tracts when surveyed took on irregular shapes and extended far onto the nonarable hillsides and ridge tops. The result was distances between farmsteads averaging $\frac{2}{5}$ of 1 mile, or approximately 2,000 feet. These families of 1804 cannot be described as isolated settlers. Rather, they formed their own community, topographically forced into a linear form.[3]

Neither the Louisiana Purchase nor land-confirmation processes affected the continuous flow of Americans into the valley, and the new arrivals did not substantially alter the linear-settlement form. They also located their farms along the creek bottom and left the adjacent hills and ridges unoccupied. In fact, as late as 1830 and well after land had been on sale in Jefferson County, only three tiny parcels of privately owned land had been added to the Plattin Creek community. These 320 acres constituted only a 2 percent increase in land in private ownership, although population in the valley had at least quadrupled. Such population growth without proportionate increase in new land can readily be accounted for by large families and subdivision of existing property.

William Conner, who lived in the adjacent and similar Joachim River Valley, provides an example. Conner had a wife, thirteen children, and six slaves on his settlement-rights claim in 1804. The children came of age during the subsequent years and, needing land of their own, found it on their parents' confirmed claim or the claims of their spouses' families or bought it from willing sellers.[4] When selling or renting his land, the original landholder could determine who his neighbors would be and keep the community homogeneous.

Among the settlers in the Plattin Valley shortly after the cession was a group of Scotch-Irish Methodists who were relatives or followers of earlier-arriving Peter McCormack. None of these families—McMullin, McClain, Wines, and others—obtained title to new land during our period of consideration. They apparently occupied land in the valley by squatting on it or by buying or leasing the improvements abandoned by Humphrey Gibson and his son. Peter McCormack himself, the leader of this small but growing, interrelated Protestant Irish community, sold his original tract and moved onto the Gibson claim. Thus, relocations, in a

---

3. For comparison, farmsteads were an average of one and one-half miles apart in the nearby Current River region of the Ozarks in 1840 (Clendenen, "Settlement Morphology," 144, 151–52).

4. *Régistre d'Arpentage*, 637.

sifting and sorting process, took place in the valley, and the technicalities of landownership could catch up later.[5]

Plattin Valley did experience its share of restless settlers. The majority of landholders before 1804 do not appear in records for the valley after 1815. Some of them obviously made profits by selling their improvements with or without the land, much of which was still unconfirmed at time of sale, to the post-1804 immigrants. Humphrey Gibson and his son left their Spanish concessions for the Boonslick country after the War of 1812. Thomas Bear, who had settled in the Plattin Valley in 1802, removed within a year to the Bellevue Valley, and John Doolin to the Farmington Plain. Following the death of her husband, Thomas, in 1801, Eliza Carlin took her family back to Illinois, where one son later became a governor of that state. Of the original eighteen family names in the valley before 1804, only two appear in the 1830 census for Plattin Township of Jefferson County and two elsewhere in Jefferson County.[6]

Nevertheless, the valley had enough families that stayed on the land as well as new families that the valley continued to grow in population and develop economically. Cleared and fenced fields increased in acreage as markets developed in the mining country and in St. Louis, or as river exports through nearby Herculaneum increased. A school was set up around 1818 in the McCormack compound in the middle part of the valley. George C. Sibley, retiring from his many years of service for the federal government as Indian factor in the Missouri Territory, searched for land to set himself up as a "gentleman farmer." He bought land around 1813 in the Plattin Valley five miles up from Herculaneum and the Mississippi, writing to his brother, "This Tract possesses according to my taste all the requisites for an elegant and pleasant country residence. . . . It is surrounded by some of the most . . . opulent farmers in the Territory."[7]

While Plattin Valley and neighboring valleys were filling in with farmers, discovery of surface lead ore in the headwaters of Joachim and Plattin Creeks around 1810 occasioned a rush to that area and prompted land speculation. Two major assemblages of properties were made in the Plat-

---

5. Z. B. Rutledge, *Our Jefferson County Heritage*, 20. Boyer, *Jefferson County, Missouri*, 17, 43.

6. Houck, *History of Missouri*, 1:372, 380; Lois Stanley, George F. Wilson, and Maryhelen Wilson, comps., *Missouri Taxpayers, 1819–1826;* R. V. Jackson, G. R. Teeples, and D. Schaefermeyer, eds., *Missouri 1830 Census Index*.

7. Peck, *Forty Years of Pioneer Life*, 89; George C. Sibley to Saml. H. Sibley, September 25, 1813, Sibley Papers, Missouri Historical Society, St. Louis.

tin Valley for lead ore, one by the Horines, the other by James Bryan, Moses Austin's son-in-law and associate who had accumulated lead lands elsewhere in the district. In 1818, Bryan's two thousand acres of land in the Plattin Valley passed into the hands of lead investor James Cox of Baltimore. Thus, East Coast money entered the agricultural valley as it did in other parts of the mineral district to the south and west. Surface lead was also produced lower in the valley beginning in 1824, although its production was quite small and never great enough to encourage a boom in land values and speculation.[8]

The valley felt the impact of the lead industry in other ways. Lead hauling, or "teaming," provided an opportunity for valley farmers to earn supplemental or even primary income and reach the condition of "opulence" that Sibley noticed. John Smith T, whose base of operations was at Shibboleth to the south, developed his own lead-shipping port, Selma, on the Mississippi, five miles south of Austin's Herculaneum. Selma, which competed with Herculaneum, was also used by the several smaller Plattin mines. Smith T's lead-hauling road from Shibboleth, which passed down the upper part of the Plattin Valley and then crossed one high ridge to descend to the Mississippi, carried heavy traffic for about a decade from both Smith T's Shibboleth and the Plattin mines, probably much of it in the wagons of Plattin farmers.[9]

Despite the flow of people up and down the twelve miles of the valley corridor and lead mining, the Plattin Valley never developed a geographical focus. The Plattin neighborhood, composed of various upland southern groups but lacking a dominant kinship relation or widespread religious bonding, remained economically and culturally unorganized. A minor focus developed in the middle of the valley around the Methodist church and the McCormack property, but this place, later called Plattin, was hardly more than a way station for travel up and down the valley. The valley was, however, distinctly set apart from other neighborhoods by vacant, paralleling ridges and early comprised its own township of Jef-

---

8. On discovery of surface lead ore: *Missouri Gazette*, February 26, 1814; Winslow, *Lead and Zinc Deposits*, 6:277; and Boyer, *Jefferson County, Missouri*, 107–8; on Horine and Bryan properties: LRB 447 and LRC 9; on Cox: LRB 594 and LRC 29, 149, 186, 187; on lower valley: Boyer, *Jefferson County, Missouri*, 107–8.

9. On Sibley: Boyer, *Jefferson County, Missouri*, 107–8; on Smith T: Steward, *Frontier Swashbuckler*, 52–65; Dorothy Heinze, presentation at the annual meeting of the Center for French Colonial Studies, Old Mines, Missouri, October 14, 1995; and Boyer, *Jefferson County, Missouri*, 18, 72. Lt. Martin Thomas calculated the distance from Potosi to Selma as thirty-three miles, which made it shorter than the distance to Herculaneum by three miles and shorter than the distance to Ste. Genevieve by twelve miles and therefore the closest point on the Mississippi River to Potosi (*ASP-PL* 4:377).

ferson County with its own local law enforcement. The Plattin Valley's simple linear organization of space with its lack of central-place development characterized all of the hilly region between Ste. Genevieve and the Meramec Valley and, in a broader sense, much of the rough-hills region of the eastern Ozarks.

Lower St. Francis River Valley

Formation of linear valley settlements such as the Plattin Valley by upland southerners took a variant form in the eastern Ozarks where physical geography prevented continuous settlement the length of a valley. Interrupted, or discontinuous, valley settlement is illustrated by the lower St. Francis River Valley in southern Madison and Wayne Counties.

The St. Francis River, south of the Fredericktown basin, flows through more accidented terrain than the Plattin does. Though the district's largest river after the Mississippi, the St. Francis is unnavigable for its course through the Missouri Ozarks. At several places, it flows through boulder-strewn igneous shut-ins that effectively block movement up and down the valley, and the straggles of settlers on arable lands trapped among the knobs could not easily communicate with each other. In this way, physical geography severely retarded the development of social and economic institutions.

The adjacent St. Francis and Black River drainages are the settlement areas in the district farthest from Ste. Genevieve and those with the least linkage with the Mississippi River. Lacking attractions such as minerals and a navigable river, they were settled by two kinds of people. One was the frontier vanguard of American hunters, trappers, and Indian traders, and the other was a group of upland southern farmer-hunters, some with slaves, a cultural group similar to those who first entered the Plattin Valley. Actually, it was not always possible to distinguish between the two kinds of people.

The hunter-trapper-trader group, it is thought, customarily preceded a farm-based settlement almost everywhere throughout the upland South. It is interesting that this group was even captured, albeit slightly, in the Spanish and American land records of the Ste. Genevieve District, because these people usually do not seek title to their lands. However, the lower St. Francis was slow to move into farm-based settlement, and hunters and trappers characterized it for quite a long period of time, before and after the Louisiana Purchase.[10]

10. Clendenen's three stages for settlement of the Ozarks are for the structure of the settlement of the region, not for type of people ("Settlement Morphology," 82–84).

If individuals of the hunter-trapper-trader group did not themselves leave much documentary record, they were different enough to be noted and described by others. The observant and critical reverend John Mason Peck, as he circulated on the far side of the Ste. Genevieve District in 1818, encountered a family that he called "a specimen of the squatter race found on the extreme frontiers." The specimen consisted of "the patriarchal head and his wife, two married daughters and their husbands, with three or four little children, and a son and daughter grown up to manhood and womanhood." These eleven or twelve persons dwelled in a single log cabin situated within a corn field. Though they had lived in the territory for some eight or ten years, having come from Appalachian Kentucky or the Carolinas, their cabin still had no table or chair, no article of furniture. The reverend was served his meal on a box, presumably used to carry cooking utensils and bedding when the family emigrated on pack horses. Except for the "old lady," only men entered the cabin during Peck's visit. The reverend saw "backwardness," "slothfulness," or "habitual neglect" in everything about the family and its surroundings. Peck thought that this family was representative of hundreds of families scattered over the extreme frontier in 1818–1819.[11]

Another description of a hunter-trapper-trader of this sort was provided in 1800 by Henri Peyroux, commandant at New Madrid, the Spanish having shifted him there from Ste. Genevieve in 1799. It seems that a person named John Braun, his two daughters, and a son-in-law had been living on the St. Francis River for twenty-two years. A trader of sorts, Braun had supposedly made five trips to New Orleans during those years. Braun and his family dug lead on the upper St. Francis for themselves and for trade with the Cherokee for making gunshot. Braun also sold horses to the Cherokee, who, in turn, sold them to Americans in the United States. Because the Osages repeatedly plundered his horses, Braun would periodically replenish his supply by stealing from the settlers of the Spanish Illinois Country. When Peyroux went to arrest him for horse theft, he found that Braun had left the St. Francis because he had heard that it would soon be settled. No one knew where the family went. Braun and his family led a peripatetic life, disdaining law. They exploited local resources as they could and interacted freely with people regardless of race and culture, and then moved on when conditions threatened their unrestrained way of life or threatened to introduce malaria into their space.[12]

---

11. Peck, *Forty Years of Pioneer Life,* 101–3. In general, the moralistic Peck was quite critical of frontier life, and his descriptions of it were largely negative.

12. Henri Peyroux to the Marquis de Casa Calva, July 6, 1800, PC 217b-121. Peyroux heard this account from Indian interpreter Jean-Baptiste Barseloux. The spelling

Settlers into this portion of the St. Francis Valley came both westward from Cape Girardeau and southward from Ste. Genevieve, encircling the Shawnee-Delaware Indian tract on its back side. Because Commandant Louis Lorimier's charge at Cape Girardeau was primarily to work with Indians and not to handle land requests (Lorimier could not read or write), commandants at Ste. Genevieve and New Bourbon oversaw settlement.

Nevertheless, shortly before the cession, Lorimier awarded three hundred arpents of land to 164 men, mostly Americans, as payment for their service to the Spanish for fighting Indians in the New Madrid District. The American board later confirmed all these lands, reasoning that Lorimier's action was in accordance with the Spanish practice of granting *mercedes* for service and that he had been ordered by Lieutenant Governor Delassus to form the militia.[13] Some of these Indian-fighting Americans took out their three hundred arpents of land on the remote St. Francis River.

They most certainly were not full-time farmers. Isaac Kelly, a self-identified trapper and Indian trader, arrived on the St. Francis in 1801 with his fifteen-year-old brother, Jacob Jr. They built a cabin and used it as a base for their hunting and trapping expeditions. Father Jacob Kelly Sr. arrived in 1804 and lived on son Jacob Jr.'s improvement until his death in 1808. Isaac Kelly did raise some crops, including one hundred pounds of cotton, and he collected fifty pounds of maple syrup. His father Jacob had sixteen cattle, four horses, a slave, and a "large production of grain."[14] The Kellys and other first Americans on the St. Francis may have built their cabins in small openings in the oak woodlands earlier cleared and maintained by Indians.

Another woodsman, Andrew Reed, built an isolated cabin on the St. Francis four miles below Isaac Kelly's in 1801, but left the country, returning in 1803. According to testimony given in 1806 by a nearby land claimant before the Board of Land Commissioners, Reed would go "out hunting, but always left in said house or cabin his said utensils or implements of husbandry, and always considered said house as his actual residence. . . . [H]e raised on said land seventy or eighty bushels of corn in 1803."[15] This testimony downplayed Reed the hunter and emphasized

---

of Braun, which would seem to make the family of German ancestry, may be incorrect, if the family itself was illiterate. For an account of an Indian trader working westward in the Ste. Genevieve District, see Morrow, "Trader William Gillis."

13. On payment: Houck, *History of Missouri,* 2:191–92. Certificates are listed in Bates's report, *ASP-PL* 3:275–317. "Lorimier's list" of 164 names is in *ASP-PL* 7:847–48.

14. *ASP-PL* 2:492, 510, 588, 3:290, 303, 7:847; Rose Fulton Cramer, *Wayne County, Missouri,* 34, 583–84.

15. *ASP-PL* 2:395.

his residence on his land and his farm activities in order to justify being awarded a settlement right.

As large game animals became less plentiful and malaria spread into the lower St. Francis, and then the great earthquakes of 1811–1812 dammed rivers with fallen timber, these hunters moved farther into the interior. The Kellys pushed westward, deeper into the Ozarks to the Current River and from there south into Arkansas. James James went to the Missouri River region in 1804. When leaving, hunters sold their improvements to their replacements when they could, and when they could not, they simply abandoned the land, which was then eventually sold for delinquent taxes. The abandoned improvements of Andrew Steel, George Stamps, and James Fisher were sold in 1820 for delinquent taxes of $31\frac{1}{4}$ cents, $37\frac{1}{2}$ cents, and $56\frac{1}{4}$ cents, respectively.[16]

Farmers, the other group on the St. Francis, were more likely to persist and prosper. One small cluster of farmers settled along the lower stretches of Twelve Mile Creek and Dry Fork, tributaries of the St. Francis, beginning in 1799, about the same time as the establishment of St. Michel and Owsley's Settlement a dozen miles up the St. Francis River. Several of the farmers were veterans of the War of 1812 and had earlier lived in the same area of the Bois Brule bottom.[17]

Another minicluster of farmers settled during the Spanish period on Clark's Fork and Ring Creek tributaries. They were Joseph Parish, a Virginian, with his son-in-law Thomas Ring and three Logan brothers who had married Parish women. They, with their families and slaves, formed a scattered but distinct kinship-based community, which attracted other Virginians and constituted the "Virginia Settlement." They had come through Kentucky and Cape Girardeau and obtained settlement permits, but no concessions, from Lorimier. All had their settlement-rights claims later confirmed.[18]

A third minicluster of related settlers came from North Carolina. The Bettis Settlement began in 1806 by occupying hunter Isaac Kelly's abandoned improvement, six miles south of Virginia Settlement on Clark's Fork. Dr. Elijah Bettis Sr., his three sons, and his three sons-in-law most certainly had prior knowledge of the St. Francis region, because David

---

16. On hunters and Kellys: R. Cramer, *Wayne County, Missouri,* 68, 73, 584; on James: Houck, *History of Missouri,* 1:376, 2:70; on improvements sold: *Missouri Gazette,* January 3, 1821.

17. Houck, *History of Missouri,* 1:376; Schoolcraft, *Journal of a Tour,* 77; ASP-PL 2:424, 454, 448, 533, 576, 579, 591, 3:290, 308, 309, 318, 322, 8:99, 152.

18. *History of Southeast Missouri: Embracing an Historical Account,* 283; R. Cramer, *Wayne County, Missouri,* 571; Houck, *History of Missouri,* 1:376, 2:152; LCC 100, 102, 151; LCS 61, 89; *Régistre d'Arpentage,* 540D, 543, 544.

Logan (of the Virginia Settlement) had signed Dr. Bettis's will while still in North Carolina. The Parish, Kelly, and Bettis families intermarried after settling on the St. Francis. The Bettis family was not poor: before leaving North Carolina directly for the St. Francis area, Elijah Jr. had twenty-six slaves; in the 1820 census for Wayne County, Missouri, the extended Bettis family accounted for one-third of the county's 204 slaves. The Bettis clan dominated the public affairs of the area. The route from Potosi and St. Michael's–Fredericktown southward to the Red River and Texas crossed the wide and barely fordable St. Francis River at their cluster of cabins. They operated a ferry, the Bettis Ferry, and by 1819, the beginnings of a town had developed at the ferry, with a gristmill and ten to fifteen houses. This ferry function provided a centralizing catalyst for the region of otherwise widely scattered miniclusters of farmers. Agriculture had become the chief occupation of the settlement region, which by then extended into the next watershed to the west, the Black River, but included were mechanics, merchants, and a few professional men. Settlers raised wheat and corn as subsistence food crops and cotton for homespun cloth. Their economic mainstay was cattle, which were driven to markets as far distant as Ste. Genevieve and Kaskaskia (75 miles) and even to St. Louis (125 miles). Baptists had organized near Bettis Ferry by 1816 and also in the Dry Fork–Twelve Mile Creek area in 1818. A Methodist circuit included Bettis Ferry in 1817.[19]

The scattered but growing miniclusters on the St. Francis and its tributaries and the Black River attained a large-enough combined population in 1818 for the creation of Wayne County. County commissioners selected the only really populated place, Bettis Ferry, for the county seat, which added to its simple central-place functions. The frontier image of these remoter Ozark settlements was memorialized in the selection of Wayne as the county name in honor of Gen. Anthony Wayne, a famous Indian fighter of the Northwest Territory, and the selection of the name Greenville as the new county-seat name to replace Bettis Ferry, to commemorate General Wayne's Treaty of Greenville of 1795, which signaled the defeat and expulsion of Indians from the Northwest Territory. The people in charge of the county paid homage to the frontier spirit and Indian fighting at the expense of patriotism to founding fathers, as the oth-

---

19. R. Cramer, *Wayne County, Missouri*, 65–66, 90–91. R. Cramer recounts a local Wayne County joke that early settlers found that they were related to so many other local families that they changed their names in order to get wives (81). On beginnings of town and economics: Schoolcraft, *Journey of a Tour*, 86–87; on Baptists: Peck, *Forty Years of Pioneer Life*, 107, 121; and R. Cramer, *Wayne County, Missouri*, 68; on Methodists: Houck, *History of Missouri*, 3:239.

er new counties in the territory at that time bore names of presidents: Washington, Jefferson, and Madison.[20]

French influence on the lower St. Francis was all but nonexistent. Though French trappers had included the region in their expeditions, they did not establish residences or farms. There is little evidence that French Creole Joseph Deblois occupied his isolated concession on the St. Francis. Thomas Ring of the Virginia Settlement, who had earlier lived in Ste. Genevieve where he met Deblois, built a cabin on the concession, but it was never inhabited and burnt a few years later. Deblois, a prized mechanic at New Bourbon and Ste. Genevieve, may never have had much opportunity to live on his concession, having been driven off it by Indians in 1804. Some members of the Deblois family, if not Joseph himself, must have come to Wayne County, because the surname persists in the county in the twenty-first century, variously spelled Deblo, Doublege, Doublaye, and Doublewye.[21]

---

20. On county seat: *History of Southeast Missouri: Embracing an Historical Account,* 458. When English naturalist George W. Featherstonhaugh passed through Greenville in 1844, a quarter century after its founding, he described it as "a poor wretched collection of four or five wooden cabins, where the miserable inhabitants die by inches of chills and fever. . . . [At this season] the poor people, feeble, emaciated, and sallow, are just beginning to recover from the malaria of the country" (*Excursion through the slave states,* 80–81). The poverty of the county precluded erection of a courthouse until sometime in the 1840s. Until that time, the county court met in rented facilities (Ohman, *Encyclopedia of Missouri Courthouses,* 221–22). Lake Wappapello, constructed in 1938–1941, submerged Greenville. A new Greenville was platted two miles to the north and is the current county seat of Wayne County.

21. On Deblois concession: *ASP-PL* 2:516, 3:318, 8:51–52. The concession was confirmed by the second board in 1835. On Ring: *ASP-PL* 8:51–52 and R. Cramer, *Wayne County, Missouri,* 49, 618; on Deblois family: R. Cramer, *Wayne County, Missouri,* 618. When the U.S. rectangular survey came through Wayne County in 1819–1821, the surveyors made no mention of Greenville, but fields and buildings may have been well away from survey lines. The surveyors did note "Swain's improvement" on the Deblois concession (U.S. General Land Office, Field Notes, Missouri, 228:6–9, 229:293–97, 301–3).

12

Private Settlement Colony

The European imperialist strategy of awarding large grants of land to private colonizers who would be responsible for populating them with settlers was tried only once in the Ste. Genevieve District. The sole colonization effort was the grant to James Maxwell on the Black River.

Father James Maxwell, an Irishman by birth, received a theological education at the Irish College of the University of Salamanca in Spain. Fluent in the Spanish language, Maxwell became an instrument of Spanish policy to send English-speaking priests to Louisiana with the expectation that they would minister to American Catholics and help in the conversion of Protestant Americans entering the province. Maxwell arrived at Ste. Genevieve in 1796, just when doors were opening wide to American immigrants, and began an eighteen-year tenure overlapping Spanish and American regimes as its parish priest, officially as "vicar general of the Province of Louisiana." Beyond his role as priest, he involved himself, not at all reluctantly, in business, in the civil administration of the territory of Missouri, and in purchasing land for speculation in and around Ste. Genevieve and New Bourbon, where he lived, and in the mining district. He spent as much time and energy on secular affairs as on religious affairs.[1]

Among his many projects, Maxwell conceived a scheme of rescuing Irish Catholics, as he put it, "from the British tyranny and persecution to which they [were] exposed on account of their religion," and bringing them to Upper Louisiana and settling them under a benign Spanish Catholic regime and his own pastoral care.[2] The scheme may not have been Maxwell's alone; Catholic immigrant–hungry Governor Carondelet could have suggested it when Maxwell passed through New Orleans and introduced himself to officials.

1. Rothensteiner, "Father James Maxwell"; Rothensteiner, *Archdiocese of St. Louis,* 1:198–209.
2. LCS 219.

This was not the first case of Irish Catholic families attempting to settle in Spanish Upper Louisiana. In December 1796, 4 Irish families asked Lieutenant Governor Trudeau for permission for them and 172 other Irish families to form a village "to protect one another," build a church and have a priest "of their own nation," and build six mills.[3] The 4 applicants had already arrived with their animals, the rest to come in April 1797. There is no record that the large contingent ever came to Upper Louisiana or what happened to the project—a proposed group of approximately 700 would have changed substantially the district's demography—but Father Maxwell was certainly aware of the proposal.

Using his close personal connections with Spanish clergy and administrators in both New Orleans and Spain that he developed when he lived in those places, Maxwell petitioned Lieutenant Governor Delassus on October 15, 1799, for an enormous grant of 4 leagues square, or 16 square leagues (112,896 arpents, or 150.8 square miles). Delassus unhesitatingly approved it on November 3, 1799.[4] *Settlement,* not *colony,* was the term used in both the petition and the concession: a "settlement" of "faithful subjects and affectionate to the Spanish Government, on account of their religion." Settlers were to own their lands themselves, not share them in any way with a proprietor as expected in a "colony," as the term was used then. Maxwell's request identified the lands as lying between the Black and Current Rivers, 30 to 35 leagues from Ste. Genevieve. His intentions, as stated in the petition, were to settle destitute Irish Catholics on the land and thus help populate the interior parts of the district. He promised to have a church built for the new immigrants. The huge tract was not surveyed until February 9, 1806, after the transfer of government.[5]

---

3. Zenon Trudeau to Francisco Baron de Carondelet, December 2, 1796, PC 33-699.

4. LCS 219, which includes a copy of Soulard's plat. The petition and its approval (grant) are also in Rothensteiner, "Father James Maxwell," 147–48. The petition is also in *ASP-PL* 8:167–69 and in *Hunt's Minutes,* 1:384–85.

Just the month before, Maxwell had been awarded another large concession of 3,000 arpents on the Mississippi River between the Saline and St. Laurent Rivers (present town of St. Mary's) for his two brothers "and others" from Ireland who were seeking to leave persecution there, and it was promptly surveyed by Soulard. Was this also a colonization project? Apparently, the land did not receive any settlers from Ireland (LCS 68; Basler, *District of Ste. Genevieve,* 222–25; *ASP-PL* 2:438).

It is worth noting that the Maxwell grant of 4 leagues square corresponds to the size of Spanish pueblos under the Laws of the Indies and the size of pueblos in Spanish municipalities in Texas (Reps, *Forgotten Frontier,* 26). Is this a coincidence, or were land-granting authorities in New Orleans thinking in terms of the Laws of the Indies?

5. Quote: Maxwell Claim Application 1875, p. 94, Missouri Historical Society, St. Louis; on petition: *Hunt's Minutes,* 1:384; on survey: LCS 219.

The scheme looked good on paper. The government sorely wanted Catholic settlers, and it had a perfect point man in the multilingual priest who had ideal connections in New Orleans, Spain, and Ireland to put the pieces together.

However, the scheme had two insurmountable problems. Foremost, Maxwell's understanding of geography was poor. The location consisted of some of the roughest Ozark land and worst soils for agriculture, and it was exceptionally remote. Maxwell may not have even seen the location and chose it on secondhand reports he had heard during his ministering trips to Mine la Motte. Perhaps Maxwell was intentionally searching for a location far enough beyond the mining country to prevent his immigrants from being tempted into working mines and to remain peasant-based agriculturists. Alternatively, businessman Maxwell's interest in the tract may have been based on its potential mineral value, situated as it was near the lead districts, yet unexplored. Was he hoping to strike it rich with a colony of laborers? Maxwell may well have expected financial gain from his enterprise, just as he did from his numerous land dealings at Ste. Genevieve and New Bourbon. However, the secretary of the Department of the Interior, during litigation still being carried on in 1875 concerning the grant, concluded that the conditions of the grant implied that "the grant was made to Maxwell in his *ecclesiastical capacity and not for his personal and individual benefit.*"[6]

The immense tract for this ambitious enterprise centered on the three forks of the Black River at present Lesterville in Reynolds County (figs. 36 and 37). At this place, the forks converge to form the trunk of the Black River, and their junction creates a larger alluvial bottom than is usual in narrow Ozark valleys. Although the bottomland itself contains tracts of productive soil, it is surrounded by miles and miles of thin-soiled, chert-mantled ridges and igneous knobs. No more than 10 percent of the tract could be considered arable land even by the most liberal definition.[7] Most of the tract has since been put into national and state forests and other conservation lands, including the wild and primitive Johnson Shut-Ins State Park.

The only way to reach the Maxwell grant from Ste. Genevieve was by a long thirty-six-mile overland trip to Mine la Motte, then cross both the Little St. Francis and the Big St. Francis Rivers at wide and deep fords, cross into the southern end of the knob-and-basin country at Stout's Settlement (where Ironton and Arcadia now are), then cross a dividing ridge

---

6. Maxwell Claim Application 1875, p. 82, Missouri Historical Society, St. Louis. Emphasis in the original.

7. H. H. Krusekopf et al., *Soil Survey of Reynolds County, Missouri.*

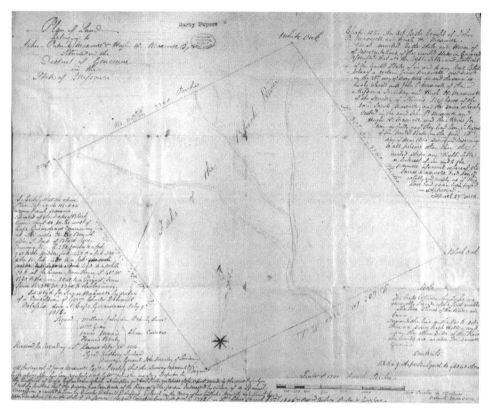

36. *Plat of Maxwell Grant on the Black River, Surveyed February 2, 1816.* The enormous sixteen-square-league grant of November 3, 1799, to Ste. Genevieve priest James Maxwell was first surveyed in 1806 and then resurveyed by the American government in 1816 during litigation over its ownership. Boundaries of the survey in the landscape are no longer known, and it is not depicted on later cadastral maps because the land was incorporated into the American rectangular land survey. The area now includes Johnson Shut-Ins State Park and other public lands. (Darby Family Papers, Missouri Historical Society, St. Louis. Courtesy of the Missouri Historical Society)

Private Settlement Colony 355

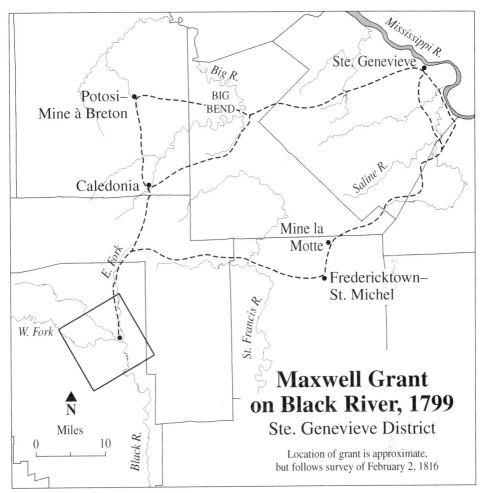

37. *Maxwell Grant on Black River, 1799, Ste. Genevieve District.* The grant lay approximately ninety miles from Ste. Genevieve across rugged hills and valleys and was thus difficult to reach.

and descend the East Fork headwaters of the Black River. The three forks' junction lay almost fifty miles from Mine la Motte and a total of eighty-six miles from Ste. Genevieve. The headwater forks of the Black River have valleys interrupted by waterfalls and rocky shut-ins. The multiday journey on simple horseback would be arduous itself, but to carry farm produce and supplies by cart or wagon would be exceptionally daunting. It was, in fact, the relative inaccessibility of the Black River country that held up its settlement for all but Indians and hunter-woodsman settlers for several decades of the nineteenth century. Madison County had been erected around Fredericktown–St. Michael's in 1818, but Reynolds County, its immediate neighbor to the west on the Black River in which most of the Maxwell grant lay, did not have enough population to be erected until 1845, twenty-seven years later.

The scheme's second problem was that the immigrant Irish never arrived. Maxwell later explained that "then existing wars and other subsequent prohibition of emigration from Ireland prevented the Irish from coming."[8] As so often, events in Europe, this time political disorder during the Napoleonic era, determined how imperial plans in the New World would unfold. There is no clear documentary evidence that any immigrant Irish Catholics ever settled on the grant.

Maxwell, apparently still intending to pursue his project after Spain left Upper Louisiana, presented his claim to the grant to the Board of Land Commissioners, which rejected it twice, June 28, 1806, and May 29, 1812, while Maxwell was still living and serving as priest at Ste. Genevieve.[9] Reasons for rejection were that the grant far exceeded the size authorized by Spanish written policy—even though Governor Carondelet himself had awarded it—and that the terms of the grant, "an obligation to bring from Ireland Roman Catholic emigrants and form a settlement of the same," were not complied with. The board agreed that the grant from the governor was valid and that there was no question of backdating or fraud, as with other large grants made during the same time.

The colonization scheme, effectively dead, lost any possibility of resurrection when Maxwell suddenly died on May 8, 1814, from a fall from a horse.[10] His reputation and leadership alone had held it together. But

---

8. *Hunt's Minutes*, 1:384–85; *ASP-PL* 2:438.

9. *ASP-PL* 2:438.

10. Irish priest James Maxwell's colonization scheme should not be confused with Irish priest John Hogan's colonization scheme in neighboring Ripley County in the late 1850s, subsequently known as the "Irish Wilderness." Hogan's project to settle destitute Irish Catholics in an agricultural colony in the interior Ozarks also failed,

even without Maxwell's death, the scheme would never have been realized because of its near-impossible physical setting.

After Maxwell's death, John P. and Hugh H. Maxwell asserted claims to the grant as his nephew heirs and American residents. By a special act on April 27, 1816, Congress "released" all the "right, title, and interest" in the land to the nephews and "vested" land in them, but this unique wording did not actually transfer the land to the heirs. They and the Roman Catholic Diocese of Missouri at St. Louis, which also claimed the tract on the basis of Maxwell's acting as an instrument of the church, tried in vain to get compensation for the land. The Maxwell claim was rejected again when brought before the second board in 1833 for the same reasons that the first board had given, while at the same time recognizing the grant document as valid and true.[11]

While all this was happening in legal, ecclesiastical, and congressional quarters, American settlers appeared on the Maxwell grant. Confirmation of the grant lacking, U.S. land surveyors had proceeded to survey the land into sections. Some of the American settlers paid the Maxwell nephews for the land in lieu of purchase from the government; others paid them to quiet land titles. When the Buford family bought land from a Maxwell nephew in 1817 or 1818, there was a clearing and improvement on the grant and several houses on it, one a stone house, where a "store" had been kept. Oral testimony in the 1870s during litigation still being pressed, much of it hearsay from the past, referred to "Maxwell's

---

mostly as a result of Civil War activities, although Irish families clearly did take up residence on the land. The post-office toponyms *Wilderness* and *Pines* mark the settlement's general location in the Mark Twain National Forest. Hogan may have been ignorant of Maxwell's proposed Irish colony a half century earlier, although records of Maxwell's project are in the St. Louis Archdiocese in which Hogan worked. For Hogan's project, see Ronald Wihebrink, "History of the Irish Wilderness Country"; Rothensteiner, *Archdiocese of St. Louis,* 2:55–56; and Murphy, "Southeastern Ozark Region," 72–76.

11. *Stats. at Large of USA* 6 (1816): 168. Ironically, Maxwell had made a will in 1802 in which he left only one dollar to his "good-for-nothing" nephew Hugh with the cryptic explanation that he "will know the reason why" (Rothensteiner, "Father James Maxwell," 152). On compensation: Rothensteiner, "Father James Maxwell," 145–47; on rejection: *ASP-PL* 8:167–69. The "heirs and legal representatives" of the two nephews, apparently never having received anything tangible from the unique action of Congress awarding them right to the land but not the land itself, applied to the U.S. General Land Office in 1873 "for land scrip in lieu of their lands sold by the United States Government," during which process testimony brought out many recollections of the early years of the Spanish grant (Maxwell Claim Application 1875, Missouri Historical Society, St. Louis).

store" or "storehouse" at the forks, and the "foreigners Maxwell had in his colony, who were very ignorant of the way to get along in a new country."[12]

These "foreigners," whoever they were (were they hired hands to build the stone store?), did not buy land in the grant as Maxwell had intended settlers to do and otherwise do not appear in county records. For years, the tract continued to be known as the Maxwell grant or reserve, and its survey lines were visible where the boundaries crossed the three forks and the Black River. Testimony in the 1870s added that "the colony and store were not continued after the death of Maxwell" in 1814. However, the Reverends Mr. Peck and Mr. Flint, who traveled the St. Francis River region in the years around 1820, made no mention of Catholic immigrants on a Maxwell grant.[13] Surely, they would have if Irish had been there, as they were so diligent in noting the existence of the farthest frontier settlements and Peck was so attentive to the presence of the Catholic religion on the frontier. Either the minuscule colony had disappeared before their travels or it had never existed.

The upper Black River valleys in and around the Maxwell grant received their first vanguard of American woodsmen in the 1810s. Some of them came from the Bellevue Valley, which by 1815 had seen its best lands taken and had begun exporting people. Henry Padgett (alias Henry Fry), a native of South Carolina, came to the upper Black in 1812 after being among the first in the Big River Bend. Other families came in the 1820s, many of them from Cumberland County, Kentucky, where they were neighbors and intermarried.[14] They scattered up and down the three forks, but carefully avoided land at the junction of the three forks, where Maxwell's store had stood. Even though the junction contained the largest contiguous tract of arable land in the vicinity, knowledge that this was the "Maxwell reserve" kept these early Americans away and therefore postponed development of the most favorable location in the upper Black River system.

---

12. Maxwell Claim Application 1875, pp. 18–20, 26–27, Missouri Historical Society, St. Louis.

13. Ibid., 26–27; Peck, *Forty Years of Pioneer Life;* Timothy Flint, *Recollections of the Last Ten Years passed in occasional residences and journeyings in the valley of the Mississippi.* . . .

14. James Bell, "Black River Township," 23–24; James Bell, *History of Early Reynolds County, Missouri,* 23–24. When the lines of the American rectangular survey were run through the forks of the Black River in the winter of 1821–1822, the surveyors made no mention of any earlier survey lines or improvements, whether from an Irish settlement or recently made by incoming Americans (U.S. General Land Office, Field Notes, Missouri, 222:276–425).

13

An Exclusive American Community

The Bellevue Valley and Caledonia

Originally called Big Lick, the Bellevue Valley reportedly received the name Bellevue in 1803 from an anonymous "old Frenchman," possibly François Azor dit Breton. The valley is a rolling, triangular lowland of some thirty square miles that lies three hundred feet below enclosing Ozarks ridges on two sides, but igneous Buford Mountain, one of the highest eminences in Missouri, rises an impressive eight hundred feet above the eastern side of the valley. Views in the valley are panoramic everywhere. The valley is not the product of river erosion. It is a structural graben, a down-dropped block of the earth's crust within the general up-domed St. Francois region. This geologic origin is highly significant, because it accounts for the complete absence of lead from the valley, although it occurs abundantly in the surrounding hills. Early settlers in the mineral district did not know this, of course, and they vainly searched the valley for lead. Entrance into the valley is everywhere abrupt. Early entrance was from the north, from Mine à Breton, and there the drop into the basin is across an ancient geologic fault line that sharply defines the basin. Early settlers must have been impressed with the grandeur of the Bellevue as they entered it; its name was aptly conferred.[1]

The valley is watered only by small perennial streams, the headwaters of the Big River system. They provided adequate water for water mills, but nowhere along them were there any tight alluvial soils to hold sur-

---

1. *Bellevue—Beautiful View: The History of the Bellevue Valley and Surrounding Area,* 9; *ASP-PL* 8:73. The official post office spelling is Belleview, and other governmental agencies use that semianglicized spelling. The local historical society, however, prefers the original spelling (Robert George Schultz, "Postal Service in Territorial Missouri, 1804–1821," 148). For the geologic origin of the Bellevue Valley, see W. A. Schroeder, "Landforms of Missouri."

face water so much feared for fevers. It was healthful country. Underlain by "purifying" limestones amid the igneous granites and rhyolites, the valley early became famous for its fertile soils, among the best of the Ozarks. At the time settlers arrived, the valley was grass covered, with timber along the water courses and scattered elsewhere. Oaks dominated the woodland, but cedars and pines grew on patches of sandy soils. The valley held all the necessary resources for initial settlement, well distributed throughout.[2] The Bellevue's benign environmental setting contrasted with the inhospitable settings for agriculture of the Maxwell grant and the adjacent lead district.

The only drawback of the Bellevue for settlement was its distance from the Mississippi River. The long fifty-mile distance from Ste. Genevieve meant that Bellevuans could not market their products there and would depend on closer purchasers. But distance had one advantage: it allowed the Bellevue to develop without much interference from the civil authorities at Ste. Genevieve.

John Stewart, a deputy surveyor for the Spanish in the mining area who claimed to have "discovered" the valley shortly after 1800 (it was, of course, known to the Osages and to Creoles searching for lead), showed it to American William Reed, who was scouting for land in Spanish Louisiana. Reed, having received permission from Lieutenant Governor Delassus in 1798 to settle with relations and "connections," led a group of Methodist Scotch-Irish in 1803 from the upper Holston River in Greene County, Tennessee, and neighbors from adjacent Rutherford County, North Carolina, to the Bellevue. Except for William Reed, who had lived temporarily in the Bois Brule bottom, apparently none of the group homesteaded, however briefly, along the way to Upper Louisiana; they came directly from Appalachia to the Spanish Bellevue. Other Scotch-Irish arrived from the upper Catawba Valley in the western Piedmont of North Carolina.[3] Some apparently came in haste to beat the December 20, 1803, deadline for expected confirmation of Spanish concessions and settlement rights. Then in 1808, more settlers, this time Presbyterians, arrived also from the Catawba region (Lincoln and Iredell Counties of North Carolina). To all these people, their religion would become im-

---

2. Schoolcraft, *View of the Lead Mines,* 51; Lewis Caleb Beck, *A Gazetteer of the States of Illinois and Missouri; containing a general view of each state—a general view of their counties—and a particular description of their towns, villages, rivers, etc.,* 264; Edmund Dana, *Geographical sketches,* 296.

3. On Stewart: ASP-PL 8:73, 108–9; on Reed: LCC 146; on rest of group: Robert Flanders, "Caledonia: Ozarks Legacy of the High Scotch-Irish," 38; and *Bellevue—Beautiful View,* 13.

portant as a cohesive force to create a kind of American covenant community.

Though the Osages repeatedly harassed them and stole their horses until 1808, the Bellevue settlers started off their community as full-time farmers who worked their lands and made them productive with the help of slaves.[4] No squatter-woodsman stage preceded the development of farming oriented to surplus production, and the settlers' direct thrust into commercial farming helped them achieve a stable and prosperous community earlier than others.

The Bellevue was completely American settled, specifically Scotch-Irish Protestant with black slaves. Not a single French, Canadian, or Creole was resident, and none was allowed to reside in the valley. This was made clear when Pascal Detchmendy, a former planter of some means who had fled Saint Domingue in the Caribbean during slave revolts there and had come to Upper Louisiana in 1796, tried to locate a 6,000-arpent (8 square miles) floating concession in the Bellevue. Detchmendy and Irish-born Thomas Maddin, the district's deputy surveyor, entered the Bellevue in January 1804, after the cession but before the formal transfer of Upper Louisiana to the United States, to locate and survey the claim, which overlapped with land already occupied by Americans. Ten armed Americans, led by William Reed, confronted the surveying party, declaring themselves and the Bellevue country no longer in the possession of the Spanish government; they would allow neither a "refugee" nor an "unprincipled Irishman" in their midst. Fearful for their lives, the surveying party retreated; Detchmendy immediately sold his claim to an American, Joshua Morrison. The Bellevue residents wrote defiantly to district commandant Vallé that they would not bow to Spain or France, only to the republic of the United States. Maddin accused Moses Austin of being an instigator of the civil disturbance and requested that the rebellious settlers be punished. Lame-duck administrators in Ste. Genevieve, effectively powerless to act, did not pursue any punishment.[5] The affair solidified the community of American republican Protestants.

4. *ASP-PL* 8:73.
5. Claim of Pascal Detchmendy, in "Final Reports of the Board of Commissioners on Private Land Claims in Missouri, Under the Act of July 9, 1832," December 15, 1835, *ASP-PL* 8:72–74; on letters to Vallé: PC 218-633, letters 77, 78, 79; and *ASP-PL* 8:73–74; on Maddin and administrators: *ASP-PL* 8:73–74; *Austin Papers*, 2:85–86, 91–92. Neither did Stoddard of the new American government punish the Bellevue settlers, observing on April 3, 1804, less than one month after he took charge, that crimes against one government are not punishable by another (Billon, *St. Louis in Its Early Days*, 376–97). Historians have noted this celebrated incident for the assertion of American citizenship before the United States had actually set up an administra-

Ten years later, a second incident affirmed the solidarity of the community when again confronted with outsiders. In 1814, Bellevuans, faced with a scarcity of salt, formed a company to develop a local salt supply. Their most promising source, intriguingly named Chicago, lay on public land, and the company learned that "designing men" from elsewhere were poised to obtain a government lease at St. Louis for the saline land as soon as the discovery's location was made known. The company requested that Frederick Bates, in charge of leases, not grant one to anyone except the local Bellevue company.[6] Although apparently nothing came of the Chicago enterprise, it demonstrated the community's vigilance in closing its space to outsiders.

Joseph Pratte, a wealthy Ste. Genevievian, received a Spanish concession in October 1797 for a gigantic tract of 20,000 arpents (26.6 square miles), chiefly in adjacent knobs and basins just to the south of the Bellevue Valley but overlapping it, for "agriculture and raising of large numbers of animals" because lands in the vicinity of the villages were "arid and uncultivable." Everyone knew, however, that the real reason for the request was the abundance of iron at Iron Mountain and adjacent knobs, but Pratte made no serious efforts to mine the ore. No Americans settled on the land for a long time, so although this episode hardly represents a French intrusion into the valley, it does indicate how large land grants in general delayed earnest settlement. Pratte's concession was eventually confirmed in 1835 by the liberal second board, but in the meantime, it was surveyed as part of the public domain and farmable tracts sold to others.[7]

A significant early addition to the basic Scotch-Irish Protestant cohort of the Bellevue Valley came from Moses Austin's coimmigrant craftsmen. Several wanted to establish themselves as farmers, and, with the land around Mine à Breton mineralized and unsuitable for agriculture, they looked to the American-settled Bellevue for the nearest tract of good land.[8] Though some were New England natives like Austin himself, they were also slave owners like the Appalachian Bellevuans, apparently having acquired the practice while living at Austin's prior lead plantation in Virginia or later in Spanish Upper Louisiana.

---

tion at St. Louis. See Flanders, "Caledonia, an Ozarks Village," 9, 52; Gracy, *Moses Austin: His Life,* 93–94; and Foley, *Genesis of Missouri,* 138–39.

6. William Stevenson to Bates, March 5, 1814, *Bates Papers,* 2:274; Schoolcraft, *View of the Lead Mines,* 54.

7. Petition for land by Joseph Pratte, September 28, 1797, LCS 85; *ASP-PL* 8:793–94.

8. Houck, *History of Missouri,* 1:373, 367, 372; LRA 115; *ASP-PL* 8:95–96.

The early economic success of the Bellevue was in large part due to its close ties with Mine à Breton, only twelve miles to the north, and other nearby mining camps. Austin was getting his lead furnace operational at the same time that settlers were entering the Bellevue Valley, and the valley farmers found in the mining camps an immediate market for grains and animals. Austin's household alone required a few hundred pounds of pork each week in 1814 for its ten white and thirty-three slave consumers, a quantity not produced on the lead-plantation grounds. The Bellevuans also found seasonal employment in jobs associated with the mines, such as timber and fuelwood cutting, hauling, construction, and working at the furnaces. Bellevuans rented their slaves to Austin.[9] In fact, Austin and colleagues probably preferred doing business with and employing Bellevue Americans over French Creoles. The markets and part-time jobs were there from the start to provide needed income during those critical initial years when farms were still being cleared to a size large enough to sustain a pioneering family. The Bellevue thus had a singular advantage over other agricultural settlement neighborhoods. Conversely, one might conclude that the immediate success of Austin's lead plantation at Mine à Breton was possible because of a local supply of food and labor that precluded the necessity to import food from and do business with more distant Ste. Genevieve.

The population of the Bellevue grew rapidly. Austin estimated that the Bellevue settlement had 20 families, or 160 persons, at the time of the cession. Another estimate based on land claims in 1804 puts the valley's population around 230. Within a few years, especially after the first contingent of Presbyterians arrived, the population doubled, and by the time Washington County was formed in 1814, the Bellevue Valley had around 700 persons and exceeded Mine à Breton in population. By then, most of the farmable land of the valley had been taken, and settlers were spilling into the next basin south, establishing Stout's Settlement, a smaller version of the Bellevue. The basin's chief distinction was a small iron furnace erected in 1815 or 1816 on Stout's Creek where shut-ins constricted the stream into a rocky gorge as it left the basin. The eco-

---

9. To Frederick Woodson, May 1, 1807, *Bates Papers*, 1:112; to Richard Bates, December 17, 1807, ibid., 238; *Austin Papers*, 2:88, 250. Gracy calculated the pork needs of Austin's household at nine hundred pounds per month (*Moses Austin: His Life*, 158). Later in the 1820s and 1830s, a similar symbiotic relationship developed between the Maramec Iron Works and nearby farmers (J. D. Norris, *Frontier Iron*). In the different economic setting of frontier North Dakota, geographer John C. Hudson also points out farmers' necessity to have additional part-time or seasonal wage work off the farm during the first critical years of establishing a frontier farm ("Migration to an American Frontier," 260).

nomic depression of 1819 caused its demise; its effect on settlement was minimal.[10]

The board confirmed virtually all of the land claims in the Bellevue, both Spanish concessions and settlement rights. There was hardly any land litigation in the valley compared to other neighborhoods. Most settlers stayed on their productive lands, confident of their right to occupy them and confident that their claims would be confirmed by a just American government. Frederick Bates, the territorial secretary, admired the quality of the Bellevue land with its community of proud republicans and considered buying land there in 1807. His intention was to be a gentleman farmer, to "cultivate with a few blacks, who may be employed at particular seasons of the year, in digging mineral." Bates observed that the location "unites two advantages, Farming & Mineral." He abandoned the idea, however, worried about the still-pending confirmation of land titles in the valley, which was a central concern of his as a member of the Board of Land Commissioners.[11] His concern was not over validity of claims or multiple claimants for the same land, as on nearby mineral lands, but whether it was ethical for him to buy any land at all before its confirmation by the board.

Other prominent Americans, however, did become gentleman farmers in the prospering Bellevue. James Bryan, Austin's son-in-law and a lead entrepreneur and land speculator, bought a settlement-rights claim and may have lived in the Bellevue at times during the years from 1808 to 1812 before turning his attention to lead mines on Hazel Run and the Terre Bleue. He may have thought the Bellevue was lead bearing. Andrew Henry, also in the lead business at that time, bought a settlement right of seven hundred acres in 1812 and became a prominent citizen of the Bellevue. His business associate Alexander Craighead established a branch store in the Bellevue and moved there from St. Louis, one of the rare businessmen who gravitated away from instead of into St. Louis.[12] The Bellevue was a prestigious place where people of substance and entrepreneurial energy wanted to live.

The dominant settlement pattern was a single compact area of dis-

---

10. Austin, "Description of Lead Mines," *ASP-PL* 1:191. The 1804 estimate is calculated according to testimony given before the Board of Land Commissioners about who was on the land by 1804 and how many children and slaves they had. On Stout's Settlement: U.S. General Land Office, Field Notes, Missouri, 197:259; on iron furnace: Cozzens, "Iron Industry of Missouri," 520–21.

11. To Frederick Woodson, May 1, 1807, *Bates Papers,* 1:112.

12. On Bryan: LPC 44; on Henry: LRB 262. Henry later achieved fame in the fur business out of St. Louis. On Craighead: Flanders, "Caledonia, an Ozarks Village," 12.

persed farms, each family living on its own farm at an average distance of one-half mile from each other. A settlement focus began when the Methodists organized in 1810 and built a church on William Reed's Spanish concession in the oldest-settled part of the valley. The Presbyterians organized with 30 families in 1816 and constituted a large rural congregation of 75 families within a few years. The log Presbyterian church, built in 1816 as the first church of that denomination anywhere in Missouri, was located within one mile of the Methodist church. The Baptists organized a Bellevue congregation by 1818.[13] Though at that time the Bellevue was one of the neighborhoods located the deepest into the Ozarks, its Protestant churches were among the largest and most vigorous in the state and even had resident ministers, accomplishments explained by the high aspirations of people in the valley and their high degree of cultural cohesion.

After 1815, when the lead boom was still on, population in the valley had become large enough and agricultural production great enough that the community was reaching the economic threshold requirements to support a town. Already, three businesses—a blacksmith, a distillery, and Henry and Craighead's store—had located in the valley. Alexander Craighead seized the opportunity and platted the town of Caledonia in 1818, and buildings went up in 1819.[14] Craighead, one of the early territorial businessmen in St. Louis, had established a store in Ste. Genevieve in 1810 and another at New Diggings next to Mine à Breton. When he and Andrew Henry set up their store in the Bellevue in 1817, he was already using the name Caledonia for a proposed town. American-born Craighead apparently chose the name to commemorate his Scottish ancestral homeland, and it was wonderfully symbolic of the Scotch-Irish culture of the valley. Caledonia was the first town in the Ste. Genevieve District to spring up purely as a central place responding to the economic needs of an agricultural community. It was not part of a larger, extraregional commercial network of lead, furs, grain, or cattle trade. Neither did it come into existence as a mandated county seat for a new county.

Caledonia was not situated centrally in the valley. It was located toward the northeast corner, where both the Methodist and Presbyterian churches and cemeteries already were and where the roads along Big River and

13. On Methodists: *Bellevue—Beautiful View,* 9; on Presbyterians: *Bellevue—Beautiful View,* 20; and Peck, *Forty Years of Pioneer Life,* 89; on Baptists: Peck, *Forty Years of Pioneer Life,* 121.

14. *Bellevue—Beautiful View,* 25. Flanders has investigated the origins and early development of Caledonia as the central place of the Bellevue Valley ("Caledonia, an Ozarks Village"; "Caledonia: Ozarks Legacy").

Cedar Creek converged.[15] Many Bellevue Valley landowners purchased lots in Caledonia for speculation and not for residences; they much preferred to stay on their spacious farms.

Caledonia succeeded but never grew much. The economic downturn of 1819–1823, felt hard in the mining country, made it difficult for the initial lot purchasers to sell their speculations or develop them. Lacking county-seat functions and with a reasonably finite population in its enclosed topographic basin (even though a stable and prosperous one), Caledonia could serve only its immediate valley population. It was not on any transportation route of consequence. The rugged Ozarks to the west and south did not allow commercial agriculture to develop there to any degree, so the Bellevue Valley remained an outpost at the edges of the sparsely settled, subsistent Ozarks. The Springfield iron furnace, which was established by Bellevuans and began operation on the northern edge of the valley in 1823, and the earlier ironworks on Stout's Creek in the basin immediately south provided local iron products, but their market range was limited because of the high cost of hauling by wagon.[16]

Some Bellevuans did move on. Members of the early Reed and Stevenson families left in 1817 to take Methodism into the Arkansas Ozarks, and the Bufords crossed over the Ozark dividing ridge onto the Maxwell grant in the headwaters of the southward-flowing Black River. Several Bellevue families joined the Austin colony in Texas after the collapse of his Missouri lead business, and thereby established a Bellevue-Texas connection that lasted for many years.[17] These special cases, however, do not

---

15. Details about the plat of Caledonia are in W. A. Schroeder, "Opening the Ozarks," 400–401.

16. On Caledonia's lack of success: Flanders, "Caledonia: Ozarks Legacy," 40; on Springfield furnace: *Missouri Gazette,* August 9, 1817; and Flanders, "Caledonia: Ozarks Legacy," 42.

17. On Reeds and Stevensons: *Bellevue—Beautiful View,* 9; and Flanders, "Caledonia, an Ozarks Village," 14–15; on Bufords: *Bellevue—Beautiful View,* 10; on Bellevue-Texas connection: Flanders, "Caledonia, an Ozarks Village," 14–15. Bellevuans were part of the large number of Missourians interested in the initial entry of Americans into the Red River country and Texas either for business or for settlement. William Ashley, who had lived in the Bellevue Valley, wanted to do business with the more than one hundred American families on the Red River in 1809, because he had heard that they were poorly provided for by New Orleans (Clokey, *William H. Ashley,* 21). In 1816, the Reverend William Stevenson (probably William O. Stevenson who lived in the Bellevue Valley) led a group of Washington County, Missouri, farmers to settle on the Red River, and in 1820, the Reverend Joseph Bays took two dozen families from Missouri to the Red River country. Austin himself had established a colony on the Red River across from Spanish Texas in 1817 before securing the grant in Texas (Gracy, *Moses Austin: His Life,* 178–82, 179–80, 210).

overshadow the geographic stability of the Scotch-Irish Protestants of the Bellevue.

The Bellevue, with Caledonia as its center, was the purest and most self-conscious Protestant American community in the Ste. Genevieve District, a distinction helped by being set apart physiographically from outside influences by encircling ridges and knobs. Without lead ore in the valley, Bellevuans put their energies into commercial farming and supplying food for the nearby mining areas. Robert Flanders, who has studied the community's settlement history, describes its society as "high Scotch-Irish" and early inclined to progressive values and strong formal institutions that were "remarkable in a region slow to develop institutions." The Bellevue, according to Flanders, embodied Jeffersonian ideals as well as any community in the district: It was "a rural neighborhood and a small village of yeomen that were Southern, middling, modest, decent (if slaveowning), striving, expansive, tasteful, Godly, patriotic, and independent. The Bellevue would have nothing to do with French or with the administration and institutions of the French and Spanish. It presaged the Americanization of the Ozarks."[18]

---

18. Flanders, "Caledonia, an Ozarks Village," 1–2.

14

Perry County

A Settlement Mosaic

The southeastern portion of the Ste. Genevieve District, now constituting Perry County, was the most geographically complex portion of the district for settlement (fig. 38). This was in large part because so much of the area was inhabitable. The Bois Brule bottom opened into the grassy karst plain of the Barrens that extended twenty miles from the Mississippi. Clusters of settlement that arose in these couple hundred square miles eventually merged into one compact settlement area.

Settlement complexity was also due to this area's position directly on the Mississippi. Left unsettled but not untouched by the French before 1796, it lay available as a first opportunity for all people who crossed the Mississippi. These included Indians of the Illinois and Cherokee Nations, and later more numerous Shawnee and Delaware; Americans of various social, economic, and religious identities, including well-organized groups of Catholics, Presbyterians, and Methodists; and individuals of Canadian, Creole, French, German, English, Irish, and Scottish "nationality." These various peoples continued to flow into the area during a long period of time, from the 1790s to 1830 and beyond.

Until the 1780s, the southeastern portion of the Ste. Genevieve District lay vacant except for occasional Indian camps and the isolated cabins of Indian traders and transients along the Mississippi River. The salt-making establishments at the mouth of the Saline River effectively marked the southern end of inhabited lands on the western side of the Mississippi River. In the 1780s, some American salt workers, having worked for three years to meet residency requirements and being pushed off the Saline River by Peyroux and the Vallés in their clash over landownership, made the first permanent settlements in the upper end of the Bois Brule bottom. They built their cabins on the concave bank of the Mississippi

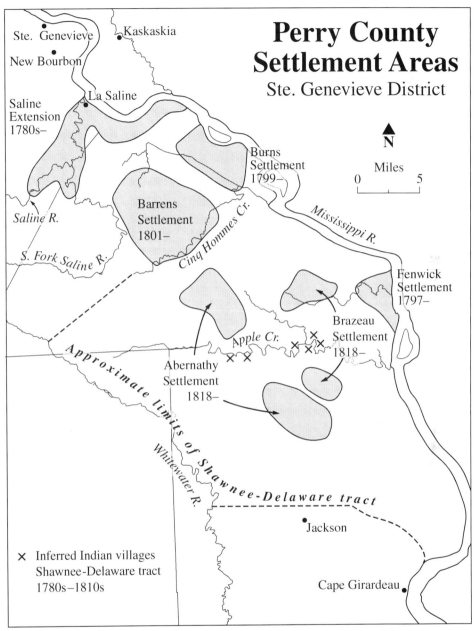

38. *Perry County Settlement Areas, Ste. Genevieve District.* Lying directly on the Mississippi River with fertile limestone and loessial soils, Perry County was settled before 1830 by a variety of immigrants, including French Creoles, Shawnee and Delaware Indians, American woodsmen, American Catholic farmers, and American Protestant farmers.

and within a few years lost them to the river, unless they disassembled them and moved them back.¹

At about the same time, people from the other side of the river began to show interest in the woods and wet prairies of the bottom. In March 1779, at a time when immigration was still restricted to Catholics and an armed revolution was taking place on the Atlantic coast of the English colonies, seven men of the "English nation" petitioned François Vallé and received permission from Lieutenant Governor Leyba to settle in the Bois Brule bottom, because, as Vallé put it in his recommendation, they would provide "safety against the incursions of Indians and means to prevent the wandering away of cattle." If these seven families did settle in the Bois Brule bottom, they soon left, for they disappeared from the record. One, Thomas Tyler, showed up on the Meramec later in 1779.²

An assortment of people appeared in these initial stages of Bois Brule settlement. Westering bands of Shawnee, Delaware, and other Indians lived in the bottoms in the 1780s. In response to their presence, Jean-Baptiste Barseloux, a Creole Indian trader and interpreter, settled in an isolated location in the lower end of the Bois Brule bottom and received a concession for it in 1787, certainly years after he had originally built there. Though no neighbors were anywhere near, the single, isolated property had the typical French long-lot shape of eight arpents front on the river by forty arpents depth. Louis Coyteux, another Indian trader who worked for the Spanish, had an isolated cabin on the bluffs overlooking the bottoms and the river. Christian Fender, known as a German and possibly a former worker at the salines, occupied land just downriver from La Saline. Fourteen other families were reported in the fourteen-mile-long bottom in the Spanish census of 1797, of which eleven were headed by Americans and three by Irish, composing a total of fifty-nine whites and eight slaves.³

---

1. On clash: LCS 243; on permanent settlement: LR 36; LCS 165, 191, 243; MN 26. The availability of work at the salines before the general opening up of Spanish Louisiana to Americans made La Saline the "port of entry" for Americans crossing the Mississippi into the Ste. Genevieve District. Apparently, the French, Creoles, and others crossed at the village of Ste. Genevieve (PC 209-598). On built cabins: PC 118-431; *ASP-PL* 2:426.

2. Petition for land by Thomas Tyler et al., March 18, 1779, LCS 100; *ASP-PL* 3:585; Houck, *History of Missouri*, 2:73.

3. On Barseloux: LCS 151; on Coyteux: Ekberg, *Colonial Ste. Genevieve*, 76–77, 313, 367–68; on Fender: *ASP-PL* 2:449. Several other persons living in the Bois Brule bottoms were of Germanic descent: William Krytz (Kreutz, Crites), George Egers (Eagen), and William Vanburken (Vanburkelow) (Houck, *History of Missouri*, 1:382). On other families: PC 2365-345.

Shawnee-Delaware Indian Grant

During the 1770s and 1780s, remnants of depleted Illinois tribes crossed into Spanish territory (did the Illinois still regard it as their territory?) as their lot on the eastern side deteriorated under British and American rule. The Peoria lived at new Ste. Genevieve village in the 1780s, and one hundred Peoria lived in a village in the lower Bois Brule bottom near Coyteux.[4] Though relations with the Creoles were peaceful, these remnant Illinois were not well integrated into the life of the Creole communities, nor were the French integrated into theirs. Marriages between the Illinois and the French, though certainly not rare, had become much less common than in the earlier years of French in the valley. The Spanish administration disregarded the Illinois in their censuses.

In the final decade of the eighteenth century, Spanish consideration of Indians changed. Although the Spanish continued to more or less ignore the Illinois remnants in Upper Louisiana, they perceived that eastern Indians could abet Spanish settlement strategies. They invited them, and the land they occupied figured prominently in directing the course of settlement of other people and in the evolving human geography of the district.

Eastern Indians had already westered into Spanish Illinois in the 1780s as part of the general displacement of native residents in front of the American frontier. Some had settled peacefully in Spanish Illinois with the implicit sanction of officials because they were expected to serve a buffering role against the Osages.[5] Along with the Shawnee in the mid-1780s came Louis Lorimier, a Métis (Canadian Shawnee) who had fought with the British against the Americans during the American Revolution. His close association with the Shawnee—he had operated an Indian trading post in Ohio—encouraged more Shawnee to follow him into new lands and better conditions in Spanish Illinois. Americans had destroyed the Shawnee villages during their imperial campaigns to conquer the Northwest Territory for white settlement and had left the Indians landless in their own country. Lorimier introduced a plan to bring the dis-

---

4. Temple, *Indian Villages*, 54–56; Timothy J. O'Rourke, *Maryland Catholics on the Frontier: The Missouri and Texas Settlements*, 2; Ekberg, *Colonial Ste. Genevieve*, 86–124; Zitomersky, *French Americans–Native Americans*.

5. Trudeau to Lorimier, May 1, 1793, in *Spanish Regime*, ed. Houck, 2:50–51. For Shawnee migration to Upper Louisiana, see PC 201-955; Temple, *Indian Villages*, 173–87; Houck, *History of Missouri*, 1:208–19. Conditions in the Northwest Territory that prompted the Shawnee to move westward are treated in Richard White, *The Middle Ground: Indians, Empires, and Republics in the Great Lakes Region, 1650–1815*; and Aron, *How the West Was Lost*. For Delaware migration, see Clinton A. Weslager, *The Delaware Indian Westward Migration*.

possessed to Spanish Illinois in 1787. Working through Lieutenant Governor Perez, he received permission on January 4, 1793, from Governor Carondelet to recruit the Shawnee and Delaware and settle them in Spanish Illinois.[6] Lorimier, who had first lived near Ste. Genevieve and La Saline, moved south to be with congregating Indians in vacant lands and finally settled at Cape Girardeau in the early 1790s. According to the agreement with Spanish officials, he was to develop a line of Indian settlements in vacant lands from Ste. Genevieve to the Tywappity Bottoms just north of the Ohio River junction. Lorimier articulated intentions to build a school, flour mill, and bridges for the Indian settlements. In 1797, the Spanish recognized his establishment at Cape Girardeau as an official post with its own administrative district, the Cape Girardeau District, with Lorimier as its commandant. An administrative post and the office of commandant were necessary because, according to Spanish policy, only a commandant could participate in Indian affairs. However, his authority in matters concerning land was limited, and he was in no way the administrative equal of the commandant at Ste. Genevieve.[7] Lorimier's new district was created within the huge expanse of unoccupied land that separated the districts of Ste. Genevieve (and New Bourbon) and New Madrid. Its northern boundary was placed along Apple Creek, which then became the southern limits of the Ste. Genevieve (including New Bourbon) District.

Although Carondelet authorized Lorimier to establish his Indian settlements for farming and hunting anywhere along the west bank of the Mississippi between the Missouri and Arkansas Rivers that no other settlers had already established, the vacant area that Lorimier chose for the Indian tract was where Indians had already gathered.[8] The tract's boundaries did not coincide geographically with Lorimier's new Cape Girardeau District. The tract was bounded on the north by Cinq Hommes Creek and on the south by the first creek above Cape Girardeau, generally assumed to be Flora Creek (to exclude Lorimier's personal land at Cape Girardeau),

---

6. PC 2365-66. A biography of Lorimier is June Cooper Stacy, *Louis Lorimer* [sic]. Lorimier's memoire of his Ohio experiences is in Paul L. Stevens, *Louis Lorimier in the American Revolution, 1777–1782*. Lorimier may have been illiterate and have had secretaries Largeau and Cousin write his journals (Houck, *History of Missouri*, 1:318; Carl J. Ekberg, "Was Louis Lorimier Literate?"). On Lorimier's plan: PC 2363-586; on permission: PC 2365-718 and *ASP-PL* 2:551. Carondelet also invited the Cherokee to settle in Louisiana (Carondelet to Gayoso de Lemos, December 18, 1792, in *Spain in the Mississippi Valley*, ed. Kinnaird, 4:106).

7. Louis Lorimier to Carondelet, June 20, 1797, PC 2365-65; Houck, *History of Missouri*, 1:318.

8. *ASP-PL* 2:551.

and reached westward to the Whitewater River. Apple Creek ran through the middle of the tract and thus put halves of the Indian tract officially in both of the two administrative districts (see fig. 38). The tract had a Mississippi River frontage of 30 miles and a depth of approximately 20 miles to the north and 30 miles to the south. On May 1, 1793, four months after Carondelet's authorization to settle, Lieutenant Governor Trudeau conceded the Apple River tract to the Indians, and he reconfirmed the award on June 19, 1797. It is curious that three years later, when the tract was to be surveyed and a plat drawn on orders of Commandant Vallé, who had in hand only a rough map of the Indian lands, the Shawnee refused to let it be surveyed, giving no reason. It never was.[9]

The concession of approximately 750 square miles included extensive prairies and woodlands, loess-mantled karst plains, and alluvial bottoms. The tract was large enough to support several Indian villages with surrounding hunting territory, although these semi-Americanized Indians were more dependent on field crops and domesticated animals than wild game. The New Bourbon census of 1797 reported 70 Shawnee and 120 Delaware families in villages on the north side of Apple Creek in the New Bourbon administrative district; possibly as many families lived in other villages south of Apple Creek.[10] At 5 persons per family, 190 families represented a population of 950, a village population similar in size to that of Ste. Genevieve and St. Louis. The intensity of land use for field crops to support a population of 1,000 must have been of nearly the same magnitude as in the celebrated common field at Ste. Genevieve and in the cultivated fields around St. Louis.

Although the tract was split between the two districts, it was Lorimier who supervised the Indians. The Ste. Genevieve administrative records, so replete with references to Osage Indian visits to villages and mines, scarcely mention the Indians on Apple Creek. Despite Lorimier's official trading privileges with the Shawnee and Delaware, the Indians looked to Ste. Genevieve to market their turkey and venison and otherwise carried on considerable trade with Ste. Genevieve merchants Pierre Menard and François Vallé, whose account books contain scores of Indian names. Traditional trading privileges did not amount to much because these Indians marketed their peltries and other products in the villages themselves,

9. On concession and reconfirmation: PC 2365-718 and *ASP-PL* 2:551; on Shawnee refusal: PC 217b-200, 206. As it turned out, the lack of survey was not a hindrance in getting the land recognized by the U.S. government, as the Indian tract did not have to be confirmed by the Board of Land Commissioners. Vallé's rough map has not been found.

10. PC 2365-345.

just as the French and Americans did. The Indians also marketed their agricultural surpluses to boatmen and travelers on the Mississippi at the mouth of Apple Creek.[11]

As many as six villages were located between Cinq Hommes and Flora Creeks, most of them along Apple Creek itself and some distance away from the Mississippi River. In his 1804 inventory of Upper Louisiana, Amos Stoddard reported one Delaware and two Shawnee villages about 20 miles up Apple Creek, supposedly built about 1794. Just a few years earlier, Finiels had described the two Shawnee villages as 5 or 6 leagues (15 to 18 miles) apart on the road between Cape Girardeau and Ste. Genevieve, and a Delaware village approximately halfway between the two, but "further inland." The largest village, a Shawnee village, was located on the north side of Apple Creek where Indian Creek joins it and approximately where the road from Cape Girardeau to Ste. Genevieve crosses Apple Creek.[12]

The villages and their locations are not identified in any official document of land concession, survey, or sale. Archaeological investigations in the 1980s failed to find any conclusive evidence of Shawnee and Delaware villages, despite careful examination of sites based on local oral history. Because these Indians lived much like other Americans on the frontier and their cabins and other tangible evidence such as clothing and implements did not differ much from that of pioneering Americans, archaeologists have found it difficult to distinguish Indian sites from frontier American sites along Apple Creek.[13]

---

11. Houck, *History of Missouri,* 1:214, 217; PC 2365-345. Shawnee on the Meramec River also supplied St. Louis with venison (Rachel Fram Vogel, "Life in Colonial St. Louis," 87). Morrow notes the importance of Indian trade in the Ozarks to merchants in Ste. Genevieve, Kaskaskia, and St. Louis from the 1780s to the 1820s in "Trader William Gillis." Cherokee farmers lived with whites on the Whitewater River in 1813 (Address of Cherokees, April 27, 1813, *Bates Papers,* 2:239–41). Official trade with Indians involved gifts (PC 2365-345). On trading and marketing: PC 2365-345, docs. 3 and 5; and Houck, *History of Missouri,* 1:215.

12. *ASP-PL* 5:800; Stoddard, *Sketches of Louisiana,* 215; Finiels, *Account of Upper Louisiana,* 34–35. Locations of Indian villages are shown on several early maps: Clark 1805 map and Forsyth 1812 map (Tucker, *Atlas,* pl. xxxi-B and xxxviii); Lewis and Clark 1806 map and Clark 1810 map (Moulton, *Journals of Lewis and Clark Expedition,* maps 123, 125); Tanner, map, Illinois and Missouri, State Historical Society of Missouri, Columbia. A map dated 1818 is unusual by showing a boundary line for the Shawnee tract (Tucker, *Atlas,* pl. xxv). For locations, also see O'Rourke, *Maryland Catholics,* 2; and Houck, *History of Missouri,* 1:213.

13. Archaeologist Duncan C. Wilkie surveyed the Apple Creek region, giving special attention to recorded locations of Indian villages, but could not find any "unquestionable evidence" of Indian occupation ("Archaeological Reconnaissance Survey of the Apple Creek Drainage: Perry and Cape Girardeau Counties, Missouri," 153–54).

Other groups of the Shawnee and Delaware had settled elsewhere in the district than on the tract the Spanish set aside for them. The most prominent of them was Rogers Town on the Meramec River in the far-northwestern edge of the district. Chief Rogers, a white man who grew up as an adoptee of the Shawnee in Ohio, was in close contact with Americans on the Meramec and especially with the Chouteaus at St. Louis. Smaller groups of the Shawnee and Delaware were on the Fourche à Courtois and Osage Fork (Huzzah) west of Mine à Breton.[14]

The well-traveled Shawnee Trace linked the Apple Creek villages southward to Lorimier's post at Cape Girardeau and northward to La Saline, New Bourbon, and Ste. Genevieve. Because the Shawnee Trace comprised a section of the royal road connecting the several administrative posts of Upper Louisiana, the Indian villages witnessed a considerable amount of travel through them by officials and outsiders. Another trail followed Apple Creek eastward to its mouth at the Mississippi where a trading post served river traffic.[15]

The Indians on Apple Creek had a significant degree of racial mixing and had adopted French and American ways. The Indians included half whites of both French and American parentage, and whites, probably orphans and captives from warfare, who had been raised as Indians according to the Shawnee practice of adopting their captives. Apple Creek villages certainly had as much "white blood" in them as French villages had "Indian blood." Some houses were log constructed in the French style, with posts set closely together into the ground and clay chinked. Other houses were built in the American style, with horizontal squared logs, some two stories high and shingled. There were granaries and barns for cattle and horses.[16] Villages gave the appearance of permanence.

14. On Rogers: Henry Harvey, *History of the Shawnee Indians from the Year 1681 to 1854, Inclusive*, 235–36. Harvey was a Quaker who lived for many years among the Shawnee and recorded events as they affected the Shawnee. Faragher, "'More Motley than Mackinaw'"; Peck, *Forty Years of Pioneer Life*, 111–13; Houck, *History of Missouri*, 1:210–12. Rogers Town is located on "Clark's Map of 1810," in *Journals of Lewis and Clark Expedition*, ed. Moulton, map 125. On smaller groups: Schoolcraft, *Journal of a Tour*, 6–7.

15. Finiels, *Account of Upper Louisiana*, 34–35; Houck, *History of Missouri*, 1:213–14.

16. Houck, *History of Missouri*, 1:209–12. C. Schultz tells an interesting account of a white woman living in the Shawnee villages who knew no English, French, or "Indian," and apparently no German (*Inland Voyage*, 2:77). On French style: Houck, *History of Missouri*, 1:214; on American style: Stoddard, *Sketches of Louisiana*, 215. Aron describes the Americanization of the Shawnee in the Ohio River country that began in the 1760s. He notes that they were not "indolent" according to Moravians and other impartial observers. The Shawnee "adopted" their enemies, including

Cultivated fields of corn, barley, pumpkins, melons, and potatoes, enclosed by rail fences, surrounded the villages. The Indians had barnyard fowl, cattle, hogs, and horses. As Lieutenant Governor Trudeau reported to Carondelet, "They are rich for Indians."[17]

Beyond the villages and enclosed fields lay the woods and prairies, the hunting grounds that all frontier Indians expected to have, just as frontier Americans required them. In his inventory of the territory, Stoddard noted that in 1804, the Indian tract was too settled to afford sufficient game. Nevertheless, ten years later, males were still leaving the villages to hunt in camps twelve miles away.[18] Thus, the land-use arrangement around the Indian villages was basically no different from that of dispersed American frontier settlements, which were part agriculture and part hunting, except that Indians lived in compact villages as the French did.

Although the Indians interpreted the entire grant as being exclusively theirs and excluding Americans, district commandants awarded concessions to white settlers within it and as close as three miles to the villages. Apparently, the commandants reckoned that it was still their right to approve locations on unoccupied land, and they could have interpreted the miles of hunting land around the villages as unoccupied in the sense that other land in the district used by Americans for hunting was considered unoccupied. Instructions to award only land not in conflict with other concessions—so rigorously adhered to everywhere else—apparently did not hold for the Indian tract.[19] Thus, at the end of the Spanish period, the tract contained approximately fifteen hundred Indians primarily in villages and approximately one hundred whites living on dispersed farms, mostly on the fringes of the unsurveyed and undemarcated tract.

The Indians protested these incursions, but there was little they could do about it, as they had hardly any influence with authorities beyond Lorimier. The first whites within the tract apparently did not disturb the Indians, and indeed some served them as useful gunsmiths and other craftsmen and others traded with them, including liquor.[20] Nevertheless, the 1804 map of land claims shows a sharp boundary along Cinq Hommes

---

Americans, by incorporating them into their community life (*How the West Was Lost*, 7–13).

17. Zenon Trudeau to the baron de Carondelet, June 8, 1794, PC 209-648; Finiels, *Account of Upper Louisiana*, 34.

18. Stoddard, *Sketches of Louisiana*, 215; Peck, *Forty Years of Pioneer Life*, 113.

19. Governor Clark to the president, January 22, 1816, *TP* 15:105–7; PC 2365-202.

20. LCS 223; PC 2365-202.

Creek that separates the compact Barrens Settlement and Bois Brule bottoms from the Indian tract with its few scattered claims of white settlers (see fig. 5).

The Shawnee did not fulfill the purposes for which the Spanish invited their immigration. They were ineffective as a buffer against the Osage Indians. From early on, the Shawnee said that Spanish Illinois was not a "tranquil" place due to the Osages, and they threatened to return to the United States.[21] The Shawnee were particularly disturbed that the Spanish expected them to fight the Osages, while at the same time, the Spanish continued to trade and bestow gifts on the Osages but offered no aid when the Osages attacked them.[22] The Osages continued their harassment of French and American settlements well past 1800. In 1807, Captain Rogers and his Meramec Shawnee requested protection from the Osages, just as American settlers did. And, as for providing a populated bank on the Mississippi to counter incursions from the United States, the Shawnee and Delaware had little interest in what they left behind in the East, except to receive more of their own who were searching for land and a better life west of the Mississippi, just like Americans.[23]

The Shawnee-Delaware tract, which was initially a haven under the Spanish administration, turned into a besieged settlement after the American cession. The people from whom the Indians had fled were now their neighbors and masters once again. Under the new government, the tract was increasingly surrounded and invaded by Americans who had little interest in being friendly, interactive neighbors. The Indians would eventually have to leave. How long would it take?[24]

By the sixth article of the Treaty of Cession of Louisiana, the American government promised to respect all treaties and "articles" that the Spanish government had made with tribes and nations of Indians. Therefore, although the Board of Land Commissioners rejected the Shawnee-Delaware land claim in 1812 for the reason that it exceeded the square-league maximum set by Congress in 1807 and the claim was never pursued after that, the Indian tract's validity did not rest on any action the board took but on the overarching federal acceptance of Spanish "articles"—the 1793 Carondelet grant—with the Shawnee and Delaware. In

---

21. Peyroux to the [governor?], April 7, 1790, PC 203-230.
22. Unnumbered letter from Lorimier to Pedro Rousseau, September 6, 1794, PC 2363-633.
23. *Bates Papers,* 1:105; Harvey, *History of the Shawnee,* 117.
24. According to Harvey's interpretation, the Shawnee realized that they would have to leave and seek a home elsewhere as soon as they learned that the United States had purchased Louisiana (*History of the Shawnee,* 117).

addition, these displaced eastern Indians fitted into the initial Jeffersonian policy of moving Indians from the Northwest Territory to lands west of the Mississippi.[25] Therefore, the federal government, fully respecting the right of the Indians to stay on the land, now had the responsibility through its territorial governor and Indian agent in St. Louis to ensure that whites stayed off designated Indian land. But at the same time, whites, many of them resolute Indian fighters, were intruding onto it with the tacit support of the rest of the local American white population. It was a situation similar to two other land problems faced by the territorial government: keeping intruders off the public domain until it could go on sale and keeping intruders off designated mineral lands that were reserved from sale. In none of these three cases was the government successful with its land policies, no matter how much public exhortation.

Before three years of the new regime had passed, Lorimier, who continued to function as a district officer enjoying the support of the territorial administration, wrote on behalf of the Indians to acting territorial governor Joseph Browne, complaining of encroachments across their boundaries. He was concerned that if such action went unchecked and unreported, "[s]ilence or inaction should be construed into a consent" to allow the interlopers in. The governor's predictable response was to order intruders off the Indian grant and to request district sheriffs to provide names of intruders so that "they may be proceeded against according to law."[26] This order did no more good than ordering the sun to stand still.

The Indians, with memories still fresh of similar treatment in the Northwest Territory, may have realized the hopeless future of their Apple Creek villages, despite their near-perfect behavior and a virtual lack of any complaint against them. Conversely, Americans were increasingly stealing from the Indians. In 1809, Shawnee and Delaware representatives visited territorial governor Meriwether Lewis in St. Louis and expressed their "long-expressed wish" to exchange their tract for lands farther west.[27] It was generally known that the terms of Carondelet's grant to the Indians included a provision that the land would revert to public domain, if they should abandon it. Perhaps white settlers were forcing

---

25. On sixth article: *LPGN* 1:1–3; on second board: *ASP-PL* 2:551; *Stats. at Large of USA* 2 (1807): 441; and Donaldson, *Public Domain,* 96; on removal: Abel, *Indian Consolidation,* 241–49.

26. Louis Lorimier to Acting Governor Browne, February 18, 1807, *TP* 14:112; "Proclamation by Governor Lewis," April 6, 1809, *TP* 14:261.

27. Harvey, *History of the Shawnee,* 163; Houck, *History of Missouri,* 1:219; to William Eustis, secretary of war, September 28, 1809, *Bates Papers,* 2:91.

this issue by their undisturbed encroachment, thereby exerting pressure on the Indians to leave it. The only outspoken champion for the Indians at any level of government had been Lorimier, and he died in 1811.

In 1815, a major Indian conclave took place at St. Louis with territorial governor William Clark and other representatives of the federal government at which Shawnee and Delaware spokesmen made passionate speeches. They described fully their worsening situation, unfulfilled promises by the American government to respect their land and property, and depredations against them by white encroachers. Within twenty-two days of the conclave, President Madison ordered all intruders to be removed from Shawnee and Delaware lands in Missouri Territory.[28] But this order, just as others that the Indians had noted, was not carried out.

Whites had reason to be perplexed by the president's directive. Had not other whites been given permission to settle in the tract, and had not those lands been confirmed by the Board of Land Commissioners? People migrating into the territory could well have believed that the large, undemarcated tract was public domain, with Americans already legally on it. Above all, everyone agreed that the tract contained land superbly fitted for agriculture. Therefore, the Missouri Territorial Assembly, strongly reacting to Madison's order and obviously reflecting the sentiments of the frontier, sent resolutions to the United States House of Representatives in January 1816, requesting that the Indians be given land elsewhere so that "people may not be Compelled to leave their Improvements" in the Indian tract.[29]

As no federal action was taken, in the following January another resolution of the Missouri Territorial Assembly noted that the Indian land grant had retarded American population growth in the vicinity. Settlers presented petitions requesting that conflicts over landownership be resolved in favor of Americans by exchanging Indian lands for land elsewhere that was "better adapted to indian persuits [sic]." It mattered little that the Indians were living nearly the same way that American frontiersmen were and that their behavior was exemplary. The Shawnee again made known their willingness to exchange their lands.[30]

In fact, the federal government had already initiated the process of finding new lands to relocate the Indians and thereby acknowledged the

---

28. *ASP-Ind* 2:11. Slightly different versions of the speeches are in Harvey, *History of the Shawnee,* 161–63; and Houck, *History of Missouri,* 1:219–20. On president's order: Harvey, *History of the Shawnee,* 163.

29. Governor Clark to the president, January 22, 1816, *TP* 15:105–7.

30. "Resolutions of the Territorial Assembly," January 24, 1817, *TP* 15:234–36; Governor Clark to the secretary of war, October 1818, *TP* 15:455.

relentless and unpreventable encroachments of whites. John Scott, a Missouri delegate to Congress in 1820, noted that conflicts took place over the intermingling of the open-range stock, during which "the safety of both is not unfrequently Jeopardized in the Contests." The tract, he said, had become "too valuable" for Indians, and they needed to be removed to lands farther west. According to Scott's argument, the governmental expense of moving the Indians and opening new farms and erecting buildings equal to those left behind would be compensated for by the sale of their abandoned lands, which would be twice as valuable as new, unimproved lands. Thus, the federal government would suffer no economic loss by paying for their removal.[31]

Actually, the Shawnee and Delaware had been in the process of leaving the tract during all these years. Some had indicated their desire for land farther west as early as 1801 under the Spanish, but others wanted to hold on to their improvements in the tract, and more Indians kept coming from the East to replace those who had gone west. By 1806, the Delaware, including newly arriving Indians passing through the Apple Creek tract, had, on their own initiative, established villages in Osage territory on the White River, the James Fork, Wilson's Creek, the Current River, and the upper Gasconade. In 1808, the Osages ceded the lands that all these Delaware villages were on to the United States, placing these Delaware once again on U.S. public domain open for survey and white settlement. The Shawnee had also moved westward in bands and in stages, first to the St. Francis and Black Rivers, then to the Current.[32]

Despite these earlier removals westward, when the Reverend Jedidiah Morse counted "souls" during a tour in 1820 for his detailed report on Indians to the secretary of war, he counted 1,383 Shawnee on Apple Creek, near the swamps to the south of Cape Girardeau, and on the Meramec, and 1,800 Delaware already moved to the Current and White Rivers. He also counted smaller groups of Peoria (97 on the Current River) and Piankashaw (207 on the St. Francis River), both of whom never received any land grants, Spanish or American.[33]

---

31. Delegate Scott to the secretary of war, September 21, 1820, *TP* 15:645–46.

32. On land farther west: PC 218-360; on improvements: PC 218-633, doc. 17; on Delaware villages: *Jackson Independent Patriot*, June 12, 1824; Houck, *History of Missouri*, 1:217–18; and Flint, *Recollections of the Last Ten Years*, 149–50. The 1818 map "Missouri Territory Formerly Louisiana" shows the Delaware on the White River. On Osage cession: Kappler, *Indian Affairs*, 95–99; and *Stats. at Large of USA* 7 (1808): 107–11; on Shawnee: Houck, *History of Missouri*, 1:235–36.

33. Jedidiah Morse, *A report to the Secretary of War of the United States, on Indian Affairs, comprising a narrative of a tour performed in the summer of 1820 . . . for the purpose of ascertaining . . . the actual state of the Indian tribes in our country*, 366; Tem-

When the Jackson land office opened in 1821, settlers were ready to file for land in the Indian tract that they had occupied as would-be preemptors after the War of 1812. Their intentions prompted new Missouri senator Thomas Hart Benton to describe the remaining village Indians as occupying an "island-like" tract, "pressed upon" by Americans, and he urged Congress to remove them both for their benefit and for "the progress of development of the state of Missouri."[34]

Finally, well after most Indians had vacated their confirmed Spanish grant, abandoning their homes, barns, and fields, the Shawnee signed a treaty on November 7, 1825, at St. Louis with William Clark, now the U.S. Indian commissioner. By this treaty, the Shawnee exchanged their Apple River tract for lands near the junction of the Kansas and Missouri Rivers, just beyond the western Missouri state border, and the sum of $14,000. The Delaware had earlier been awarded land in southwestern Missouri, and then later in Indian Territory. They formally surrendered their share of the title to the Apple Creek tract in 1829. Because questions lingered over who owned the cabins and other improvements on the Apple Creek tract, representatives of both Indian nations signed the Treaty of Castor Hill in St. Louis County in 1832 by which the Indians relinquished all claims to their nations' lands and improvements in the Apple Creek tract.[35]

Thus, in a protracted series of official acts, the settlement of eastern Indians in the Ste. Genevieve District, which began with a Spanish invitation to live there on a large land grant that was subsequently recognized by the U.S. government and therefore treated as valid a land title as any could be, ended with their removal and resettlement onto an American-style reservation. By the mid-1820s, the Indian villages once populated by as many as 1,500 Indians were completely gone. When Timothy Flint

---

ple, *Indian Villages*, 54–56. Maj. Stephen Long reported in 1820 one remaining village of four hundred Shawnee and Delaware, apparently on Apple Creek, and that the government was trying to move them west (Temple, *Indian Villages*, 181). Also see the special census of Indians in "Indian Tribes in Missouri Territory," August 24, 1817, *TP* 15:305.

34. *Jackson Independent Patriot,* June 12, 1824.

35. On 1825 treaty: Richard Peters, ed., *Treaties between the United States and the Indian Tribes,* 284–86; Kappler, *Indian Affairs,* 262–64; and *Jackson Independent Patriot,* March 4, 1826; on 1829 surrender and 1832 treaty: Kappler, *Indian Affairs,* 303–5, 370–72. Castor Hill was the name for General Clark's prairie land northwest of St. Louis city, now along Natural Bridge Road where it passes from St. Louis city into Pine Lawn; Council Grove is the name of one of the neighborhood streets. The conclave was a large, long, and momentous meeting involving the Shawnee, Delaware, various Illinois tribes, and others.

passed through the villages, he described "every house deserted, the deer browsed upon their fields, and the red-bird perched upon their shrubs and flowers." It was a "desolate contrast" to what he had seen but a few months before: "villages . . . all bustle and life."[36]

Bois Brule

Among the first Americans to cross the river in 1796 as a direct result of the liberalization of Spanish immigration policy were members of the Burns family. Two members, Michael and Barnabas, had earlier taken the oath of allegiance in July 1795 at New Madrid, but the clan came as a group to the Bois Brule bottom in 1796 and 1797. Among them were Michael Burns, his three sons, eight sons-in-law, and their families. Patriarch Michael was Irish born and listed as Catholic in the 1797 New Bourbon census, but he more likely was Scotch-Irish and only nominally Catholic in order to enter Spanish Illinois in 1795. The others were born in Virginia, Pennsylvania, Kentucky, and Illinois. As the clan westered, they probably heard about the advantages of Spanish Illinois from the Luzières advertisement. They received individual concessions in the middle Bois Brule bottom totaling some four thousand arpents (see fig. 38). Most of the concessions were in the form of French-style long lots, like Barnabas Burns's 1797 concession of six arpents frontage by forty arpents depth, but they changed in size and shape through the confirmation process.[37]

Part hunters and part farmers, the Burnses fit the general notion of the first wave of Americans into a region. Although identified and held together by kinship ties, the clan lived on dispersed improvements, apparently had no slaves, and had no church or other kind of community institution. They probably arrived with few dollars and material belongings, and their economic system did not lead to wealth accumulation. As subjects of the Spanish regime, thirteen of them dutifully contributed money in 1799 to support an expected war, albeit meagerly: their average contribution per family was only a few piastres, contrasted to ten or more piastres from other settlers. Their family names, even allowing for difficulty in Span-

---

36. Flint, *Recollections of the Last Ten Years*, 149–50.
37. Timothy J. O'Rourke, *Perry County, Missouri: Religious Haven in the Trans-Mississippi West*, 3; *History of Southeast Missouri: Embracing an Historical Account*, 253. On 1797 census: PC 2365-345; LCC 135; and LCS 32; on advertisement: request for land by Jacques Dodson, April 15, 1798, LCS 168; on individual concessions: O'Rourke, *Perry County*, 3; on Barnabas Burns concession: request for land by Barn Burns, March 4, 1797, LCS 152; and *Régistre d'Arpentage*, 196–98. Barnabas Burns also appears in records as Barn Burns.

ish spelling, appeared in the records in numerous ways: for example, McConochie, McConnachen, and McConohue; and Murdock, Mordough, Moredock, and Mordock. They did not have enough money to pay surveyor fees to Maddin in 1799, even when Delassus ordered them to pay, and everything tangible they had—their oxen, goods, and improved properties—was sold in public auction.[38] Interested in living as self-reliant, independent, free persons on the land, they chose to live apart from the public life of the district, dominated as it was at the turn of the century by Creoles and the French.

They continued their restlessness as settlers. Most did not stay in the Bois Brule bottom. During the period from 1804 to 1820, members of this clan sold their Spanish concessions, their settlement rights, or their improvements, and left the neighborhood for the mining country and virgin lands in the Ozarks. Thomas Allen, who had been convicted of the rape of a twelve-year-old girl in 1800, fled the neighborhood. Francis Clark Sr., who was from Ireland, was murdered, and his son Francis Clark Jr. went to live on the St. Francis River and was joined by relatives.[39] The Burnses were sellers and out-migrants, as time went on. They did not build a community, yet they left a legacy. By clearing land and building cabins, they added value to land, and it could be sold to later arrivers for profit. As chronic land clearers, their future lay in unoccupied lands farther west, not in the Bois Brule.

Fenwick Settlement

Not all of the Americans arriving after 1796 were Protestant. A group of Kentucky Catholics also responded to the opening up of Spanish territory and received permission to settle in the closing years of the eighteenth century. Known as Maryland Catholics or English Catholics to distinguish them from the resident French-speaking Catholics, they consisted of two subgroups: the Fenwick Settlement at the mouth of Brazeau Creek and the Tucker Settlement in the Barrens.

In the mid-1700s, Catholics in Maryland were unhappy with their lot and sought land elsewhere for a better life. Among the places was Catholic

38. Letter from the [governor] to Carlos de Lassus [sic], August 28, 1799, PC 134a-400; "Patriotic Donations and Loans Made by the Residents of Upper Louisiana to Aid Spain in the War—1799," in *Spanish Regime,* ed. Houck, 2:292–94; LPS 16.

39. On Allen: Ekberg, *Colonial Ste. Genevieve,* 374. Ekberg speculates that Allen fled to the American side of the Mississippi River to escape punishment, but if he did, he returned later and had his land claim approved (*ASP-PL* 2:591). He also testified before the second board in 1833 about events in the Bois Brule during 1799–1803 (*ASP-PL* 8:66). On Clarks: Houck, *History of Missouri,* 1:381–82; and *History of Southeast Missouri: Embracing an Historical Account,* 628.

French Louisiana, which had been made known to them by the passage of Acadians through Maryland on their way to Louisiana. Nothing came from this colonization idea, but correspondence resumed in 1767 with the new Spanish governor of Louisiana, Antonio de Ulloa. However, conditions for Catholics in the Chesapeake eased, and emigration fever subsided.[40]

In the meantime, several families of Maryland Catholics had gone to Kentucky after the Revolution, about 1785, and their presence there came to the attention of Spanish authorities in Upper Louisiana. As Catholics, they fitted well the description of prospective immigrants. Joseph Fenwick, head of one of these Kentucky families who had investigated land at New Bourbon, received an invitation in 1797 from New Bourbon District commandant Luzières to bring with him his doctor son, Walter Fenwick. Ste. Genevieve and New Bourbon were without the services of either a doctor or a surgeon. Joseph Fenwick responded and received a generous floating concession for 6,000 arpents, plus an additional 20,000 arpents for fellow Catholics whom he could persuade to come with him. This was a large amount of land for Americans (a total of 34.6 square miles), and by far the largest land award to any newcomer up to that time, except for those to the Luzières family itself. Apparently, the bait to come live under an authoritarian regime of coreligionists was stronger than Fenwick's desire to continue to live in a republic. Luzières proudly wrote to Carondelet that Joseph Fenwick had arrived in Spanish Illinois on April 18, 1797, with twenty other families with seventy slaves to come shortly. Indeed, before 1803, twenty-five or more mostly slave-owning Catholic families from the White Sulphur area of Kentucky had migrated.[41]

Joseph Fenwick lived first at Luzières's New Bourbon village, buying Israel Dodge's "great house" and three pieces of farmland. The 1797 New Bourbon census listed him as the largest slaveholder in the village, with sixteen slaves. But Fenwick got into controversy with Dodge over the purchased land, and this legal issue must have discouraged the newcomer

---

40. James A. Robertson, "A Projected Settlement of English-Speaking Catholics from Maryland in Spanish Louisiana, 1767, 1768"; several docs., Kinnaird, *Spain in the Mississippi Valley,* 2:36–43; O'Rourke, *Perry County,* 12–13.

41. O'Rourke, *Maryland Catholics,* 2, 5; PC 204-699, 213-905, 213-906; Ekberg noted the need for a physician at Ste. Genevieve (*Colonial Ste. Genevieve,* 263). On floating concession: *ASP-PL* 2:518, 600, 8:118–21, 229–30, 137; LCS 185; LRA 34; and PC 34-717; on Luzières: PC 213-903, 906; on White Sulphur area families: O'Rourke, *Maryland Catholics,* 588–89; and O'Rourke, *Perry County,* 16.

from staying around.⁴² Fenwick had already used his floating concession to select a tract of 3,000 arpents on Apple Creek, 30 miles southeast of New Bourbon village.⁴³ This was excellent loessial and limestone land for full-time farmer Fenwick and his slaves, away from any Creole village. He began building on his concession on Apple Creek at the mouth of Indian Creek in 1797, close to the villages that the Shawnee were erecting at the same time. The presence of so many Indians probably caused Fenwick to give up plans to live at Apple Creek, and he shifted eastward to the Mississippi River, 3 miles above the mouth of Brazeau Creek, still on the Indian grant but 10 miles removed from the Indian villages. As the years went on, the river location became quite famous. After Fenwick's death, his widow, Chloe, and slaves turned the place into a well-known stop for river travelers. River men prominently noted it, and Bishop DuBourg and his party stopped by this Catholic woman's house on Christmas Day 1817 en route upriver to assume the diocesan leadership at St. Louis. Curiously, the American board confirmed Joseph Fenwick's large concession in the midst of the Indian villages at Apple Creek, but not his farm and residence on the Mississippi; apparently, his widow filed no settlement-rights claim for it. Joseph's son Ezekiel also established himself on the Mississippi River, south of his parents at the mouth of Brazeau Creek, where he maintained a "house of entertainment," or facilities for river travelers.⁴⁴

Another son, Thomas Fenwick, received a concession on June 1, 1797, for several thousand arpents (reduced by the American board to 640 acres and split into two tracts when surveyed) on Prairie Spring Creek on the karst plain upland between Apple and Cinq Hommes Creeks, also within the Indian tract, but he did not live there. He was a merchant at La Saline in 1799 and in 1807 an attorney at Mine à Breton. In 1811, he also bought land on the Mississippi just downriver from his mother's place and near his brother Ezekiel. Thomas Fenwick sold his Prairie Spring

---

42. PC 218-516, doc. L; PC 2365-345. The extended Dodge-Fenwick property issue at New Bourbon may be traced in PC 215b-635-77.

43. O'Rourke, *Maryland Catholics*, 590; Houck, *History of Missouri*, 2:317. The Irish-born priest at Ste. Genevieve, James Maxwell, had outspoken contempt for the Maryland (or English) Catholics, which may have been a factor in Joseph Tucker's departure from New Bourbon, where Maxwell lived ("Two Maxwell Letters").

44. On Fenwick building and moving: PC 215b-669 and O'Rourke, *Maryland Catholics*, 589; on DuBourg's stop: Houck, *History of Missouri*, 2:317; and Z. Cramer, *Navigator*, 136; on nonconfirmation of farm and residence: PC 215b-669 and O'Rourke, *Maryland Catholics*, 590; on Ezekiel: *ASP-PL* 8:119; LRB 250, 506; and Houck, *History of Missouri*, 3:61.

concession in 1813 to Isidore Moore, perhaps the only American who went to the trouble to obtain written permission from the Shawnee to settle within their tract.[45]

Only the doctor son, Walter Fenwick, stayed a resident in New Bourbon, where he fulfilled Luzières's expectations by serving the residents of the village and of Ste. Genevieve; in fact, a high flood during the year of his arrival, 1797, brought him considerable business, with fevers and other ailments. Walter Fenwick also invested heavily in land there and elsewhere. With lead businessman Andrew Henry, he bought in 1806 the small but important original mine at Mine à Breton awarded to discoverer François Azor dit Breton. He also bought land on the Mississippi in 1809, called the "Grand Tower tract," near his mother's and brothers' places, but he did not live there. As a professional, he incorporated himself quickly and smoothly into the society and business affairs of Creole New Bourbon–Ste. Genevieve. He married Julie Vallé, the commandant's daughter. Clearly the most prominent and successful one of the family, the doctor unfortunately lost his life in a duel in 1811 while defending the honor of his brother Ezekiel.[46]

Other Kentucky Catholic immigrants of the Fenwick party and some Revolutionary War veterans also settled along the Brazeau and at its mouth on the Mississippi. By 1803, the Brazeau (or Obrazo) was known as the Fenwick Settlement; it had twenty families in 1809 (see fig. 38). This land was an unusual choice for settlement in several ways. First, although located near the Mississippi River and highly accessible in that important regard, it was far removed from population and administrative centers, being thirty-five miles below Ste. Genevieve and twenty-five miles above Lorimier's post at Cape Girardeau. Did Joseph Fenwick intend to distance his group from Creole society and administrators? Sec-

---

45. On Thomas's concession: *ASP-PL* 6:605. The site was an ideal location combining the natural grass of prairie, timber, sinkhole ponds, and a never-failing spring. Fenwick was unable to settle on it, however, because of the usufructuary claim by the Shawnee who were granted it two weeks later, and he turned to other occupations (*ASP-PL* 5:202–3). After the U.S. government later surveyed it and put it on sale and claimant Isidore Moore purchased it by preemption, the U.S. government confirmed the Spanish concession to him in 1830, and Moore then asked for compensation for having purchased the confirmed Spanish concession that he inhabited (*ASP-PL* 6:605). On Thomas as merchant and attorney: Houck, *History of Missouri*, 1:367; and LRA 99; on 1811 purchase: LRB 395; on 1813 sale: *ASP-PL* 5:202–3; and PC 215b-669.

46. On Walter as physician: PC 204-699 and 204-717; on Walter and Henry: *ASP-PL* 3:590; LR 14; and William Mathers to Bates, January 29, 1808, *Bates Papers*, 1:270–71; on "Grand Tower tract": LRB 19; on marriage: Ekberg, *Colonial Ste. Genevieve*, 263; on death: Houck, *History of Missouri*, 3:76.

ond, the land was in the process of being set aside for the Shawnee and Delaware. What was Luzières thinking when he permitted the Fenwicks, with multithousand-arpent land concessions, to settle on land that his superior had already set aside for Indian villages and hunting territory? Two other smaller concessions along Apple Creek had been given to Pierre Menard and François Berthiaume, but Menard was an Indian trader, and Berthiaume was a gunsmith for the Shawnee and had an Indian wife.[47] The Fenwick group had no relations with Indians. The only explanation for the location is that the Fenwicks could move anywhere they wanted with their floating concessions and did not need permission for a specific site.

The site was unusual also because the land on the Mississippi and Brazeau is poorly endowed for agriculture. It was an incongruous choice for former residents of smooth tidewater St. Mary's County, Maryland, and even the White Sulphur area of Kentucky. The Mississippi River washes the base of rocky cliffs in this reach, and, except for Brazeau and Apple Creeks, the tiny creeks that break the bluff line have too narrow a valley for any but the smallest-size corn patches. The bluff lands, though capped with loess, are among the steepest and most rugged along the Mississippi between Ste. Genevieve and Cape Girardeau.

Members of the Fenwick Settlement built cabins and cleared and fenced land, but the settlement did not prosper and eventually disintegrated. No subsequent Kentucky Catholic immigration reinforced the group, and many of the Fenwick group drifted away beginning in 1807–1808.[48] Perhaps they heeded calls to get off the Indian grant, or perhaps they were uncomfortable having a thousand Indians, no matter how Americanized, as neighbors. Perhaps they tired of being so distant from a Catholic church, mills, stores, and other necessities of life. But whatever the reasons, land unsuitable for farming had to be one.

47. LCC 55; Rev. John Rothensteiner, "Father Charles Nerinckx and His Relations to the Diocese of St. Louis," 157–75; O'Rourke, *Maryland Catholics,* 588; Houck, *History of Missouri,* 1:387. Houck credits George Morgan for bringing many of his Revolutionary War companions to Spanish Louisiana and lists veterans by county. Among them are the Harrisons and Hinkstons on Brazeau Creek in the Fenwick Settlement (*History of Missouri,* 3:83–86). When English traveler Thomas Ashe passed by the mouth of "Happle Creek" in 1806, he said that the "people live in the manner of Indians, that is by hunting, and in bartering. . . . They cultivate very little ground." It is unclear whether he was referring to the Shawnee on Apple Creek, who were usually referred to as Indians, not people; to the Fenwick colony between the mouths of Apple and Brazeau Creeks; or to some other settlers (*Travels in America,* 288).

The second board confirmed both Berthiaume's and Menard's concessions in November 1833. *ASP-PL* 5:729, 801; LRA 86; Houck, *History of Missouri,* 2:191, 232.

48. LRA 135; LRB 87; *Bates Papers,* 1:285–93; and O'Rourke, *Perry County,* 17.

When it came time for confirming land claims in the Fenwick Settlement, something strange happened. The second board, which was extremely liberal in confirming lands still unresolved in the 1830s, rejected several of the Fenwick Settlement claims because of alterations on the original concessions. "Brazeau" had been obliterated and in its place was "à la Viande," a local name for Cinq Hommes Creek. The board concluded that the claimants were attempting fraudulently to shift the location of their lands from the hilly, nonarable Brazeau to the much better lands of the karst plain, outside the Indian grant.[49]

Protestant Americans from Appalachia with farming systems more adapted to rough hills moved into the Brazeau Valley after the War of 1812 and bought up the Catholics' settlement rights and Spanish concessions. Thus, despite the financial resources of the slave-owning families and a working network of family and friends, and despite initial personal help from and connections with the commandants of both New Bourbon and Ste. Genevieve, the Fenwick Settlement strategy was unsuccessful and the community disappeared.

The Barrens and Perryville

Like the Fenwick Settlement, the Tucker Settlement of the Barrens was primarily the responsibility of a single person who was the leader of a large group of extended family and friends. But in contrast to the Fenwick Settlement, the Tucker Settlement prospered and expanded. Both were composed of English-speaking Catholics who had westered from Maryland through Kentucky and arrived in the Ste. Genevieve District during the last six years of the Spanish regime. Both had organization and adequate financial resources. What made the difference? Explanations are found in the different physical environments they selected, the contrasting goals of leaders, demographic developments, the role of the Catholic Church in community cohesion, and differences in commitment to agriculture.

Isidore Moore was the person most responsible for establishing the Tucker Settlement, or the Barrens, although the more numerous Tucker family gave it its first name. Moore's and other Catholic families from Maryland had earlier settled in Marion, Nelson, and Washington Counties in Kentucky in the 1780s. Learning about Spanish Upper Louisiana and perhaps dissatisfied with his new Kentucky home, twenty-year-old Isidore Moore scouted both sides of the Illinois Country in 1792, at which time his visit to Kaskaskia discouraged him, and then again in 1797, when

---

49. *ASP-PL* 8:119–20, 232–33.

he visited the St. Louis region and the unoccupied grasslands of the karst plain south of Ste. Genevieve. He paid particular attention to the American Bottoms between Kaskaskia and Cahokia, and lived there for a while. However, slaveholder Moore and his family, like the Fenwicks, could not accept residency on the east side of the Mississippi, where slavery had just been prohibited. The American Bottoms were even less inviting after disastrous Mississippi River floods of 1785, 1788, and 1794; stagnant ponds and lagoons in the bottoms and associated fevers dissuaded Moore. On yet a third scouting trip in 1800, he again explored the grasslands, stopping to visit with his Kentucky coreligionist Joseph Fenwick on his three-year-old farm. Moore liked what he saw and chose the grassy barrens. He received a concession on February 9, 1801, and returned to Kentucky for his family and other families already closely interrelated by marriage: the Tuckers, Haydens, Cissells, Hagans, Laytons, Brewers, and others.[50] Some of these families arrived early enough to receive Spanish concessions. Those who arrived during the winter of 1803–1804 and built cabins established settlement rights (see fig. 38).

Isidore Moore's choice of location for his concession and for his extended group is instructive for understanding how settlement decisions that affect numerous families can be made by a single individual for personal and exceptional reasons. During Moore's visit with Joseph Fenwick in 1800 at Fenwick's home on the Mississippi, Moore realized that he had major differences of opinion with his fellow Catholic. In an 1837 letter to Bishop Rosati at St. Louis in which he recalled his past, Moore wrote that "a difference of opinion in worldly matters, and some other circumstances, prevented our wishing to be very close to each other." Therefore, these two leaders had to distance themselves from each other, and Moore,

---

50. The name Barrens is indigenous to Missouri and was not transferred from the Barrens of Kentucky (O'Rourke, *Maryland Catholics*, 2). The term *Barrens*, referring to the community as distinct from the open landscape, was used as early as 1805 by surveyor John Hawkins (*Régistre d'Arpentage*, 638).

Benedict Joseph Webb has accounts of the Tucker and other Kentucky Catholic families in *The Centenary of Catholicity in Kentucky*. Two different Tucker families, apparently unrelated until after arrival, settled in the Barrens. The "Short Tuckers," led by Josephus Tucker, came from Georgia (earlier from Virginia). The "Long Tuckers," led by Joseph Tucker, were part of the St. Mary's, Maryland, contingent. "Long" and "Short" derived from the heights of family members (*History of Southeast Missouri: Embracing an Historical Account*, 253).

On Moores: O'Rourke, *Maryland Catholics*, 48–50; O'Rourke, *Perry County*, 17–18. Moore's concession and subsequent land dealings are summarized in O'Rourke, *Perry County*, 35–36, 40–41. On other families: O'Rourke, *Maryland Catholics*; *History of Southeast Missouri: Embracing an Historical Account*, 688, 693, 710, 733; Larry Hoehn, *The Genealogist and Historian's Guide to Perry County, Missouri*.

the later arriver, chose a location in the grassy barrens. Moore's choice was on Cinq Hommes Creek, on the northern boundary of the Shawnee-Delaware tract, which was well populated with Indians by that time. All of the accompanying families also selected sites in the barrens near him. One might also like to think that Moore recognized that the Fenwick Settlement was in an environment difficult for agriculture and thereby avoided the Fenwicks' error. The group was used to clear-field agriculture as practiced in Maryland, not the patch cropping of hill people, and their short stay in the bluegrass region of Kentucky had not transformed them into upland southerners.[51]

Choice of the barrens at the headwaters of Cinq Hommes Creek was an excellent one for clear-field agriculture. Big bluestem growing on rich loessial soils blanketed the surface.[52] Farms could be started immediately with no necessity to wait for time-consuming tree felling and stump removal. But it was not pure prairie. Adequate timber was available in every shallow valley and in karst depressions. Numerous limestone springs provided a water supply entirely adequate for household purposes. Settlers cut prairie hay in abundance for winter provisioning, even in the initial year; there was no need to drive horses and cattle to cane breaks along the Mississippi. Probably no other natural setting in the entire Ste. Genevieve District was more conducive to rapid agricultural development. The barrens spread over a couple hundred square miles and provided abundant land for expansion of the community; there was no need to string out cramped farms for miles along a narrow valley.

By 1804, settlement in the barrens was compact and contiguous over many square miles (see fig. 5). Because so much of the karst plain was usable, little land was left unclaimed. Neighbors could readily see each

---

51. Moore letter: O'Rourke, *Maryland Catholics,* 590. Father Charles Nerinckx of Kentucky described the two separate settlements of Kentucky Catholics during a visit in 1810. Tucker's Settlement had about sixty families and the Fenwick Settlement about twenty (Rothensteiner, "Father Charles Nerinckx," 157–75).

52. Festervand, *Soil Survey of Perry County,* 13–21; W. A. Schroeder has mapped the extent of the prairie of Perry County from information in the field notes of the U.S. General Land Office rectangular survey in 1817–1818 (*Presettlement Prairie of Missouri,* 13–14). For the Kentucky Barrens as also a desirable environment for settlement, see John Opie, "Frontier History in Environmental Perspective."

In clear-field agriculture, farmers do the extra labor to remove stumps from fields, keep them reasonably weed free, and extend the cropland to property lines, where feasible. The result is a visual neatness to the farm landscape and, on a larger scale, a slower westward movement of the settlement frontier. Upland southerners usually did not go to such lengths, leaving their fields with stumps and a "messy" appearance, which resulted in a faster westward movement of the frontier (Jordan and Kaups, *American Backwoods Frontier,* 100).

other's houses and fireplace smoke across the plain. Seeing one's neighbors, as opposed to being walled in by trees and hills on individual clearings, helped further the sense of community.

The Barrens Settlement, as it was known, already with enough population to have its own constable in the first year of American rule, grew and prospered in succeeding years, despite a slack in immigration attributed to the uncertainty of land titles until the War of 1812 was over. Families were substantially larger than the average even for pioneering families. Leader Isidore Moore had nineteen children. Of Michael Tucker's sixteen children, fifteen grew to adulthood. Joseph Manning had twelve children; Clement Hayden, ten children; Joseph Tucker, nine sons; James Tucker, who married his second cousin, ten children. To a large degree, these progeny found spouses within the settlement group, which meant, in some cases, marrying second and third cousins or cousins once removed, because the families had already intermarried before arrival in Missouri. Three of Michael Tucker's children married Joseph Manning's children. Four of Isidore Moore's children married Cissell children, and three other Moore children married Manning children.[53] In those few cases in which Barrens settlers married outside the settlement, they more often than not married other English Catholics, French Catholics, or converted Catholics. Despite the personal differences between group leaders Isidore Moore and Joseph Fenwick, marriage tended to fuse the two settlements, with the Barrens being the clear choice of residence of new couples.

Both newcomers and the numerous children found land for farms within the settlement. Some found unoccupied public domain, but most of the new-farm expansion was accommodated by subdividing the large Spanish concessions and settlement rights. Because a high fraction of any tract in the karst plain was arable, economic impoverishment of families did not take place with land division even into fifths, which resulted in still sizable farms of two hundred acres. Thus, John Layton Sr. carved out five farms from his concession for his children and the orphaned children of his first wife by her first marriage, who lived with him, when the children married and began farming on their own.[54] It was the prac-

---

53. Houck, *History of Missouri,* 2:384. A separate constable, Thomas Donahoe, patrolled the Bois Brule settlement. On slack in immigration: O'Rourke, *Perry County,* 6; on families: O'Rourke, *Maryland Catholics,* 45; O'Rourke, *Perry County,* 40, 46–47; and *History of Southeast Missouri: Embracing an Historical Account,* 253.

54. LRC 66, 66a. Whether land is passed on to children at the time of their marriage and getting started in farming or held by the parents until their death has consequences for whether the children remain in the community (usually the result in

tice in the Barrens community to give children a share of the property when they began their own households rather than wait until the father's death and subsequent subdivision of his estate. Similarly, Joseph Tucker, while living, provided room and "good homes" on his land for seven sons and a daughter, all married. Even when property went to children upon their father's death, wills were written to ensure their continued residence in the community. For example, Isidore Moore's will stipulated that his unmarried son Lewis (the younger of Moore's nineteen children were not yet at marrying age when their father was aging) would receive his inheritance "on condition that he reside at home with us while single, and on the farm near us when married . . . to see that his mother and me are not misused by the slaves." Such a practice of land division encouraged the younger generation to remain and raise their own families in the neighborhood and led to the development of a stable and culturally homogeneous community. Tax rolls for the first years of the 1820s indicate at least eighty Catholic families lived in the Barrens community, or a population of approximately eight hundred, including slaves.[55]

Of paramount importance to the distinctiveness of the Barrens community was the farmers' general disinterest in the lead mines as a place of quick wealth. Apparently, not a single Barrens resident moved to the mining communities, although Isidore Moore did form a partnership with Andrew Henry to invest in lead mining. Neither did these agriculturists get involved in fur or river trade or in salt making. Their commitment to agriculture with slaves was near total.

Above all, it was the Catholicism of these "pious planters, so simple and so free from malice," as noted by Father Marie Joseph Durand, a Kentucky priest who visited the Barrens in 1809, that gave them their greatest distinction. Treated as a minority in Maryland and later again in Kentucky, they had developed a strong sense of community from their

---

the former case) or move out of it (usually the result in the latter). See Robert C. Ostergren, "Land and Family in Rural Immigrant Communities." Despite the remarkable persistence of families in the Barrens, some progeny pressed on to Texas in the 1830s with other Missourians in the wake of Moses Austin. Isidore Moore had served in the Missouri legislature with Moses Austin (O'Rourke, *Maryland Catholics,* 1–9).

55. LRB 325–31; LRC 80, 80a; Rothensteiner, *Archdiocese of St. Louis,* 1:294; O'Rourke, *Maryland Catholics,* 839–40; Stafford Poole and Douglas J. Slawson, *Church and Slave in Perry County, Missouri, 1818–1865,* 3–4. In a detailed study of colonial North Carolina in the 1720s that used multiple regression analysis, Robert E. Gallman showed that community stability (similar to that of the Barrens) is associated with economic prosperity and the ability of households to participate broadly in the market and associated much less with the rules of land disposal ("Influences on the Distribution of Landholdings in Early Colonial North Carolina").

Catholic religion on a Protestant frontier. It is highly unlikely that Isidore Moore would have brought his group to the Barrens but for the Catholic Spanish administration in Upper Louisiana. Their zealous practice of Catholicism contrasted with the adjacent Creoles' much less spirited practice in Ste. Genevieve and New Bourbon. "These people are the most devout of any in the United States," commented Father Joseph Rosati, later to become bishop at St. Louis. The Reverend James Maxwell initially served these Anglo Catholics from Ste. Genevieve and in 1812 blessed the Barrens log church, the first Catholic church erected in the entire district beyond the village of Ste. Genevieve.[56]

The best testimony to religious devotion was the community's successful effort to establish St. Mary's of the Barrens, the first seminary west of the Mississippi. That this seminary, which had a significant role in the early development of Catholicism in the St. Louis Archdiocese, was established in the agricultural Barrens and not at Ste. Genevieve or even St. Louis speaks to the confidence of the church hierarchy in the future of this rural community, its economic prosperity, its Catholic faith, and its discipline in accomplishing results.[57]

The seminary resulted directly from the initiatives of the families of the Barrens themselves. William Louis DuBourg, the new bishop for Louisiana after it became American, introduced the Vincentian order to the territory when he arrived at New Orleans in 1812. In 1815, Bishop DuBourg and his assistant, Father Felix de Andreis, received permission from Rome to establish a seminary of the Order of St. Vincent in Upper Louisiana. A logical choice for its location would have been the administrative center of St. Louis, the largest town and clearly the one with the most potential for growth. Yet, the French Creole Catholics of St. Louis and Ste. Genevieve, according to Bishop Benoit Joseph Flaget of Kentucky who had come to the new territory to help in its Catholic organization, initially did not take any more interest in Bishop DuBourg "than about that of the Emperor of China." Bishop Flaget, incidentally, was guided around the territory by "Mr. Tucker" from the Barrens. When residents of Missouri Territory learned that a seminary was to be built in Louisiana, a deputation from the Barrens seized the opportunity and went to St. Louis in 1817 to approach the new bishop and persuade him to locate it at the

---

56. "Epistle or Diary of the Reverend Father Marie Joseph Durand," 45; Janet, "St. Mary's," 21; Rothensteiner, "Father James Maxwell," 146.

57. Accounts of the origin and early years of St. Mary's of the Barrens Seminary are in Janet, "St. Mary's"; O'Rourke, *Perry County,* 272–33; Poole and Slawson, *Church and Slave,* 141–89, esp. on the ownership and use of slaves at the seminary; and Rothensteiner, *Archdiocese of St. Louis,* 1:292–99.

Barrens. On a second trip in 1818, they formally offered DuBourg a 640-acre tract, an act that convinced the bishop to visit the site. He liked it and also the ardor and energy of the people—he called them "honest and industrious: the best set I ever knew"—and resolved to construct his seminary there. The Barrens Catholics agreed to share the construction work and to raise seventy-five hundred dollars over the next five years for maintenance of clergy, impressive undertakings for the farming community. The numerous slaves of the community, called "servants" in church records, helped in the construction, begun in 1818.[58] Losing no time, seminarians began arriving the same year, lodging in residents' homes until buildings were finished.

Presence of a resident priest at the Barrens seminary encouraged a renewal of Catholic migration from Kentucky to the Barrens, beginning in 1818. In 1822, a lay "college" for boys was added and the next year a girls' school started. In 1823, sixty-six persons were studying at the seminary, and in 1827 more than one hundred lay students were enrolled, including some sons of Protestants, in addition to the seminarians. Pupils at the seminary and lay college came from great distances, as far as the state of Louisiana, and paid tuition in the form of sugar, molasses, coffee, or cotton. Seminary fathers exchanged these imported items for butter, eggs, lard, bacon, and grain of local farmers.[59] This cashless commercial system, common on the frontier, soon turned into a true store at the seminary. Thus, the seminary and lay college were not just educational institutions, but also, in the dearth of merchants and other stores between Ste. Genevieve and Jackson, became a commercial center and the true central place of the Barrens, before the county seat of Perryville functioned as one.

By 1821, the population of the Barrens and surrounding neighborhoods warranted separation of the southeastern portion of Ste. Genevieve County into a new county, Perry County. Because of its centrality, population, and location on the historic road between Ste. Genevieve and Cape Girardeau, the prosperous Barrens became the obvious location for the

---

58. On Flaget: Houck, *History of Missouri,* 2:316; on offer to DuBourg: Janet, "St. Mary's," 19–20; and Houck, *History of Missouri,* 2:323–24; quote: Rothensteiner, *Archdiocese of St. Louis,* 1:294; on money agreement: "The Centenary of the Foundation of the St. Louis Diocesan Seminary," 47–48; on slaves: O'Rourke, *Perry County,* 19.

59. On renewal of Catholic migration: O'Rourke, *Perry County,* 7; on girls' school: Janet, "St. Mary's," 23. For example, Reuben Smith of Washington County sent an orphan girl whom he was raising to the Catholic School in the Barrens in 1828 (Eunice Bernard, *Wills and Administrations, Washington County, Missouri, 1814–1870,* 47); on enrollment: S. P. Hueber, ed., *Centennial History of Perry County, Missouri, 1821–1921;* on barter system: *Missouri Republican,* October 14, 1828; and Hueber, *Centennial History.*

county-seat town. Bernard Layton donated fifty-one acres of his land with a good spring, one and one-half miles east of the seminary for the new town, Perryville, platted in 1822 using the Shelbyville square design.[60] Initial sale of lots was brisk, and, as at Farmington and Caledonia, they were mostly bought by local landowners for speculation. The town existed for a few years in name only, then grew only slowly because the store at the nearby seminary continued to serve as the commercial-exchange place for the area. Because the community was fully engaged in building the seminary and church, in both labor and money, the courthouse had lower priority and was not built until 1825–1826. The delay in building a courthouse was also attributed to Perry Countians' continuing obligation to help build a courthouse at Ste. Genevieve, as they had been a part of that county. When built, the Perry County Courthouse had a cross above the doorway, bearing witness, Bishop Rosati admiringly noted, that the residents "glory in being Catholics." The Barrens community of 160 families in 1823 was the demographic, economic, and geographic core of the new county.[61]

We cannot conclude the story of the initial years of the Barrens without returning to leader Isidore Moore. Moore's achievements went far beyond the Barrens. He was elected in 1815 to the Territorial General Assembly from Cinq Hommes Township of Ste. Genevieve County and continued to serve in it until statehood, and then was elected to the first state legislature as the representative from Perry County and served there through 1826. He thus was in influential positions to promote the interests of the Barrens. Moore was a major land speculator, first by dealing in New Madrid earthquake certificates and Spanish land claims and later by purchasing land from the General Land Office. His interests remained fundamentally with the welfare of the people in the Barrens: according to his will, this founder and patriarch of the cohesive Barrens community would allow the sale of his slaves only to his fellow Roman Catholics residing in Perry County.[62]

---

60. Douglas, *History of Southeast Missouri: A Narrative Account*, 270; O'Rourke, *Perry County*, 23; Ohman, *Counties, County Seats, and Courthouse Squares*, 33; Ohman, *Encyclopedia of Missouri Courthouses*, 157–58.

61. Hueber, *Centennial History*. Surprisingly, no courthouse was erected in Ste. Genevieve until 1821. Court sessions took place in homes or the parish house (Steward, *Frontier Swashbuckler*, 124). Quote: Rothensteiner, *Archdiocese of St. Louis*, 1:299. The Roman Catholic influence in Perryville is also recorded in the original street names of St. Mary, St. Joseph, St. Francois, St. Augustine, St. [sic] Genevieve, and St. Valentine.

62. On political life: O'Rourke, *Perry County*, 34–42; on other interests: Hueber, *Centennial History*.

Brazeau and Abernathy Settlements

Among the reasons for the decline in the rate of immigration between 1805 and 1815 into the southeastern part of the Ste. Genevieve District were reports, perhaps exaggerated, of the effects of the New Madrid earthquakes, national events like the War of 1812, and the confusion over land titles and inability to obtain clear title to new land. Following the War of 1812, however, the rate of immigration increased significantly as it did elsewhere in the Missouri Territory. In addition to the renewed Kentucky Catholic immigration to the Barrens, two more groups arrived: the Brazeau Presbyterians and the Longtown Methodists or Abernathy Settlement.

Beginning in 1817 and continuing for several years, interrelated families from Rowan, Iredell, Cabarras, and Mecklenburg Counties in North Carolina arrived and opened land in the Brazeau Creek drainage basin, just northwest of the fading Fenwick Settlement (see fig. 38). They were part of the large exodus from the Piedmont of North Carolina following the War of 1812. They may have heard about the eastern Ozarks from German Lutherans from adjacent Lincoln County, North Carolina, who had settled in the Cape Girardeau District some fifteen years earlier and prospered. Of Presbyterian Scotch-Irish ancestry, the new settlers quickly organized a church in 1819 in the Brazeau area.[63] They fitted broadly into the pattern of settlement of much of the eastern Ozarks by kinship groups from Tennessee and North Carolina.

At the same time that the Brazeau neighborhood was being formed, other Americans of Scotch-Irish ancestry arrived from Lincoln County in the North Carolina Piedmont. The first arrivals occupied land in 1818 south of Apple Creek at Pocahontas, then in 1820 more settled in the productive karst plain south of the Barrens Settlement and filed for their land as soon as they could after the Jackson land office opened (see fig. 38). Both locations lay within the still active Indian grant. Initially called the Abernathy Settlement after its largest family, the settlement was later identified as Longtown and York Chapel. As a group of interrelated families, they distinguished themselves from their fellow North Carolinians on the Brazeau and others primarily by their Methodist religion, which provided a strong community focus, spiritual and social.[64] Both the Brazeau and Longtown communities had slaves.

---

63. *History of Southeast Missouri: Embracing an Historical Account*, 731–32, 739; *A History of the Brazeau Presbyterian Church, Brazeau, Missouri, 1819–1970*, 4.

64. Elizabeth Denty Abernathy, *The Abernathy Family;* Anne B. Rutledge, *York Chapel Graveyard, 1821–1971*. The settlement of Pocahontas was named in honor of the group's leader, Joseph Abernathy, who was, according to tradition, a descendant

Perry County received still other groups that must be mentioned to complete the picture of settlement diversity. All of these groups were part of the westering of people from the Carolinas, Virginia, Tennessee, and Kentucky, and they shared the broad outlines of upland southern culture and settlement strategies. Among them before 1830 were English-speaking Lutherans of German descent from Lincoln County, North Carolina. These "English Lutherans" followed their Scotch-Irish neighbors to Missouri and settled on both sides of Apple Creek in the former Indian grant, now legally open for whites. Some merged later into the much larger numbers of Lutherans who arrived directly from Germany beginning in 1839.[65]

During the 1820s, a contingent of immigrant Flemings bought land in the Bois Brule bottom along the Mississippi, and some French-speaking Swiss and French moved into the Barrens. Their presence introduced French-language services into the large English Catholic church of the Barrens.[66] Finally, we should note that all of the various settlement groups, with the exception of the immigrants who came directly from Europe, accepted slavery.

By the end of the 1820s, people had spread into all of the readily occupiable lands of the Ste. Genevieve District; the initial phase of settlement was closing. The process produced some two dozen settlement neighborhoods, each with its own cultural and economic distinctiveness.

---

of Pocahantas (Abernathy, *The Abernathy Family*, 30). The name has nothing to do with the Shawnee and Delaware in the midst of whose land Pocahantas was located. Abernathy, not Pocahantas, is the name of the settlement on Johnson's 1862 Map of Missouri and Kansas by Johnson and Ward. On group: *History of Southeast Missouri: Embracing an Historical Account*, 699, 726. Although the families were of Scotch-Irish ancestry, some of German ancestry were included, such as the Quick families.

65. Hoehn, *Genealogist and Historian's Guide*, 5, 22. Houck, *History of Missouri*, 2:187–89, 3:169–70. These English Lutherans were not part of the "Whitewater Dutch" of Cape Girardeau District who were also from Lincoln County, N.C., but arrived much earlier during the Spanish period. Beginning in 1839, a large number of German Lutherans from Saxony, totally unassociated with earlier German immigrants into Perry County, settled in the hills and valleys of the Brazeau and Apple Creek watersheds where the Fenwick colony had been, and they have distinguished the population of southeastern Perry County since (Hildegard Binder Johnson, "The Location of German Immigrants in the Middle West," 12–16; Sauer, *Ozark Highland*, 164–65; Walter O. Forster, *Zion on the Mississippi*; E. F. Stegen, "The Settlement of the Saxon Lutherans in Perry County, Missouri"; P. E. Kretzmann, "Saxon Immigration to Missouri, 1838–1839").

66. Poole and Slawson, *Church and Slave*, 3–6. The post office name of Belgique, in the Bois Brule bottom eight miles northeast of Perryville, commemorates the Flemings who arrived in Perry County before an independent Belgium was recognized in 1839.

Such a geographic and historical complexity can be understood by examining the physical form that settlements took, ranging from dispersed farms and amorphous mining camps to planned French villages and American towns. For some of these forms, the major explanation must be sought in the cultural practices of the various groups that entered the district. Thus, Americans from Appalachia settled on individual farms and immigrant Indians and French Creoles in compact villages. Extended kinship networks engendered solidarity and expansion in some communities, whether clustered or dispersed.

In other communities, physical geography became more important. In the difficult topography of the Ozarks, the traditional French village failed owing to a scarcity of tillable land within walking distance, and American dispersed farms occurred only in narrow valleys. But in the more open basins and plains, the same kinds of people as in the hills could spread and form prosperous communities.

In still other localities, economy was the most important factor in shaping it. The chief differentiation here was lead, the great impetus for moving away from the Mississippi River. Sometimes mining concentrated people, as with Austin's several land developments, and sometimes it delayed settlement, as with Mine la Motte and other large tracts awarded for mining, but undeveloped.

Finally, variety in settlement form was due to constraints imposed by administrative regulations, Spanish and American. In general, settlers were free to arrange themselves on the land as they wanted under Spain, producing an irregular cadastral geometry, but under the Americans, they were restricted to squares and rectangles, both on farms and in towns.

Part IV
Organizing Space

15

"Connecting Tissues"

French historian Fernand Braudel called the total interaction among people of a region its "connecting tissues."[1] More than just making it possible for governmental and economic institutions to carry on their functions, connecting tissues relentlessly mold the people through daily, weekly, and monthly spatial interactions into communities and larger societies. France, according to Braudel, was created over the centuries by pervasive and continuous interactions among villages and villagers. And so it has been with the Ste. Genevieve District, except that the result is not the forging of a great nation but the creation of a much smaller society of people of diverse provenances. Spatial interaction first brought the resident Indians, French Canadians, and African slaves into functioning exchange and social systems and then allowed the French villages of Upper Louisiana to operate within the larger Spanish administrative system. Later, interaction brought together French Creoles, Americans, more blacks and Indians, and others into a spatial system. How did the system

---

1. Braudel, *The Identity of France*, 32. In a similar vein, Donald W. Meinig calls for study of the United States as a "varying set of regional parts developing through a recognizable sequence of interrelationships" ("American Wests: Preface to a Geographical Interpretation," 184). Examples of recent historical geographic studies that employ the concept of central-place systems and city systems are Robert D. Mitchell, "The Shenandoah Valley Frontier," in which Mitchell emphasizes that how central places historically evolved is important, not just the relationship of hierarchical centers at any one time in history; John A. Jakle, "Transportation and the Evolution of River Settlements along the Middle Mississippi River"; Michael P. Conzen, "The Maturing Urban System in the United States, 1840–1910"; Hudson, *Plains Country Towns*; and Farmer, *Absence of Towns*. A strong argument for understanding regions by their "organization" in addition to environment and culture is in Timothy R. Mahoney, *River Towns in the Great West: The Structure of Provincial Urbanization in the American Midwest, 1820–1870*. A review of the literature of the "functionalist" approach is in James O'Mara, *An Historical Geography of Urban System Development: Tidewater Virginia in the 18th Century*.

of linkages evolve and function, and what was the changing role of Ste. Genevieve village as the primary node in that spatial system?

At the smallest spatial level, a landowner organized his own property. He selected where to put the house and where to carry out different activities in relation to distance and essential resources, always with a guarded eye to "salubrity." He put his house close to the water source and, if village-dwelling French, close to others. He put his vegetable garden and other truck adjacent to the house, as well as stables and shelter for those barnyard fowl and animals that needed daily attention. Fenced field crops and orchards were farther distant, and open range for cattle and hogs the farthest, as they needed the least attention and could utilize unimproved woodland and nonarable lands.

People also organized space at higher levels, where it becomes the aggregate of many decisions by many persons. The few dozen farm families in the Bellevue Valley placed their Presbyterian and Methodist churches close to each other at the place of maximum aggregate access. Then, building on that focus, Alexander Craighead chose an adjacent site for his store and the valley's first town, Caledonia.

The Bellevue farmers also had linkages beyond their valley. They worked part-time in nearby mining districts. They took their farm produce and animals north to Mine à Breton for sale and bought goods there that were not available at Craighead's small store, thereby forging economic ties with that community. Because Mine à Breton–Potosi also served farmers from other neighborhoods, it was a higher-order place for a much larger area than the Bellevue.

Until 1814, Ste. Genevieve village was a notch above Mine à Breton–Potosi in the district. Everyone in the district with any official business had to transact it in Ste. Genevieve, first through the Spanish commandant and later through various American county officers. The village also served the entire district as its economic center. At a still broader geographic scale, Ste. Genevieve was subordinate to St. Louis as the increasingly dominant focus of the middle Mississippi Valley.

The Administrative Perspective

The process of organizing space by administrative areas began with the earliest inhabitants, Indians, who had a sense of territoriality no less keen than Europeans. Auguste Chouteau, an Indian trader and agent at St. Louis, provided a depiction of the territories of the Illinois, the Osages, and other Indians in areas relating to St. Louis.[2] Indian geographic iden-

---

2. The map drawn from Chouteau's account is in Temple, *Indian Villages*, pl. 41, and the account is in Auguste Chouteau, "Notes of Auguste Chouteau on Boundaries

tities, however, succumbed to European and American invasion of the middle Mississippi Valley.

For most of the eighteenth century, actual French-controlled space remained close to the Mississippi River and focused in villages, despite maps made in Europe that depicted French domain extending throughout the million-square-mile Mississippi River basin (fig. 39). In 1766, only five hundred colonial people were counted in the Ste. Genevieve District, and a total of less than one thousand in the entire Spanish Illinois. It was the new Spanish regime that created the first geopolitical structure for the west side of the Mississippi River by establishing geographical districts of Upper Louisiana and made St. Louis the administrative center (fig. 40).[3]

In the 1780s, permanent settlement spread into the mining camps of the interior and was followed in the 1790s by American farmers surging in after the change in immigration policy. By 1804, the area occupied included most of the mining country and all of the major good-soil tracts of the district (fig. 41). Dispersal of settlement required the division of the district into cantons, which were created around the pockets of population (fig. 42). Population in the Ste. Genevieve District (including the New Bourbon District) rose considerably by 1804 to 3,610, plus an additional 1,540 Indians, mostly Shawnee and Delaware (fig. 43). In all of Upper Louisiana, population rose to 10,340.[4] Essentially, all the neighborhoods of the district had been established by 1804.

---

of Various Indian Nations." Indians marked their boundaries in the landscape with posts and symbols just as Europeans did (Patricia J. O'Brien, "Prehistoric Politics: Petroglyphs and the Political Boundaries of Cahokia").

3. Population counts made by the Spanish, as would be expected in colonial areas in general, leave a lot to be desired because of their incompleteness and their lack of geographic precision. The problems with Spanish and early American censuses, as applied to the Ste. Genevieve District and Indians, are discussed in W. A. Schroeder, "Opening the Ozarks," 455–57. Spanish censuses in general are summarized and discussed in Viles, "Population and Extent of Settlement."

The evolution of the administrative organization of the Ste. Genevieve District from 1770 to 1830 is traced in detail in W. A. Schroeder, "Opening the Ozarks," 507–47. "The Lewis and Clark Map of 1804" is probably the first map to show the administrative districts of Spanish Illinois (Moulton, *Journals of Lewis and Clark Expedition*, 7, map 6). The map is discussed, but not shown, in Donald Jackson, "A New Lewis and Clark Map."

4. The figure for the Ste. Genevieve District is an independently calculated figure based on testimony before the American Board of Land Commissioners. It is a higher figure, by 26 percent, than the figure of 2,870 estimated by Stoddard and commonly used by historians. Stoddard based his estimate on earlier, "incomplete" Spanish censuses (W. A. Schroeder, "Opening the Ozarks," 459–65). Using board testimony to calculate population also provides a high degree of geographical resolution, down to the neighborhood level, whereas the Stoddard figure and U.S. censuses for 1810–1830 provide totals only for counties. For Upper Louisiana figure: Stoddard, *Sketches of Louisiana*, 226.

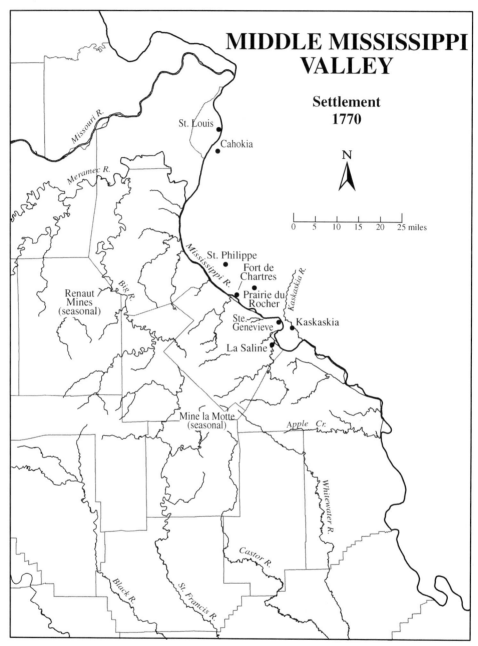

39. *Middle Mississippi Valley Settlement, 1770.* In 1770, seven French villages occupied the middle Mississippi Valley, two of them in newly created Spanish territory on the west side and five in newly created British territory on the east side. La Saline was a salt-mining camp and the Renault and Mine la Motte mines were worked seasonally.

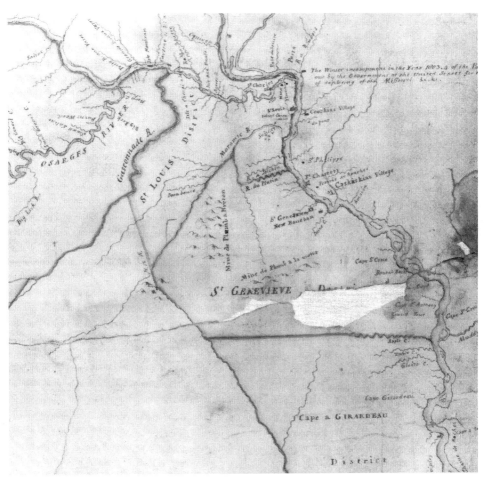

40. *Part of "Lewis and Clark Map of 1804."* Meriwether Lewis and William Clark made this map before leaving on their expedition up the Missouri River. It shows Lewis's recommended subdivision of Upper Louisiana into administrative districts, which closely follows the existing Spanish districts. It is the first map to show the Spanish subdivisions, which became the first subdivisions of territorial Missouri. (Center for Cartographic and Architectural Archives, National Archives, IR 20, Record Group 77. Courtesy of the National Archives)

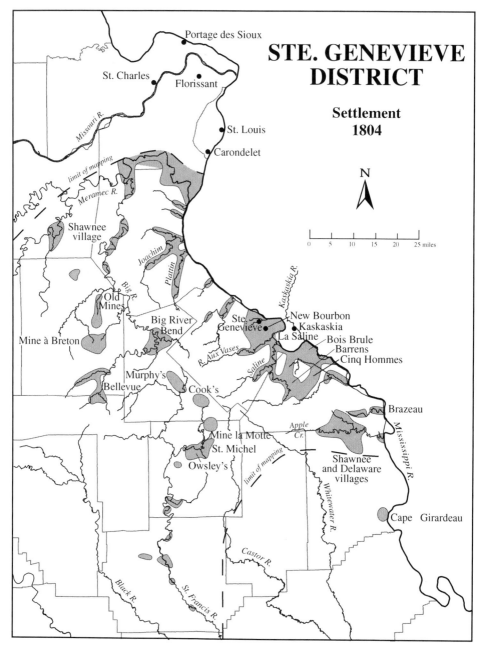

41. *Ste. Genevieve District Settlement, 1804.* At the end of the Spanish period, permanent settlement had spread away from the Mississippi River onto the karst plain and into the mining district thirty miles into the interior. Americans arriving in the preceding ten years occupied narrow valleys and selected tracts of good soils in the interior.

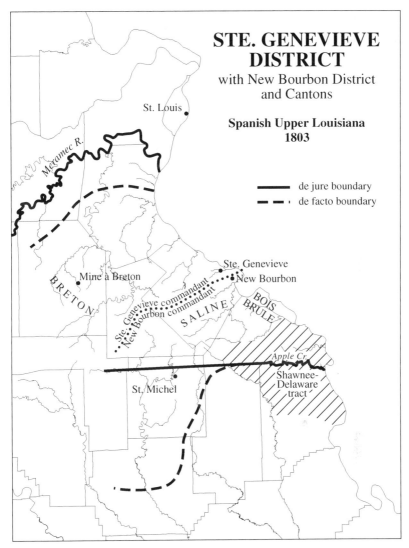

42. *Ste. Genevieve District with New Bourbon District and Cantons, 1803.* The Spanish administration defined the Ste. Genevieve District as the land between the Meramec River and Apple Creek, but in practice the district commandant administered land on the St. Francis River well into Wayne County, and the St. Louis lieutenant governor directly administered land in northern Jefferson County. When the New Bourbon District was carved from the Ste. Genevieve District in 1797, a line separating the two (reconstructed on the basis of lands conceded by commandants) ran from the Mississippi River into the interior, splitting the historic common field and the mining district. The Spanish established three cantons in the two districts for local policing away from the villages.

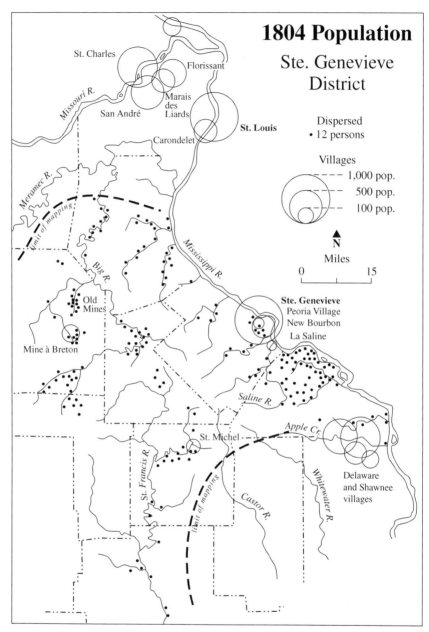

43. *1804 Population, Ste. Genevieve District.* Distinct neighborhoods were in place by 1804. Some were primarily French Creole, some American, and some Indian, and the mining neighborhoods were mixed. The Indian villages probably contained as many residents as the French villages of Ste. Genevieve and New Bourbon.

Though the purchase of Louisiana offered the opportunity to reconfigure its political geography, the incoming American government opted to carry over without change the Spanish districts and boundaries (except the reincorporation of the New Bourbon District into the Ste. Genevieve District). The districts were renamed counties in 1812. Population of the district reached 4,620 in the third U.S. census of 1810, and the whole territory of Louisiana reached 19,783. Due to exceptional growth in the mining district, the estimated population of the district in the special census of 1817 was 7,947. Growth and dispersal resulted in the creation of new counties within the district, beginning with Washington in 1814, followed by Jefferson, Madison, and Wayne in 1818.[5] The administrative fragmentation of the historic district was under way, and Ste. Genevieve village was losing its administrative influence over the most productive parts of the district.

In the fourth U.S. census of 1820, the counties that comprised the historic district had a total population of 13,056, an average annual growth rate of 18 percent since 1810, which was high, but only three-quarters as fast as the state as a whole (fig. 44). The karst plain and the mineral district were well settled, but the rough hills still held farmers only sparsely in narrow valleys. West, beyond the mineral district, the eastern Ozarks still lay virtually untouched by American settlers. Slaves made up 20 percent of the population of shrunken Ste. Genevieve County, the highest percentage of all counties in Missouri Territory. By the end of 1821, seven counties constituted the historic district, each of them subdivided into civil townships for local policing, administration, and militia purposes (fig. 45).[6] In general, both county and township boundaries went through unoccupied areas between the settlement clusters and were poorly marked, if at all, on the ground. All the occupiable tracts of the district had been located and settled, although population densities would increase in subsequent years. Further spreading onto new land would have

---

5. On renaming districts as counties: "Proclamation by Governor Howard," October 1, 1812, *TP* 13–599–601; and *ASP-Misc* 2:201–3; on 1810 figures: *Aggregate Amount of Persons within the United States in the Year 1810;* on 1817 figures: "An Act supplementary to the acts, providing for taking the enumeration of free white male inhabitants," February 1, 1817, *LPGN* 1:550–51; and "Enumeration of White Males in 1818," *St. Louis in Its Territorial Days,* by Billon, 51. The special censuses of 1814 and 1817 were conducted to apportion the territorial assembly among the counties. For estimating total population from the number of white males counted, see W. A. Schroeder, "Opening the Ozarks," 466–68. On new counties: *LPGN* 1:554, 567, 576–80.

6. Iron County, the final county to be created from the historic district, was created in 1857.

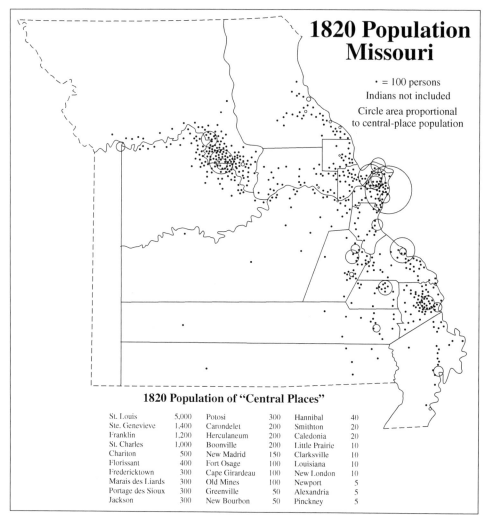

44. *1820 Population, Missouri.* Growth of population in Missouri in the first two decades of the nineteenth century was proportionately much greater along the Missouri and upper Mississippi Rivers than in the historic Ste. Genevieve District. Consequently, the eastern Ozarks was reduced to a less influential position in the new state of Missouri by 1821.

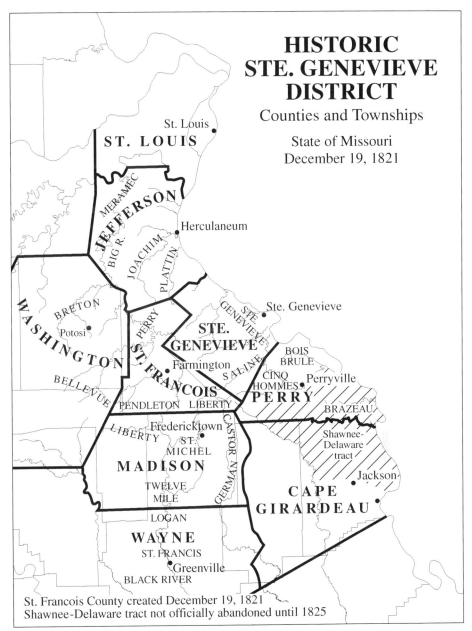

45. *Historic Ste. Genevieve District, Counties and Townships, State of Missouri, December 19, 1821.* By the end of 1821, the historic district had been subdivided into seven counties (Iron County was created later). Each county was subdivided into civil townships for policing, voting, and militia purposes in local neighborhoods. The townships had no official boundaries, as much of the space between neighborhoods was unoccupied.

to await the development of new agricultural technologies or the immigration of different groups.

A Restless Population

The initial years of populating a new land are restless ones, as those arriving search for the best place to put down roots, and then, learning of new opportunities in the apparent limitless expanse of unexploited land around them, consider relocating to another place, then yet another place within the general region. Thus, during the time it takes for a region to achieve a measure of residential stability, movement is one of the connecting tissues that helps form a geographic consciousness. People moving from neighborhood to neighborhood within the Ste. Genevieve District, whether seasonally or permanently, acquired knowledge of its various parts and other people. Places that attracted migrants more than others emerged as incipient economic centers, and the routes that people chronically moved over evolved into the basic infrastructural network of the district.[7]

Among the motives mentioned for moving within the district in testimony before the Board of Land Commissioners were the lure of short-term profits in mining; the replacement of eroded and unproductive soils with virgin soils; the use of larger expanses of land for larger herds of cattle and swine; the necessity to have a supplementary income in the first few critical years while the farm was getting established; dissatisfaction with the initial, hurried choice of land; married children, especially daughters, leaving their home places; and the frontier practice of selling a cabin and clearing for profit. To these we can add the concern with unhealthy locations and the desire to distance oneself from others to escape the fevers. Even skilled tradesmen made seasonal or temporary moves when they were not otherwise occupied by their farms. Thus, the blacksmith Etienne Govereau moved about the district, working first in Ste. Genevieve and New Bourbon, then at La Saline in 1799, and at busy Mine à Breton in 1802, ending up on the Big River.[8]

The rate of property turnover serves as a measure of the mobility of the population. In Ste. Genevieve County in 1819 (which still included

7. Mahoney argues forcefully that an environmental explanation to settlement in the American interior is inadequate and that it is necessary to consider access, lines of movement, and "patterns of human activity organized systematically across space" to understand where people settled and where towns grew up (*River Towns,* 5–9, 45–46).

8. Houck, *History of Missouri,* 1:338. Dorrance used the Vallé Papers to describe how skilled workmen moved about from settlement to settlement as their services were needed (*Survival of French,* 26–27).

present Perry County), 327 land sales had been recorded since the 1790s for 215 pieces of rural property, which is an average rate of 1.52 sales per property.[9] Many properties changed owners four and five times. The analysis is heavily weighted against the most mobile element, the Americans, as they entered the district much later than the French and could buy only land that was already in private ownership.

The analysis reveals significant differences between French and American neighborhoods. Despite the bias against Americans, French neighborhoods averaged only 1 turnover per property, and the American neighborhoods averaged 1.71 turnovers. The major exception to this pattern was the Barrens, which had a turnover rate of 0.95, similar to that of French neighborhoods. The American-populated Bois Brule riverbank properties, first occupied by the land-clearing Burns clan, had the highest rate, 3.41, which suggests its function as a kind of continuous staging area for those arriving from the East before moving into the interior. Thus, French-settled areas in general were relatively stable compared to American-settled areas, but significant differences occurred among the American-settled areas.

The 1789–1819 land records of the entire historic Ste. Genevieve District, supplemented by Houck's record of who lived where in the district, provide another insight into population movement within the district.[10] From these two sources combined, a total of 368 internal moves of residence could be identified over the thirty-year record but concentrated in the last half, from 1804 to 1819. It is not known how close to the actual number of moves this total might be, but it is certainly a minor fraction,

---

9. The number of recorded sales used in this analysis is obviously not the complete total for the county since the 1790s, but it is usable as a sample if the percentage of recorded sales is reasonably the same in all geographic neighborhoods of the county. A more complete analysis is in W. A. Schroeder, "Opening the Ozarks," 481–83, 605.

The land sales are recorded in Ste. Genevieve County Deed Books A, B, and C, from 1804 to 1819, and in the Ste. Genevieve Archives for sales before 1804. Aron warns that land sold for speculation may confound the use of land sales for reckoning population movement (*How the West Was Lost,* 74–75). No evidence of speculators buying up numerous confirmed land claims in the Ste. Genevieve District has been uncovered.

10. "Land records" refers to the complete database of landownership accumulated by the Ste. Genevieve Project at the University of Missouri–Columbia from a variety of sources. Houck lists residents by neighborhood in many places throughout his three-volume *History of Missouri,* but most of the names for the Ste. Genevieve District are near the end of vol. 1. Duplications from combining the two sources were omitted. An early version of this analysis was Walter A. Schroeder, "Local Population Movement on the Eastern Ozarks Frontier, 1770–1820." It is more fully presented in W. A. Schroeder, "Opening the Ozarks," 483–85, 606.

and the results should be a conservative reckoning of population movement.

Four neighborhoods, all of them along the Mississippi, were significant net losers by migration: Ste. Genevieve, New Bourbon, the Bois Brule bottom, and the Brazeau neighborhood of Fenwick connections. The aggregate ratio of immigration to emigration for the four is 1:2.5, or five families leaving for every two entering. None of the four was growing toward the end of the period; New Bourbon and the Brazeau were already depopulating in the decade of the 1810s.

Two kinds of neighborhoods were significant gainers by migration and thus were seen as offering more opportunities to district residents. The mining areas of Old Mines, St. Michael's–Fredericktown, and Mine à Breton–Potosi, with its farming adjunct at the Bellevue, attracted strong intradistrict movement, especially Mine à Breton–Potosi. A second consisted of those neighborhoods on the extreme edges of the district where settlers were just beginning to enter after 1820: the lower St. Francis in Wayne County, Fourche à Courtois west of Potosi, and the valleys of the Black and Current Rivers. For all these gaining neighborhoods, the ratio of immigration to emigration was 2.4:1, or almost five families entering for two families leaving. Conversely, the stable Barrens community, although well populated, participated little in population mobility: only seven families entered, and only five left over the two decades of record.

The mining neighborhoods accounted for one-half of the total intradistrict moves, and the fraction would be much higher if persons who did not own property were included. Residents of all kinds, American and Creole, rich and poor, saw the mines as opportunities. The mining district in general attracted people from all over the district, but none drew more widely than Mine à Breton–Potosi, which received migrants from fifteen of the twenty-one district neighborhoods. In contrast, persons from only nine neighborhoods moved into Ste. Genevieve village, the district's historic center.

French neighborhoods in general were not attractive to migrants from the district. Neither Ste. Genevieve nor New Bourbon offered new agricultural land or new economic opportunities to persons already resident in the district. In this sense, they tended to stay "closed" communities through the 1820s, in contrast to Potosi and other American towns that were growing and fast diversifying in population and economy. Ste. Genevieve did, however, receive new American residents arriving from the American East, who entered civil administration and various economic activities undeveloped by Creole residents.

Thus, the restlessness of the district's residents during the period of initial settlement reveals differences among neighborhoods and population groups. Some parts of the district grew, and some lost, from internal population redistribution, a process that set in motion the emergence of a hierarchy of towns and central places. But a mobile population also led to interactive farm and mining economies and to integrated populations, as the French moved into American neighborhoods, and Americans moved into French neighborhoods.

Pathways of Integration

Over the years, repeated movements of people hauling lead and farm products and journeying to stock farms, sugar camps, and lumber camps eventually created a network of geographic "desire lines": routes that linked where people were and where they wanted to go. The network consisted of a dense pattern of local routes linking individual farms with each other and neighborhoods with each other, supplemented by routes linking the district with the world beyond.[11]

Land travel was usually on foot, and it was not uncommon for people and troops to walk the length of Upper Louisiana, from St. Louis to Cape Girardeau, or even walk the greater part of the distance to Louisville. The French had small, strong horses for riding, plowing, and pulling *charettes,* the sturdy all-wooden, two-wheeled carts used universally for both farm and long-distance lead hauling. Oxen also drew plows and carts. The *charette* had a limited capacity; it held only about double the contents of the bed of a large wheelbarrow, but when solid sides were put on it, it became a *tombereau* with increased capacity. A *charette* carried Madame Chouteau and her children the forty miles from Fort de Chartres to Cahokia in 1764.[12] With such minimal capacities for land transport, no wonder the French used the river as much as they did and had so many difficulties hauling lead ore to the river, sometimes spurning *charettes* for packhorses.

Four-wheeled wagons with iron rims on the wheels came with American settlement, replacing the French *charettes* rapidly in all but the most traditional French locales. They had much greater capacities but needed

---

11. Various terms were used for land routes during 1760–1830: *route, trail, trace, sentier, chemin, route* (French), *rue, street, alley,* and *camino.* These terms are discussed in W. A. Schroeder, "Opening the Ozarks," 489.

12. Houck, *History of Missouri,* 2:235–36; Ekberg, *Colonial Ste. Genevieve,* 152–53. For *charette, calèche, berline,* and *tombereau,* see Dorrance, *Survival of French,* 21–22. Citing French sources, Houck says that the horses of the Illinois Country were originally from Canada (as were the cattle), and were crossed with wild horses of the plains from Indians who had gotten them from the Spanish (*History of Missouri* 2:236–37). On Madame Chouteau: Houck, *History of Missouri,* 2:8.

more horses or oxen to pull them.[13] Individuals continued to travel on horseback or on foot; that was the mode of travel for circuit-riding preachers and priests and the postal service. Horse-drawn coaches for passenger travel increased in number with Americans, but poor roads in the district limited their use to the immediate vicinity of towns and villages.

The prevailing ridge and valley pattern of the district made it easy for local people to conceptualize geographic patterns of movement and keep track of where things were and how to get to places without the aid of maps. Routes regularly followed the ridges. Small ridges led to larger and larger ridges in a predictable way; it was the exact landscape obverse of the stream pattern of smaller streams leading into larger ones. For example, once one ascended to a ridge near the mouth of the Joachim at Herculaneum, one could travel continuously, with no fear of disorientation, by passing onto larger and broader (not necessarily higher) ridges to the Joachim's headwaters, then cross the watershed divide into the Big River drainage and follow the dendritically branching ridge pattern in reverse onto smaller and narrower ridges until one was forced to descend to the gravel-bottomed Big River. After fording it, the ridge-route scheme would be repeated on the other side. American immigrants from similar stream-dissected ridge and valley landscapes of Kentucky and Tennessee brought this mental visualization of space with them to the district and felt quite comfortable moving around in it.

The advantages of ridge locations for routes are obvious. Trees were scattered and smaller on the ridges, and a road was easy to lay out and maintain. In fact, hardly any tree clearing was necessary. "If a person in carriage is dissatisfied with a beaton [sic] road," wrote Timothy Flint of ridge roads, "he can travel through untrodden forest at ease." Openness of the woodland made visibility much better, and travelers were less likely to be surprised by Indians. On the other hand, ridges were more water deficient for travelers and their animals.[14]

Ridge roads were the main roads serving both interdistrict and intradistrict traffic and the ones used by travelers whose written descriptions have evoked images of what traveling was like. Farms in the hilly parts of the district, however, occupied valleys, and a separate local road network

---

13. Dorrance notes that some French had iron-rimmed wheels before Americans entered the district (*Survival of French*, 21–22).

14. Flint, *Condensed Geography*, 82; Schoolcraft, *View of the Lead Mines*, 52; Schoolcraft, *Travels in the Mississippi Valley*, 252; U.S. Senate, "Letter from the Secretary of War, in Relation to Leases of Lead Mines and Salt Springs," 19th Cong., 1st sess., S. Doc. 45, 22 February 1826, 9–10; Kedro, "Three-Notch Road Frontier," 191.

developed for them, mostly unused and unrecorded by visitors and travelers.[15] In any valley neighborhood, such as the Brazeau or the Aux Vases, trails or crude roads on the valley floors, often taking advantage of the gravel creek bed itself, connected the farms. This intimate component of the total route system helped socially bind neighborhoods together as one kind of connecting tissue, but it did not directly contribute to the evolution of a districtwide system and the emergence of central places.

The Mississippi River, inflexibly positioned in its valley, provided an umbilical connection with the rest of the world. It tethered the district to St. Louis, Canada, New Orleans, and the American East. Its monopoly in the district's external connections was not broken until overland transport developed after the cession, but even then, the river continued as the primary connector, especially for heavy cargoes.

A variety of riverboats served different purposes. Simple canoes sufficed for people, and it was by canoe that seven-year-old Henry Marie Brackenridge traveled down the Ohio in 1793 to Ste. Genevieve to learn French ways. The pirogue, a hollowed log in the shape of a canoe, but less tippy, transported larger loads. Bateaux, which could carry up to fifteen tons or more, however, were the mainstay of Mississippi River colonial commerce. A bateau was about ten feet wide by forty feet long, flat bottomed, and shallow drafted for low water and maneuvering into the muddy mouths of tributaries as small as the Gabouri at Ste. Genevieve. Its rear portion could be roofed over. Convoys of bateaux were the chief method of transporting products to New Orleans.[16]

Keelboats replaced bateaux after the turn of the century. Usually larger than a bateau, a keelboat had a cargo box or enclosed room for freight and sometimes also rooms for passengers. Its distinguishing features were a more substantial construction, with thicker gunwales than a bateau and a keel running the length of the boat. Though keelboats could be moved upstream by a variety of methods—rowing, cordelling, poling, or wind sailing—moving them upstream was one of the most difficult and hazardous tasks in the Mississippi Valley.[17] Flatboats, simpler than

---

15. Some of these routes are in the field notes of the land surveyors of the General Land Office.

16. Brackenridge, *Recollections of Persons and Places*, 22; William Steinbacher-Kemp, "Henry Marie Brackenridge and the Illinois Country"; Houck, *History of Missouri*, 2:261–67; Finiels, *Account of Upper Louisiana*, 30; Ekberg, *Colonial Ste. Genevieve*, 164–70; Ekberg, *French Roots*, 273–82; Nancy Maria Surrey, *The Commerce of Louisiana during the French Regime, 1699–1763*, 55–76.

17. For keelboat travel, see Leland D. Baldwin, *The Keelboat Age on Western Waters*. Stoddard described keelboats and "Kentucky flats" or "arcs" and their role in the

keelboats, were the cheapest way to move freight downriver, and, not worth bringing back upriver, they were sold for lumber at New Orleans. Many of these various types of river craft continued in both local usage and traffic to New Orleans even after steam-powered vessels entered the Mississippi River.

A late-eighteenth-century trip from the Illinois Country to New Orleans in large cargo boats usually took from fourteen to sixteen days, but the return, in large twenty-ton boats with twenty or twenty-two oars, took from fifty to ninety days depending on the speed of the current. Boats, often in convoys, that left New Orleans in August or September traveled the thousand miles upstream at one or two miles per hour and arrived in the Illinois Country in November or December. Boats that left the Illinois Country in February or March, aided by high water and a fast current at five to six miles per hour, reached New Orleans half a month later. Major trips of these large boats to and from New Orleans were made usually twice a year, a frequency of connection similar to that of seventeenth-century Atlantic colonies with England. Smaller boats made the river trip much more frequently. Col. George Morgan reported seeing three or four every day in 1766. American keelboats could make eight or ten miles per day going upstream, significantly faster than bateaux. Going downstream, a keelboat would float in twenty-four hours from forty-five to fifty miles at low water and from ninety to one hundred miles at high water. Steamboats arrived on the Mississippi above the Ohio in the summer of 1817 and radically shortened the time for a round-trip from St. Louis to New Orleans and back to nine days.[18] The volume of cargo increased, costs decreased, and the efficiency and dependability of service improved dramatically.

The Mississippi also linked the district to the Great Lakes and Canada by way of the Illinois River, but this water connection closed in winter. The linkage, essential to those in the fur trade and with family ties in Canada, declined in usage after 1804. The time required to go to Canada

---

economy of Upper Louisiana (*Sketches of Louisiana,* 374–75). Hunt described her keelboat trip up the Mississippi from Cape Girardeau to St. Louis in the summer of 1805 ("Early Recollections," 44).

18. Alvord and Carter, *New Regime,* 302, 540; Berquin-Duvallon, *Vue de la colonie espagnole,* 10; Pittman, *Present State,* lxxxi; Frederick J. Dobney, *River Engineers on the Middle Mississippi: A History of the St. Louis District, U.S. Army Corps of Engineers,* 13–14; Usner, *Indians, Settlers, and Slaves,* 233; on Morgan: "Voyage down the Mississippi, November 21, 1766–December 18, 1766," in *New Regime,* ed. Alvord and Carter, 440–41; Peck, *Forty Years of Pioneer Life,* 80–83. Stoddard reckons upstream progress by keelboats at fourteen to twenty miles a day (*Sketches of Louisiana,* 371, 375). On steamboats: Dobney, *River Engineers,* 13–14.

and back was approximately the same as for travel to New Orleans and back.[19]

The first recognizable overland routes used by Europeans in the Ste. Genevieve District were two that connected the Mississippi River with the mines at Mine à Renault and Mine la Motte (fig. 46). The former, described in 1743 as a "well beaten trail" used by the Osages, went up the Establishment Creek Valley opposite Fort de Chartres.[20] Renaut had properties near Fort de Chartres, and this route was the shortest distance to his mines. With the abandonment of the mines, Renaut's departure, and the demise of the fort, the route fell into disuse.

The other desire line connected Kaskaskia and Mine la Motte. When Lamothe visited the location in 1715, he noted a trail already in existence between the mines and La Saline on the Mississippi River, where Frenchmen were working. When mining began in earnest in the 1720s, this route crossed the Mississippi at La Saline, but later, after the founding of Ste. Genevieve in the Pointe Basse, it crossed at that location. Both that village and La Saline served as depots for Mine la Motte lead and obviated the necessity of moving heavy lead across the wide river to Kaskaskian depots. The route between old Ste. Genevieve and Mine la Motte, called the Three-Notch Road, apparently after the way it was blazed on trees, followed the smooth-crested ridges and forded only three streambeds.[21] The distance of thirty-three miles took two days to travel with a *charette* and horses.

Before 1769, the French moved between their several villages on the eastern side of the Mississippi both by way of the river and by a land route between Kaskaskia and Cahokia. The British inherited that land route after the partition of the Illinois Country, but the Spanish, lacking one on their side to connect their two fast-growing villages of refugees, could use only the river. It took at least two days for the seventy-eight-mile trip upriver from Ste. Genevieve to St. Louis, and sometimes as long as four days. In winter, the river was not always usable. Therefore, the residents of the Spanish side were often obliged to ferry across the river at the two ends of the trip, Ste. Genevieve and St. Louis, and rent horses and conveyances

---

19. Walter B. Douglas, *The Spanish Domination of Upper Louisiana*, 77.

20. Ekberg, "Antoine Valentin de Gruy," 143. The corners of two of the earliest French properties along Establishment Creek were marked with lead-ore stones hauled forty miles from the mines, providing evidence that lead hauling must have passed along the creek (*Régistre d'Arpentage*, 93). The name *Etablissement* probably derived from its function as the route beginning opposite the *établissement* of Fort de Chartres.

21. On Lamothe: Rothensteiner, "Earliest History," 199–202; Kedro, "Three-Notch Road Frontier."

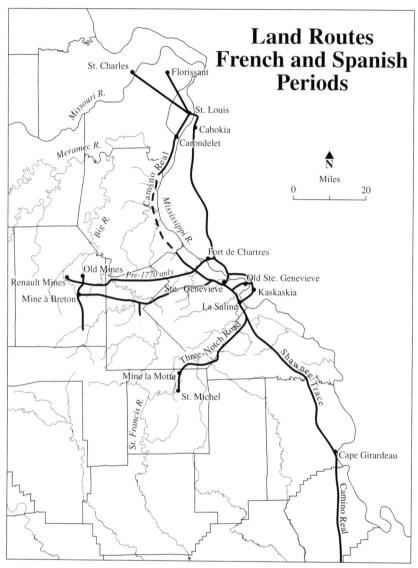

46. *Land Routes, French and Spanish Periods.* Three land routes were used during the French period: the Kaskaskia–Ste. Genevieve to Cahokia–St. Louis route on the east side of the river, and two routes to the lead mines at Mine la Motte and Renault mines. During the later part of the Spanish period, a new route to Mine à Breton replaced the earlier route to the Renault mines. The Spanish authorized a *camino real* (royal road) parallel to the Mississippi River to connect administrative posts, but it did not exist as a trafficable route through Jefferson County. (The map is not complete for areas not part of the historic Ste. Genevieve District.)

on the British side, as those could not easily be ferried.²² To the local Creoles and merchants, this was no annoyance because of relatives and friends on the British side. Spanish administrators and military, however, accepted such travel arrangements embarrassingly, as they were supposedly defending their colony against the British, against whom they ultimately declared war in 1779.

In 1776, the Spanish lieutenant governor realized that the two principal villages of St. Louis and Ste. Genevieve needed a land connection, a *camino real,* and in 1779 wrote to his superior that such a road had been constructed. Spain later reportedly extended the road southward from Ste. Genevieve to Cape Girardeau and New Madrid when they were established as provincial posts and northwestward to the post of St. Charles. The sole rationale for the *camino real* was as a military road to connect the several district posts for defense and administrative purposes, and it received limited usage by settlers except in close proximity to villages. The government road was the *chemin du roi* to French speakers and has been translated grandiosely as "king's highway" in English.²³

Like other unfulfilled Spanish government intentions for development of Upper Louisiana, the *camino real* never became a road throughout much of its length, despite the official report that it had been finished. From just north of Ste. Genevieve to the Meramec River, the route existed at best as a simple trace for horses and foot travelers and probably was not even locatable on the ground during many years. As late as 1796, anything large enough to be transported by wagon was still being sent on the river between Ste. Genevieve and St. Louis or by ferry at the two ends and moved by wagon on the American side.²⁴ The carefully constructed map that Finiels made at the end of the century shows no route of any kind

---

22. On length of trip: Finiels, *Account of Upper Louisiana,* 51; and Collot, "Journey in North America," 284–85; on ferrying: Leyba to Gálvez, July 13, 1779, in *Spain in the Mississippi Valley,* ed. Kinnaird, 2:348; and Collot, "Journey in North America," 295.

23. *ASP-PL* 2:550; Martha May Wood, "Early Roads in Missouri," 34; Leyba to Galvez, July 13, 1779, in *Spain in the Mississippi Valley,* ed. Kinnaird, 2:348–49. Any colonial road ordered by higher authorities was a "royal road," but that designation did not necessarily translate into a more trafficable and trafficked road. Other roads in Upper Louisiana were much more important than the *camino real.* Kingshighway continues as a street name in present-day St. Charles, St. Louis, Cape Girardeau, Sikeston, and New Madrid, but curiously not in Ste. Genevieve.

24. Finiels, *Account of Upper Louisiana,* 53. Several surveys of concessions contain the location of the *camino real* on them, but neither that term nor *chemin du roi* is written on the surveys. Any name along the road carried the name of the village to which it led, as *chemin à St. Louis* (Missouri Land Petitions no. 92, Missouri Historical Society, St. Louis).

between the two chief villages on the west side of the river. Finiels, whose object was to inventory features in the middle Mississippi Valley for military purposes, contrasted in detail the land routes on both sides between St. Louis and Ste. Genevieve and concluded that the Spanish-side route was usable only by horses and traveled only occasionally in summer and autumn. If in fact the bulk of the traffic in goods and people moved between the two chief places of Upper Louisiana either by water or on the alien British-American side, one doubts that the river really functioned as an international boundary in local economic and social matters, as depicted on political maps of empires.

The *camino real* south of Ste. Genevieve to Cape Girardeau did serve local populations, and its usage increased with their steady growth. From Ste. Genevieve to New Bourbon, it followed the base of the bluff, which was the back side of the *grand champ* common field, and was lined with houses. From New Bourbon to the River Aux Vases and on to the Saline River, the road was on the low bluff top, passing close by Commandant Luzières's house. South of the Saline, the road crossed the open, grassy barrens of the karst plain to the Indian villages on Apple Creek, then south through more open country to Lorimier's post at Cape Girardeau. For the twenty miles across the Indian tract, from Cinq Hommes Creek to near Cape Girardeau, the route bore the name Shawnee Trace. Finiels described the sixty-mile Ste. Genevieve–Cape Girardeau segment as an Indian trail that "can easily be accomplished in thirty-six hours, and some *voyageurs* in a hurry have done it in twenty-four."[25]

Discovery of lead ore at Mine à Breton opened a major road around 1791 from Ste. Genevieve due westward some forty-five miles to the mine (see fig. 46). As it developed in the ensuing years, the road went up the North Gabouri Valley and ascended the ridge between Establishment Creek and Fourche à du Clos, passing by Belle Fontaine (Price's Spring) at eighteen miles, probably a common rest stop. The road then passed into the limestone basin and took a smoother route across large *vacheries* of the Villars and Delassus families, reaching the young neighborhood of American farms and mines of the Big Bend of the Big River at twenty-six miles. An alternate route across the northern portion of the limestone basin developed after 1800 and passed through the neighborhood of Petit Canada (now French Village), which became an overnight stop. From the major ford across the Big River, the road mounted a sterile ridge and

25. Houck, *History of Missouri*, 2:150; M. M. Wood, "Early Roads in Missouri," 46–47; Finiels, *Account of Upper Louisiana*, 43–44. The road crossing of Apple Creek is preserved at the McClain House (1820) at Appleton on the south side of the creek (Wilkie, "Reconnaissance Survey," 26).

followed it westerly for the remaining eighteen miles to Mine à Breton.[26] This road to the mines, widened and beaten down under countless heavy wagon loads of lead, also served the newly established agricultural neighborhoods at Murphy's and the Bellevue by connecting wagon roads and was, by far, the chief way to access the interior of the district at the end of the Spanish period.

In his description of Upper Louisiana given to Congress upon his purchase of Louisiana, President Jefferson reported that the settlements of the newly annexed lands were "separated from each other by immense and trackless deserts, having no communication by land."[27] This was an overstatement, of course, but the generalization divulged an eastern American's perception of the primitive state of road development in the territory and implied the necessity of doing something about it.

When the Americans took over the west side of the Mississippi, there was no longer a reason to develop a separate road on that side parallel to the Mississippi, either for defense or for administrative purposes. Nevertheless, as soon as 1808, the territorial legislature passed two acts to lay out and mark a public wagon road from St. Louis to Ste. Genevieve, to Cape Girardeau, and on to New Madrid, thus connecting the administrative posts of all those districts (fig. 47).[28] The wagon road would be "near the old king's trace," implying that the *camino real* was essentially defunct where it had existed and an entirely new road had to be built.

This act did not result in a road being built the entire length. According to merchant Christian Wilt, wagon traffic between St. Louis and Ste. Genevieve was still using the Illinois side, with ferries on both ends in 1813. When lead manufacturer John Maclot's wife fell sick at Herculaneum in 1815, the "only way" to get her to a doctor in Ste. Genevieve was by water, which was probably a smoother journey in any instance. In

---

26. LRA 237. The alternate route can be traced in part on Soulard plats. The mileages given here have been measured on modern topographic maps and indicate that the distance from Ste. Genevieve to Mine à Breton was forty-four miles. Soulard's plat of Austin's concession at Mine à Breton gives that location as fifty miles west of Ste. Genevieve (*Régistre d'Arpentage*, 85). The Mine à Breton road today is witnessed mostly by gravel county roads and minor ridge-top trails. It is a highway only on the west side of Ste. Genevieve and near Potosi and Bonne Terre.

27. *ASP-PL* 1:345.

28. *LPGN* 1:188–89, 225–26. M. M. Wood points out the distinction between this territorial road that linked all the districts and the district roads that could be authorized separately by the districts ("Early Roads in Missouri," 46). One plat of this road, dated 1809 for the segment in Cape Girardeau County, remains in existence in the office of the Cape Girardeau County clerk of the circuit court. On the plat are these words: "This road follows the Shawnee trail the whole distance without any deviation from Cape Girardeau to the Indian town."

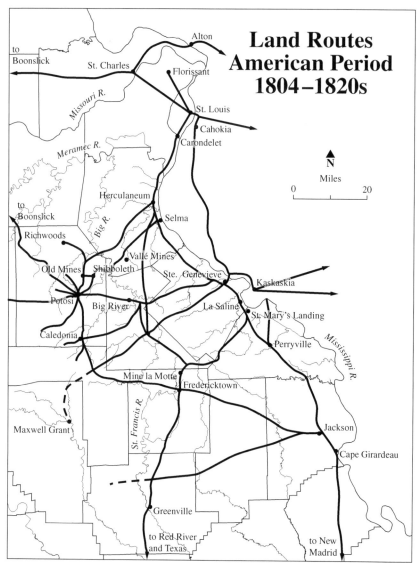

47. *Land Routes, American Period, 1804–1820s.* Land routes greatly expanded in the first two decades of the American period. Several routes connected lead-mining areas with the Mississippi River. The Potosi–St. Louis route became especially important as the mines were reoriented to St. Louis. Potosi became the district's chief route focus. A route led from Potosi northwestward directly to the Boonslick region on the Missouri River. Another route led southward to the Red River and Texas and was used by emigrants who followed the Austins to colonize Texas. (The map is not complete for areas not part of the historic Ste. Genevieve District.)

1817, the Reverend John Mason Peck denigrated the land connection between Herculaneum and Ste. Genevieve as nothing but a one-horse "bridle trail," and even later in the early 1820s, Duke Paul Wilhelm referred to that same distance as "over most difficult roads" that had to be done on horseback. The state designated the St. Louis–Ste. Genevieve route a "state road" in late 1821, surveyed it, then left it to the counties to build and maintain. No road of any kind between Herculaneum and Ste. Genevieve was depicted on early state maps until 1826.[29] Thus, though the Spanish and American bureaucrats' idea of a road to connect administrative centers was logical to them, it could not be backed up by a desire line based on local demand until the mid-1820s.

Expansion of lead mining in the years after the purchase had a great impact on the route system of the district. A rather dense network of wagon roads converged on Mine à Breton. Moses Austin built his own road to the Mississippi at Herculaneum, which also served as the southern half of the major road from St. Louis to Potosi. It carried increasing amounts of traffic between St. Louis and the mining region, reflecting growing St. Louis economic involvement in the mining industry. The thirty-four miles between Potosi and Herculaneum could be traveled in eleven hours, or about three miles per hour, and the entire trip between Potosi and St. Louis took only one and one-half days on horseback. Overnight accommodations could be had in 1820 at several places, and at statehood the road functioned as close as any in Missouri to a real "highway."[30]

Local roads were developing elsewhere in the district as neighbor-

---

29. Sister Marietta Jennings, *A Pioneer Merchant of St. Louis, 1810–1820: The Business Career of Christian Wilt,* 158; on Maclot's wife: Billon, *St. Louis in Its Territorial Days,* 178; Peck, *Forty Years of Pioneer Life,* 117; Wilhelm, *Travels in North America,* 174–75; on 1821 designation: *LPGN* 1:782–83; D. H. Vance, Map of Missouri and Territory of Arkansas. But the St. Louis–Ste. Genevieve road was not shown on a state map of 1827 (H. C. Carey and I. Lea, Geographical, Statistical, and Historical Map of Missouri). It reappeared on state maps of 1832, 1836, and subsequently (S. Augustus Mitchell, Map of Missouri and Territory of Arkansas, 1832; Mitchell, Map of the State of Missouri and Territory of Arkansas, 1836). Mapmakers probably obtained information on state roads from official acts of the legislature, not by actually observing what was in place in the landscape.

30. Schoolcraft, *Travels in the Mississippi Valley,* 250–54; Brackenridge, *Recollections of Persons and Places,* 215. The road from Old Mines to St. Louis is marked as the "Old St. Louis Road" on the 1873 plat of the Chouteau claim. It crosses Mineral Fork above Old Mines Creek and is clearly a separate and older road than the one through Shibboleth, which is also marked on the plat as a road to St. Louis (Shumard, *Chouteau League Tract*). The road rivaled the Boonslick road, which linked St. Charles and the Boonslick region, in traffic and trafficability.

hoods became more populated in the 1820s. Virtually every ridge radiating out of Potosi–Mine à Breton had some trail, if not a wagon road, such was the determined effort to find lead in the vicinity in the 1820s. Both prongs of the Bellevue Valley had wagon roads connecting farms with Caledonia, and from there north to Potosi. Another road led south from the Bellevue into Stout's Settlement and into the Black River drainage. Stout's Settlement was also connected to Fredericktown by way of the iron furnace at the Stout's Creek shut-in. Yet another road connected the Big River Bend with Fredericktown and Mine la Motte by way of the new county seat of Farmington (Murphy's) and Cook's Settlement. Another road led from Farmington due north into Jefferson County. Thus, by the end of the decade, a rather well defined road network interconnected all the populated neighborhoods of the district.

Ambitious Moses Austin, who observed after the War of 1812 that the swelling numbers of immigrants headed for the Boonslick region on the Missouri River were funneling through St. Louis and bypassing Ste. Genevieve and his town of Potosi, reckoned that he could divert at least part of the heavy Boonslick-bound traffic through Potosi. To this end, he persuaded the territorial legislature in 1816 to enact a law providing for the opening of a public wagon road from Potosi, then the third largest town in the territory, to the Boonslick "by the most direct and best route." Through newspapers, Austin encouraged immigrants headed for the Boonslick in 1818 to cross the Mississippi at Ste. Genevieve, then travel to Potosi and on to the Boonslick, because "this route saves 80 miles, and is a very good wagon road." Although built and used, the road was never in as good a shape as the Boonslick road west from St. Charles, and Potosi was never able to rival St. Louis and St. Charles as a way to reach the Boonslick. Austin's plan to make Potosi a competitive outfitting point in westward migration across Missouri failed, and no one championed the scheme after his death in 1821, although the road continued as a mail route until 1823. Except for short segments, the road has since been lost in the landscape.[31]

Another connection of the Ste. Genevieve District with new lands opening in the West was the road to Arkansas Territory, the Red River, Natchitoches, and Texas, known at different times by all those destination

---

31. On Austin: *LPGN* 1:479–80; Van Zandt, *Full description,* 158; and *Missouri Gazette,* March 23, 1816. M. M. Wood speculated that the road's fate was one of two possibilities. Either it disappeared because of lack of repair and nonuse, or it could have been given the status of a county road ("Early Roads in Missouri," 66–67). One portion that continues in use is Missouri Highway 185 that leads diagonally northwestward out of Potosi through Latty.

names. This road, which fulfilled an earlier intention by Spanish authorities to link Upper Louisiana by land with their posts on the Arkansas and Red Rivers and farther into Texas, ironically took form shortly after the Spanish left Upper Louisiana. It followed a route that had been traveled from "time immemorial" by Indians as a hunting and "war trail." More trafficable after 1815, it was the military road used by the United States Army for deployment of soldiers. William Darby predicted that it would become a great thoroughfare and carefully noted distances along it from St. Louis to Natchitoches in his 1818 emigrant's guide.[32]

The Red River road began at the outfitting places of Potosi and Fredericktown–St. Michael's. The road crossed the St. Francis River at Bettis Ferry–Greenville and gave that place its initial reason for existence. It crossed the Black, then the Current at Hix's Ferry (Currenton), likewise giving birth to that platted town, and then continued into Arkansas. Emigrants who left the district with wagons and animals to settle on the Red River in 1818–1820 and those who emigrated to Austin's colony in Texas traveled this road. Missouri interests early realized the necessity of land connections southward and by territorial resolution in 1815 had asked the federal government for an overland connection from St. Louis to New Orleans and the lower Mississippi Valley, essentially along the Mississippi River.[33] Such a river-bordering road, however, never reached farther than New Madrid, south of which were unoccupied swamps not expected to be soon occupied, if ever. The Red River road replaced the plan for a land road to New Orleans along the Mississippi, leaving New Madrid as a dead end on the old *camino real*. Ste. Genevieve village was also bypassed by the Red River road and did not share in the economic activity generated by its Texas-bound emigrants and military traffic.

After the *Zebulon M. Pike* first entered the Mississippi River above the Ohio in 1817 en route to St. Louis, steamboats stopped at Ste. Genevieve, Herculaneum, and other landings along the river and brought the river towns frequent and dependable service.[34] The advent of steamboats, however, did not appreciably affect lead mining, the district's most important, basic economic activity, the relative productivity of which re-

---

32. William Darby, *The emigrant's guide to the western and southwestern states and territories, comprising a geographical and statistical description of the states . . .* , 149, 161. See also descriptions of the road in Gracy, *Moses Austin: His Life,* 193–207; Schoolcraft, *Journal of a Tour,* 80–89; and M. M. Wood, "Early Roads in Missouri," 69–75.

33. "Resolution of the Territorial Assembly," January 15, 1815, *TP* 15:38–39.

34. *Missouri Gazette,* August 2, 1817; Darby, *Emigrant's guide,* 154–55; Z. Cramer, *Navigator,* 135.

mained bottlenecked by the high overland costs of wagon hauling to the river. The geographic effect of steamboats was to concentrate even more movement and economic activity in the largest places, especially St. Louis. To operate efficiently, the vastly larger boats required more cargoes and more people, which were concentrated in larger places. The volume of river business at Ste. Genevieve did not grow appreciably in the 1820s, with so much lead being diverted to Herculaneum and Selma and with its immediate agricultural hinterland small and relatively unproductive. Compared to St. Louis and Cape Girardeau, the economic effect of steamboats on Ste. Genevieve and its district was slight.

As the 1820s came to a close, the route network, as connecting tissue, had become reasonably dense within the district, and the district had reasonably good connections with regions to the east, southwest, north, and northeast.[35] However, the most important lines of movement into and out of the middle Mississippi Valley were no longer passing through the district but bending around it, especially to the north where traffic coming from the Ohio Country headed directly to St. Louis and the Missouri River. Similarly, most traffic bound southwestward to the Red River Country and Texas no longer passed through Potosi and St. Michael's, but went from source regions in Tennessee directly into Arkansas and Texas. The reason was simple geography. No navigable river such as the Missouri or Arkansas led westward from the Mississippi to provide a corridor for movement. The Ste. Genevieve District was backed up on the west by an endless succession of rough hills impenetrable for long-haul commercial transport.

Rapid decline in freight rates that began in the New West in the 1820s and produced radical changes elsewhere in settlement geography had less effect in the Ste. Genevieve District.[36] The high cost of hauling lead the thirty to fifty miles to the river continued through the first decade of statehood as a seemingly intractable problem and hindered development of the district's most important resource. Additionally, the district had not become a significant agricultural producer and exporter to the degree that other sections of the New West had become.

---

35. Evolution of the network of land routes after 1804 is traceable through the evolution of post routes. These are mapped in W. A. Schroeder, "Opening the Ozarks," 502–3 and figs. 5–7, and discussed in detail in R. G. Schultz, "Postal Service." Information on post offices, postal routes, postmasters, and frequency of service is in the well-indexed volumes of the *Territorial Papers of the United States*.

36. Carville V. Earle, "Regional Economic Development West of the Appalachians, 1815–1860," 176.

The Economic Perspective

Concurrent with the evolving organization of its area into administrative jurisdictions of districts and cantons, counties and townships, the Ste. Genevieve District underwent a continuing economic transformation that also led to the establishment of economic space organized around towns. Much more than administrative jurisdictions of counties, economic activities bound residents into interactive communities and eventually linked them and the district to a larger American system anchored in the great cities in the American East.

The district's basic economy rested heavily on a broad base of farmers, both French and American. Economically self-sufficient, they produced most of what they ate and needed to operate their farmsteads. As farmers became more numerous and more productive, they required more middlemen in villages and towns to handle their growing surpluses and provide goods and services that they could buy with their profits. This simplistic model of agricultural settlement by itself cannot explain the emergence of an interactive economic system, because the villages and towns from their initial years also handled lead, furs, wood, and other nonfarm commodities, which were equally important in establishing economic centers and the network of connecting routes.[37]

The early economic base of the district was agriculture, lead, furs and skins, and salt. As settlement spread into the interior and population increased, the economy diversified and developed by adding new functions such as merchants and professional services (lawyers, teachers, doctors, and the like) and the processing of raw products into articles of more value (hemp into rope, live hogs into preserved pork, lead into shot, wood into lumber, and so on). Those towns that added new functions—as distinguished from doing more of the same—were the towns that rose in the incipient hierarchy of central places.[38] Thus, early decisions made by

---

37. In his analysis of economic statistics of the United States in 1810, Tench Coxe takes note of the "considerable home industry in frontier settlements that goes unreported in censuses" (*A Statement of the Arts and Manufactures of the United States of America for the Year 1810*, vi). Ekberg has a full discussion of the economic life in his community study of colonial Ste. Genevieve, but the various constituents of the economy are not spatially patterned in the district (*Colonial Ste. Genevieve*, 126–77). Lewis E. Atherton discusses the Missouri economy at the time of statehood, but his discussion also lacks a spatial perspective ("Missouri's Society and Economy in 1821").

38. Aron points out that the coming of hemp into Kentucky in the 1790s as a cash crop for rope making helped transform a subsistent society into a commercial one (*How the West Was Lost*, 129–33, 143). Also see Lewis E. Atherton's comments on the economic transformation from subsistent to "established" communities (*The Frontier Merchant in Mid-America*, 9).

individuals in local communities planted the seeds for future economic development.

During the French regime, the geographic pattern of economic activity in the Ste. Genevieve District consisted simply of movement on the Mississippi River, a few short overland routes to lead mines, and contact points where these routes met the river. It was a typical colonial pattern for the extraction of resources with a modicum of attention to local agriculture that also provided exports in some years.

Little economic expansion or development occurred during the Spanish period until the 1790s; consequently, the basic economic spatial pattern underwent no change except for the loosening of ties between Ste. Genevieve and Kaskaskia. But economic freedom from Kaskaskia did not mean that Ste. Genevieve became the economic dominant on the west side, for under Spanish administration St. Louis rose faster.

Lead activity throughout the Spanish period centered in the Ste. Genevieve District; all mines were linked to Ste. Genevieve in respect to workers, owners, and lead traffic. Ste. Genevieve received the lead from the mines, warehoused it, and sent it out unprocessed. At the end of the century, Moses Austin helped turn Creole Mine à Breton into a permanent settlement, but it continued as an outlier of Ste. Genevieve. Revenue from lead (and salt from La Saline) enriched Ste. Genevieve and helped develop a small class of reasonably wealthy people who were then able to expand into other activities such as stock raising, wheat production, and lumbering. They purchased more slaves for increased production of these items. They built better houses and thereby stimulated the trades. Toward the end of the Spanish period, they used their wealth and social standing to obtain large land concessions before the Americans took over. Wealthier residents of St. Louis also benefited from large land concessions from Delassus at the end of the period and, aware of profits to be made from lead, located some of them in the mining areas of the Ste. Genevieve District. In this way, St. Louisans invaded the lead activity theretofore controlled by Ste. Genevievians.

Ste. Genevieve, like all settlements of French Canadians in the middle valley, participated early on in trapping and the lucrative fur trade with Indians. Many of the men of these villages, perhaps as many as half of them, at some time plied the rivers. However, the fur trade concentrated at St. Louis from the day of its founding by Gilbert Antoine Maxent, a New Orleans merchant, and his junior partner, Pierre de Laclède. Believing they had exclusive rights to fur trading on the Missouri River, which was then becoming the chief source for furs for the Mississippi Valley, they seized the boats of Ste. Genevieve and other fur traders already operating on the Missouri River and effectively closed the river to them. A license was required to conduct trade with Indians, and Ste. Genevievians who held licenses argued successfully with the Spanish governor for their rights, but they were unable to resume operations on the rivers above St. Louis. A few moved to St. Louis to stay in the fur trade. By 1795, the Missouri River fur trade included only three Ste. Genevieve traders (one inactive) out of twenty-eight licensed in the trade by Lieutenant Governor Trudeau.[39]

But human decisions cannot be the complete explanation. The hydrographic pattern manifestly helped St. Louis in competition with any other village for supremacy in fur trading on the immense inland river network. Because furs came down both the Missouri and Mississippi River systems (although much of the far-upper Mississippi traffic was diverted east into the Great Lakes), St. Louis could geographically co-opt the traffic before it floated farther downstream to Ste. Genevieve. After collection at St. Louis, the larger portion of furs was exported through the Great Lakes via the Illinois River, which could be easily canoed upstream due to its exceptionally low gradient.[40] Exporters avoided shipping through New Orleans because of the difficulty of keeping furs in the high humidity of the Gulf Coast. Thus, the fur trade would likely have devolved to St. Louis because of its location in the Mississippi Valley drainage network, regardless of manipulation of traffic by St. Louis merchants. Those living in other places who wished to continue as major participants in the fur business found it to their advantage, if not their necessity, to move their operations to St. Louis. One notable exception was the partnership of Pierre Menard of Kaskaskia and the Vallé family of Ste. Genevieve, established in 1817, which carried on an extensive and profitable Indian trade, but mostly in the Ozarks. Menard, who continued to live all this time at Kaskaskia, also became a primary investor in the St. Louis consortium that founded the American Fur Company.[41]

---

39. Jacqueline Trenfel-Treantafeles, "Spanish Occupation of the Upper Mississippi Valley, 1765–1770," 257–58, 262–65; "Distribution of Missouri Trading Posts, May 1–3, 1794," in *Before Lewis and Clark,* ed. Nasatir, 1:209–11; Trudeau to Carondelet, St. Louis, October 26, 1795, in *Before Lewis and Clark,* ed. Nasatir, 1:366–69.

40. Stoddard, *Sketches of Louisiana,* 297–306. The gradient of the Illinois River in some places is less than one inch per mile, which is not much different from lake water. In contrast, the lower Missouri River drops one foot per mile, or is twelve times as steep.

41. Rozier, *Rozier's History,* 131; Richard E. Oglesby, *Manuel Lisa and the Opening of the Missouri Fur Trade,* 65–75. Frequent and prolonged absences by men from Ste. Genevieve for trading, trapping, hunting, and commerce had implications on the role and status of their wives who remained in Ste. Genevieve and had to manage busi-

St. Louis derived vastly more economic benefit from trade in furs and skins than Ste. Genevieve did from lead and salt. In 1799, Spain reported the value of deerskins alone at New Orleans sent from Upper Louisiana at $70,000, but the value of lead at only $2,160 (before Austin's production began).[42] Little wonder St. Louis grew richer faster and became a more important economic center during the Spanish period.

The profits, real and potential, to be made from furs and lead constantly detracted from agriculture, not to mention the attractiveness of the life of *coureurs de bois* and voyageurs over that of cultivators. "The passion for the hunt and for trading has always caused the French settlements to languish and to neglect what is essential, the cultivation of the soil," observed Berquin-Duvallon, a New Orleans resident who traveled in Upper Louisiana. In his 1791 annual report to Carondelet, Trudeau lamented a "so-so" wheat harvest, then explained that no one in Upper Louisiana had any reason to grow more than they needed because of the exorbitant prices of barrels, boats, and rowers. Consequently, concluded the lieutenant governor, people turned to the fur trade to make money. Later, in 1797, the New Bourbon commandant was still claiming that agricultural surpluses could not be sold because of the high cost of transportation. Curious it is, though, that Americans in Kentucky, just as far away from New Orleans, in the same years and with the same farm commodities, were able to market their surpluses at New Orleans quite profitably.[43] Were the Kentuckians getting higher yields? Were their products of higher quality and more marketable? Did they have better marketing

---

ness and social affairs in their husbands' absences. Boyle concluded that Ste. Genevieve women enjoyed more financial autonomy under French and Spanish authority than later under Anglo-American. Other conditions that Boyle identified that promoted a prominent role and status of colonial French women were the demographic imbalance of significantly more men than women and the provisions of French customary law that protected women's economic and social interests ("Did She Generally Decide?").

42. *ASP-Misc* 1:383. Another account of the colony's economy for the period 1796–1802 concluded that the Illinois Country hardly mined enough lead for the needs of the colony and sent any excess to American settlements along the Ohio and Kentucky Rivers (James Pitot, *Observations on the Colony of Louisiana from 1796 to 1802*, 70). For other comparisons between the value of fur and lead, see Stoddard, *Sketches of Louisiana*, 297–306; and Finiels, *Account of Upper Louisiana*, 78, 93–107. Finiels virtually ignores lead, as if it was not worth considering.

43. Berquin-Duvallon, *Vue de la colonie espagnole*, 61–62; Report of Trudeau, 1791, November 28, 1791, in *Spanish Regime*, ed. Houck, 1:322; "Extract of the deliberation of the habitants of the post of Nouvelle Bourbon by Delassus de Luzières," December 17, 1797, PC 215b-608. Produce from the Ohio River Valley outstripped that from Upper Louisiana by nearly twenty to one in 1801–1802 (Ekberg, *French Roots*, 234).

connections? Were they more aggressive, persuasive, and competitive in their transactions? Was the "cut" that Spanish officials in Upper Louisiana allegedly took from shipping so great that it priced Upper Louisiana products out of the market?

Be that as it may, the habitants at Ste. Genevieve were successful farmers as measured in their terms. The Pointe Basse common field drew widespread acclaim for its productivity, despite the alleged nonchalant attitude toward farming of the French farmers. Wheat in the bottoms yielded five to eight times what was planted, and corn yielded "a hundredfold," despite poor care of the fields. When fields were opened up in the bluff lands, wheat yields increased. St. Louis, however, regularly fell short in supplying itself with grain, thereby acquiring the nickname "Paincourt," and had to rely on Ste. Genevieve surpluses. Ste. Genevieve clearly outproduced St. Louis in wheat and corn both in total production and in per capita production. In 1800, Ste. Genevieve produced more than twelve *minots* of wheat and twenty-four *minots* of corn per capita (or 216 and 432 bushels, respectively, per household of six), contrasted to St. Louis's production of less than eight *minots* of wheat and only seven *minots* of corn per capita. Ste. Genevieve also far outproduced St. Louis in tobacco, flax, hemp, cotton, and maple sugar, and had twice as many horses per capita.[44]

When the Americans took over in 1804, St. Louis was clearly the primary economic center of Upper Louisiana. Ste. Genevieve was the granary, but wheat, tobacco, cattle, hogs, and even lead and salt could not bring in the value per capita that fur and skins did. Furthermore, Ste. Genevieve itself was being strongly challenged by Mine à Breton as the center of lead activity in the Ste. Genevieve District.

The economic quickening after 1804 cannot be explained simply by more people occupying more land and exploiting more resources. To a

---

44. On yield in the bottoms: Delehanty, "Livelihood in Fort de Chartres," 41; on yield in the bluff lands: Stoddard, *Sketches of Louisiana,* 228; John Francis McDermott, "Paincourt and Poverty"; Finiels, *Account of Upper Louisiana,* 55–56; "Trudeau's Report," in *Spanish Regime,* ed. Houck, 2:250; Flint, *Condensed Geography,* 2:103. Production is one thing, quality is another. The French villages constantly struggled for better-quality flour, in large part due to their inadequate mills (Ekberg, *French Roots,* 265). One study of watermills in the rural American South concluded that they functioned to bolster the local subsistence economy rather than contribute to the market economy, and thus quality was less a concern than supply (Larry Hasse, "Watermills in the South: Rural Institutions Working against Modernization"). "General Census of 1800 of Upper Louisiana," in *Spanish Regime,* ed. Houck, 2:414. The census may lack accuracy, but comparison among districts can still be made if errors are systematic across districts. Some quantities are obviously estimated.

far greater degree than the French or Spanish could ever muster, the new residents had access to financial resources and credit of a vigorous economic system of the American East. It was an economic system with tentacles over hundreds of miles that soon superceded local entrepreneurs in Upper Louisiana regardless of their business acumen, and it unleashed new forces of economic development, as opposed to simple economic growth. External money and credit flowed across the Mississippi River. The infusion enabled an enormous expansion of fur-trading companies at St. Louis and shipping services on the Mississippi River. By 1820, federal government activity itself was contributing hundreds of thousands of dollars annually to the economy of St. Louis through Indian agents and traders (worth a phenomenal six hundred thousand dollars annually) and the outfitting and wages of numerous land surveyors and army personnel.[45] The territory was being integrated into a dynamic national economic system, and St. Louis was the hinge.

Business ties with eastern investment capital blossomed after 1804 in three chief arenas of speculation: furs, lead, and land. East-coast investors, such as John Jacob Astor, entered the trans-Mississippi fur business, although St. Louis and Kaskaskia money still largely financed the St. Louis Missouri Fur Company in 1808–1809. Eastern money also flowed abundantly into the mining district. The firm of Seth Craig and Company of Philadelphia bought three lots in the Old Mines concession in September 1804, barely six months after Stoddard took charge at St. Louis, and other eastern firms followed them into the district.[46] Eastern money also speculated heavily in land not only in and around St. Louis but also throughout the territory, driving up land prices and otherwise injecting money and credit into the local economy. Easterners bought up large tracts of land that Delassus had doled out in the closing years of his administration, precarious as they were in a state of land-confirmation limbo. Easterners showed great interest in purchasing New Madrid certificates that allowed holders virtual freedom to locate claims on the choicest, highest-priced land. Although the fur and lead money was concentrated in St. Louis and a few selected smaller places, economic ripples from land speculation were widespread, as far as the western and northern borders of the state by the end of the 1820s. The Ste. Genevieve District, however, participated much less in land speculation than other parts of the state, especially in the mining area where the most coveted

45. John A. Paxton, *The St. Louis Directory and Register, containing the Names, Professions, and Residence of All the Heads of Families and Persons in Business; together with Descriptive Notes on St. Louis: etc.*, 263.
46. Foley, *Genesis of Missouri*, 208–9; LRB 8.

lands of the district were locked up as reserved and unpurchasable mineral land.

Basic economic development for most residents continued to depend on exploitation of local natural resources. Whether loessial and alluvial soils, natural grasslands, pines and hardwoods, the various minerals, or the river systems, the natural resources were exactly the same for French entrepreneurs and their Spanish administrators as for the arriving Americans. If their development were to take place, natural resources had to be exploited with the aid of credit and marketing institutions and linked to the rest of the world by competitive transportation. The territory's new position in a larger economic system under a more responsive government enabled the economy to accelerate after 1804, even though its basic physical geography had not changed one iota.

Therefore, economic development of the region after 1804 must be interpreted within a much more extensive geographical framework than before the transfer, one that is provided by zones of economic activities organized around a core region in the eastern United States and shown modeled in Figure 48. At the core were the large cities of the East where goods were consumed, business transactions made, strategy laid out, and wealth transferred. Surrounding the core was intensively used land of high value that provided products for direct sale in the cities, including more perishable products. Successively outward were zones of land use of correspondingly less intensively used land, because the value of land, in general, diminished with distance from the core. Beyond was unoccupied and little-used land, viewed as part of this centralized economic system. Upper Louisiana in the first decades of the nineteenth century lay in one of the outermost zones of occupation. The reason land had less value was certainly not its inherent natural endowment but geographic and economic distance from the core. The cost of overcoming that distance was so great—the economic friction of distance—that whatever was produced in these farther regions had to have a high value per weight, such as gold and furs, or have a low cost of production, such as open-range cattle (for hides), or be unobtainable in places nearer the core, such as lead.[47]

47. By use of a simple model, geographer Jordan shows how open-range cattle ranching in frontier regions can be interpreted as part of a world system. By extension, the model, which is basically the von Thünen model of rings around a center, can also be applied to other economic activities of a region being settled (*Cattle-Ranching Frontiers*, 7–17). Also see the transatlantic model developed by Meinig around his center-core-periphery concept (*Shaping of America*, 1:258–67). Historians have more recently incorporated spatial models to understand the American

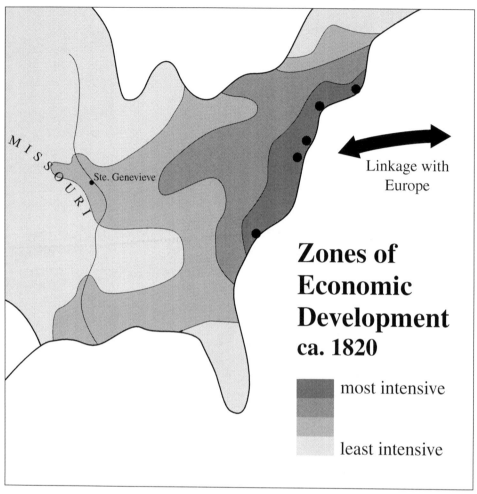

48. *Zones of Economic Development, ca. 1820.* The map is an idealized depiction of the intensity of economic development around 1820 and suggests that land use in Missouri was directly affected by access to water transportation that was needed for economic development linked to the American East.

Different degrees of access distorted the model of concentric zones. For example, navigable rivers such as the Ohio pushed any one zone outward, and less accessible locations away from rivers and brought the zones inward toward the core. Along the Mississippi River, residents used land more intensively because agricultural products could reach markets, but thirty miles away from it, land dropped into the next outward zone of range cattle and lead and fur exploitation. The constraint of distance, or, better, accessibility, determined what could be successfully produced for export, not the quality of the land and soil and not entrepreneurship of whatever ethnicity, Creole or American. Commodities that were produced beyond the zone of access had a market limited to the local population only and could not be directly tied into the larger regional and continental economy. Such was the case of the Bellevue farmers who had a market in the adjacent mining region but could not market to the Mississippi River. The major products of the region that fitted into the larger system of U.S. and world economic organization were wheat, corn, and tobacco near the river, and furs, peltries, lead, and animal products from less accessible lands. Return flows, or imports into the region, were no less sensitive to accessibility, if the imports were tangible, such as manufactured goods. Some imports such as cash and credit from government sources, Indian and military payments, and the capitalization of lead companies, fur trading, and river shipping were relatively insensitive to physical accessibility.

To integrate the economic activities of the various zones into the national economic system required human coordination of economic activity by countless individual decisions, which was articulated through a system of control points, or central places. They existed in hierarchical arrangement, from the highest order in the American East down to the lowest order in a local neighborhood. For example, a Baltimore or Philadelphia wholesaler-supplier sold to a merchant in St. Louis, such as Christian Wilt, a certain lot of goods, usually on credit. There may have been middlemen at Pittsburgh, Cincinnati, or Louisville. Because the wholesaler-supplier also sold to merchants in other towns in the East and West, his shipment to Christian Wilt constituted only part of a much larger shipment divided among merchants in other places. Wilt then resold the goods to the local St. Louis population, but he also consigned or sold a certain amount to smaller stores in the region. Several other St. Louis

---

frontier. See, for example, Gitlin's interpretation of the frontier as a hinterland of Europe that required geographic linkages and "connectedness" ("Boundaries of Empire," 72).

merchants joined Wilt in the redistribution of goods through their own branch stores in the Missouri Territory, where they were purchased by people in local neighborhoods without having to travel to St. Louis. For example, the Philipson brothers, William Shannon, and H. M. Shreve, all St. Louis merchants, had branch stores in Ste. Genevieve.[48]

The process requires contacts, contracts, credit arrangements, business-management skills, and knowledge of how to move goods by different conveyances. The process was repeated for goods and services from many places, including Canada and New Orleans, but the Ohio Valley and the American East dominated Missouri imports as time went on. This process firmly welded the new territory onto the overall American economic system before the 1820s.[49]

While the economic system of Missouri was geographically organizing itself hierarchically downward from capital-rich control centers in the East, the system was also geographically organizing itself upward from local neighborhoods into villages and towns. Commodities that came from individual farms, individual mines, and individual trades with Indians moved to the local grain mill, smelting furnace, and trade-goods buyer, and then, if the quantity exceeded what the local community needed, to a port for shipment to still higher-order places in the American East or elsewhere. Exporting from the district required coordination through people in all these articulation points of the system no less than importing did.[50]

As the economy became more sophisticated and complex in the 1820s, so did the transactional and geographic pathways. In 1787, Ste. Genevieve merchants annually procured their merchandise directly from Canada, which was the simplest and most direct pathway, usually making the trip themselves. After 1804, Ste. Genevieve merchant Ferdinand Rozier himself still traveled to Philadelphia for goods to sell locally, but this was a dying practice that middlemen in the district and along the route of movement were replacing. In the 1820s, wholesaling flourished in St. Louis both for redistribution of goods brought into the region and for increased warehousing and processing of locally produced goods for export.[51]

48. Jennings, *Pioneer Merchant: Christopher Wilt.* Atherton explains how local merchants had to have credit standing with wholesalers in larger centers and thereby created a version of geographic networks (*Frontier Merchant,* 18). Walter Ehrlich, "The First Jews of St. Louis."
49. Atherton, *Frontier Merchant,* 59–60, 80–86.
50. Meinig, *Shaping of America,* 2:248–49.
51. Alvord, *Kaskaskia Records, 1778–1790,* 5:411; Atherton, *Frontier Merchant,* 96–98. Both the wholesalers and any individual merchant who refused to buy through them made incredibly long trips to obtain goods. Morris Birkbeck, an Englishman

By the 1820s, a hierarchical pattern of economic central places was taking form in Missouri. The central places included both the collection and distribution of goods and their processing and, extremely important, the institutional control of the business transactions. These centers ranged from St. Louis at the top, although still low order in the national system, down to rural neighborhood stores and other dispersed economic activities.

St. Louis's unrivaled position was justified by such measures as the volume of its economic activity both in value and in tonnage, the diversity of its economic activities, the number of persons employed in nonfarm activities, the number and range of professional services of an economic nature, the number of buildings specifically devoted to commerce (wholesaling and retailing) and manufacturing, and the focus of transportation routes upon the town.[52]

Ste. Genevieve, Potosi, Jackson, and Franklin (in the Boonslick) composed the next lower level of economic places in Missouri in 1821. They had fewer economic establishments and a much narrower range of kinds than St. Louis.[53] Their basic services were mercantile stores, lawyers, and physicians. Still lower-order places, such as Herculaneum, Caledonia, Fredericktown, and St. Mary Seminary at Perryville, had only the most essential economic activities and services for local neighborhoods, consisting of a store, blacksmith, gunsmith, boot and shoemaker, mill, postal service, and church.

---

who traveled in the Mississippi Valley in 1818, marveled that people "start on an expedition of three thousand miles by boats, on horseback or on foot, with as little deliberation or anxiety as [English] should set out on a journey of three hundred" (*Notes of a journey,* 102). Birkbeck further observed that many people go to New Orleans by boat, seventeen hundred miles from Cincinnati, then walk back a thousand. Shopkeepers travel to Baltimore, New York, and Philadelphia once a year to buy goods, then return home to Illinois. Truly, the frontiersman's concept of distance was different from that of a resident of a long-settled region.

52. Both Lewis Atherton *(Frontier Merchant)* and Halvor Gordon Melom ("The Economic Development of St. Louis, 1803–1846") describe the economic primacy of St. Louis in the first third of the nineteenth century, but neither uses the concept of geographic linkages of central-place analysis. A geographic, but outdated, analysis of the rise of St. Louis to preeminence in the interior of the United States is Semple, "Geographic Influences." St. Louis's first business directory in 1821 provides an account of the range and number of business activities in St. Louis (Paxton, *St. Louis Directory and Register*).

53. The economic functions of lower-order centers were identified and counted from records of them in newspapers (especially business licenses and advertisements), travelers' accounts, community and county histories, and in Houck, *History of Missouri,* 3:186–97. An analysis of places in the Ste. Genevieve District by economic functions is in W. A. Schroeder, "Opening the Ozarks," 548–71, 610.

Some economic-exchange functions and manufactories were located in rural areas of dispersed settlement and do not fit into the emerging central-place system. Among these isolated activities were gunsmiths and taverns, such as William James's "house of entertainment" (or tavern) at the mineral springs on his farm on the Saline River, six miles south of Ste. Genevieve. Introduced by Americans into the district as a staple service to travelers, taverns were superfluous in French society, according to Henry Marie Brackenridge, because the hospitality of French homes filled that need. Boatbuilders operated at the mouth of Apple Creek. Charles Ellis Jr. opened an isolated store in his house on the Mississippi River in the Bois Brule bottom to serve a population relatively distant from stores at the Barrens, Jackson, Ste. Genevieve, or across the river at Kaskaskia.[54]

All of the several types of connecting tissues of the region may be combined and measured by the relative volume of business of post offices. The American postal system reflected administrative activity of the federal and state governments in the region; spatial relationships of the social and cultural activities of people through their personal correspondence; and economic activity of stores, wholesaling, manufacturing, and financing. Consequently, post office business serves both as a sensitive measure of the relative position of places in the emerging hierarchy of places in the region and as a measure of the interaction between the region and the rest of the country.

Additionally, a map of location of post offices is a good representation of the extent of effective settlement. The post office was the government service closest to individual citizens. Every person was to have reasonable access to it, and virtually every neighborhood in Missouri of more than a few hundred people had a post office in the 1820s. Presence of postal service served as a geographic indicator of a certain critical mass of population and economic activity on the frontier.[55]

Figure 49 shows the volume of gross receipts of all seventy-seven post offices in Missouri (and fifteen in Illinois within twenty-five miles of Missouri) in 1828.[56] The map powerfully shows population and economic

54. On James: *Missouri Gazette,* June 28, 1817; on Brackenridge: Wish, "French of Old Missouri," 185; on boatbuilders: *Missouri Herald,* August 20, 1819; on Ellis: *Jackson Independent Patriot,* January 4, 1821.

55. James R. Shortridge used the year and location of the establishment of post offices to trace the spread of settlement in Missouri ("The Expansion of the Settlement Frontier in Missouri").

56. The map is based on dollar receipts reported annually from individual post offices. No expenses have been subtracted. The data come from U.S. House, "Postage

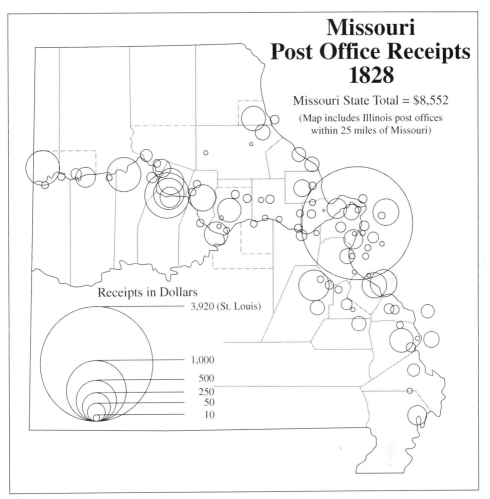

49. *Missouri Post Office Receipts, 1828.* St. Louis's dominance in Missouri may be measured by its accounting for almost half of all post office receipts. Eight other towns, most on the Missouri River, had surpassed Ste. Genevieve and Potosi as economic centers in Missouri. By the end of the 1820s, the historic Ste. Genevieve District was falling economically behind the rest of the populated state. (20th Cong., 2d sess., House of Rep. Doc. 71, 1829)

development restricted to a zone no greater than thirty miles from the Mississippi and Missouri Rivers, except for the mining country of the Ste. Genevieve District. It also shows an enormous concentration in St. Louis that is disproportionately much greater than its population. St. Louis accounted for 46 percent of the total Missouri post office receipts. St. Louis had been designated the western regional center for the national post office system at the beginning of the decade, an early indication of its imminent rise into the national system of central places.[57] Its centrality went beyond state boundaries to include federal and private-business interests in all the lands opening up to the west, including Santa Fe in Mexican territory, the Red River Country along the Texas border and Mexican Texas, and the Missouri and upper Mississippi River traffic, which still focused on Indian relations and the fur trade. Washington funneled its communication links with its vast interests in the great West, whether part of the United States or not, through the St. Louis post office.

Second-order places (those with receipts of more than $250) are most numerous along the Missouri River. The Boonslick region, which had no centers before 1815, had been rapidly filling in with settlers since then, towns platted, and counties erected. Steamboats encouraged commercial economic development, and Missouri located its state capital at Jefferson City, just above the mouth of the Osage River.

In the eastern Ozarks, much less had happened since statehood. Ste. Genevieve, Potosi, and Jackson (with a land office) continued as second-order centers, with Herculaneum, Fredericktown, Farmington, Caledonia, and Perryville–St. Mary Seminary somewhat behind them. Growth in the district lagged behind other settled regions of the state. It is curious that the post office in Greenville in Wayne County reported no receipts during the 1820s (it reported $19 in 1830), although it was on the mail route into Arkansas Territory. Although the lower St. Francis River area was well-enough populated to have its own county by 1818 and was on a post office route and a well-traveled road to the Southwest, it was

---

Accruing for Year Ending 31st March, 1828," 20th Cong., 2d sess., H. Doc. 71, 1829, 73–74. Comparable data for other years of the decade are in U.S. House, "Letter from the Postmaster General Transmitting Statements Shewing the Amount of Postage Received at Each of the Post Offices in the United States and the Territories Thereof during the Year 1822," 18th Cong., 1st sess., H. Doc. 95, 1824, 1–118; U.S. House, "Postage Accruing in the U.S.—1826," 19th Cong., 2d sess., H. Doc. 35, 1827, 71; and U.S. House, "Postage Accruing for Year Ending March 31, 1831," 22nd Cong., 1st sess., H. Doc. 262, 1832, 79. For maps and data for other years in the 1820s, see W. A. Schroeder, "Opening the Ozarks," 564–66, 611, 684.

57. R. G. Schultz, "Postal Service," 118, 128.

not an administrative center of any consequence and apparently had not developed economically enough to generate much mail.[58]

During the 1820s, Ste. Genevieve village could not maintain its relative rank in the evolving hierarchical system. Its share of postal receipts in the state dropped from 4.1 percent in 1822 to only 1.2 percent in 1830. Furthermore, the decline in real dollars in post office business at Ste. Genevieve was continuous through the decade: from $268 in 1822 to $167 in 1826, $159 in 1828, and down to $125 in 1830, or less than half as much business as eight years earlier! A *relative* decline was to be expected as population spread throughout the state and Ste. Genevieve became only one town of dozens, but an absolute decline in dollars indicated economic stagnation or even decline and contrasted sharply with growth in almost all other towns. The whole historic Ste. Genevieve District was also declining relative to other parts of the state. In 1822, the counties of the historic district accounted for 12 percent of postal receipts, and in 1830, only 7.5 percent.

Thus, while the district was maturing in its settlement geography—arable lands being put into production and mines being exploited—and was interlinking itself with a network of roads, it was at the same time declining as a component of larger administrative and economic systems. Any hope that Ste. Genevieve, the first settlement of the territory, would become the chief central place in the middle Mississippi Valley faded in the 1760s when the Spanish designated St. Louis as the administrative center of Upper Louisiana. The Spanish selected St. Louis because of physical geography: its strategic location for defense and for controlling the fur trade in a river-oriented world, both imperial goals that had little to do with the settlement pattern of Creoles in the valley. Then, as time went on, St. Louis, though continuing its function as the

---

58. The sum of $19 does not seem to be much business; some post offices returned as little as $1 in receipts for one year. However, this small dollar amount was at a time when David Wills, of Wayne County, had his name printed in the *Jackson Independent Patriot* for delinquency in taxes for the grand amount of $0.04 to the county and $0.08 to the state (December 8, 1821). Truly, every cent counted on the frontier. Other possible explanations for the low dollar amount at Greenville are that receipts were reported through Fredericktown, that Greenville had a postmaster appointed not to receive money but only to distribute mail, or that Greenville's returns for 1828 were delayed and included in the returns for the following year, a common practice.

The cost of sending a letter in the 1820s varied according to the distance sent. It ranged from six cents for deliveries less than thirty miles to twenty-five cents for deliveries more than four hundred miles (*Historical Statistics of the United States, Colonial Times to 1957: A Statistical Abstract Supplement,* 498).

administrative center for the territory and all the West, turned to new economic activities, especially the processing of raw materials into products, which Ste. Genevieve did not do, and river commerce, which largely ignored Ste. Genevieve. New towns arose in the district and displaced Ste. Genevieve from its original hinterland.

Lucille Basler, a respected local historian, concluded in her history of the Ste. Genevieve District that by the 1820s, the town had "degenerated so much, and [become] of so little commercial importance that it could not sustain a hotel." When Timothy Flint revisited Ste. Genevieve in the 1830s, he observed that "the town does not appear to thrive, not possessing more inhabitants than it did 30 years ago." William Shannon, a principal merchant of Ste. Genevieve in 1806, left the town for greater opportunities in St. Louis and was followed by merchant Horace Austin and others. Conversely, no single business left St. Louis or any other town to move to Ste. Genevieve during this time, although St. Louis businesses set up branches there.[59] Such economic losses were compounded by the relegation of Ste. Genevieve's administrative function to a simple county seat of one of the least-populated counties in the state and one with no state or federal administrative function.

59. Basler, *District of Ste. Genevieve,* 335–37; Flint, *Condensed Geography,* 2:100; on Shannon: Houck, *History of Missouri,* 3:193–94; Walter Ehrlich, "The First Jews of St. Louis."

Conclusion

16

The Ste. Genevieve District as a Settlement Frontier

The laws of biology are exactly the same across the United States, yet, because of the uniqueness of time and place, they produce an infinitely complex landscape of plants and animals. In like manner, when the historical and geographic processes that shape the human experience interact in unique times and places, the result is also an infinitely complex cultural landscape. The Ste. Genevieve District, as it passed through the decades from 1760 to 1830, transformed itself into a variety of communities and settlement forms each the result of its distinctive time and place. But if we stopped with idiosyncratic identification and explanation, however fascinating that is, we would not get too far in understanding the human experience. The experience of the Ste. Genevieve District must be put within a larger context of settlement of American frontiers.

Cultural Diversity on the Frontier

Contemporary American concern for multiculturalism—the identification, preservation, and celebration of different peoples—raises the question whether we unthinkingly project that concern retrospectively into an exaggeration of the significance of cultural differences on past frontiers. Certainly, people of different cultural backgrounds settled in the Ste. Genevieve District, and it is a legitimate question to ask what role cultural differences had in the settlement process.

What we conveniently call "French" society during the colonial period actually comprised people whom the Spanish identified by a variety of national identities. Spanish censuses and other documents disclose the presence of Alsatians, Germans, Spanish, Flemings, Scots, Protestant Irish, Anglican English, Catholic English, "Bostoneses" or Americans, and variations among the French groups: Caribbean French, Canadian French, Indian French, Creoles, and French directly from France. French Creoles born in the valley were the most numerous of all and defined "Frenchness" in general for the district. Despite their identification as distinct nationals, all of these whites, including mixtures among them, were subsumed under the designation of French culture, and they generally interacted as one single cultural group. Later, Americans were identified by states of origin, but more commonly by religion.

Consequently, cultural differences among the white population were usually reduced to a dichotomy between "French" and "Americans." Reading documents establishes that both groups held strong opinions of the other. The French, according to the Reverend Timothy Flint, "were born for the most part in the woods, or at least far from society." Amos Stoddard believed they "mostly submitted themselves to indolence and fatuity" and spent much of their time on social pleasures. Americans used the term *French fence* derogatorily to refer to "logs laid round where the fence ought to be," that is, just a pile of logs instead of any effort to build a real rail fence. The French were creatures of habit, "strongly attached to the ways of their forefathers." They cowered before Indians.[1]

To the French, Americans were rude people. They stole. They were more preoccupied with laws than the smooth functioning of society. They were violent people, and their government was tolerant of its citizens' turbulent life. The French bishop of Louisiana in 1799 believed that Americans, including those in the Illinois Country, were "a gang of adventurers who have no religion and acknowledge no God, and they have made much worse the morals of our people."[2]

However, one soon realizes that these characterizations were made by the elite of both groups for the common folk, and that the more learned and politically active French and Americans never included themselves in assessments they made of common folk, whom they considered culturally inferior. In fact, as events in the post-1803 years proved, the elite of both groups had much in common and found it in their interest to cooperate on issues concerning land, thereby effacing supposed cultural differences between them, at least in public affairs. Viewed in this way,

---

1. Flint, *Condensed Geography*, 2:105; Stoddard, *Sketches of Louisiana*, 294; on *French fence:* testimony of Charles McDurmett, December 5, 1807, *ASP-PL* 3:598; on creatures of habit: Flint, *Condensed Geography*, 2:106. Perhaps the most extreme view of the French in Missouri was expressed by an anonymous Vermont author in 1828 who wrote, "The French are numerous, and are many of them ignorant and vicious" (as quoted in Harrison A. Trexler, "Missouri in the Old Geographies," 150 n).

2. Gayarré, *History of Louisiana*, 3:407.

group characterization is an expression of social class independent of cultural or ethnic group, and multiculturalism on this frontier, expressed as nationality, language, or religious differences, was outplaced by socioeconomic differences.

Thus, a conclusion of this account of the settlement of the Ste. Genevieve District is that one's "ethnicity" made less difference than one's station in society in matters concerning land. Aristocrat Luzières's elitism contributed to the unraveling of the New Bourbon community of habitants. In Mine à Breton and St. Michel, miners of various national origins worked together, side by side, but they had vast differences between themselves, as a class, and the large operators. The salt workers of whatever origin had serious differences with Henri Peyroux and the Vallé families. The Mine à Breton miners struggled mightily with Austin and he with them. When the ordinary Creoles of Old Mines sought arbitration for dispute over water degradation, they specifically sought out the American captain Stoddard, not their co-Creole commandant Vallé, whom they contended would not be fair to them.

This conclusion is strengthened in the American period when society became more structured along socioeconomic divisions. Land issues—the major issues of the territorial period—followed class lines, poorer French and Americans on one hand and the wealthier of both on the other. Many of the poorer French simply despaired of getting their untitled soil rights confirmed, and perhaps as many as half of the ordinary Americans chose to be squatters even after public land went on sale. Prominent French and wealthy Americans kept their land claims alive until their eventual confirmation. A few Americans highly placed in federal administrative positions, however, insisted on enforcing the rules of American democracy—that law should treat people equally, regardless of class—and they struggled to keep the district from devolving into one of even greater inequality in landownership.

Assimilation of people of diverse national origins and religions into a single culture structured by class did not necessarily extend to those of different skin color. In general, in the early years of settlement, the three racial elements did cooperatively interact, as Faragher demonstrated for the lower Missouri River Valley, although land records for the Ste. Genevieve District are less persuasive on this point, certainly in regard to the Osages.[3] In the early stages of settlement, one can imagine that sparse population lent itself to cooperation among all racial elements. A white male population's need for wives, the scarcity of labor,

---

3. Faragher, "'More Motley than Mackinaw.'"

the necessity to trade for essentials, and the search for information about local geography promoted cooperation among all people in the region. In addition, imperial governments do not much care who their subjects are, as long as they maintain allegiance, order, and a certain level of productivity.

In the Ste. Genevieve District during the Spanish period, blacks, as slaves, were forced to "cooperate," but, as Carl Ekberg has shown, more than a few seemed to be accepted as part of the larger society by marriages and property ownership.[4] From a spatial perspective, because slaves lived in cabins on the same lot as their owners—distances of less than one hundred feet—and the two interacted daily in homes, gardens, fields, and mines, one cannot help but conclude that the two shared much. They shared religion, language, food, and probably such intimate experiences as raising children. People cannot live juxtaposed long without incorporating aspects of each other's culture. Culture sharing, however, can never overcome the oppressive weight of human bondage, however humane slavery may have been.

Indians, as a single category, had equally complex interracial relationships. The immigrant Shawnee and Delaware, numerous enough to compose their own community on their own dedicated lands, lived in correct social and economic relations with the French, even though they occupied separate space. On the other hand, these Indians were enemies of others of their race in the district, the Osages. Then, when Americans came into the district, the lot of the Shawnee and Delaware changed abruptly, and they were stripped of their lands. Although such relations are interpreted as "exclusionary" in frontier terminology, one can interpret the Indians' departure from the district differently. Before Americans invaded the Indian tract for settlement, the Shawnee, who had a peripatetic reputation, were already seeking lands elsewhere and had asked the American government to relocate them. Before antagonism sharpened between Americans and these Indians, the Shawnee and Delaware were experiencing a reduction of game and a general inability to carry on their traditional lifestyle. No different from frontiering American woodsmen, these Indians were westering for land, and their departure from the Ste. Genevieve District probably resulted from the same forces that westering Americans were responding to in their search for game-filled woodlands with a modicum of patch agriculture. From its establishment, the Indian tract clearly served as a pass-through tract for Indians who never intended to live permanently on it. It was only later that

---

4. Ekberg, *Colonial Ste. Genevieve,* 197–239.

American aggressiveness forced the Indians to surrender ownership of the land. Had the Shawnee-Delaware tract been initially occupied instead by American woodsmen, they probably would have left it also, leaving their cabins and cleared land for others to occupy, just as the Burns family did in the adjacent Bois Brule.

Relations between whites and the Osages to the west of the district were uneasy at best and occasionally flared into violence during both Spanish and American administrations. The Osages entered mining camps, stealing horses and sending the white invaders fleeing back to their riverine-village refuges. The Spanish government once declared war on the Osages at the same time that it gave its blessings to St. Louisan Chouteau's negotiations with them, involving gifts and other appeasements. This action perplexed the peaceful Shawnee, who never received rewards for their cooperation. When the Osages continued their attempts to retain control of their land after Americans entered, this time they met with forceful resistance in the form of guns and pursuit. Within a short time, the Osages were gone from the district and much of Missouri. Records at St. Louis show that the Osages did incorporate French and mixed bloods into their life by awarding land to them (although the Osage concept of landownership was different from the European and American concept), but such records are lacking for the Ste. Genevieve District. Thus, as far as the Ste. Genevieve District goes, the Osages were not included in the cooperative, interactive frontier.

Frontiers, it is believed, tend to be culturally inclusive in their early encounters and become exclusive as settlement proceeds. Whether this is by administrative action, as Faragher emphasizes, or by market forces, as Usner emphasizes, the Ste. Genevieve District teaches us that it is one of degree and, quite important, a matter of which groups are involved.[5]

The early incorporation of national groups into a single functioning society continued into the American period, although increasingly stratified socioeconomically. The U.S. census, for example, did not collect information on national or ethnic origin for whites. People, including annexed French Creoles, were simply "Americans." The push to populate the frontier was officially blind to cultural variations on this issue, which it should be to a republican government, as Captain Stoddard proclaimed to his mixed audience in his inaugural speech on March 10, 1804, at St. Louis: "[D]raw the veil of oblivion over [past differences and] embrace each other as brethren of the same mighty family." The French them-

---

5. Mikesell, "Comparative Studies in Frontier History"; Faragher, "'More Motley than Mackinaw'"; Usner, *Indians, Settlers, and Slaves.*

selves were incorporated, however imperfectly and gradually, into the American cultural fabric, or, as historical geographer Donald Meinig puts it, "minorated" by being submerged in a much larger American population.[6]

But was the Americanization of the French simply a matter of demographic submergence by Americans? The larger American population and its widespread distribution into all the district neighborhoods cannot be discounted, but other factors worked against the perpetuation of a French way of life. First, one can argue that the French had a specialized way of occupying land. Their custom of a tripartite village, upon which so much of the successful functioning of the entire society was built, required cleared fields, organized animal herding, and, above all, a strong group solidarity and *mentalité*. Such specialized land strategies, encumbered as they are with land-use regulations, lose out on the frontier to nonspecialized, regulation-free land strategies, as Terry Jordan and Matti Kaups have pointed out.[7] In the Ste. Genevieve District, it was westering upland southerners who were eminently preadapted with guns, tools, and land-use techniques to move into hilly woodlands. Their "messy" fields filled with stumps, open-range animal raising, living on the land rather than commuting to it, and loosely structured communities were more efficient ways to spread into raw frontier lands and furthered the disintegration of the traditional French village system.

Second, it may be argued that the French had a smaller entrepreneurial element in their population than Americans. Although some French did engage in large-scale mining, stock raising, farming, and Mississippi River trade, Americans of this bent were more numerous. Recognizing the inadequate numbers of tradesmen and professionals among the Creoles of Upper Louisiana, French commandants and Spanish lieutenant governors welcomed them from the United States, even in defiance of official immigration policy, and when they came, the Americans had more and better market contacts for supplies and sales. Potosi, Herculaneum, Fredericktown, and Caledonia all were run by American businessmen, and Ste. Genevieve village itself became a town of American merchants and professionals. However, the works of Susan Boyle and Anna Price show that entrepreneurial families in Ste. Genevieve engaged in strategies, such as diversification of their business activities, to cope with risks on the frontier and that, in general, the community was far from being a simple agrarian colonial backwater. We need to do more work to uncov-

6. Stoddard, *Sketches of Louisiana,* 105; Houck, *History of Missouri,* 2:370–72; Meinig, *Shaping of America,* 2:173.
7. Ekberg, *French Roots;* Jordan and Kaups, *American Backwoods Frontier.*

er the degree of French entrepreneurial activity in colonial Upper Louisiana. Berquin-Duvallon, the French visitor touring the valley before the transfer, observed that the American government on the east side expected and enabled individuals to be entrepreneurs, but the Spanish government considered building mills and providing services a governmental responsibility, not one left to individual enterprise.[8]

Finally, the Creole French in the Mississippi Valley as a group descended from immigrants from seventeenth- and eighteenth-century France and thus carried the cultural legacy of a prerevolutionary peasantry since their Canadian days. Becoming free agents in the Mississippi Valley far removed from colonial centers of control did not necessarily transform them into equivalents of Americans who were transformed from British colonists. Lacking the stimulus of further immigrants from postrevolutionary France carrying new ideas, these valley Creoles retained values and attitudes echoing the ancien régime that Americans entering the valley immediately noticed: subservience to authority, both ecclesiastical and political; the habit of doing as their parents did; a reluctance to try new crops and new ideas (slavery an exception); and fondness of sitting around rather than working. Thus, their way of life, which historical geographer Carl Sauer nearly a century ago labeled inferior to American ways and probably a long-standing legacy of the Old World, receded in favor of American ways.[9]

French aristocrats who found their way into the continental interior had an opportunity to use their resources and leadership to organize land and society. In this regard, evidence strongly supports the conclusion already established for Laurentian Canada by Cole Harris and for Maryland by Carville Earle that the privileged class was unsuitable for the frontier. Spain's often-repeated desire to settle Louisiana with a European Catholic peasantry administered by a privileged elite was unsuccessful because they did not attract enough of the privileged class, and, when the few did come, they could not prevail. No document better divulges the Old World way of organizing land and society than immigrant aristocrat Pierre-Charles Delassus de Luzières's observations of 1797 in which he describes the nature of the district's habitants (that is, peasants), lays out his extensive plans for development of the New Bourbon District, and lists

---

8. Flader, "Final Narrative Report," 5; Boyle, "Did She Generally Decide?"; Anna Price, "Economic Change in Eighteenth Century Ste. Genevieve"; Berquin-Duvallon, *Vue de la colonie espagnole*, 61–62.

9. Moogk notes the persistence in Canada of aspects of the ancien régime of France and also analyzes the transformation of French immigrants into Canadians and Acadians in *Nouvelle France*, xvi, 265–81. Sauer, *Ozark Highland*, 92–93.

his complaints about his subordinates, exactly as if he were back home in prerevolutionary Flanders.[10] In the end, the lure of land for the taking, lure of the mines and fur trading for quick profits, lure of the life of a river man—all to the abject neglect of agriculture—made it impossible to keep people, except slaves, subservient to an aristocrat. The Luzières family, including his proud wife, ended up toiling in the house and fields alongside their slaves and hired help and descending into debt.

Consequently, as Cole Harris has hypothesized for frontiers of European settlement in general, the French colonial frontier in the eastern Ozarks was "simplified" from its stratified European base at the start.[11] European class differences vanished, and a new social structure had to be created from within the group itself. This came about when some were able to take advantage of markets and connections to make profits, whether in wheat, salt, lead, furs, or land speculation. These self-made local Creoles, of whom the Vallé family was the most notable at Ste. Genevieve, created the first "elite" in colonial Upper Louisiana and replaced immigrant Old World aristocracy.

The Environmental Encounter

Frontiers are thought of as zones of encounter between invaders and indigenous residents, but they are equally zones of encounter between arriving peoples and the environment, as William Cronon has shown for Europeans taking up land in New England.[12] The environmental encounter on the frontier is not one to be painted with broad strokes; it requires attention to minute geographic detail, as the Ste. Genevieve District informs us. For example, it made a difference at old Ste. Genevieve that houses were situated near the Mississippi River bank. It is conclusive that the abandonment of the village, an act that set in motion a whole host of new settlement decisions and expansion into the interior, was due more to loss of structures and land from bank caving than to flooding, which has heretofore been presented as the cause. Neighboring Kaskaskia suffered as much from flooding of its houses and fields, but because its houses on the quieter Kaskaskia River were not subject to bank caving, Kaskaskians did not have to abandon their built village, and it persisted until late in the nineteenth century.

The village of St. Michel serves as another example of the need to pay

---

10. R. C. Harris, *Seigneurial System;* Carville V. Earle, *The Evolution of a Tidewater Settlement System: All Hallow's Parish, Maryland, 1650–1783;* [Luzières's observations on the means of increasing the prosperity of the colony], 1797, PC 2365-345.
11. Harris, "Simplification of Europe Overseas."
12. Cronon, *Changes in the Land.*

attention to geographic precision, because it was the presence of a small, compact tract of alluvium, unusual in the mining country, that permitted the development of a traditional French common field of long lots. Thus, the Creole settlement at St. Michel has to be explained as much by physiographic detail as by political or economic factors. Similarly, the quiet, deep water mouths of Mississippi River tributaries, only a few hundred feet long, determined the precise sites of river landings and harborages. Settlers' sensitivity to fevers and malaria forced them to consider such microdetails as hillside exposure to the sun, sinkhole ponds, and blufftop locations when placing their homes and fields.

The closest geographic relationship of people with the environment was with soils. Settlers rapidly identified the productive soils of the district, even if only a couple dozen acres in extent, to locate their improvements. The spatial correlation of alluvial, loessial, and limestone soils with claimed lands in 1804 in the Ste. Genevieve District is nearly perfect.[13] Indeed, a map of soil quality in this hill region would be a good predictor of the location of early settlement. Frontiers, as zones of encounter with the environment, demand attention to geographic detail of the natural environment.

All migrating people visualize the new lands they are entering. They form mental images of the landscape and the spatial arrangement of its features. Where are the rivers, and how are they interconnected? What is the shortest route between the river and a mine? Visualizations of spatial patterns improve in accuracy and precision as more people share their experiences, and better mental maps make further entrance into the region less risky and administrative management of it more efficient.

Eventually, visualization takes the form of real maps, which stand as documents of how people understood the lands they were occupying. In this regard, it is instructive that the Spanish and French had maps only of the courses of the Mississippi and Missouri Rivers. The lands along them were replete with place-names, but lands away from them were toponymically poor, despite the storehouse of knowledge already accumulated. Accurate locations of lead mines and hundreds of land concessions, though already precisely surveyed on the ground, are absent from these maps. Because the Spanish administrators' focus on Upper Louisiana was on its role in international affairs and the fur trade, the Mississippi River was all important, and the rest of the region was not. Neither

13. James D. Harlan, "A G.I.S. Analysis of Early Land Choice in the Historic Ste. Genevieve District." Harlan's map of the superposition of property lines upon topography and his statistical analysis are reproduced in W. A. Schroeder, "Opening the Ozarks," 85–86, 591, 620.

the French nor the Spanish visualized European settlement ever extending completely across the landscape; hence, they had no interest in mapping it.

Entering Americans, however, visualized Upper Louisiana differently, as a region to be settled throughout. Moses Austin began immediately to map the mining country for its development, and the federal government set about to survey every square mile to enable the orderly spread of Americans across the Ozark hills.

Until the rectangular land survey was draped across the landscape, vague visualizations led to errors in settlement decisions involving distances and locations. Austin, not having the use of any map of the interior when scouting, seriously misjudged access to the mines, believing that the overland distance to the mines was shorter than it really was and concluding that the Meramec River was navigable. Although these two errors did not bring down his mining enterprise, they did cause Austin and all others intractable problems with the high cost of land transport. One wonders if Austin would have begun his Missouri adventure had he had a better geographic understanding of the region.

Lacking accurate maps, Father Maxwell misunderstood the difficult terrain and the long distance to the land he selected for his colony at the forks of the Black River, and his incorrect visualization of space must have contributed to the project's failure. In other instances, incorrect mental maps led to continuous boundary problems between landholdings (as in the overlapping claims of Peyroux and Vallé at La Saline). Frontier land decisions must be evaluated in terms of how well or how poorly people visualized the lands they were entering.

Population in the Mississippi Valley concentrated along the Mississippi River, both sides of it, and thus the river served as a unifier of human activities despite its appearance on maps as a boundary separating countries until 1804. Although Spanish immigration policy hindered trans-river migration, activities extended across the big river as if it had no jurisdictional significance. The French had relatives and businesses on both sides, a priest often served both, Spanish officers traveled British-side roads, and the fur trade disregarded the river as a boundary. A major reason that Upper Louisiana officials did not fear American immigration as authorities in New Orleans did was that both sides of the river were already functionally and profitably linked. On frontiers, administrative territorial jurisdictions, so prominently displayed on maps in contrasting colors, may mislead those not on the scene as to their significance in how the frontier region actually functioned.

The arrangement of resources—not merely their presence—must be considered in understanding the settlement process. That surface lead was widespread over thousands of acres and exploitable without capital by the crudest methods allowed hundreds of miners to dig their own pits and operate reasonably independently. Dispersal of mining into many hands delayed technological improvements to make the industry more profitable and spur its expansion, even after Austin's initial success with the reverberatory furnace and shaft mining and John Smith T's aggregation of lead land. Thus, the simple presence of lead provided the initial impetus for immigration into the eastern Ozarks, but peculiarities of its geographic distribution held the key to why the industry developed the way it did.

As on other American frontiers, the Ste. Genevieve District offers little evidence of concern for maintaining environmental systems or specific resources, although occasional voices spoke out that degradation was occurring. Loessial soils eroded badly within a few years of use. Untold acres of timber were repeatedly cut for lead-smelting fuel. Commandant Luzières decried the indiscriminate burning of woodlands. People believed that lead washing polluted stream waters and that lead smelting poisoned nearby plants and animals, but governmental attempts to do something about it met with little cooperation. As on other frontiers, the sheer abundance of land, timber, and water meant that the extra effort expended to maintain productivity was illogical and an inefficient use of labor and time. As a result, the only approach of settlers to the environmental encounter was to exploit resources with little thought for the future.

Administrative Settlement Policy and Traditional Practices

Frontier settlers experience a friction between their traditional practices of occupying land and administrative direction for occupying it. The struggle is implicit in almost all frontier studies, including Andrew Cayton's Kentucky study in which he shows how a rule of law eventually replaced early settlers' "rights to the woods." So it was in the Ste. Genevieve District where a loosely regulated and supervised population, operating according to custom, was gradually replaced by a society operating according to a rule of law and a sense of administrative "order." It was during the multidecade process of replacement, as Stuart Banner demonstrates for the Missouri Territory, that friction occurred between the two settlement approaches.[14]

14. Cayton, *The Frontier Republic: Ideology and Politics in the Ohio Country, 1780–1825*; Banner, *Legal Systems in Conflict*.

Both Spanish and American governance was promulgated from imperial and national centers and based on considerations far removed from what was occurring in the Ste. Genevieve District: international strategies and Old World settlement systems in the case of Spain and eastern views on how the public domain was to be managed in the case of the United States. Residents in the district disliked or ignored any "institutional impediments" to acquiring and occupying land.[15] As time went on, however, distant presidents and Congresses heard local voices, adapting settlement policy to conform with local practices.

Several examples illustrate this point. Spanish immigration policy to keep out Americans eventually relented to their admission, especially in Upper Louisiana where Americans refused to cease their westering habit. American policy in respect to Indian land eventually collapsed because authorities could not keep American settlers from their practice of occupying Indian lands. Settlers disregarded the prohibition to settle on the public domain not yet surveyed and put on sale. The American government attempted to limit settlement in various ways for good reasons from a Washington point of view, but settlers would suffer none of those restrictions. And in the case of lead mining, which Spanish land laws did not directly address, miners dug on the royal domain, each miner respecting each other's pits according to ancient tradition. The American government sought to control this ancient custom by a land-leasing program that made good sense to easterners, but eventually, after years of lobbying from the Missouri frontier, the government put the lands on sale. The explanation common to all these examples is American land aggressiveness, the inability of government to control where settlers went.

Both Spanish and American governments had policies to limit size of landholdings (for a peasantry in the former and to prevent speculation and monopolies by the latter), but the Spanish nevertheless awarded enormous parcels of land to favorites. And, as Paul Wallace Gates has pointed out, Americans did not always achieve their national goal of distributing land equitably to a farming yeomanry; wealthy land claimants, both French and American, generally got their way, whether by pressing successfully for confirmation of their enormous pre-American claims or by buying and amassing large acreages during the lead boom years.[16]

In cases where law and policy were lacking or inappropriate, the settlement process went on according to the customs of the people, and in

15. Term is from Meinig, *Shaping of America*, 1:361.
16. Gates, *Jeffersonian Dream*.

this respect, we see a major difference between the colonial and the American regimes. As Donald Meinig explains, during colonial times, society tended to be shaped more by the actions of individuals and local groups than by imperial policies. However, during the American period, with more specific laws and a more determined attempt to enforce them, the government exerted greater control and order to settlement. This contrast leads to the interpretation that the French in the district were more tradition bound than Americans, which may have been the case, but if it was, it was due as much to other reasons than whether authority was enforced. Authoritarian regimes commonly rely on custom and obedience. Democratic governments have more laws to regulate behavior and public processes, including how to acquire and occupy land. "Tradition and rights counted heavily in the Illinois Country," concluded Carl Ekberg, but the strong habitant *mentalité* succumbed to an American "rule of laws."[17]

Settlement Pattern

The complex spatial pattern of landholdings in the district derived from multiple sources: people who settled it, environments they encountered, land regulations under which they settled, and the economic activities they engaged in. Thus, the settlement pattern becomes a tangible and revealing record of the complexity of the total human experience, as historical geographer Edward Price has shown in his detailed study of land division in the eastern United States. To aggregate the variety of settlement forms of the Ste. Genevieve District into seven categories creates a misleading simplicity warranted only for analytical purposes. In other regions of great cultural diversity settled under different regimes, such as Lower Louisiana, the settlement pattern is also complex, but in regions of basic cultural homogeneity, the settlement pattern is simple.[18]

Tradition determined the first land-division systems implanted on the frontier. The earliest French occupied the western side of the Mississippi River (as well as the eastern side) according to their customs. Officials may have formalized the landholdings by granting concessions to the land, but the Creoles arranged themselves on the land and determined how they used it according to ancient practices. Then, when old Ste. Genevieve broke up and habitants looked for other places to live, they initially took out lands in various locations by repeating the traditional long-

17. Meinig, *Shaping of America*, 1:249; Ekberg, *Colonial Ste. Genevieve*, 460.
18. E. T. Price, *Dividing the Land;* Hall, "Louisiana Survey Systems."

lot shape of a few arpents width by 40 arpents length. They did this in the Gabouri Valley, in the Bois Brule, at the Big River Bend, and at St. Michel. Even the partition of Old Mines into private properties later in the American period reproduced the long-lot pattern, though it was not done for farming purposes.

Topography shaped settlement patterns everywhere through natural lines and curves. Even after straight cadastral lines were laid down in the landscape, topography determined where houses and fields were located, as landownership parcels were initially much larger than necessary for a farm. Economic activities differentiated mining settlements from agricultural ones and American commercial towns from French agricultural villages. Eventually, of course, administrative regulations regulated the settlement pattern; the American land survey determined size, shape, and orientation of properties and forced a rigid compliance with it no matter what one's cultural practice was in land division or what the topographical circumstances were.

Another conclusion regarding settlement pattern is that large landholdings had a pronounced negative effect on the frontier settlement process, consonant with the conclusions of Robert D. Mitchell for the Shenandoah Valley and Cole Harris for Laurentian Canada.[19] Joseph Pratte's enormous 20,000-arpent Iron Mountain concession not only lay undeveloped for iron for a generation, but also delayed its agricultural settlement by being in litigation so long. The Luzières square league, supposedly a *vacherie* and seasonally inhabited by miners, lay adjacent to thriving Murphy's Settlement of ordinary farmers but remained undeveloped while the county seat of Farmington was laid out on its border on fast-appreciating American land. Frontier Americans valued land more as a commodity to be subdivided and developed than as ancestral property to be kept in the family. In addition to denying settlers ownership of land, many of these large tracts had their timber resource destroyed for fuel demands, and thus lost their value as farmland.

The large size of even those tracts occupied by ordinary farmers—the standard size was 640 acres American or 800 arpents Spanish—affected the settlement process. Because 1 square mile was greatly oversized for the needs of one family, its size forced farms to be spaced farther apart than they would have been otherwise. Farms spaced at mile or half-mile intervals promoted dispersed settlement and hindered the buildup of adequate populations for the development of community institutions.

---

19. Mitchell, *Commercialism and Frontier: Perspectives on the Early Shenandoah Valley*; R. C. Harris, *Seigneurial System*.

Thus, it follows that the organization of community life in the eastern Ozarks was initially harder to achieve than in the New England rural villages, where rural property sizes were significantly smaller and local populations greater in any given neighborhood.[20]

Conversely, the large property sizes per farm family—so astonishing to European visitors—in time contributed to community stability and cohesion because parents could readily subdivide their properties for their numerous children and keep them at home. Where children and grandchildren could find land near their families, the neighborhood would develop more cultural cohesion than where offspring had to move away to find land. Communities thus grown more cohesive as time went on, such as the Barrens, contrast with the experience of Kenneth Lockridge's Dedham, Massachusetts, where deterioration of the original covenant community of noticeably smaller-size properties set in when its children had to leave to find land.[21]

We further learn that the frontier does not have to pass through "stages" of settlement. Although a hunter-woodsman stage did precede farming in many neighborhoods of the district, the French villages, the American Bellevue and Barrens communities, and the mining camps prove that this earlier stage was not necessary.

Settlement Systems

Geographer James Vance thought that the historical geography of North America most fundamentally is a study of the forces that cause people to move. Others have proposed that "Americanization by motion" be a replacement for Turner's frontier thesis.[22] It should not surprise us then that residents of the Ste. Genevieve District moved about as frequently as they did. We should imagine a dynamic settlement frontier in motion, not one fixed in space as static maps portray it.

Although the early French community is often presented in the literature as residentially stable, which it certainly was relative to other more mobile people, many French made frequent seasonal and short-term moves to mines, or longer-term moves to rivers for fur and Indian trading and as rowers and boatmen to New Orleans and other distant places. Their houses and families may have stayed put in the villages, but the men did not.

20. J. S. Wood, *The New England Village*.
21. Lockridge, *New England Town*.
22. Vance, "California and the Search for the Ideal," 194. See also John C. Hudson, "North American Origins of Middlewestern Frontier Populations"; and George W. Pierson, *The Moving Americans*.

For their part, Americans moved their residences so readily that it is often difficult to keep track of them. As with the French and slaves, a special version of movement was seasonal employment at the mines and selling trade skills in order to supplement their marginal farm subsistence and make frontier life economically successful. Movement also helped the district achieve a degree of social and economic cohesion.

Chronic movement of settlers in the district makes one wonder whether it is valid to separate white settlers from Indians on this basis, as is usually done. Were, in fact, the Osages, the Shawnee and Delaware, and the Peoria any more transient on the land than the French and Americans? One could argue that the hunting forays of the Osages and Shawnee in the eastern Ozarks were little different in purpose and frequency from the hunting and trading forays of the French and Americans in that same space. Calls for Indians to cease their wanderings and settle down to live like American settlers appear hypocritical. Of course, as settlement proceeded into the nineteenth century and the region became better populated and its residents had more capital and labor invested in property and developed local institutions, most people did become more geographically put. A reduction of people in motion could well be a defining characteristic in the passing of a frontier.

According to Timothy R. Mahoney and many others, a frontier needs to be studied in terms of the organization of space and linkages between places. In this regard, the Turner frontier thesis that social and economic institutions with towns as markets and service centers developed from a dispersed, productive farm population has long been challenged by those such as Richard C. Wade who maintain that towns were already in place as spearheads of frontier settlement.[23] The experience of the Ste. Genevieve District supports both views, the evidence for the two coming from opposite ends of the geographic scale.

There is no question that the villages of Kaskaskia, Ste. Genevieve, and St. Louis, each in turn, served as hinges connecting the middle Mississippi Valley with the world in the pre-American period. Though they were not "cities," they were central places through which people, goods, credit, and ideas funneled into and out of the valley. They were as much the product of external forces as they were of the internal forces of local residents. As time went on, externally directed activities in St. Louis (Indian and military affairs, fur trading, postal services, the Catholic bish-

---

23. Mahoney, *River Towns;* Wade, *The Urban Frontier: The Rise of Western Cities, 1790–1830.* The argument among historians whether "cities" preceded or followed the settlement frontier will not proceed toward a resolution until historians agree on what constitutes a "city."

opric, credit services, river shipping, wholesaling, and mercantile activity) influenced, if they did not control, the development and organization of economic, political, and social life throughout the whole middle Mississippi Valley, including the subordinate Ste. Genevieve District. In the prerailroad era, organization of spatial activities was tied to river networks, and in this respect, river ports functioned as hinges to the world, just as colonial seaports on the Atlantic Coast functioned as hinges to the European world.

At the other end of the geographic scale, in the countryside, where life was lived daily by the vast majority of the people, one finds evidence of a Turnerian organization of space at work. At this lower level, central places had to arise from the needs of the neighborhood and grow as the neighborhood grew. Eighteenth-century French society, centered on its agricultural villages, was spatially organized in tracts some six miles in diameter determined by three-mile daily commuting from village to fields. Services and markets had to be local. American-organized local space was larger. For people living on dispersed farms, counties were twenty-four miles across or twelve miles from the center (assuming a half-day horseback ride at three miles per hour). County seats, the local central places for Americans, came into existence as population growth warranted the erection of counties. The county-seat towns remained basically administrative in function into the 1830s, adding only a few economic activities as demand gradually arose from the county neighborhoods (as in Perryville, Farmington, and Greenville). Only where lead mining supplemented the basic farm economy was the county seat as much economic as administrative (as in Potosi, Fredericktown).

Agriculture exerted less influence on the rise of local centers than in frontiers of the American West better endowed for it. The French and Spanish expectation of a granary in the middle Mississippi Valley could not be fulfilled, as colonists found fur trading, hunting, and lead mining more attractive economically and psychologically. African slaves were also diverted into other activities than full-time agriculture. And, faced with no navigable rivers in the district, settlers more than thirty miles from the Mississippi had no way to get grain cheaply through the hills to the river. Thus, commercial agriculture could not be extended westward. Agriculture did succeed in the few "covenant" communities built upon strong kinship and religious solidarity.

The externally related hinge "cities" and internally produced local centers merged into a single hierarchical system as settlement spread in the district and the export economy grew. Early on, raw lead moved outward from Ste. Genevieve, but as Americans dominated the industry and began manufacturing lead into products, first Herculaneum, then St. Louis served as the links between the lead mines and the world. Necessary changes in the network of roads and postal routes accompanied the reorientation of the incipient central-place system from Ste. Genevieve to St. Louis. We also learn from the Ste. Genevieve District that any route infrastructure imposed by external authorities for reasons not associated with the reality of local settlement cannot long persist and will be replaced by desire lines created by the local Turnerian-style farmer-settlers and, in the special case of the Ste. Genevieve District, by the mining industry.

A frontier stays in flux and continually reorganizes itself during fast population growth and economic development. During this period of dynamism, it is no surprise that the incipient hierarchical central-place system rearranged itself. Ste. Genevieve declined in relation to Potosi, Jackson, and St. Louis, and the whole district declined in importance in relation to the Boonslick and other faster-growing regions of Missouri. The reasons for the rise of St. Louis to the top of the hierarchy according to this analysis of the Ste. Genevieve District are its designation as the Spanish administrative center for defense, fur trading, and Indian affairs; its infusion with streams of new people and ideas; the economic and social cooperation of St. Louis French with Americans; the enormous tributary hinterland of rivers focusing on St. Louis; the early venture of St. Louisans into manufacturing and processing of raw materials; and St. Louis's access to eastern credit and business ties.

Settlement of the Ste. Genevieve District did not end in 1830. The district received further immigration, especially from large numbers of German immigrants with more intensive farming practices, who expanded settlement onto loess-capped ridges and onto underused loessial soils of the karst plain. In many respects, the cultural landscape of Perry and Ste. Genevieve Counties now reflects more their contribution than that of earlier French and Appalachian Americans. The mining area, resuscitated by rail transport, introduction of the diamond-bit drill to reach deep-seated lead and iron, and an enormously expanded market that brought it more investment capital, evolved into the world's largest lead-mining district. The energized industry, centered in the big bend of the Big River in St. Francois County, attracted new population elements, including European immigrants, and radically altered the natural and cultural landscape. More recently, the rough hills and igneous knobs have become a major recreation region. All the public domain of the historic district had been alienated just after the turn of the twentieth century, but since then, the reverse process of returning private land to public ownership has set in, accompanied by widespread abandonment and migration from formerly occupied lands. This reversal, in effect, constitutes a "desettlement" of the historic Ste. Genevieve District.

These events after 1830, however, have not erased landscape features from earlier settlement. The first people to effectively occupy new land leave imprints that remain indelible long after their time of impression—a palimpsest from which others cannot completely remove earlier signatures. Few actions are more permanent in the cultural landscape than the first land-division decisions. Still plainly visible after two centuries of elapsed time are property lines in the common fields of Ste. Genevieve, New Bourbon, and St. Michel and the long-lot subdivision of the Old Mines group concession. Also visible are bounds of the large pre-American land concessions, many still only slightly developed, which were so determinedly fought for during the land-confirmation process. Because Ste. Genevieve's population and economic growth stagnated by 1830, the village grid and buildings as far back as the relocation in the 1790s remain remarkably intact, presenting us an unexcelled example of the French tripartite village in North America. Less appreciated are the unusual rural settlements of extended Creole families in the hardscrabble mined-out lands at Old Mines, the first planned American towns west of the Mississippi River with their courthouse squares, eighteenth-century lead-hauling roads and furnace remnants, and still unrecovered woodlands from repeated cuts before 1830. The townscapes of Potosi and Fredericktown still display their original conjoined Creole and American neighborhoods.

The historic Ste. Genevieve District was diverse in environmental niches and opportunities, in the cultural groups that occupied it, in its changing administration, and in the way people acquired land and formed settlements. Nevertheless, residents organized themselves into functioning social, political, and economic systems, and their success points to the necessity of investigating the smallest details of local physiography and human geography. As we work toward generalizations about how American frontiers formed and operated, the lower local end of the geographic scale holds great promise of uncovering new ideas.

Appendix A

Spanish Governors-General of Louisiana

From Francis P. Burns, "The Spanish Land Laws of Louisiana":

Antonio de **Ulloa** (never actually took possession), 1766–1768
Alejandro **O'Reilly**, arrived in Louisiana on July 24, 1769; commission dated April 16, 1769
Luis **Unzaga** y Amezaga, appointed August 17, 1772
Bernardo de **Galvez**, appointed May 8, 1778
Esteban **Miró**, appointed July 31, 1786
Francisco Luis Hector, baron de **Carondelet**, appointed March 17, 1791
Manuel **Gayoso** de Lemos, appointed October 28, 1796
Sebastian Calvo de **Casa Calvo**, appointed 1799
Manuel Juan de **Salcedo**, appointed October 24, 1799; Salcedo put Louisiana in the hands of Americans at New Orleans, December 20, 1803

The Intendant Juan Ventura **Morales** (Juan Bonaventure Morales) had the authority to issue land grants beginning October 22, 1798

Appendix B

Lieutenant Governors of Upper Louisiana during the Spanish Period (in St. Louis)

Louis **St. Ange** de Bellerive, provisional commandant at St. Louis, 1765–1770

Pedro **Piernas**, took possession at St. Louis, May 20, 1770; had come earlier, but left
Francisco **Cruzat**, May 19, 1775
Fernando de **Leyba**, June 17, 1778
Silvio Francisco de **Cartabona**, acting, June 8, 1780
Francisco **Cruzat**, reappointment, September 24, 1780
Manuel **Perez**, November 27, 1787
Zenon **Trudeau**, July 21, 1792
Charles de Hault **Delassus**, July 29, 1799, arrived from commandant post at New Madrid

Amos **Stoddard** took charge from Delassus at St. Louis for the United States, March 10, 1804

Appendix E

U.S. Government Officials Involved in the Land-Confirmation Process in Missouri

Territorial Governors

Indiana Territory

William Henry **Harrison** (October 1, 1804–March 3, 1805)

Louisiana Territory

James A. **Wilkinson** (March 3, 1805–last part of 1806)
Joseph **Browne** (secretary and acting governor, last part of 1806–April 1, 1807)
Frederick **Bates** (secretary and acting governor, April 1, 1807–March 1808)
Meriwether **Lewis** (appointed March 1807, arrived March 1808, served until September 1809)
Frederick **Bates** (secretary and acting governor, September 1809–September 19, 1810)
Benjamin **Howard** (September 19, 1810–November 29, 1812)
Frederick **Bates** (secretary and acting governor, November 29, 1812–December 7, 1812)

Missouri Territory

Frederick **Bates** (secretary and acting governor, December 7, 1812–July 1813)
William **Clark** (July 1813–1820)

Territorial Secretary

Frederick **Bates** (May 1807–1812)

Board of Land Commissioners (1805–1812)

John B. C. **Lucas**, commissioner (1805–1812)
Clement Biddle **Penrose**, commissioner (1805–1812)
James Lowry **Donaldson**, recorder of land titles (1805–1807)
Frederick **Bates**, replaced Donaldson (1807–1812)
Charles **Gratiot**, clerk for the board (1806–1807)
William **Christy**, clerk for the board (1807)
Thomas F. **Riddick**, clerk for the board (1807–1812)
Philippe Marie **Leduc**, translator for the board (1806–1812)

Second Board of Land Commissioners (1832–1834)

A. G. **Harrison**, commissioner, resigned
Lewis Fields **Linn**, commissioner, resigned
F. H. **Martin**, recorder
James H. **Relfe**, commissioner
Frederick Rector **Conway**, commissioner

Surveyors (at St. Louis)

Antoine **Soulard** (continued from Spanish regime until 1806)
Silas **Bent**, principal deputy (1806–1813)
William **Rector**, principal deputy (1813–1816); surveyor general (1816–1824)
William **Clark**, acting surveyor general (1824–1825)
William **McRee**, surveyor general (1825–1832)

Others (at St. Louis)

William **Clark**, U.S. Indian agent at St. Louis
Pierre **Chouteau**, U.S. Indian agent specifically for the Osages
William C. **Carr**, U.S. territorial land agent, 1806–1811, representing the U.S. secretary of treasury in public land matters
Martin **Thomas**, U.S. superintendent of mines, 1824–1830
Amos **Stoddard**, acting civil commandant of the Province of Upper Louisiana, March 10, 1804–October 1, 1804
Theodore **Hunt**, recorder of land titles, 1820s

Bibliography

Abbreviations

The Ste. Genevieve Project at the University of Missouri-Columbia amassed, organized, translated, and computerized an immense collection of documents during the 1980s, funded in large part by the National Endowment for the Humanities. The bulk of documents came from the Ste. Genevieve Archives, a microfilmed collection of materials from Ste. Genevieve, Mo., located in the Western Historical Manuscripts Collections, State Historical Society of Missouri, University of Missouri-Columbia, under the title "Ste. Genevieve (Mo.) Archives, 1756-1930, microfilm collection C3636." Other documents in the collection came from the Ste. Genevieve Catholic Church; various Ste. Genevieve County offices; the Missouri Historical Society, St. Louis; the Minute Books of the U.S. Board of Land Commissioners; the *American State Papers;* the Missouri State Archives (U.S. government land sales); the Bureau of Land Management (eastern office) in Washington, D.C. (land sales); and the Archivo General de Indias: Papeles Procedentes de Cuba. The Papeles de Cuba, as they are usually called, include Spanish government documents from Louisiana first removed to Cuba, then to Seville, Spain. Documents in the Papeles de Cuba that pertain directly or indirectly to Ste. Genevieve were selected for microfilming, translated, and transcribed or summarized by Anna Price through joint funding by the Missouri Historical Society, St. Louis; the Saline Creek Valley Project in the Division of American Archaeology at the University of Missouri-Columbia; and the Ste. Genevieve Project. A microfilm copy of the documents of the Papeles de Cuba that were selected for study of Ste. Genevieve is in the Missouri Historical Society, St. Louis.

The various documents cited in this study have numbers assigned by the Ste. Genevieve Project at the University of Missouri, preceded by abbreviations that are a key to the specific source of that document. For example, LRB 122 refers to land record 122 (a Ste. Genevieve Project number), which is keyed to Ste. Genevieve County Deed Book B.

Bibliography

STE. GENEVIEVE PROJECT ABBREVIATIONS

- LCC: Confirmed Claims, 1806–1808, nos. 1–184. Minute Books of the Board of Land Commissioners. Vols. 1–3. Missouri State Archives.
- LCS: Land Concessions, 1759–1805, nos. 1–255. Nos. 1–139, Ste. Genevieve Archives, folders 86–96. Nos. 140–255, Land Petitions Collection of the Missouri Historical Society, St. Louis.
- LEX: Land Exchanges, 1759–1805, nos. 1–17. Ste. Genevieve Archives, folders 97–98.
- LL: Land Leases, 1759–1805, nos. 1–8. Ste. Genevieve Archives, folder 99.
- LPC: Confirmed Claims, 1833–1835, nos. 1–57. *American State Papers: 8, Public Lands.* Vols. 6–7.
- LPS: Public [Land] Sales, 1759–1805, nos. 1–19. Ste. Genevieve Archives, folders 102–5.
- LR: Land Records (Sales), 1759–1805, nos. 1–308. Ste. Genevieve Archives, folders 33–75, 88.
- LRA: Land Records, Ste. Genevieve District/County, 1804–1809, nos. 1–242. Ste. Genevieve County Deed Book A.
- LRB: Land Records, Ste. Genevieve County, 1809–1818, nos. 1–608. Ste. Genevieve County Deed Book B.
- LRC: Land Records, Ste. Genevieve County, 1818–1823, nos. 1–315. Ste. Genevieve County Deed Book C.
- Minutes: J. B. C. Lucas, Minute Books of the Board of Land Commissioners. Vols. 1–5, 1805–1812. Second Board: vols. 6–7, 1832–1835.
- MN: Mining File, 1770–1806, nos. 1–40. Ste. Genevieve Archives, folders 380–90.
- PC: Archivo General de Indias: Papeles Procedentes de Cuba (Seville), nos. 1–6 through 2365–523. The first number of a pair is the folder number, and the second is the page number within the folder.

OTHER ABBREVIATIONS

ASP-Ind: American State Papers: 2, Indian Affairs. 2 vols. Washington, D.C.: Gales and Seaton, 1832–1834.

ASP-Misc: American State Papers: 10, Miscellaneous. 2 vols. Washington, D.C.: Gales and Seaton, 1834.

ASP-PL: American State Papers: 8, Public Lands. 9 vols. Washington, D.C.: Gales and Seaton, 1832–1861.

Austin Papers: The Austin Papers. Ed. Eugene C. Barker. Vol. 2. Annual Report of the American Historical Association for 1919. Washington, D.C.: GPO, 1924.

Bates Papers: The Life and Papers of Frederick Bates. Ed. Thomas Maitland Marshall. 2 vols. St. Louis: Missouri Historical Society, 1926.

Hunt's Minutes: Testimony before the Recorder of Land Titles. Comp. Theodore Hunt. St. Louis, 1825. Missouri State Archives, Jefferson City.

LPGN: Laws of a Public and General Nature, of the District of Louisiana, of the Territory of Louisiana, of the Territory of Missouri, and of the State of Missouri, up to the Year 1824. Vol. 1. Jefferson City: W. Lusk and Son, 1842. *Laws of a Public and General Nature of the State of Missouri, Passed between the Years 1824 and 1836, Not Published in the Digest of 1825, nor in the Digest of 1835.* Vol. 2. Jefferson City: W. Lusk and Son, 1842.

Stats. at Large of USA: Statutes at Large of the United States of America, 1789–1873. 17 vols. Washington, D.C., 1850–1873.

TP: The Territorial Papers of the United States. Ed. Clarence E. Carter. Vol. 13, *Louisiana-Missouri, 1803–1806* (1948). Vol. 14, *Louisiana-Missouri, 1806–1814* (1949). Vol. 15, *Louisiana-Missouri, 1815–1821* (1954); Vol. 17, *Territory of Illinois, 1814–1818* (1950). Washington, D.C.: GPO.

Collections and Papers

Library of Congress, Washington, D.C.

Overby, Osmund, dir. Ste. Genevieve Project of the Historic American Buildings Survey, 1985. Historic American Buildings Survey Collections.

Missouri State Archives, Jefferson City

Abstract of Land Sales. St. Louis and Jackson Land Offices, 1818–1834.
Record of Surveys. Mine à Breton, Village à Robert, and St. Ferdinand.
Record of Surveys. Recorder of Land Titles.
Record of Surveys. Sainte Genevieve and New Bourbon Lands.
Record of Surveys. Ste. Genevieve and New Bourbon Commonfield and Outlots.
Régistre d'Arpentage. Collection of survey plats drawn by Antoine Soulard and his deputies, 1798–1806. 2 vols. Abstract, *American State Papers: 8, Public Lands* 8:848–65.
Sundry Surveys, by J. C. Brown.
U.S. Surveyor General. Missouri Plats and Surveys. Miscellaneous plats.

Missouri Department of Natural Resources, Division
of Geology and Land Survey, Rolla

Miscellaneous plats.
U.S. General Land Office. Field Notes, Missouri, 1815–1862. 650 vols.

Missouri Historical Society, St. Louis

Amoureux-Bolduc Collection. 1754–1844.
Darby Family. Papers. 1785–1919.
Glimpses of the Past. Serial publication (1933–1943) of "miscellaneous extracts from rare old manuscripts, books, newspapers, and public records."
Guibourd Papers. 1752–1832.
Journal. "General Merchandize." Sales journal for unidentified general stores at New Diggings (1809–1810) and Belleview (1817–1820).
Maxwell Claim Application, 1875.
Missouri Historical Society Collections. Serial publication of various documents, 1880–1931.
Sibley, George C. Papers. 1803–1828.
Vallé, Francois. Papers. 1742–1939.

State Historical Society of Missouri, Columbia

Ste. Genevieve County Plat Book [ca. 1861].

Maps

Browne, E., and E. Barcroft. Map of the States of Missouri and Illinois and Territory of Arkansas. N.p., 1827.
Carey, H. C., and I. Lea. Geographical, Statistical, and Historical Map of Missouri. N.p., 1827.
Cram, George F. New Sectional Map of the State of Missouri. Chicago, 1869.
Johnson and Ward. Johnson's Map of Missouri and Kansas. N.p., 1862.
Lloyd's Official Map of Missouri. New York, 1861.
Map of Missouri Territory, formerly Louisiana. Philadelphia, 1818.
Map of Northern Part of Missouri Territory. N.p., 1815.
Mitchell, S. Augustus. Map of Missouri and Territory of Arkansas. Philadelphia, 1832.
———. Map of the State of Missouri and Territory of Arkansas. Philadelphia, 1836.
[Official Map of] Missouri. N.p., 1822.
Tanner, H. S. A New Map of Missouri with Its Roads and Distances. N.p., 1833.

———. Illinois and Missouri. N.p., 1823.
Vance, D. H. Map of Missouri and Territory of Arkansas. N.p., 1826.

WESTERN HISTORICAL MANUSCRIPTS COLLECTION, STATE HISTORICAL SOCIETY, UNIVERSITY OF MISSOURI–COLUMBIA

Ste. Genevieve (Mo.) Archives, 1756–1930, microfilm collection C3636. Originals in Ste. Genevieve.

ELLIS LIBRARY, UNIVERSITY OF MISSOURI–COLUMBIA

Jefferson, Thomas. Papers. Microfilm.
Sanborn Fire Insurance Company maps.

NATIONAL ARCHIVES, WASHINGTON, D.C.

Land Entries. St. Louis and Jackson land districts, 1818–1830.
Office of the Recorder of Deeds. St. Louis Deed Books C, G.

Missouri Newspapers

Fredericktown Bee. 1869.
Fredericktown Democrat News. 1999.
Jackson Independent Patriot. 1820–1826.
Jackson Missouri Herald. 1819–1820.
Jefferson City Daily Tribune. 1926.
Nile's Weekly Register. 1815–1830.
St. Louis Missouri Gazette. 1808–1822.
St. Louis Missouri Republican. 1822–1830.

Government Documents

Aggregate Amount of Persons within the United States in the Year 1810. Washington, D.C., 1811.
Annals of Congress of the United States, 1789–1824. 42 vols. Washington, D.C.: Gales and Seaton, 1834–1856.
Bureau of the Census. *Census of 1820.* Washington, D.C.: Gales and Seaton, [1821].
———. *Fifth Census; or, Enumeration of the Inhabitants of the United States, 1830.* Washington, D.C.: Duff Green, 1832.
Congressional Debates, 1824–1837. 14 vols. Washington, D.C., 1825–1838.
Congressional Globe, 1833–1873. 46 vols. Washington, D.C., 1834–1873.
General Public Acts of Congress, Respecting the Sale and Disposition of the Public Lands, with Instructions Issued, from Time to Time, by the Secretary of the Treasury and Commissioner of the General Land Office, and

Official Opinions of the Attorney General on Questions Arising under the Land Laws. 2 vols. Washington, D.C.: Gales and Seaton, 1838.

Journal of the House of Representatives of the State of Missouri at the Second Session of the First General Assembly. St. Charles: Robert McCloud, 1821.

Report of the U.S. Secretary of the Treasury, 1790–1828. Vols. 1–2. Washington, D.C.: Blair and Rives, 1837.

Reports of the U.S. General Land Office, 1826–1863. 15 vols. Washington, D.C., 1826–1863.

Other Sources

Abel, Annie Heloise. *The History of Events Resulting in Indian Consolidation West of the Mississippi.* Vol. 1. Annual Report of the American Historical Association for the Year 1906. Washington, D.C.: GPO, 1908.

Abernathy, Elizabeth Denty. *The Abernathy Family.* [Pulaski, Tenn.]: Pulaski Citizen Print, 1929.

Abramoske, Donald J. "The Federal Lead Leasing System in Missouri." *Missouri Historical Review* 54 (1959): 27–38.

Ackerknecht, Erwin Heinz. *Malaria in the Upper Mississippi Valley, 1760–1900.* Supplements to the Bulletin of the History of Medicine, no. 4. Baltimore: Johns Hopkins University Press, 1945.

Adams, David P. "Malaria, Labor, and Population Distribution in Costa Rica: A Biohistorical Perspective." *Journal of Interdisciplinary History* 27 (1996): 75–85.

Allain, Mathé. "Slave Policies in French Louisiana." *Louisiana History* 21 (1980): 127–37.

Alvord, Clarence Walworth. "Genesis of the Proclamation of 1763." Paper read before the Michigan Pioneer and Historical Society, December 13, 1907. Copy in the State Historical Society of Missouri, Columbia.

———. *The Illinois Country, 1673–1818.* Centennial History of Illinois, vol. 1. Springfield: Illinois Centennial Commission, 1920.

———. *The Mississippi Valley in British Politics.* 2 vols. Cleveland: Arthur H. Clark, 1917.

———, ed. *Cahokia Records, 1778–1790.* Collections of the Illinois State Historical Library, vol. 2. Springfield: Illinois State Historical Library, 1907.

———. *Kaskaskia Records, 1778–1790.* Collections of the Illinois State Historical Library, vol. 5. Springfield: Illinois State Historical Library, 1909.

Alvord, Clarence Walworth, and Clarence Edwin Carter, eds. *The Critical Period, 1763–1765.* Collections of the Illinois State Historical Library, vol. 10. Springfield: Illinois State Historical Library, 1915.

———. *Invitation serieuse aux habitants des Illinois by un habitant des Kaskaskias.* Reprinted in facsimile from the original edition published at Philadelphia in 1772 with an introduction by Clarence Walworth Alvord and Clarence Edwin Carter. Providence: Club for Colonial Reprints of Providence, Rhode Island, 1908.

———. *The New Regime, 1765-1767.* Collections of the Illinois State Historical Library, vol. 11. Springfield: Illinois State Historical Library, 1916.

———. *Trade and Politics, 1767-1769.* Collections of the Illinois State Historical Library, vol. 16. Springfield: Illinois State Historical Library, 1921.

Ambrose, Stephen E. *Undaunted Courage: Meriwether Lewis, Thomas Jefferson, and the Opening of the American West.* New York: Simon and Schuster, 1996.

Anderson, Hattie M. "Frontier Economic Problems in Missouri, 1815-1828." *Missouri Historical Review* 34 (1939-1940): 38-70, 182-203.

———. "Missouri, 1804-1828: Peopling a Frontier State." *Missouri Historical Review* 31 (1937): 150-80.

Archibald, Robert R. "Honor and Family: The Career of Lt. Gov. Carlos de Hault de Lassus [sic]." *Gateway Heritage* 12:4 (1992): 32-41.

Arensburg, Conrad M. "American Communities." *American Anthropologist* 57 (1955): 1143-62.

Arnold, Morris S. *Colonial Arkansas, 1686-1804: A Social and Cultural History.* Fayetteville: University of Arkansas Press, 1991.

———. *Unequal Laws unto a Savage Race: European Legal Traditions in Arkansas, 1686-1836.* Fayetteville: University of Arkansas Press, 1985.

Aron, Stephen. *How the West Was Lost: The Transformation of Kentucky from Daniel Boone to Henry Clay.* Baltimore: Johns Hopkins University Press, 1996.

Ashe, Thomas. *Travels in America, performed in . . . 1806, for the purpose of exploring the rivers Alleghany, Monongahela, Ohio, and Mississippi, and ascertaining the produce and condition of their banks and vicinity.* London: R. Phillips, 1809.

"Ashley's Cave." *Missouri Historical Review* 24 (1930): 607-8.

Atherton, Lewis E. *The Frontier Merchant in Mid-America.* Columbia: University of Missouri Press, 1971.

———. "Missouri's Society and Economy in 1821." *Missouri Historical Review* 55 (1971): 450-77.

Austin, Moses. "'A Memorandum of M. Austin's Journey from the Lead Mines in the County of Wythe in the State of Virginia to the Lead Mines in the Province of Louisiana West of the Mississippi,' 1796-1797." *American Historical Review* 5 (1900): 518-42.

Babeau, Albert. *Le Village sous l'ancien régime.* 1878. Reprint, New York: AMS Press, 1972.

Baker, William. "Land Claims as Indicators of Settlement in Southwestern Illinois, circa 1809–13." *Bulletin of the Illinois Geographical Society* 16:1 (1974): 29–42.

Baldwin, Leland D. *The Keelboat Age on Western Waters.* Pittsburgh: University of Pittsburgh Press, 1941.

Balesi, Charles J. *The Time of the French in the Heart of North America, 1673–1818.* Chicago: Alliance Française Chicago, 1991.

Banner, Stuart. *Legal Systems in Conflict: Property and Sovereignty in Missouri, 1750–1860.* Norman: University of Oklahoma Press, 2000.

Bannon, John Francis. *The Spanish Borderlands Frontier, 1513–1821.* New York: Holt, Rinehart, and Winston, 1970.

Barrette, Gérald. "Contribution de l'arpenteur-géomètre à la géographie du Quebec." *Canadian Geographer* 2 (1952): 67–71.

Barrow, Elmer Cave. "The Early History of Iron Mining in Southeast Missouri (1815–1861)." Master's thesis, University of Missouri–Columbia, 1940.

Basler, Lucille. "The District of Ste. Genevieve, 1725–1980." Ste. Genevieve, Mo.: by the author, 1980.

Bassett, Thomas J. "Cartography and Empire Building in Nineteenth-Century West Africa." *Geographical Review* 84 (1994): 316–35.

Batek, Michael J. "Presettlement Vegetation of the Current River Watershed in the Missouri Ozarks." Master's thesis, University of Missouri–Columbia, 1994.

Baulig, Henri. *Vocabulaire Franco-Anglo-Allemand de géomorphologie.* Publications de la Faculté des Lettres de l'Université de Strasbourg, fascicule 130. Paris: Société d'Edition: Belles Lettres, 1956.

Beahan, Gary W. "Missouri's Public Domain: United States Land Sales, 1818–1922." Archives Information Bulletin 2, no. 3 (1980). State of Missouri, Office of Secretary of State, Records Management and Archives Service, Jefferson City.

Beaulne, Kent, Judith Escoffier, and Patricia Weeks. *Les Noms des vielle famille de la region de la Vielle Mines* [sic] (Old family names of the Old Mines area in Missouri). Old Mines, Mo.: Old Mines Area Historical Society, 1982.

Becherer, Floyd Oliver. "The History of St. Louis, 1817–1826." Master's thesis, Washington University, 1950.

Beck, Lewis Caleb. *A Gazetteer of the States of Illinois and Missouri; containing a general view of each state—a general view of their counties—and a particular description of their towns, villages, rivers, etc.* Albany, N.Y.: Charles R. and G. Webster, 1823.

Beers, Henry Putney. *The French and British in the Old Northwest: A Bibliographical Guide to Archive and Manuscript Sources.* Detroit: Wayne State University Press, 1964.

Behnen, Robert J. "Coleman Family Origins Found—200 Year Old Mystery Is Solved!" *Diggin's* (Old Mines Area Historical Society, Cadet) 3:4 (1997): 42–43.

Bell, James. "Black River Township." In *Bellevue—Beautiful View: The History of the Bellevue Valley and Surrounding Area,* 23–24. Potosi, Mo.: Bellevue Valley Historical Society, 1983.

———. *History of Early Reynolds County, Missouri.* Paducah, Ky.: Turner Publishing, 1986.

Bell, Ovid. *Cote Sans Dessein, a history.* Fulton, Mo.: Ovid Bell Press, 1930.

Bellevue—Beautiful View: The History of the Bellevue Valley and Surrounding Area. Potosi, Mo.: Bellevue Valley Historical Society, 1983.

Bellovich, Steven J. "The Establishment and Development of the Main Colonial Territorial Routes of Lead Movement in Eastern Missouri, 1700–1965." Master's thesis, Southern Illinois State University, 1970.

Belote, Theodore Thomas. *The Scioto Speculation and the French Settlement at Gallipolis.* University of Cincinnati, University Studies, ser. 2, vol. 3, no. 3. Cincinnati: University of Cincinnati Press, 1907.

Belting, Natalia Maree. *Kaskaskia under the French Regime.* Illinois Studies in the Social Sciences, vol. 29, no. 3. Urbana: University of Illinois Press, 1948.

Bernard, Eunice. *Wills and Administrations, Washington County, Missouri, 1814–1870.* 1939. Copy in the State Historical Society of Missouri, Columbia.

Berquin-Duvallon. *Vue de la colonie espagnole du Mississippi, ou des provinces de Louisiane et Floride Occidentale; en l'année 1802, par un observateur résident sur les lieux.* Paris: Imprimerie expéditive, 1803.

Beuckman, Frederick. "The Commons of Kaskaskia, Cahokia, and Prairie du Rocher." *Illinois Catholic Historical Review* 1 (1919): 405–12.

Beveridge, Thomas R. *Geologic Wonders and Curiosities of Missouri.* Educational Series, no. 4. 2d ed. Rolla: Missouri Department of Natural Resources, Division of Geology and Land Survey, 1990.

Billon, Frederic L. *Annals of St. Louis in Its Early Days under the French and Spanish Dominations, 1764–1804.* 1886. Reprint, New York: Arno Press and New York Times, 1971.

———. *Annals of St. Louis in Its Territorial Days, from 1804–1821.* 1888. Reprint, New York: Arno Press and New York Times, 1971.

Bining, Arthur Cecil. *Pennsylvania Iron Manufacture in the Eighteenth Century.* Harrisburg: Pennsylvania Historical Commission, 1938.

Birkbeck, Morris. *Notes on a journey in America, from the coast of Virginia* to the territory of Illinois. 1819. Reprint, New York: Augustus M. Kelley, 1971.

Blume, Helmut. *The German Coast during the Colonial Era, 1722–1803: The Evolution of a Distinct Cultural Landscape in the Lower Mississippi Delta during the Colonial Era.* Trans. and ed. Ellen C. Merrill. Kiel, Germany: Geographisches Institut der Universität Kiel, Germany, 1956; reprint, Destrehan, La.: German-Acadian Coast Historical and Genealogical Society, 1990.

Bossu, Jean Bernard. *Travels in the Interior of North America, 1751–1762.* Norman: University of Oklahoma Press, 1962.

Boyer, Mary Joan. *Jefferson County, Missouri, in Story and Pictures.* Imperial, Mo.: by the author, 1958.

Boyle, Susan. "Did She Generally Decide? Women in Ste. Genevieve, 1750–1805." *William and Mary Quarterly,* 3d ser., 44 (1987): 775–89.

———. "Economic Change in the Later Eighteenth and Nineteenth Centuries." Paper delivered at the Ste. Genevieve Seminar on French Colonial Studies, November 1–3, 1985, Ste. Genevieve, Mo.

———. *Los Capitalistas: Hispano Merchants and the Santa Fe Trade.* Albuquerque: University of New Mexico Press, 1997.

Brackenridge, Henry Marie. *Recollections of Persons and Places in the West.* 1834. Reprint, Philadelphia: J. B. Lippincott, 1868.

———. *Views of Louisiana, together with a Journal of a Voyage up the Missouri River, in 1811.* 1814. Reprint, Chicago: Quadrangle Books, 1962.

Bradbury, John. *Travels in the Interior of America in the Years 1809, 1810, and 1811.* 2d ed. London: Sherwood, Neely, and Jones, 1819.

Brassieur, C. Ray. "Expressions of French Identity in the Mid-Mississippi Valley." Ph.D. diss., University of Missouri–Columbia, 1999.

Braudel, Fernand. *The Identity of France.* Vol. 1, *History and Environment.* New York: Harper and Row, 1988.

———. *On History.* Chicago: University of Chicago Press, 1980.

Breckenridge, William Clark. "Early Gunpowder Making in Missouri." *Missouri Historical Review* 20 (1926): 85–95, with commentaries pp. 124–28, 338–40.

Bretz, J. Harlen. *Geomorphic History of the Ozarks of Missouri.* Missouri Geological Survey and Water Resources, 2d ser., vol. 41. Rolla, Mo., 1965.

Briggs, Winstanley. "The Forgotten Colony: *Le Pays des Illinois.*" Ph.D. diss., University of Chicago, 1985.

———. "Le Pays des Illinois." *William and Mary Quarterly,* 3d ser., 47 (1990): 30–56.

Brod, Raymond M. "The Art of Persuasion: John Smith's *New England* and *Virginia* Maps." *Historical Geography* 24 (1995): 91–106.

Brown, Burton L. *Soil Survey of St. Francois County, Missouri.* U.S. Department of Agriculture, Soil Conservation Service and Forest Service. Washington, D.C.: GPO, 1981.

Brown, Burton L., and James D. Childress. *Soil Survey of Ste. Genevieve County, Missouri.* U.S. Department of Agriculture, Soil Conservation Service and Forest Service. Washington, D.C.: GPO, 1985.

Brown, Burton L., and Kenneth L. Gregg. *Soil Survey of Iron County, Missouri.* U.S. Department of Agriculture, Soil Conservation Service. Washington, D.C.: GPO, 1991.

Brown, Lawrence A., Rodrigo Sierra, Scott Digiacinto, and W. Randy Smith. "Urban-System Evolution in Frontier Settings." *Geographical Review* 84 (1994): 249–65.

Brown, Margaret Kimball, and Lawrie Cena Dean. *The Village of Chartres in Colonial Illinois, 1720–1765.* New Orleans: Polyanthos Press, 1977.

Brown, Ralph H. *Historical Geography of the United States.* New York: Harcourt, Brace, 1948.

Brown, Samuel R. *The Western Gazetteer; or, Emigrant's Directory.* Auburn, N.Y.: H. C. Southwick, 1817.

Bryan, William S. "Daniel Boone in Missouri." *Missouri Historical Review* 3:4 (1909): 293–99.

Buisseret, David, ed. *Monarchs, Ministers, and Maps.* Chicago: University of Chicago Press, 1992.

Burghardt, A. F. "A Hypothesis about Gateway Cities." *Annals of the Association of American Geographers* 61 (1971): 269–85.

Burnham, J. H. "Destruction of Kaskaskia by the Mississippi River." In *Transactions of the Illinois State Historical Society for 1914.* Illinois State Historical Society Publication 20, pp. 95–112. Springfield: Illinois State Historical Society, 1915.

Burns, Francis P. "The Spanish Land Laws of Louisiana." *Louisiana Historical Quarterly* 11 (1928): 557–81.

Butzer, Karl W. "The Indian Legacy in the American Landscape." In *The Making of the American Landscape,* ed. Michael P. Conzen, 27–50. New York and London: Routledge, 1994.

Call, Steven Rene. "French Slaves, Indian Slaves: Slavery and the Cultural Frontier in the Illinois Country, 1703–1756." Master's thesis, University of Missouri–Columbia, 1988.

Callender, Charles. "Illinois." In *Northeast,* ed. Bruce G. Trigger, 673–80. Vol. 15 of *Handbook of North-American Indians,* ed. William C. Sturtevant. Washington, D.C.: Smithsonian Institution, 1978.

———. "Shawnee." In *Northeast,* ed. Bruce G. Trigger, 622–35. Vol. 15 of *Handbook of North-American Indians,* ed. William C. Sturtevant. Washington, D.C.: Smithsonian Institution, 1978.

Calvert, John H. "Fencing Laws in Missouri: Restraining Animals." *Missouri Law Review* 32 (1967): 519–42.

Carstensen, Vernon, ed. *The Public Lands: Studies in the History of the Public Domain.* Madison: University of Wisconsin Press, 1968.

Cayton, Andrew R. L. *The Frontier Republic: Ideology and Politics in the Ohio Country, 1780–1825.* Kent, Ohio: Kent State University Press, 1986.

Cayton, Andrew R. L., and Peter S. Onuf. *The Midwest and the Nation: Rethinking the History of an American Region.* Bloomington: Indiana University Press, 1990.

Cazier, Lola. *Surveys and Surveyors of the Public Domain, 1785–1975.* U.S. Department of the Interior. Washington, D.C.: GPO, 1976.

"The Centenary of the Foundation of the St. Louis Diocesan Seminary." *St. Louis Catholic Historical Review* 1:1 (1918): 40–49.

Chapman, Carl H. "The Indomitable Osage in the Spanish Illinois (Upper Louisiana), 1763–1804." In *The Spanish in the Mississippi Valley, 1762–1804,* ed. John F. McDermott, 287–313. Urbana: University of Illinois Press, 1974.

Chapman, Carl H., and Eleanor F. Chapman. *Indians and Archaeology of Missouri.* Columbia: University of Missouri Press, 1964.

Chardon, Roland. "The Linear League in North America." *Annals of the Association of American Geographers* 70 (1980): 129–53.

Chester, William W. "'The Barrens': A Study in Geographic Terminology." *Transactions of the Missouri Academy of Science* 6 (1972): 3–5.

Chouteau, Auguste. "Narrative of the Settlement of St. Louis." In *The Early Histories of St. Louis,* ed. John Francis McDermott, 45–59. St. Louis: St. Louis Historical Documents Foundation, 1952.

———. "Notes of Auguste Chouteau on Boundaries of Various Indian Nations." *Glimpses of the Past* 7:9–12 (1940): 119–40.

Cleary, Patricia. "Contested Terrain, Environmental Agendas, and Settlement Choices in Colonial St. Louis." In *Common Fields: An Environmental History of St. Louis,* ed. Andrew Hurley, 58–72. St. Louis: Missouri Historical Society, 1997.

Clemson, Thomas G., E. F. Pratte, C. C. Vallee, and Lewis L. Linn. *Observations on the La Motte Mines and Domain, in the State of Missouri, with some account of the advantages and inducements there promised to capitalists and individuals desirous of engaging in mining, manufacturing, or farming operations.* Washington, D.C.: Blair and Rives, 1838.

Clendenen, Harbert L. "Settlement Morphology of the Southern Courtois Hills, Missouri, 1820–1860." Ph.D. diss., Louisiana State University, 1973.

Clokey, Richard M. *William H. Ashley: Enterprise and Politics in the Trans-Mississippi West.* Norman: University of Oklahoma Press, 1980.

[Collot, Georges-Victor]. "Victor Collot: A Journey in North America." *Transactions of the Illinois State Historical Society, 1908.* Illinois State Historical Society Publication 13, pp. 269–98. Springfield: Illinois State Historical Society, 1909.

Comeaux, Malcolm L. "An Archaeological Perspective on Animal Exploitation at French Colonial Sites in the Illinois Country." *P.A.S.T. (Pioneer America Society Transactions)* 13 (1990): 94.

———. "Origin and Evolution of Mississippi River Fishing Craft." *Pioneer America* 10 (1978): 72–97.

Conrad, Glenn R. "Administration of the Illinois Country: The French Debate." In *The French Experience in Louisiana.* Vol. 1 of *The Louisiana Purchase Bicentennial Series in Louisiana History,* 442–59. Lafayette: University of Southwestern Louisiana, Center for Louisiana Studies, 1995.

Conzen, Michael P. "The Maturing Urban System in the United States, 1840–1910." *Annals of the Association of American Geographers* 67 (1977): 88–108.

Cook, Edward M., Jr. "Geography and History: Spatial Approaches to Early American History." *Historical Methods* 13 (1980): 19–28.

Courville, Serge. "Contribution à l'étude de l'origine du rang au Québec: La Politique Spatiale des cent-associés." *Cahiers de Géographie du Québec* 25:65 (1981): 197–236.

Coxe, Tench. *A Statement of the Arts and Manufactures of the United States of America for the Year 1810.* Philadelphia: A. Cornman Jr., 1814.

Cozzens, Arthur B. "The Iron Industry of Missouri." Parts 1–3. *Missouri Historical Review* 35 (1941): 509–38; 36 (1941): 48–60; 36 (1942): 214–20.

———. "The Natural Regions of the Ozark Province." Ph.D. diss., Washington University, 1937.

Cramer, Rose Fulton. *Wayne County, Missouri.* Cape Girardeau, Mo.: Ramfre Press, 1972.

Cramer, Zadok [or Zadoc]. *The Navigator, containing directions for navigating the Monongahela, Allegheny, Ohio, and Mississippi waters.* . . . 10th ed. Pittsburgh: Cramer and Spear, 1818.

Crane, G. W. *The Iron Ores of Missouri.* 2d ser. Vol. 10. Jefferson City: Missouri Bureau of Geology and Mines, 1912.

Cronon, William. *Changes in the Land: Indians, Colonists, and the Ecology of New England.* New York: Hill and Wang, 1983.

———. *Nature's Metropolis: Chicago and the Great West.* New York: W. W. Norton, 1991.

Crouch, Dora P., Daniel J. Garr, and Axel I. Mundigo. *Spanish City Planning in North America.* Cambridge: MIT Press, 1982.

Cruchet, René. "La Vie en Louisiane de 1752 à 1756, d'après un manuscrit bordelais inédit." In *France et Louisiane*. Romance Language Series, no. 2, pp. 45–86. Baton Rouge: Louisiana State University Press, 1939.

Cutter, Charles R. *The Legal Culture of Northern New Spain, 1700–1810*. Albuquerque: University of New Mexico Press, 1995.

Dale, Harrison Clifford, ed. *The Ashley-Smith Explorations and the Discovery of a Central Route to the Pacific, 1822–1829, with the Original Journals*. Rev. ed. Glendale, Calif.: Arthur H. Clark, 1941.

Dalton, Mary Louise. "Notes on the Genealogy of the Valle Family." *Missouri Historical Society, Collections* 2:7 (1906): 54–82.

Dana, Edmund. *Geographical sketches on the western country, designed for emigrants and settlers . . . including a particular description of all the unsold public lands, collected from a variety of authentic sources. Also, a list of the principal roads*. Cincinnati: Looker, Reynolds, 1819.

Darby, William. *The emigrant's guide to the western and southwestern states and territories, comprising a geographical and statistical description of the states. . . .* New York: Kirk and Mercein, 1818.

DeConde, Alexander. *This Affair of Louisiana*. New York: Scribner, 1976.

Deffontaines, Pierre. "Le Rang, type de peuplement rural du Canada français." *Cahiers de Géographie de Québec*, old ser., no. 5 (1953): 3–32.

Delehanty, James M. "Livelihood in the Region of Fort de Chartres under the French, 1720–1763." Unpublished paper submitted to the Department of Geography, University of Chicago, August 1979. Copy in the State Historical Society of Missouri, Columbia.

Demangeon, Albert. "L'Habitat rural: Villages, hameaux, et fermes." In *La France, Part 2: France économique et humaine*. Vol. 6 of *Géographie Universelle*, ed. Paul Vidal de la Blache and Lucien Gallois, 186–207. Paris: Armand Colin, 1946.

Denevan, William M. "Adaptation, Variation, and Cultural Geography." *Professional Geographer* 35 (1983): 399–407.

Denman, David D. "French Peasant Society in Flux and Stress: The Reintegration of Traditional Village Communal Activity in Ste. Genevieve, 1703–1830." Master's thesis, University of Missouri–Columbia, 1980.

———. "History of 'La Saline': Salt Manufacturing Site, 1675–1825." *Missouri Historical Review* 78 (1979): 307–20.

Derruau, Max. "À l'origine du 'rang' Canadien." *Cahiers de Géographie de Québec*, new ser., 1 (1956): 39–47.

Diamond, Sigmund. "An Experiment in Feudalism: French Canada in the Seventeenth Century." *William and Mary Quarterly*, 3d ser., 18 (1961): 3–34.

Dick, Everett. *The Lure of the Land: A Social History of the Public Lands from the Articles of Confederation to the New Deal.* Lincoln: University of Nebraska Press, 1970.

Dickason, Olive Patricia. "A Legend Reconsidered: The French and Indians in the New World." *France Magazine* 24 (fall 1992): 18–23.

Din, Gilbert C. "Captain Francisco Ríu y Morales and the Beginnings of Spanish Rule in Missouri." *Missouri Historical Review* 94 (2000): 121–45.

———. "The Immigration Policy of Governor Estaban Miró in Spanish Louisiana." *Southwestern Historical Quarterly* 73 (1969): 155–75.

———. "Proposals and Plans for Colonization in Spanish Louisiana, 1787–1790." *Louisiana History* 9 (1970): 197–213.

———. "Spain's Immigration Policy and Efforts in Louisiana during the American Revolution." *Louisiana Studies* (Northwestern State College, Natchitoches, La.) 14 (1975): 241–57.

Din, Gilbert C., and Abraham P. Nasatir. *The Imperial Osages: Spanish-Indian Diplomacy in the Mississippi Valley.* Norman: University of Oklahoma Press, 1983.

Di Piazza, Daniel. "A History of Federal Policy toward the Public Mineral Lands, 1785–1866." Master's thesis, University of Missouri–Columbia, 1957.

Dobney, Frederick J. *River Engineers on the Middle Mississippi: A History of the St. Louis District, U.S. Army Corps of Engineers.* [St. Louis?]: U.S. Army Corps of Engineers, 1978.

Donaldson, Thomas, comp. *The Public Domain, Its History with Statistics.* . . . 1884. Reprint, New York: Johnson Reprint Corporation, 1970.

Dorrance, Ward A. *The Survival of French in the Old District of Ste. Genevieve.* University of Missouri Studies, vol. 10, no. 2. Columbia: University of Missouri, 1935.

———. *We're from Missouri.* Richmond: Missourian Press, 1938.

Dorsey, Dorothy P. "The Panic of 1819 in Missouri." *Missouri Historical Review* 29 (1935): 79–91.

Dosse, François. *New History in France.* Urbana: University of Illinois Press, 1994.

Douglas, Walter B. *The Spanish Domination of Upper Louisiana.* Proceedings of the State Historical Society of Wisconsin for 1913. Madison: State Historical Society of Wisconsin, 1914.

Douglass, Robert Sidney. *History of Southeast Missouri: A Narrative Account of Its Historical Progress, Its People, and Its Principal Interests.* 1912. Reprint, Cape Girardeau, Mo.: Ramfre Press, 1961.

Dowd, Gregory Evans. *A Spirited Resistance: The North American Indian*

Struggle for Unity, 1745–1815. Baltimore: Johns Hopkins University Press, 1992.

Drake, Daniel. *Systematic treatise, historical, etiological, and practical, on the principal diseases of the interior valley of North America, as they appear in the Caucasian, African, Indian, and Esquimaux varieties of its population.* Cincinnati: W. B. Smith, 1850; New York, Mason and Law, 1850.

Duclos, Zachary. "Hypolite Robert—His Origin?" *Le Journal* (Center for French Colonial Studies) 10, nos. 3–4 (1994): 6.

Dugger, Harold H. "Reading Interests of the Book Trade in Frontier Missouri." Ph.D. diss., University of Missouri–Columbia, 1951.

Earle, Carville V. *The Evolution of a Tidewater Settlement System: All Hallow's Parish, Maryland, 1650–1783.* Department of Geography Research Paper, no. 170. Chicago: University of Chicago, 1975.

———. "The First English Towns of North America." *Geographical Review* 67 (1977): 34–50.

———. "Regional Economic Development West of the Appalachians, 1815–1860." In *North America: The Historical Geography of a Changing Continent,* ed. Robert D. Mitchell and Paul A. Groves, 172–97. London: Hutchinson, 1987.

East, Ernest E. "Lincoln and the Peoria French Claims." *Journal of the Illinois State Historical Society* 42 (1949): 41–56.

Eccles, W. J. *France in America.* East Lansing: Michigan State University Press, 1990.

Echeverria, Durand. "General Collot's Plan for a Reconnaissance of the Ohio and Mississippi Valleys, 1796." *William and Mary Quarterly,* 3d ser., 9 (1952): 512–20.

Ehrlich, Walter. "The First Jews of St. Louis." *Missouri Historical Review* 38 (1988): 57–76.

Ekberg, Carl J. "Antoine Valentin de Gruy: Early Missouri Explorer." *Missouri Historical Review* 76 (1982): 136–50.

———. *Colonial Ste. Genevieve: An Adventure on the Mississippi Frontier.* Gerald, Mo.: Patrice Press, 1985.

———. *French Roots in the Illinois Country: The Mississippi Frontier in Colonial Times.* Urbana: University of Illinois Press, 1998.

———. "Was Louis Lorimier Literate?" *Le Journal* (Center for French Colonial Studies) 15:1 (1999): 5.

Ekberg, Carl J., Charles R. Smith, William D. Walters Jr., and Frederick W. Lange. *A Cultural Geographical and Historical Study of the Pine Ford Lake Project Area: Washington, Jefferson, Franklin, and St. Francois Counties, Missouri.* Archeological Survey Research Report no. 2. Normal: Illinois State University, 1981.

"Epistle or Diary of the Reverend Father Marie Joseph Durand." Parts 1–2. *American Catholic Historical Society of Philadelphia Records,* 26 (1915): 328–46; 27 (1916): 45–64.

Ernst, Joseph, and Roy Merrens. "'Camden's Turrets Pierce the Skies!': The Urban Process in the Southern Colonies during the Eighteenth Century." *William and Mary Quarterly,* 3d ser., 30 (1973): 549–74.

Espinosa, J. Manuel. "Spanish Louisiana and the West: The Economic Significance of the Ste. Genevieve District." *Missouri Historical Review* 32 (1938): 287–97.

Evans, E. Estyn. "Culture and Land Use in the Old West of North America." In *Heidelberger Studien zur Kulturgeographie. Festgabe zum 65. Geburtstag von Gottfried Pfeifer. Heidelberger Geographische Arbeiten,* 15:72–80. Wiesbaden: Franz Steiner Verlag GmBH, 1966.

Faragher, John Mack. "Americans, Mexicans, Métis: A Community Approach to the Comparative Study of North American Frontiers." In *Under an Open Sky: Rethinking America's Western Past,* ed. William Cronon, George Miles, and Jay Gitlin, 90–109. New York: W. W. Norton, 1992.

———. *Daniel Boone: The Life and Legend of an American Pioneer.* New York: Henry Holt, 1992.

———. "'More Motley than Mackinaw': From Ethnic Mixing to Ethnic Cleansing on the Frontier of the Lower Missouri, 1783–1833." In *Contact Points: American Frontiers from the Mohawk Valley to the Mississippi, 1750–1830,* ed. Andrew R. L. Cayton and Fredrike J. Teute, 304–26. Chapel Hill: University of North Carolina Press, 1998.

———. *Rereading Frederick Jackson Turner.* New York: Henry Holt, 1994.

———. *Sugar Creek: Life on the Illinois Prairie.* New Haven: Yale University Press, 1986.

Farmer, Charles J. *In the Absence of Towns: Settlement and Country Trade in Southside Virginia, 1730–1800.* Lanham, Md.: Rowman and Littlefield, 1993.

Featherstonhaugh, G. W. *Excursion through the slave states, from Washington on the Potomac, to the frontier of Mexico; with sketches of popular manners and geological notices.* New York: Harper and Brothers, 1844.

———. *Geological report of an examination made in 1834 of the elevated country between the Missouri and Red rivers.* Washington, D.C.: Gales and Seaton, 1835.

Festervand, D. F. *Soil Survey of Perry County, Missouri.* U.S. Department of Agriculture, Soil Conservation Service. Washington, D.C.: GPO, 1986.

Finiels, Nicolas de. *An Account of Upper Louisiana by Nicolas de Finiels.* Ed. Carl J. Ekberg and William E. Foley. Columbia: University of Missouri Press, 1989.

Finnie, W. Bruce. *Topographic Terms in the Ohio Valley, 1748–1800*. Publication of the American Dialect Society, no. 53. Tuscaloosa: University of Alabama Press, 1970.

Flader, Susan L. "Final Narrative Report, Ste. Genevieve: An Interdisciplinary Community Study." Final Report to General Research Program, National Endowment for the Humanities. March 1986.

———. "Settlement History of Old Ste. Genevieve." Paper delivered at the Ste. Genevieve Seminar on French Colonial Studies, November 1–3, 1985, Ste. Genevieve, Mo.

———. "Ste. Genevieve: An Interdisciplinary Community Study." Renewal proposal to General Research Program, National Endowment for the Humanities, for grant period March 1, 1983, to August 31, 1984.

Flagg, Edmund. *The Far West; or, A tour beyond the mountains.* . . . 2 vols. 1838. Reprinted in vols. 26–27 of *Early Western Travels, 1748–1846,* ed. Reuben Gold Thwaites. Cleveland: Arthur H. Clark, 1906.

Flanders, Robert. "Caledonia: Ozarks Legacy of the High Scotch-Irish." *Gateway Heritage* 6:4 (1986): 34–52.

———. "Caledonia, an Ozarks Village: History, Geography, Architecture." Springfield: Southwest Missouri State University, Center for Ozark Studies, 1984.

Flatrès, Pierre. "Hamlet and Village." In *Man and His Habitat: Essays Presented to Emyr Estyn Evans,* ed. R. H. Buchanan, Emrys Jones, and Desmond McCourt, 165–85. New York: Barnes and Noble, 1971.

Flint, Timothy. *A Condensed Geography and History of the Western States, or the Mississippi Valley.* 2 vols. Cincinnati: William M. Farnsworth, 1828.

———. *Recollections of the Last Ten Years passed in occasional residences and journeyings in the valley of the Mississippi.* . . . 1826. Reprint, New York: DaCapo Press, 1968.

Foley, William E. "The American Territorial System: Missouri's Experience." *Missouri Historical Review* 55 (1971): 403–26.

———. *The Genesis of Missouri: From Wilderness Outpost to Statehood.* Columbia: University of Missouri Press, 1989.

Foley, William E., and C. David Rice. *The First Chouteaus: River Barons of Early St. Louis.* Urbana: University of Illinois Press, 1983.

Ford, Thomas. *A History of Illinois.* New York: Ivison and Phinney, 1854.

Forster, Walter O. *Zion on the Mississippi.* St. Louis: Concordia Publishing House, 1953.

Fortier, Alcée. *A History of Louisiana.* 4 vols. New York: Goupil and Co. of Paris; Manzi, Joyant and Co., successors, 1904.

Fram, Rachel. "Social Life in St. Louis, 1764–1804." *Glimpses of the Past* 1 (1933): 85–91.

Franke, Judith A. *French Peoria and the Illinois Country, 1673–1846.* Illinois State Museum Popular Science Series, vol. 12. Springfield: Illinois State Museum Society, 1995.

Franzwa, Gregory M. *The Story of Old Ste. Genevieve.* Gerald, Mo.: Patrice Press, 1967.

French, Carolyn O. "Cadastral Patterns in Louisiana: A Colonial Legacy." Ph.D. diss., Louisiana State University, 1978.

Friis, Herman R. "A Series of Population Maps of the Colonies and the United States, 1625–1790." *Geographical Review* 30 (1940): 463–70.

Fuller, Myron L. *The New Madrid Earthquake.* U.S. Geological Survey Bulletin 494. Washington, D.C.: GPO, 1912.

Gallman, Robert E. "Influences on the Distribution of Landholdings in Early Colonial North Carolina." *Journal of Economic History* 42 (1982): 549–75.

Gardner, James Alexander. *Lead King: Moses Austin.* St. Louis: Sunrise Publishing, 1980.

Gates, Paul Wallace. *Fifty Million Acres.* Ithaca: Cornell University Press, 1954.

———. *Frontier Landlords and Pioneer Tenants.* Ithaca: Cornell University Press, 1945.

———. *History of Public Land Law Development.* Washington, D.C.: Public Land Law Review Commission, 1968.

———. *The Jeffersonian Dream: Studies in the History of American Land Policy and Development.* Ed. Allan G. Bogue and Margaret Beattie Bogue. Albuquerque: University of New Mexico Press, 1996.

———. "Tenants of the Log Cabin." *Mississippi Valley Historical Review* 49 (June 1962): 3–31.

Gayarré, Charles. *Histoire de la Louisiane.* 2 vols. New Orleans: Magne and Weisse, 1846–1847.

———. *History of Louisiana.* 4 vols. New York: Redfield, 1854.

Gentilcore, R. Louis. "Vincennes and French Settlement in the Old Northwest." *Annals of the Association of American Geographers* 47 (1957): 285–97.

Gerlach, Russel L. *Immigrants in the Ozarks: A Study in Ethnic Geography.* Columbia: University of Missouri Press, 1976.

———. "The Ozark Scotch-Irish." In *Cultural Geography in Missouri,* ed. M. O. Roark, 11–29. Cape Girardeau: Southeast Missouri State University, 1983.

Ghorra-Gobin, Cynthia. "La 'Frontière,' espace et formation de l'identité nationale américaine: Comment réinterpreter la thèse de Turner un siècle après?" *Hérodote: Revue de Géographie et de Géopolitique,* nos. 72–73 (January–June 1994): 170–79.

Giraud, Marcel. *A History of French Louisiana*. 5 vols. Baton Rouge: Louisiana State University Press, 1991.

Gitlin, Jay. "'Avec bien du regret': The Americanization of Creole St. Louis." *Gateway Heritage* 9:4 (1989): 2–11.

———. "On the Boundaries of Empire." In *Under an Open Sky: Rethinking America's Western Past,* ed. William Cronon, George Miles, and Jay Gitlin, 71–89. New York: W. W. Norton, 1992.

Glacken, Clarence J. *Traces on the Rhodian Shore: Nature and Culture in Western Thought from Ancient Times to the End of the Eighteenth Century.* Berkeley and Los Angeles: University of California Press, 1967.

Goff, Frederick R. "A Rare Pamphlet about Missouri." *Bulletin of the Missouri Historical Society* 13 (1957): 400–402.

Gold, Gerald L. "Lead Mining and the Survival and Demise of French in Rural Missouri (les gens qui ont pioché le tuf)." *Cahiers de Géographie de Québec* 23:59 (1979): 331–41.

Goodrich, James W., and Lynn Wolf Gentzler, eds. "'I Well Remember': David Holmes Conrad's Recollections of St. Louis, 1819–1823." *Missouri Historical Review* 90 (1995): 1–37.

Gracy, David B., II. *Moses Austin: His Life.* San Antonio: Trinity University Press, 1987.

Green, Mary Susan. "The Material Culture of a Pre-enclosure Village in Upper Louisiana: Open Fields, Houses, and Cabinetry in Colonial Ste. Genevieve, 1750–1804." Master's thesis, University of Missouri–Columbia, 1983.

Greene, Evarts B., and Virginia D. Harrington. *American Population before the Federal Census of 1790.* New York: Columbia University Press, 1932.

Gums, Bonnie L., William R. Iseminger, Molly E. McKenzie, and Dennis D. Nichols. "The French Colonial Villages of Cahokia and Prairie du Pont, Illinois." In *French Colonial Archaeology: The Illinois Country and the Western Great Lakes,* ed. John A. Walthall, 85–122. Urbana: University of Illinois Press, 1991.

Guyette, Richard P., and E. E. Cutter. "Tree-Ring Analysis of Fire History of a Post Oak Savanna in the Missouri Ozarks." *Natural Areas Journal* 11 (1991): 93–99.

Haggett, Peter. *Geography: A Modern Synthesis.* 2d ed. New York: Harper and Row, 1975.

Hall, John W. "Louisiana Survey Systems: Their Antecedents, Distribution, and Characteristics." Ph.D. diss., Louisiana State University, 1970.

———. "Sitios in Northwestern Louisiana." *North Louisiana Historical Association Journal* 1 (1970): 1–9.

Hanley, Lucy Elizabeth. "Lead Mining in the Mississippi Valley during the Colonial Period." Master's thesis, St. Louis University, 1942.

Harlan, James D. "A G.I.S. Analysis of Early Land Choice in the Historic Ste. Genevieve District." Research paper, Department of Geography, University of Missouri–Columbia, 1995.

Harley, J. B. "Maps, Knowledge, and Power." In *The Iconography of Landscape*, ed. Denis Cosgrove and Stephen Daniels, 277–312. Cambridge: Cambridge University Press, 1988.

———. "Silences and Secrecy: The Hidden Agenda of Cartography in Early Modern Europe." *Imago Mundi* 40 (1988): 57–76.

Harris, Richard Colebrook. "French Landscapes in North America." In *The Making of the American Landscape*, ed. Michael P. Conzen, 63–79. New York: Routledge, 1994.

———. *The Seigneurial System in Early Canada: A Geographical Study.* Madison: University of Wisconsin Press, 1966.

———. "The Simplification of Europe Overseas." *Annals of the Association of American Geographers* 67 (1977): 469–83.

Harvey, Henry. *History of the Shawnee Indians from the Year 1681 to 1854, Inclusive.* Cincinnati: Ephraim Morgan and Sons, 1855.

Hasse, Larry. "Watermills in the South: Rural Institutions Working against Modernization." *Agricultural History* 58 (1984): 280–95.

Henderson, John R. "The Cultural Landscape of French Settlements in the American Bottom." Master's thesis, Illinois State University, 1966.

Hero, Alfred Olivier, Jr. *Louisiana and Quebec: Bilateral Relations and Comparative Sociopolitical Evolution, 1673–1993.* New Orleans: Tulane University Press, 1995.

Hibbard, Benjamin Horace. *A History of the Public Land Policies.* New York: Peter Smith, 1939.

Hibbert, Wilfrid. "Major Amos Stoddard: First Governor of Upper Louisiana and Hero of Fort Meigs." *Quarterly Bulletin of the Historical Society of Northwestern Ohio* 2:2 (1930): 1–11.

Hill, Roscoe R. *Descriptive Catalogue of the Documents relating to the History of the United States in the Papeles Procedentes de Cuba deposited in the Archivo General de Indias at Seville.* Washington, D.C.: Carnegie Institution of Washington, 1916.

Hilliard, Sam B. "Headright Grants and Surveying in Northeastern Georgia." *Geographical Review* 72 (1982): 416–29.

———. "An Introduction to Land Survey Systems in the Southeast." Geoscience Reprint Series no. 752. Baton Rouge: Louisiana State University, School of Geoscience, 1973.

Historical Statistics of the United States, Colonial Times to 1957: A Statis-

tical Abstract Supplement. Washington, D.C.: U.S. Bureau of the Census, 1960.

History of Franklin, Jefferson, Washington, Crawford, and Gasconade Counties, Missouri. Chicago: Goodspeed Publishing, 1888.

History of Southeast Missouri: Embracing an Historical Account of the Counties of Ste. Genevieve, St. Francois, Perry, Cape Girardeau, Bollinger, Madison, New Madrid, Pemiscot, Dunklin, Scott, Mississippi, Stoddard, Butler, Wayne, and Iron. Chicago: Goodspeed Publishing, 1888.

A History of the Brazeau Presbyterian Church, Brazeau, Missouri, 1819–1970. N.p., [1972?]. Copy in the State Historical Society of Missouri, Columbia.

Hodge, Frederick Webb, ed. *Handbook of American Indians North of Mexico.* 2 vols. Washington, D.C.: GPO, 1907–1910.

Hoehn, Larry. *The Genealogist and Historian's Guide to Perry County, Missouri.* St. Louis: Lineage Press, 1985.

Hoffman, Richard C. "Medieval Origins of the Common Fields." In *European Peasants and Their Markets,* ed. William N. Parker and Eric L. Jones, 24–71. Princeton: Princeton University Press, 1975.

Hofstra, Warren R. "Land Policy and Settlement in the Northern Shenandoah Valley," In *Appalachian Frontiers: Settlement, Society, and Development in the Preindustrial Era,* ed. Robert D. Mitchell, 105–26. Lexington: University Press of Kentucky, 1991.

Holmes, Jack D. L. *Docomentos Ineditos para la Historia de la Luisiana, 1792–1810.* Madrid: Ediciones Jose Porrua Turanzas, 1963.

———. "The Historiography of the American Revolution in Louisiana." *Louisiana History* 19 (1978): 309–26.

———. "The Value of the Arpent in Spanish Louisiana and West Florida." *Louisiana History* 24 (1983): 314–20.

Holweck, F. G. "Public Places of Worship in St. Louis, before Palm Sunday 1843." *St. Louis Catholic Historical Review* 4 (1922): 5–12.

Hornbeck, David. "Spanish Legacy in the Borderlands." In *The Making of the American Landscape,* ed. Michael P. Conzen, 51–62. New York and London: Routledge, 1994.

Houck, Louis. *History of Missouri, from the Earliest Explorations and Settlements until the Admission of the State into the Union.* 3 vols. Chicago: R. R. Donnelley and Sons, 1908.

———, ed. *The Spanish Regime in Missouri.* 2 vols. Chicago: R. R. Donnelley and Sons, 1909.

Hudson, John C. "A Location Theory for Rural Settlement." *Annals of the Association of American Geographers* 59 (1969): 365–81.

———. "Migration to an American Frontier." *Annals of the Association of American Geographers* 66 (1976): 242–65.

———. "North American Origins of Middlewestern Frontier Populations." *Annals of the Association of American Geographers* 78 (1988): 395–413.

———. *Plains Country Towns*. Minneapolis: University of Minnesota Press, 1985.

Hueber, S. P., ed. *Centennial History of Perry County, Missouri, 1821–1921*. Perryville, Mo.: Centennial History Committee, 1921.

Hunt, Anne Lucas. "Early Recollections." *Glimpses of the Past* 1 (1933–1934): 41–51.

Hurley, Andrew, ed. *Common Fields: An Environmental History of St. Louis*. St. Louis: Missouri Historical Society, 1997.

Hutchins, Thomas. *A Historical and Topographical Description of Louisiana and West Florida, 1784*. Edited by Joseph G. Tregle Jr. Facsimile ed. Gainesville: University of Florida Press, 1968.

Ingalls, Walter Renton. *Lead and Zinc in the United States: Comprising an Economic History of the Mining and Smelting of the Metals and the Conditions Which Have Affected the Development of the Industries*. New York: Hill Publishing, 1908.

Ives, David J., J. Alan May, and David D. Denman. "Pine Ford Lake: Phase I Archeological Survey." American Archeology Division, Department of Anthropology, University of Missouri–Columbia, 1982.

Jackson, Donald. "A New Lewis and Clark Map." *Bulletin of the Missouri Historical Society* 17 (1961): 117–32.

Jackson, John Brinckerhoff. "The Movable Dwelling and How It Came to America." In *Discovering the Vernacular Landscape*, 88–100. New Haven: Yale University Press, 1984.

Jackson, R. V., G. R. Teeples, and D. Schaefermeyer, eds. *Missouri 1830 Census Index*. Bountiful, Utah: Accelerated Indexing Systems, 1976.

Jaenen, Cornelius J. "Colonisation compacte et colonisation extensive aux XVIIe et XVIIIe siècles en Nouvelle-France." In *Colonies, territoires, sociétés: L'Enjeu français,* ed. Alain Saussol and Joseph Zitomersky, 14–22. Paris: L'Harmattan, 1996.

Jakle, John A. "Salt on the Ohio Valley Frontier, 1770–1820." *Annals of the Association of American Geographers* 59 (1969): 687–709.

———. "Transportation and the Evolution of River Settlements along the Middle Mississippi River." Master's thesis, Southern Illinois University, 1963.

James, Eugene LeRoy. "The History of Lead Mining in Southeastern Missouri." Master's thesis, Northeast Missouri State Teacher's College, 1957.

"James Clemens, Jr. to Isachar Pawling." *Glimpses of the Past* 3 (1936): 135–37.

Janet, Richard Joseph. "St. Mary's of the Barrens Seminary and the Vin-

centians in Southeast Missouri, 1818–1843." Master's thesis, Southeast Missouri State University, 1979.
Jefferson, Thomas. *The Writings of Thomas Jefferson*. Ed. Paul Leicester Ford. 10 vols. New York: G. P. Putnam's Sons, 1892–1899.
Jennings, Sister Marietta. *A Pioneer Merchant of St. Louis, 1810–1820: The Business Career of Christian Wilt*. New York: Columbia University Press, 1939.
Johnson, Hildegard Binder. "French Canada and the Ohio Country: A Study in Early Spatial Relationships." *Canadian Geographer* 3 (1958): 1–10.
———. *French Louisiana and the Development of the German Triangle*. A German-American Tricentennial Publication. Minneapolis: Associates of the James Ford Bell Library, University of Minnesota, 1983.
———. "A Historical Perspective on Form and Function in Upper Midwest Rural Settlement." *Agricultural History* 48 (1974): 11–25.
———. "The Location of German Immigrants in the Middle West." *Annals of the Association of American Geographers* 41 (1951): 1–41.
Johnson, Hugh N. "Sequent Occupance of the St. Francis Mining Region." Ph.D. diss., Washington University, 1950.
Johnson, Jerah. "La Coutume de Paris: Louisiana's First Law." *Louisiana History* 39 (1989): 145–55.
Jones, J. Wyman. "A History of the St. Joseph Lead Company from Its Organization, in 1864, to January 1, 1892." St. Joseph: St. Joseph Lead Company, n.d.
Jordan, Terry G. "Antecedents of the Long-Lot in Texas." *Annals of the Association of American Geographers* 64 (1974): 70–86.
———. "Between the Forest and the Prairie." *Agricultural History* 38 (1964): 205–16.
———. *North American Cattle-Ranching Frontiers: Origins, Diffusion, and Differentiation*. Albuquerque: University of New Mexico Press, 1993.
———. "Preadaptation and European Colonization in Rural North America." *Annals of the Association of American Geographers* 79 (1989): 489–500.
Jordan, Terry G., and Matti Kaups. *The American Backwoods Frontier: An Ethnic and Ecological Interpretation*. Baltimore: Johns Hopkins University Press, 1989.
Kain, Roger J. P., and Elizabeth Baigent. *The Cadastral Map in the Service of the State: A History of Property Mapping*. Chicago: University of Chicago Press, 1992.
Kappler, Charles J. *Indian Affairs: Laws and Treaties*. Vol. 2, *Treaties*. Washington, D.C.: GPO, 1904.
Kedro, Milan J. "The Three-Notch Road Frontier: A Century of Social and

Economic Change in the Ste. Genevieve District." *Bulletin of the Missouri Historical Society* 29 (1975): 189–204.

Keefe, James F., and Lynn Morrow, eds. *The White River Chronicles of S. C. Turnbo: Man and Wildlife on the Ozarks Frontier.* Fayetteville: University of Arkansas Press, 1994.

Keene, David. "Fort de Chartres: Archaeology in the Illinois Country." In *French Colonial Archaeology: The Illinois Country and the Western Great Lakes*, ed. John A. Walthall, 29–42. Urbana: University of Illinois Press, 1991.

Keslin, Richard O. "Archaeological Implications of the Role of Salt as an Element of Cultural Diffusion." *Missouri Archaeologist* 26 (1964): 1–181.

King, Brian. "The Delimitation and Demarcation of the State Boundary of Missouri." Master's thesis, University of Missouri–Columbia, 1995.

King, Leslie J. "A Quantitative Expression of the Pattern of Urban Settlements in Selected Areas of the United States." *Tijdschrift voor Economische en Sociale Geografie* (Rotterdam) 53:1 (1962): 1–7.

Kinnaird, Lawrence. "American Penetration into Spanish Louisiana." In *New Spain and the Anglo-American West,* ed. Charles W. Hackett, George P. Hammond, and Lloyd Mecham, 1:211–38. Lancaster, Pa.: Lancaster Press, 1932.

———, ed. *Spain in the Mississippi Valley, 1765–1794.* Annual Report of the American Historical Association for the Year 1945. Vols. 2–4. Washington, D.C.: GPO, 1946–1949.

Klein, Ada Paris, ed. "Ownership of the Land under France, Spain, and United States." *Missouri Historical Review* 44 (1950): 274–94.

Koehler, Kevin J. "French New Bourbon: A Study of a Spanish Missouri Community during the 1790s." Master's thesis, University of Missouri–Columbia, 1990.

Kretzmann, P. E. "Saxon Immigration to Missouri, 1838–1839." *Missouri Historical Review* 33 (1939): 157–70.

Krusekopf, H. H., W. DeYoung, W. I. Watkins, and C. E. Deardorff. *Soil Survey of Reynolds County, Missouri.* U.S. Department of Agriculture, Bureau of Soils. Washington, D.C.: GPO, 1921.

Ladd, Doug. "Re-examination of the Role of Fire in Missouri Oak Woodlands." In *Proceedings of the Oak Woods Management Workshop,* ed. J. E. Ebinger and G. S. Wilhelm, 67–80. Charleston: Eastern Illinois University, 1991.

La Flesche, Francis. *The Osage Tribe.* Annual Report of the Bureau of American Ethnology 36. Washington, D.C.: GPO, 1921.

Larimore, Ann Evans. *The Alien Town: Patterns of Settlement in Busoga,*

Uganda. Department of Geography Research Paper no. 55. Chicago: University of Chicago, 1958.

Latrobe, Benjamin Henry Boneval. *Impressions Respecting New Orleans: Diary and Sketches, 1818–1820.* Ed. Samuel Wilson. New York: Columbia University Press, 1951.

Latrobe, Charles Joseph. *The Rambler in North America.* 2 vols. London: R. E. Seeley and W. Burnside, 1835.

LeBeau, Pierre. "The French Colonial Period in the Illinois Country, 1673–1765." *Le Journal* (Center for French Colonial Studies) 16:3 (summer 2000): 6–9.

———. "The French Colonial Period in the Teaching of Illinois History." *Le Journal* (Center for French Colonial Studies) 16:3 (summer 2000): 1–5.

Lebergott, Stanley. "The Demand for Land: The United States, 1820–1860." *Journal of Economic History* 45 (1985): 181–212.

LeBlanc, Gabriel. "A Closer Look at the September 16, 1732, Montreal Earthquake." *Canadian Journal of Earth Sciences* (Ottawa) 18 (1981): 539–50.

Lemon, James T. *The Best Poor Man's Country.* Baltimore: Johns Hopkins University Press, 1972.

———. "Colonial America in the Eighteenth Century." In *North America: The Historical Geography of a Changing Continent,* ed. Robert D. Mitchell and Paul A. Groves, 121–46. London: Hutchison, 1987.

Leopold, Luna B., M. Gordon Wolman, and John P. Miller. *Fluvial Processes in Geomorphology.* San Francisco: W. H. Freeman, 1964.

Le Page du Pratz, [Antoine Simon]. *The History of Louisiana.* 1774. Facsimile reproduction, Baton Rouge: Louisiana State University Press, 1975.

Lewis, Donald. "Economic and Social Life of the French Villages of Missouri." Master's thesis, University of Missouri–Columbia, 1936.

Lewis, G. Malcom, ed. *Cartographic Encounters: Perspectives on Native American Mapmaking and Map Use.* Chicago: University of Chicago Press, 1998.

Lewis, Kenneth E. *The American Frontier: An Archaeological Study of Settlement Pattern and Process.* Orlando: Academic Press, 1984.

Liming, F. G. *The Range and Distribution of Shortleaf Pine in Missouri.* Technical Paper no. 106. Columbus: Central States Forest Experiment Station, U.S. Department of Agriculture, 1946.

Lockridge, Kenneth A. *A New England Town: The First Hundred Years.* New York: W. W. Norton, 1970.

Loeb, Isidor. "The Beginnings of Missouri Legislation." *Missouri Historical Review* 1 (1906): 53–71.

Long, John H., ed. *Historical Atlas and Chronology of County Boundaries, 1788–1980.* Vol. 4, *Iowa, Missouri,* comp. Adele Hast and John H. Long. Hermon Dunlap Smith Center for the History of Cartography, the Newberry Library, Chicago. Boston: G. K. Hall, 1984.

Long, Stephen H. *Account of an expedition from Pittsburgh to the Rocky Mountains, performed in the years 1819 and '20.* Comp. Edwin James. Philadelphia: H. C. Carey and I. Lea, 1822–1823.

Louder, Dean R., and Eric Waddell, eds. *French America: Mobility, Identity, and Minority Experience across the Continent.* 1983. Reprint, Baton Rouge: Louisiana State University Press, 1993.

Lucas, John B. C., ed. *Letters of Hon. J. B. C. Lucas from 1815 to 1836.* St. Louis: Lippincott, 1905.

Luzières, Pierre-Charles Delassus de. *An Epoch in Missouri history; being a facsimile of a brochure printed in Lexington, Kentucky in the year 1796, with a twentieth century view of it.* Commentary by Frances L. S. Dugan and Jacqueline P. Bull. Keepsake no. 5. Lexington: University of Kentucky Library Associates, 1958.

Lyon, E. Wilson. *Louisiana in French Diplomacy, 1759–1804.* Norman: University of Oklahoma Press, 1934.

Mahoney, Timothy R. *River Towns in the Great West: The Structure of Provincial Urbanization in the American Midwest, 1820–1870.* New York: Cambridge University Press, 1990.

Marbut, Curtis Fletcher. "Physical Features of Missouri." In *Surface Features of Missouri,* 11–109. Missouri Geological Survey, vol. 10. Jefferson City, 1896.

———. *Soils of the Ozark Region: A Preliminary Report on the General Character of the Soils and the Agriculture of the Missouri Ozarks.* Research Bulletin no. 3, pp. 151–273. Columbia: University of Missouri, Agricultural Experiment Station, 1910.

———. "Soil Reconnaissance of the Ozark Region of Missouri and Arkansas." In *Field Operations of the Bureau of Soils, 1911,* 1717–1872. Map No. 41 in separate folio. U.S. Department of Agriculture, Bureau of Soils. Washington, D.C.: GPO, 1914.

March, David D. *The History of Missouri.* 4 vols. New York: Lewis Historical Publishing, 1967.

Margry, Pierre. *Mémoires et documents pour servir à l'histoire des origines français des pays d'outre-mer; découvertes et établissements des français dans l'ouest et dans le sud de l'Amérique septentrionale (1683–1724).* Vol. 5, *Première formation d'une chaîne de postes entre le fleuve Saint-Laurent et le Golfe du Mexique (1683–1724).* Paris: Maisonneuve Frères et Ch. Leclerc, 1887.

Marshall, Howard Wight. *Folk Architecture in Little Dixie: A Regional Culture in Missouri.* Columbia: University of Missouri Press, 1981.

Martin, François-Xavier. *The History of Louisiana.* 1927. Reprint, New Orleans: James A. Gresham, 1882.

Martin, Gaston. *Histoire de l'esclavage dans les colonies françaises.* Paris: Presses Universitaires de France, 1948.

Mathews, John J. *The Osages: Children of the Middle Waters.* Norman: University of Oklahoma Press, 1961.

McCandless, Perry. *A History of Missouri: Volume 2, 1820 to 1860.* Columbia: University of Missouri Press, 1972.

McCloskey, David N. "The Persistence of English Common Fields." In *European Peasants and Their Markets,* ed. William N. Parker and Eric L. Jones, 73–119. Princeton: Princeton University Press, 1975.

McCormick, E. M. "The Coleman Family History." Typescript, [1930s?]. Available as annotated by Patricia Weeks (1979) from Old Mines Area Historical Society, Cadet, Mo.

McDaniel, Chad. "Aspects of the Demography of Ste. Genevieve." Paper delivered at the Ste. Genevieve Seminar on French Colonial Studies, November 1–3, 1985, Ste. Genevieve, Mo.

———. Presentation on preliminary findings from a demographic study of Ste. Genevieve. Workshop on Ste. Genevieve History, Architecture, and Archaeology, sponsored by Foundation for the Restoration of Ste. Genevieve, Ste. Genevieve, Mo., October 30, 1981.

McDermott, John .F, ed. *Before Mark Twain: A Sampler of Old, Old Times on the Mississippi.* Carbondale: Southern Illinois University Press, 1968.

———. *A Glossary of Mississippi Valley French, 1673–1850.* Washington University Studies, New Series, Language and Literature, no. 12. St. Louis: Washington University, 1941.

———. "Myths and Realities Concerning the Founding of St. Louis." In *The French in the Mississippi Valley,* 1–15. Urbana: University of Illinois Press, 1965.

———. "Paincourt and Poverty." *Mid-America,* new ser., 5 (1934): 210–12.

———, ed. *The French in the Mississippi Valley.* Urbana: University of Illinois Press, 1965.

———, ed. *Frenchmen and French Ways in the Mississippi Valley.* Urbana: University of Illinois Press, 1969.

———, ed. *Old Cahokia: A Narrative and Documents Illustrating the First Century of Its History.* St. Louis: St. Louis Historical Documents Foundation, 1949.

———, ed. *The Spanish in the Mississippi Valley, 1762–1804.* Urbana: University of Illinois Press, 1965.

McMahon, David Francis. "Tradition and Change in an Ozark Mining Community." Master's thesis, St. Louis University, 1958.
McManis, Douglas R. *The Initial Evaluation and Utilization of the Illinois Prairies, 1815-1840*. Department of Geography Research Paper no. 94. Chicago: University of Chicago, 1964.
McQuillan, D. Aidan. *Prevailing over Time: Ethnic Adjustment on the Kansas Prairies, 1875-1925*. Lincoln: University of Nebraska Press, 1990.
McReynolds, Edwin C. *Missouri: A History of the Crossroads State*. Norman: University of Oklahoma Press, 1962.
McWhiney, Grady. *Cracker Culture: Celtic Ways in the Old South*. Tuscaloosa: University of Alabama Press, 1988.
Meade, Melinda S. "The Rise and Demise of Malaria: Some Reflections on Southern Settlement and Landscape." *Southeastern Geographer* 20 (1980): 77-99.
Meinig, Donald W. "American Wests: Preface to a Geographical Interpretation." *Annals of the Association of American Geographers* 62 (1972): 159-85.
———. "The Continuous Shaping of America: A Prospectus for Geographers and Historians." *American Historical Review* 83 (1978): 1186-1205.
———. *The Shaping of America: A Geographical Perspective on 500 Years of History*. Vols. 1-3. New Haven: Yale University Press, 1986-1993.
Meitzen, August. *Siedlung und Agrarwesen des Westgermanen und Ostgermanen, der Kelten, Römer, Finnen und Slawen*. 3 vols. Berlin: W. Hertz, 1895.
Melom, Halvor Gordon. "The Economic Development of St. Louis, 1803-1846." Ph.D. diss., University of Missouri-Columbia, 1947.
Merrens, Roy. "The Physical Environment of Early America: Images and Image Makers in Colonial South Carolina." *Geographical Review* 59 (1969): 530-56.
Meyer, Duane G. *The Heritage of Missouri*. 3d ed. St. Louis: River City Publishers, 1982.
Mikesell, Marvin. "Comparative Studies in Frontier History." *Annals of the Association of American Geographers* 50 (1960): 64-74.
———. "Tradition and Innovation in Cultural Geography." *Annals of the Association of American Geographers* 68 (1978): 1-16.
Miller, William Marion. "The Bicentennial of Ste. Genevieve, Missouri." *French Review* 10 (1936): 30-34, 109-14.
Miner, Horace. *St. Denis: A French-Canadian Parish*. Chicago: University of Chicago Press, 1939.

Mires, Peter B. "Relationships of Louisiana Colonial Land Claims with Potential Natural Vegetation and Historic Standing Structures: A GIS Approach." *Professional Geographer* 45 (1993): 342–50.

Mitchell, Robert D. "The Colonial Origins of Anglo-America." In *North America: The Historical Geography of a Changing Continent*, ed. Robert D. Mitchell and Paul A. Groves, 93–120. London: Hutchinson, 1987.

———. *Commercialism and Frontier: Perspectives on the Early Shenandoah Valley.* Charlottesville: University Press of Virginia, 1977.

———. "The Shenandoah Valley Frontier." *Annals of the Association of American Geographers* 62 (1972): 461–86.

Monette, John W. *History of the Discovery and Settlement of the Valley of the Mississippi . . . until the year 1846.* 2 vols. New York: Harper and Brothers, 1846.

Moneymaker, B. C. "Some Early Earthquakes in Tennessee and Adjacent States, 1699 to 1850." *Journal of the Tennessee Academy of Science* 29 (1954): 224–33.

Moogk, Peter. *La Nouvelle France: The Making of French Canada—a Cultural History.* East Lansing: Michigan State University Press, 2000.

Moore, Glover. *The Missouri Controversy, 1819–1821.* Lexington: University Press of Kentucky, 1953.

Morrow, Lynn. "New Madrid and Its Hinterland, 1783–1826." *Bulletin of the Missouri Historical Society* 36 (1980): 241–50.

———. "Ozark/Ozarks: Establishing a Regional Term." *White River Valley Historical Quarterly* 35:2 (1996): 4–11.

———. "Trader William Gillis and Delaware Migration in Southern Missouri." *Missouri Historical Review* 75 (1981): 147–67.

Morse, Jedidiah. *A report to the Secretary of War of the United States, on Indian Affairs, comprising a narrative of a tour performed in the summer of 1820 . . . for the purpose of ascertaining . . . the actual state of the Indian tribes in our country.* Reprint, St. Clair Shores, Mich.: Scholarly Press, 1972.

Moulton, Gary E., ed. *The Journals of the Lewis and Clark Expedition.* Vol. 1, *Atlas of the Lewis and Clark Expedition.* Lincoln: University of Nebraska Press, 1983.

Muller, Edward K. "Early Urbanization in the Ohio Valley: A Review Essay." *Historical Geography Newsletter* 3:2 (1973): 19–30.

Murphy, James Lee. "A History of the Southeastern Ozark Region of Missouri." Ph.D. diss., St. Louis University, 1982.

Musick, James B. *St. Louis as a Fortified Town.* St. Louis: R. F. Miller, 1941.

Nagel, Paul C. *Missouri, a Bicentennial History.* New York: W. W. Norton, 1977.

Nasatir, Abraham P. *Borderland in Retreat: From Spanish Louisiana to the Far Southwest.* Albuquerque: University of New Mexico Press, 1976.

———. "St. Louis during the British Attack of 1780." In *New Spain and the Anglo-American West*, ed. Charles W. Hackett, George P. Hammond, and J. Lloyd Mecham, 1:239–61. Los Angeles: n.p., 1932.

———, ed. *Before Lewis and Clark: Documents Illustrating the History of Missouri, 1785–1804.* 2 vols. St. Louis: St. Louis Historical Documents Foundation, 1952; reprint, University of Nebraska Press, Lincoln, 1990.

———. *Spanish War Vessels on the Mississippi, 1792–1796.* New Haven: Yale University Press, 1968.

Nash, Roderick. *Wilderness and the American Mind.* 3d ed. New Haven: Yale University Press, 1982.

Nason, Frank L. *A Report on the Iron Ores of Missouri from Field Work Prosecuted during the Years 1891 and 1892.* Geological Survey of Missouri, vol. 2. Jefferson City: Tribune Printing, 1892.

Nigh, Tim, et al. *The Biodiversity of Missouri: Definition, Status and Recommendations for Its Conservation.* Jefferson City: Missouri Department of Conservation and the National Forest Service, 1992.

Nobles, Gregory H. "Straight Lines and Stability: Mapping the Political Order of the Anglo-American Frontier." *Journal of American History* 80 (1993): 9–35.

Norall, Frank. *Bourgmont: Explorer of the Missouri, 1698–1725.* Lincoln: University of Nebraska Press, 1988.

Norris, F. Terry. "Old Cahokia: An 18th Century Archaeological Site Model." *Le Journal* (Center for French Colonial Studies) 2:1 (1984): 1–22.

———. "Ste. Genevieve: A French Colonial Village in the Illinois Country." In *French Colonial Archaeology: The Illinois Country and the Western Great Lakes,* ed. John A. Walthall, 133–48. Urbana: University of Illinois Press, 1991.

———. "Where Did the Villages Go? Steamboats, Deforestation, and Archaeological Loss in the Mississippi Valley." In *Common Fields: An Environmental History of St. Louis,* ed. Andrew Hurley, 73–90. St. Louis: Missouri Historical Society, 1997.

Norris, James D. *Frontier Iron: The Maramec Iron Works, 1826–1876.* Madison: State Historical Society of Wisconsin, 1964.

Nostrand, Richard. "The Century of Hispano Expansion." *New Mexico Historical Review* 62 (1987): 361–86.

Nuttall, Thomas. *A Journal of Travels into the Arkansas Territory during the Year 1819.* Ed. Savoie Lottinville. Norman: University of Oklahoma Press, 1980.

Nuttall, Zelia. "Royal Ordinances Concerning the Laying Out of New Towns." *Hispanic American Historical Review* 5 (1922): 249–54.

Nuttli, Otto W. *The Effects of Earthquakes in the Central United States.* 2d ed. Cape Girardeau, Mo.: Center for Earthquake Studies, 1990.

Oberly, James W. *Sixty Million Acres: American Veterans and the Public Lands before the Civil War.* Kent, Ohio: Kent State University Press, 1990.

O'Brien, Michael J. *Grassland, Forest, and Historical Settlement: An Analysis of Dynamics in Northeast Missouri.* Lincoln: University of Nebraska Press, 1984.

———. *Paradigms of the Past: The Story of Missouri Archaeology.* Columbia: University of Missouri Press, 1996.

O'Brien, Patricia J. "Prehistoric Politics: Petroglyphs and the Political Boundaries of Cahokia." *Gateway Heritage* 15:1 (1994): 30–47.

O'Callaghan, Mary M. "Ste. Genevieve in the Spanish Regime, 1770–1804." Master's thesis, St. Louis University, 1936.

Official Manual of the State of Missouri, 1935–1936. Jefferson City: Midland Printing, 1936.

Oglesby, Richard E. *Manuel Lisa and the Opening of the Missouri Fur Trade.* Norman: University of Oklahoma Press, 1963.

Ohman, Marian M. "Diffusion of Foursquare Courthouses to the Midwest, 1785–1885." *Geographical Review* 72 (1982): 171–89.

———. *Encyclopedia of Missouri Courthouses.* Columbia: University of Missouri Extension Division, 1981.

———. *A History of Missouri's Counties, County Seats, and Courthouse Squares.* Columbia: University of Missouri Extension Division, 1983.

———. *Twenty Towns: Their Histories, Town Plans, and Architecture.* Columbia: University of Missouri Extension Division, 1985.

O'Mara, James. *An Historical Geography of Urban System Development: Tidewater Virginia in the 18th Century.* Geographical Monographs no. 13. Downsview, Ontario: York University, Department of Geography, 1983.

150th Anniversary of Saxon Lutheran Churches in Perry County, Missouri. Perryville, Mo.: 150th Joint Anniversary Committee, 1989.

Onuf, Peter S. *Statehood and Union: A History of the Northwest Ordinance.* Bloomington: Indiana University Press, 1987.

Opie, John. "Frontier History in Environmental Perspective." In *The American West: New Perspectives, New Directions,* ed. Jerome Steffen, 3–4. Norman: University of Oklahoma Press, 1979.

———. "Learning to Read the Pioneer Landscape: Braudel, Eliade, Turner, and Benton." *Great Plains Quarterly* 2:1 (1982): 5–19.

O'Rourke, Timothy J. *Maryland Catholics on the Frontier: The Missouri and Texas Settlements*. Parsons, Kans.: Brefney Press, 1973.

———. *Perry County, Missouri: Religious Haven in the Trans-Mississippi West*. Parsons, Kans.: Brefney Press, 1979.

———. *A Reconstructed Census of Perry County, Missouri, 1821*. Parsons, Kans.: Brefney Press, 1979.

Ostergren, Robert C. "Land and Family in Rural Immigrant Communities." *Annals of the Association of American Geographers* 71 (1981): 400–411.

Otto, John Solomon. "Migration of the Southern Plain Folk: An Interdisciplinary Synthesis." *Journal of Southern History* 51 (1985): 183–200.

Otto, John Solomon, and Nain Estelle Anderson. "The Diffusion of Upland South Folk Culture, 1790–1840." *Southeastern Geographer* 20 (1982): 89–98.

Overby, Osmund. "House and Home in French Ste. Genevieve." Paper delivered at the Ste. Genevieve Seminar on French Colonial Studies, November 1–3, 1985, Ste. Genevieve, Mo.

Palm, Sister Mary Borgia. *The Jesuit Missions of the Illinois Country, 1673–1763*. Cleveland: Sisters of Notre Dame, 1933.

Pattison, William D. *Beginnings of the American Rectangular Land Survey System, 1784–1800*. Department of Geography Research Paper no. 50. Chicago: University of Chicago, 1957.

Paul Wilhelm, duke of Württemberg. *Travels in North America, 1822–1824*. Ed. Savoie Lottinville. Norman: University of Oklahoma Press, 1973.

Paxton, John A. *The St. Louis Directory and Register, containing the Names, Professions, and Residence of All the Heads of Families and Persons in Business; together with Descriptive Notes on St. Louis: etc.* St. Louis: St. Louis Directory and Register, 1821.

Peck, John Mason. *Forty Years of Pioneer Life: Memoir of John Mason Peck, DD, Edited from His Journals and Correspondence*. Ed. Rufus Babcock. 1864. Reprint, Carbondale: Southern Illinois University Press, 1965.

———. *A Guide for Emigrants, Containing Sketches of Illinois, Missouri, and the Adjacent Parts*. Boston: Lincoln and Edmands, 1831.

Pellican, Lou Hudson. *Some of the Stories and Legends of the Burnt Mill*. Perryville, Mo.: Perry County Historical Society, 1984.

Pelzer, Louis. *Henry Dodge*. Iowa City: State Historical Society of Iowa, 1911.

———. "The Spanish Land Grants of Upper Louisiana." *Iowa Journal of History and Politics* 11 (1913): 3–37.

Penick, James Lal, Jr. *The New Madrid Earthquakes*. Rev. ed. Columbia: University of Missouri Press, 1981.

Penn, Dorothy, ed. "The French in the Valley." Parts 1–2. *Missouri Historical Review* 40 (1945): 90–112; (1946): 245–75, 407–30, 562–78.

Perrin du Lac, François Marie. *Voyage dans les deux Louisianes et chez les nations sauvages du Missouri par les Etats-Unis, d'Ohio, et les provinces qui le bordent, en 1801, 1802, et 1803.* Paris: Capelle et Renaud, 1805.

Peters, Richard, ed. *Treaties between the United States and the Indian Tribes.* Vol. 7 of *The Public Statutes at Large of the United States of America.* Boston: Charles C. Little and James Brown, 1848.

Peterson, Charles E. *Colonial St. Louis: Building a Creole Capital.* Tucson: Patrice Press, 1993.

———. "Early Ste. Genevieve and Its Architecture." *Missouri Historical Review* 35 (1941): 207–32.

———. *A Guide to Ste. Genevieve, with Notes on Its Architecture.* [St. Louis?]: U.S. Department of the Interior, National Park Service, Jefferson National Expansion Memorial, 1940.

Pierson, George W. *The Moving Americans.* New York: Alfred A. Knopf, 1972.

Pitot, James. *Observations on the Colony of Louisiana from 1796 to 1802.* Trans. Henry C. Pitot. Baton Rouge: Louisiana State University Press, 1979.

Pittman, Philip. *The Present State of the European Settlements on the Mississippi, with a geographical description of that river.* 1770. Bicentennial Floridiana Facsimile Series. Facsimile reproduction, with introduction and index by Robert R. Rea. Gainesville: University of Florida Press, 1973.

Platt, Robert S. "A Detail of Regional Geography: Ellison Bay Community as an Industrial Organism." *Annals of the Association of American Geographers* 18 (1928): 81–126.

Poole, Stafford, and Douglas J. Slawson. *Church and Slave in Perry County, Missouri, 1818–1865.* Studies in American Religion, vol. 22. Lewiston, N.Y., and Queenston, Ont.: Edwin Mellen Press, 1986.

Price, Anna. "Economic Change in Eighteenth Century Ste. Genevieve." Paper delivered at the Ste. Genevieve Seminar on French Colonial Studies, November 1–3, 1985, Ste. Genevieve, Mo.

———. "The Three Lives of Fort de Chartres: French Outpost on the Mississippi." *Historic Illinois* 3 (1980): 1–4.

Price, Cynthia R. "Report of Construction Monitoring at Durham Hall, 23WA77, in the Town of Potosi, Washington County, Missouri, 1981." Prepared for Missouri Engineering Corporation and Associates, Rolla, by the Center for Archaeological Research, Southwest Missouri State

University, Springfield. Copy in the Division of State Parks and Historic Sites, Missouri Department of Natural Resources, Jefferson City.

Price, Edward T. "The Central Courthouse Square in the American County Seat." *Geographical Review* 58 (1968): 29–60.

———. *Dividing the Land: Early American Beginnings of Our Private Property Mosaic.* University of Chicago Geography Research Paper no. 238. Chicago: University of Chicago Press, 1995.

Price, James E., and Cynthia R. Price. *An Archaeological Survey of Selected Portions of Cape Lacroix, Goose, and Hubble Creeks in Cape Girardeau County, Missouri.* Prepared for the U.S. Army Corps of Engineers, St. Louis District. Columbia: University of Missouri, American Archaeology Division, 1977.

Primm, James Neal. *Lion of the Valley: St. Louis, Missouri.* Boulder: Pruett Publishing, 1981.

———. "Locational Factors in the Development of St. Louis: The First Century." *Gateway Heritage* 4:2 (1983): 10–17.

"Proceedings of the Centennial Anniversary of the City of Gallipolis, Ohio." *Ohio Archaeological and Historical Publications* 3 (1891): 1–81.

Puckett, William Touches Deer. "Settlement History Development of the Cahokia Canal Drainage Area." *Bulletin of the Illinois Geographical Society* 21 (1979): 115–27.

Rabbitt, Mary C. *Minerals, Lands, and Geology for the Common Defense and General Welfare.* Vol. 1. U.S. Geological Survey. Washington, D.C.: GPO, 1979.

Reps, John W. *The Forgotten Frontier: Urban Planning in the American West before 1890.* Columbia: University of Missouri Press, 1981.

———. "New Madrid on the Mississippi." *Journal of the Society of Architectural Historians* 18 (1959): 21–26.

———. *Town Planning in Frontier America.* Princeton: Princeton University Press, 1969; reprint, Columbia: University of Missouri Press, 1980.

Reynolds, John. *The Pioneer History of Illinois.* Belleville, Ill.: N. A. Randall, 1852.

Reynolds County, Missouri: Sesquicentennial Year, 1845–1995. Ellington, Mo.: Reynolds County Genealogy and Historical Society, 1995.

Rice, Howard Crosby. *Barthélémi Tardiveau, a French Trader in the West.* Baltimore: Johns Hopkins University Press, 1938.

Rice, John G. "The Effect of Land Alienation on Settlement." *Annals of the Association of American Geographers* 68 (1978): 61–72.

Richardson, Lemont K. "Private Land Claims in Missouri." *Missouri Historical Review* 50 (1956): 132–44, 271–86, 387–99.

Rickey, Don, Jr. "The British-Indian Attack on St. Louis, May 26, 1780." *Missouri Historical Review* 55 (1960): 35–45.

Robbins, Roy M. *Our Landed Heritage: The Public Domain, 1776–1970*. 2d ed. Lincoln: University of Nebraska Press, 1976.

———. "Preëmption: A Frontier Triumph." *Mississippi Valley Historical Review* 18 (1931): 331–49.

Robertson, James A. "A Projected Settlement of English-Speaking Catholics from Maryland in Spanish Louisiana, 1767, 1768." *American Historical Review* 16 (1911): 319–27.

Rohrbough, Malcolm J. *The Land Office Business: The Settlement and Administration of American Public Lands, 1789–1837*. New York: Oxford University Press, 1968.

———. *The Trans-Appalachian Frontier: People, Societies*. New York: Oxford University Press, 1978.

Rollings, Willard M. *The Osage: An Ethnohistorical Study of Hegemony on the Prairie-Plains*. Columbia: University of Missouri Press, 1992.

Ronnebaum, Sister Chelidonia. "Population and Settlement in Missouri, 1804–1820." Master's thesis, University of Missouri–Columbia, 1936.

Rose, Allen Henry. "The Extension of the United States Land System to Missouri, 1804–1817." Master's thesis, Washington University, 1941.

Rose, Gregory S. "Information Sources for Nineteenth Century Midwestern Migration." *Professional Geographer* 37 (1985): 66–72.

———. "Reconstructing a Retail Trade Area: Tucker's General Store, 1850–1860." *Professional Geographer* 39 (1987): 33–40.

Rothensteiner, Rev. John. *Chronicles of an Old Missouri Parish. Historical Sketches of St. Michael's Church, Fredericktown, Madison County, Missouri. A Souvenir of the Centenary of Its Erection into a Parish, 1827–1927*. Cape Girardeau, Mo.: n.p., 1928. Copy in the State Historical Society of Missouri, Columbia.

———. "Earliest History of Mine La Motte." *Missouri Historical Review* 20 (1926): 199–213.

———. "Father Charles Nerinckx and His Relations to the Diocese of St. Louis." *St. Louis Catholic Historical Review* 1 (1919): 157–75.

———. "Father James Maxwell of Ste. Genevieve." *St. Louis Catholic Historical Review* 4 (1922): 142–54.

———. *History of the Archdiocese of St. Louis*. 2 vols. St. Louis: Blackwell Wielandy, 1928.

Rout, Leslie B., Jr. *The African Experience in Spanish America, 1502 to the Present Day*. Cambridge: Cambridge University Press, 1976.

Royce, Charles C., comp. *Indian Land Cessions in the United States*. Eighteenth Annual Report of the Bureau of American Ethnology to the Secretary of the Smithsonian Institution, 1896–1897. 2 pts. Washington, D.C.: GPO, 1899. Printed as U.S. House, 56th Cong., 1st sess., H. Doc. 736.

Rozier, Firmin A. "Rev. James Maxwell, Missionary at St. [sic] Genevieve." *United States Catholic Historical Magazine*, no. 3 (July 1887): 283–86.

———. *Rozier's History of the Early Settlement of the Mississippi Valley.* St. Louis: G. A. Pierrot and Son, 1890.

———. "Ste. Genevieve: Its Early History." *Western Journal* (St. Louis) 3 (1850): 239–43.

Rush, Benjamin. "An Inquiry into the Causes of the Increase of Bilious and Intermitting Fevers in Pennsylvania." *Medical Inquiries and Observations* (Philadelphia) 11 (1797): 265–76.

Rutledge, Anne B. *York Chapel Graveyard, 1821–1971.* Perryville, Mo.: Perry County Historical Society, 1971.

Rutledge, Zoe Booth. *Our Jefferson County Heritage.* Cape Girardeau, Mo.: Ramfre Press, 1970.

Rutman, Darrett B., and Anita H. Rutman. "Of Agues and Fevers: Malaria in the Early Chesapeake." *William and Mary Quarterly*, 3d ser., 33 (1976): 31–60.

Saalberg, Gloria. "The New Madrid Land Claims in Howard County, Missouri." *Missouri Mineral Industry News* (Missouri Department of Natural Resources at Rolla) 7 (1967): 69–79.

Sauer, Carl O. *The Geography of the Ozark Highland of Missouri.* Chicago: University of Chicago Press, 1920.

———. *Geography of the Pennyroyal.* Kentucky Geological Survey, ser. 6, vol. 25. Frankfort, Ky., 1927.

———. "Homestead and Community on the Middle Border." *Landscape* 12 (1962): 3–7.

Saussol, Alain. "Nouvelle Calédonie: Stratégies coloniales et organisation de l'espace. Le Mythe d'une colonie de peuplement." In *Colonies, territoires, sociétés: L'Enjeu français*, ed. Alain Saussol and Joseph Zitomersky, 183–215. Paris: L'Harmattan, 1996.

Savelle, Max. "The Founding of New Madrid, Missouri." *Mississippi Valley Historical Review* 19 (1932): 30–56.

———. *George Morgan: Colony Builder.* New York: Columbia University Press, 1932.

Savitt, Todd L. *Medicine and Slavery: The Diseases and Health Care of Blacks in Antebellum Virginia.* Urbana: University of Illinois Press, 1978.

Schaaf, Ida M. "The First Roads West of the Mississippi." *Missouri Historical Review* 29 (1935): 92–105.

———. "Henri Pratte: Missouri's First Native Born Priest." *St. Louis Catholic Historical Review* 5 (1923): 129–48.

Schafer, Joseph. *The Wisconsin Lead Region.* Madison: State Historical Society of Wisconsin, 1932.

Scharf, J. Thomas. *History of Saint Louis City and County, from the earliest periods to the present day: including biographical sketches of representative men.* 2 vols. Philadelphia: L. H. Everts, 1883.

Schiavo, Giovanni E. *The Italians in Missouri.* Chicago: Italian Press, 1929.

Schmudde, Theodore H. "Some Aspects of Land Forms of the Lower Missouri River Floodplain." *Annals of the Association of American Geographers* 53 (1963): 60–73.

Schoolcraft, Henry Rowe. *Journal of a Tour into the Interior of Missouri and Arkansaw, from Potosi, or Mine a Burton, in Missouri Territory, in a South-West Direction, toward the Rocky Mountains; performed in the years 1818–1819.* London: Richard Phillips, 1821.

———. *Scenes and Adventures in the Semi-Alpine Regions of the Ozarks Mountains of Missouri and Arkansas.* Philadelphia: Lippincott, Grambo, 1853.

———. *Travels in the Central Portions of the Mississippi Valley.* New York: Collins and Hannay, 1825.

———. *A View of the Lead Mines of Missouri.* 1819. Reprint, New York: Arno Press, 1972.

Schroeder, Adolf E., and Carla Schulz-Geisberg, eds. *Hold Dear, As Always: Jette, a German Immigrant Life in Letters.* Columbia: University of Missouri Press, 1988.

Schroeder, Walter A. "The Environmental Setting of the St. Louis Region." In *Common Fields: An Environmental History of St. Louis,* ed. Andrew Hurley, 13–37. St. Louis: Missouri Historical Society Press, 1997.

———. "Landforms of Missouri: A Geography of the Configuration of the Land Surface of Missouri." In preparation.

———. "Local Population Movement on the Eastern Ozarks Frontier, 1770–1820." Paper delivered at the annual meeting of the Association of American Geographers, Minneapolis, May 6, 1986. [Abstract in *AAG '86, Twin Cities, Abstracts.* Washington D.C.: Association of American Geographers, 1986, sess. 182].

———. "Opening the Ozarks: Historical Geography of the Ste. Genevieve District (Missouri), 1760–1830." Ph.D. diss., University of Missouri–Columbia, 2000.

———. "Order in the Orientation of Pre-American Land Grants in the Ste. Genevieve District (Missouri)." In *French and Germans in the Mississippi Vallley: Landscape and Cultural Traditions,* ed. Michael Roark, 125–36. Cape Girardeau, Mo.: Southeast Missouri State University, Center for Regional History and Cultural Heritage, 1988.

———. *Presettlement Prairie of Missouri*. Natural History Series no. 2. Jefferson City: Missouri Department of Conservation, 1981.

———. "Settlement Patterns and Land Use in the Ste. Genevieve District." Paper delivered at the Ste. Genevieve Seminar on French Colonial Studies, November 1–3, 1985, Ste. Genevieve, Mo.

———. "Spread of Settlement in Howard County, Missouri, 1810–1859." *Missouri Historical Review* 63 (1968): 1–37.

Schroeder, Walter A., and Tim Nigh. "Atlas of Missouri Ecoregions." Sponsored by the Missouri Department of Conservation, Jefferson City. In preparation.

Schultz, Christian, Jr. *Travels on an Inland Voyage*. 2 vols. New York: Isaac Riley, 1810.

Schultz, Robert George. "Postal Service in Territorial Missouri, 1804–1821." Master's thesis, Northeast Missouri State University, 1980.

Scott, James. *The Illinois Nation*. 2 pts. Streator, Ill.: Streator Historical Society, 1973, 1976.

Searcy, J. K. *Floods in Missouri: Magnitude and Frequency*. U.S. Geological Survey Circular 370. Washington, D.C.: GPO, 1955.

Sellars, Richard West. "Early Promotion and Development of Missouri's Natural Resources." Ph.D. diss., University of Missouri–Columbia, 1972.

Semple, Ellen Churchill. *American History and Its Geographic Conditions*. Boston: Houghton Mifflin, 1903.

———. "The Anglo-Saxons of the Kentucky Mountains: A Study in Anthropo-geography." *Journal of Geography* 17 (1901): 588–623.

———. "Geographic Influences in the Development of St. Louis." *Journal of Geography* 3 (1904): 290–300.

Shinn, Charles Howard. *Mining Camps: A Study in American Government*. New York: Alfred A. Knopf, 1948.

Shoemaker, Floyd Calvin. "Madison County." *Missouri Historical Review* 53 (1958): 1–9.

———. *Missouri and Missourians*. 5 vols. Chicago: Lewis Publishing, 1943.

———. *Missouri's Struggle for Statehood, 1804–1821*. Jefferson City: Hugh Stephens Printing, 1916.

Shortridge, James R. "The Expansion of the Settlement Frontier in Missouri." *Missouri Historical Review* 75 (1980): 64–90.

Shortt, Adam, and Arthur S. Doughty, eds. *Documents Relating to the Constitutional History of Canada, 1759–1791*. Canadian Archives, Sessional paper, no. 18, A.1907. Ottawa, 1907.

Showalter, George. "Potosi, Missouri, Bicentennial, 1763–1963: A Bicentennial Scrapbook Commemorating the 200th Anniversary of Potosi at the Heart of the Historic Mining Country of Missouri." N.p., 1963. Copy in the State Historical Society of Missouri, Columbia.

Shumard, Benjamin F. *A Geological Report on the Old Mines Property of William Long, Esq., in Washington County, Missouri.* St. Louis: R. P. Studley, 1867.

———. *Report on the Chouteau League Tract, or Spanish Mineral Land Grant, Known as U.S. Survey no. 2066, in Washington County, Mo., supplemented by a report on the same tract, now the property of the St. Louis Lead Mining Company, by N. W. Bliss.* St. Louis: Democrat Lithography and Printing, 1873.

Smith, J. Calvin, ed. *Harper's Statistical Gazetteer of the World.* New York: Harper and Brothers, 1855.

Smith, Russell E. "The Towns of the Missouri Valley: An Element of the Historical Geography of Missouri." Master's thesis, University of Missouri–Columbia, 1957.

Soltow, Lee, and Margaret Soltow. "A Settlement That Failed: The French in Early Gallipolis, an Enlightening Letter, and an Explanation." *Ohio History* 94 (1985): 46–67.

"The Spanish Forts at the Mouth of the Missouri River." *Missouri Historical Society Collections* 3:3 (1911): 269–74.

The Spanish Land Grants of Our Own Perry County, Missouri. Perryville, Mo.: Perry County Historical Society, 1984.

Sparke, Matthew. "A Map That Roared and an Original Atlas: Canada, Cartography, and Narration of Nation." *Annals of the Association of American Geographers* 88 (1998): 463–95.

St. Joseph Lead Company. *Herculaneum: Lead Smelting Division of St. Joseph Lead Company.* Centennial brochure [1964]. Copy in the State Historical Society of Missouri, Columbia.

"St. Joseph Lead Company." Lead Smelting Division of St. Joseph Lead Company. N.p., n.d. Copy in the State Historical Society of Missouri, Columbia.

Stacy, June Cooper. *Louis Lorimer [sic].* Cape Girardeau, Mo.: n.p., 1978. Copy in the State Historical Society of Missouri, Columbia.

Stanislawski, Dan. "The Origin and Spread of the Grid-Pattern Town." *Geographical Review* 36 (1946): 105–20.

Stanley, Lois, George F. Wilson, and Maryhelen Wilson, comps. *Missouri Taxpayers, 1819–1826.* Decorah, Iowa: privately published, 1979.

Steck, Francis Borgia. *A Tentative Guide to Historical Materials on the Spanish Borderlands.* Philadelphia: Catholic Historical Society of Philadelphia, 1943.

Stegen, E. F. "The Settlement of the Saxon Lutherans in Perry County, Missouri." Master's thesis, Indiana University, 1930.

Steinbacher-Kemp, William. "Henry Marie Brackenridge and the Illinois Country." *Le Journal* (Center for French Colonial Studies) 16:4 (fall 2000): 1–7.

Stercula, Beverly M. *Heads of Families, 1830 Census of Missouri*. Fullerton, Calif.: Genealogems Publications, 1966.

Stevens, Donald L., Jr. *A Homeland and a Hinterland: The Current and Jacks Fork Riverways*. Historic Resource Study, Ozark National Scenic Riverways. Omaha: National Park Service, Midwest Region, 1991.

Stevens, Paul L. *Louis Lorimier in the American Revolution, 1777–1782: A Mémoire by an Ohio Indian Trader and British Partisan*. Extended Publication Series no. 2. Naperville, Ill.: Center for French Colonial Studies, 1997.

Steward, Dick. *Frontier Swashbuckler: The Life and Legend of John Smith T*. Columbia: University of Missouri Press, 2000.

———. "John Smith T and the Way West: Filibustering and Expansion on the Missouri Frontier." *Missouri Historical Review* 89 (1994): 48–74.

———. "'With the Scepter of a Tyrant': John Smith T and the Mineral Wars." *Gateway Heritage* 14:2 (1993): 24–37.

Stilgoe, John R. *Common Landscape of America, 1580–1845*. New Haven: Yale University Press, 1982.

Stoddard, Amos. *Sketches, Historical and Descriptive, of Louisiana*. Philadelphia: A. Carey, 1812.

———. "Transfer of Upper Louisiana: Papers of Captain Amos Stoddard." *Glimpses of the Past* 2 (1935): 78–122.

Strauser, Claude N. "Restoration of the Middle Mississippi River." St. Louis District, U.S. Army Corps of Engineers, April 1986.

Strickland, Arvarh E. "Aspects of Slavery in Missouri, 1821." *Missouri Historical Review* 55 (1971): 505–26.

Strickland, Rex Wallace. "Anglo-American Activities in Northeastern Texas, 1803–1845." Ph.D. diss., University of Texas, 1937.

Surrey, Nancy Maria. *The Commerce of Louisiana during the French Regime, 1699–1763*. New York: Columbia University Press, 1916.

Swank, James M. *History of the Manufacture of Iron in All Ages*. Philadelphia: American Iron and Steel Association, 1892.

Swartzlow, Ruby J. "The Early History of Lead Mining in Missouri." Parts 1–3. *Missouri Historical Review* 28 (1934): 184–94, 287–95; 29 (1934): 27–34; (1935): 109–14, 195–205.

Swenson, Robert W. "Legal Aspects of Mineral Resources Exploitation."

In *History of Public Land Law Development,* ed. Paul W. Gates, 699–764. Washington, D.C.: Public Land Law Review Commission, 1968.

Tarr, William Arthur. *The Barite Deposits of Missouri and the Geology of the Barite District.* University of Missouri Studies, Science Series 3, no. 1. Columbia, 1918.

Taylor, Alan. *William Cooper's Town: Power and Persuasion on the Frontier of the Early American Republic.* New York: Alfred A. Knopf, 1995.

Taylor, J. W. "The Leasing of the Mineral Lands of the United States up to 1850." Ph.D. diss., University of Wisconsin, 1915.

Taylor, Peter J. *Quantitative Methods in Geography.* Boston: Houghton Mifflin, 1977.

Temple, Wayne C. *Indian Villages of the Illinois Country.* Rev. Scientific Papers, vol. 2, pt. 2. Springfield: Illinois State Museum, 1966.

Tesreau, Mary Ellen C. "Search and Find: Tesreau Genealogy." N.p., 1969. Copy in the State Historical Society of Missouri, Columbia.

Thom, Richard H., and James H. Wilson. "The Natural Divisions of Missouri." *Transactions of the Missouri Academy of Science* 14 (1980): 9–23.

Thomas, Rosemary Hyde. *It's Good to Tell You: French Folktales from Missouri.* Columbia: University of Missouri Press, 1981.

Thompson, Henry Clay, II. *A History of Madison County, Missouri.* 1940. Reprint, Fredericktown, Mo.: McMinn Printing, 1992.

Thompson, Wilson. *The Autobiography of Elder Wilson Thompson, embracing a sketch of his life, travels, and ministerial labors.* 1873. Conley, Ga.: Old School Hymnal, 1978.

Thorndale, William, and William Dollarheide. *Map Guide to the U.S. Federal Census, 1790–1920.* Baltimore: Genealogical Publishing, 1987.

Thorne, Tanis C. *The Many Hands of My Relations: French and Indians on the Lower Missouri.* Columbia: University of Missouri Press, 1996.

Treat, Payson Jackson. *The National Land System, 1785–1820.* New York: E. B. Treat, 1910.

Tregle, Joseph, Jr. "On That Word 'Creole' Again: A Note." *Louisiana History* 23 (1982): 192–98.

Trenfel-Treantafeles, Jacqueline. "Spanish Occupation of the Upper Mississippi Valley, 1765–1770." Master's thesis, University of California–Berkeley, 1941.

Trewartha, Glenn T. "Types of Rural Settlement in Colonial America." *Geographical Review* 36 (1946): 568–96.

———. "The Unincorporated Hamlet." *Annals of the Association of American Geographers* 33 (1943): 32–81.

Trexler, Harrison A. "Missouri in the Old Geographies." *Missouri Historical Review* 32 (1938): 148–55.

———. *Slavery in Missouri, 1804–1865.* Baltimore: Johns Hopkins University Press, 1914.

Trimble, Michael K., Teresita Majewski, Michael J. O'Brien, and Anna L. Price. "Frontier Colonization of the Saline Creek Valley." In *French Colonial Archaeology: The Illinois Country and the Western Great Lakes,* ed. John A. Walthall, 165–88. Urbana: University of Illinois Press, 1991.

Tucker, Sara J. *Atlas.* In *Indian Villages of the Illinois Country.* Scientific Papers, vol. 2, pt. 1. Springfield: Illinois State Museum, 1942.

Turnbull, David. *Maps Are Territories: Science Is an Atlas.* Chicago: University of Chicago Press, 1994.

Turner, Frederick Jackson. *The United States, 1830–1850: The Nation and Its Sections.* 1935. Reprint, Gloucester, Mass.: Peter Smith, 1958.

"Two Maxwell Letters." *St. Louis Catholic Historical Review* 4 (1922): 231–34.

Uhlig, Harald. "Fields and Field Systems." In *Man and His Habitat: Essays Presented to Emyr Estyn Evans,* ed. R. H. Buchanan, Emrys Jones, and Desmond McCourt, 93–125. New York: Barnes and Noble, 1971.

Unklesbay, A. G., and J. D. Vineyard. *Missouri Geology.* Columbia: University of Missouri Press, 1992.

Usner, Daniel H., Jr. "An American Indian Gateway: Some Thoughts on the Migration and Settlement of Eastern Indians around Early St. Louis." *Gateway Heritage* 11:3 (1991): 42–51.

———. "From African Captivity to American Slavery: The Introduction of Black Laborers to Colonial Louisiana." *Louisiana History* 20 (1979): 25–48.

———. *Indians, Settlers, and Slaves in a Frontier Exchange Economy: The Lower Mississippi Valley before 1783.* Chapel Hill: University of North Carolina Press, 1992.

Vance, James E., Jr. "California and the Search for the Ideal." *Annals of the Association of American Geographers* 62 (1972): 185–210.

———. "Democratic Utopia and the American Landscape." In *The Making of the American Landscape,* ed. Michael P. Conzen, 204–20. New York and London: Routledge, 1994.

———. *The Merchant's World: The Geography of Wholesaling.* Englewood Cliffs, N.J.: Prentice-Hall, 1970.

———. *This Scene of Man.* New York: Harper and Row, 1977.

Van Ravenswaay, Charles. *St. Louis: An Informal History of the City and Its People, 1764–1865.* St. Louis: Missouri Historical Society, 1991.

Van Zandt, Nicholas Biddle. *A full description of the soil, water, timber, and prairies of each lot, or quarter section of the military lands between the Mississippi and Illinois rivers.* Washington, D.C.: P. Force, 1818.

Viles, Jonas. "Population and Extent of Settlement in Missouri before 1804." *Missouri Historical Review* 5 (1911): 189–213.

Villiers du Terrage, Marc de. *The Last Years of French Louisiana.* Ed. Carl A. Brasseaux and Glenn R. Conrad. Lafayette: University of Southwestern Louisiana, Center for Louisiana Studies, 1982.

Villmer, Natalie. "History of the Old Mines Area, Washington County, Missouri: A Resume." Old Mines, Mo.: Old Mines Area Historical Society, 1987.

[———]. "250th Anniversary Historical Program Pageant Book of Old Mines, Missouri, 1973." Old Mines, Mo.: Old Mines Area Historical Society, 1973.

Violette, Eugene M. "Early Settlements in Missouri." *Missouri Historical Review* 1 (1906): 38–52.

———. *Spanish Land Claims in Missouri.* Washington University Studies, Humanistic Series, vol. 8, no. 2, pp. 167–200. St. Louis, 1921.

Vogel, Rachel Fram. "Life in Colonial St. Louis." *Glimpses of the Past* 1 (1933): 85–91.

Voight, Eric E., and Michael J. O'Brien. "The Use and Misuse of Soils-Related Data in Mapping and Modeling Past Environments: An Example from the Central Mississippi River Valley." *Contracts Abstracts and CRM Archaeology* 2:3 (1982): 24–31.

Volney, Constantin François Chasseboeuf, comte de. *A View of the Soil and Climate of the United States of America.* Philadelphia: J. Conrad, 1804.

Voss, Stuart F. "Town Growth in Central Missouri, 1815–1860." Parts 1–2. *Missouri Historical Review* 64 (1969): 64–80; (1970): 197–217, 322–50.

Wade, Richard C. *The Urban Frontier: The Rise of Western Cities, 1790–1830.* Cambridge: Harvard University Press, 1959.

Wallace, Agnes. "The Wiggins Ferry Monopoly." *Missouri Historical Review* 42 (1947): 1–19.

Walter, Joseph B. "Distribution of African Americans in Missouri from 1860 to 1990." Master's thesis, University of Missouri–Columbia, 1996.

Walthall, John A., ed. *French Colonial Archaeology: The Illinois Country and the Western Great Lakes.* Urbana: University of Illinois Press, 1991.

Warhus, Mark. *Another America: Native American Maps and the History of Our Land.* New York: St. Martin's Press, 1997.

Watrin, François Philibert. *Banissement des Jésuites de la Louisiane.* Paris, 1865.

Webb, Benedict Joseph. *The Centenary of Catholicity in Kentucky.* Louisville, Ky.: C. A. Rogers, 1884.

Weber, David J. *The Spanish Frontier in North America.* New Haven: Yale University Press, 1992.

———, ed. *New Spain's Far Northern Frontier: Essays on Spain in the American West.* Albuquerque: University of New Mexico Press, 1979.

Webre, Stephen. "The Problem of Indian Slavery in Spanish Louisiana, 1769–1803." *Louisiana History* 25 (1984): 117–35.

Weeks, Patricia. *The 1830 Federal Census of Washington County, Missouri.* Old Mines, Mo.: Old Mines Area Historical Society, 1998.

Weller, Stuart, and Stuart St. Clair. *Geology of Ste. Genevieve County, Missouri.* 2d ser., vol. 22. Rolla: Missouri Bureau of Geology and Mines, 1928.

Weslager, Clinton A. *The Delaware Indian Westward Migration.* Wallingford, Penn.: Middle Atlantic Press, 1978.

Westfall, Frank W. "Politics in Territorial Missouri, 1804–1821." Master's thesis, State University of Iowa, 1938.

Westover, John Glendower. "The Evolution of the Missouri Militia, 1804–1919." Ph.D. diss., University of Missouri–Columbia, 1948.

———. *Selective Memories.* Vol. 1. N.p.: by the author, 1989. Copy in the State Historical Society of Missouri, Columbia.

Wheatley, Paul. *The Pivot of the Four Quarters: The Origin of the Ancient City.* Chicago: Aldine Publishing, 1971.

Whitaker, Arthur Preston. *The Spanish-American Frontier, 1783–1795: The Westward Movement and the Spanish Retreat in the Mississippi Valley.* 1927. Reprint, Gloucester, Mass.: Peter Smith, 1962.

White, C. Albert. *A History of the Rectangular Survey System.* U.S. Department of the Interior, Bureau of Land Management. Washington, D.C.: GPO, 1984.

White, Francis Harding. "The Administration of the General Land Office, 1812–1911." Ph.D. diss., Harvard University, 1912.

White, Joseph M. *A new collection of laws, charters, and local ordinances of the governments of Great Britain, France, and Spain, relating to the concessions of land in their respective colonies. . . .* 2 vols. Philadelphia: T. and J. W. Johnson, 1839.

White, Richard. *The Middle Ground: Indians, Empires, and Republics in the Great Lakes Region, 1650–1815.* Cambridge: Cambridge University Press, 1991.

Wihebrink, Ronald. "History of the Irish Wilderness Country." Rolla, Mo.: U.S. Department of Agriculture, Forest Service, Mark Twain National Forest, 1970.

Wilkie, Duncan C. "Archaeological Reconnaissance Survey of the Apple Creek Drainage: Perry and Cape Girardeau Counties, Missouri." Report prepared for the Historic Preservation Program, Division of Parks and Historic Preservation, Missouri Department of Natural Resources, Jefferson City, December 1984.

Williams, Michael. *Americans and Their Forests: A Historical Geography.* Cambridge: Cambridge University Press, 1989.

Williams, Walter, ed. *The State of Missouri.* Columbia: E. W. Stephens, 1904.

Williamsen, Tora Lorraine. "An Analysis of Colonial Town Planning in North America and Its Influence on Ste. Genevieve, Missouri." Master's thesis, University of Missouri–Columbia, 1990.

Willms, Welton Lyle. "Lead Mining in Missouri, 1700–1811." Master's thesis, Washington University, 1935.

Winslow, Arthur. *Lead and Zinc Deposits.* Missouri Geological Survey, vols. 6–7. Jefferson City, 1894.

Winsor, Justin. "Cartography of Louisiana and the Mississippi Basin under the French Domination." In *Narrative and Critical History of America,* ed. Justin Winsor, 5:79–86. New York: Houghton Mifflin, 1887.

Wish, Harvey. "The French of Old Missouri, 1804–1921: A Study in Assimilation." *Mid-America,* new ser., 12 (1941): 167–89.

Wishart, David J. *The Fur Trade of the American West, 1807–1840: A Geographical Synthesis.* Lincoln: University of Nebraska Press, 1979.

Wood, Joseph S. "'Build, Therefore, Your Own World': The New England Village as Settlement Ideal." *Annals of the Association of American Geographers* 81 (1991): 32–50.

———. "Elaboration of a Settlement System: The New England Village in the Federal Period." *Journal of Historical Geography* 10 (1984): 331–56.

———. *The New England Village.* Baltimore: Johns Hopkins University Press, 1997.

———. "Village and Community in Early Colonial New England." *Journal of Historical Geography* 8 (1982): 333–46.

Wood, Martha May. "Early Roads in Missouri." Master's thesis, University of Missouri–Columbia, 1936.

———. "Traces in Early Missouri, 1700–1804." *Missouri Historical Review* 38 (1943): 18–24.

Wood, Peter H. *Black Majority: Negroes in Colonial South Carolina from 1670 through the Stono Rebellion.* New York: Alfred A. Knopf, 1974.

Wood, W. Raymond. "The Missouri River Basin on the 1795 Soulard Map: A Cartographic Landmark." *Great Plains Quarterly* 16 (1996): 183–98.

Works Progress Administration. *Missouri: A Guide to the "Show Me" State.* American Guide Series. Sponsored by the Missouri State Highway Department. New York: Duell, Sloan, and Pearce, 1941.

Wright, Henry C. "Report of the Committee on Medical Topography and Endemics and Epidemics." *Transactions of the Missouri State Medical Association* 1 (1851): 29–40.

Wright, James E. *The Galena Lead District: Federal Policy and Practice, 1824–1847.* Madison: State Historical Society of Wisconsin for the Department of History, University of Wisconsin, 1966.

Wyckoff, William. *The Developer's Frontier: The Making of the Western New York Landscape.* New Haven: Yale University Press, 1988.

Yealy, Francis J. *Sainte Genevieve: The Story of Missouri's Oldest Settlement.* Ste. Genevieve, Mo.: Bicentennial Historical Committee, 1935.

Zelinsky, Wilbur. *The Cultural Geography of the United States.* Englewood Cliffs, N.J.: Prentice-Hall, 1973.

Zimmer, Gertrude M. "Place Names of Five Southeast Counties of Missouri." Master's thesis, University of Missouri–Columbia, 1944.

Zitomersky, Joseph. *French Americans–Native Americans in Eighteenth-Century French Colonial Louisiana. The Population Geography of the Illinois Indians, 1670s–1760s. The Form and Function of French-Native Settlement Relations in Eighteenth-Century Louisiana.* Lund Studies in International History, no. 31. Lund, Sweden: Lund University Press, 1994.

———. "Ville, état, implantation, et société en Louisiane française: La Variante 'mississipienne' du modèle colonial français en Amérique du Nord." In *Colonies, territoires, sociétés: L'Enjeu français,* ed. Joseph Zitomersky and Alain Saussol, 23–48. Paris: L'Harmattan, 1996.

Index

Page numbers in italics refer to maps and other illustrations.

Abernathy, Joseph, 396–97n1164
Abernathy Settlement, 204, *369*, 396
Able, Ezekiel, 160
Acadia and Acadians, 3, 79, 88, 210n11, 384. *See also* Canada; French territories and settlers
Adams, David, 57n71
African slaves. *See* Slavery
Agriculture: by Americans, 29–30, 69–70, 105, 218–21, 220n24, 265, 340–45, 361, 362, 367, 429; at Barrens Settlement, 390; in Bellevue Valley, 41, 295, 361, 362–63, 365, 367; in Big River Bend, 295, 329–30, 336; clear-field agriculture, 390, 390n52; and climate, 28–30; cost of transportation for agricultural surpluses, 432–33, 463; in Current River region, 342n3; farmers' need for wage work, 336, 363, 363n9, 402, 462; in Farmington Plain, 337, 338; by French, 28–30, 207–9, 243n37, 245, 265, 429, 433; and Herculaneum, 326; by Indians, 68, 70, 373, 376; influence of, on rise of local centers, 463; in karst plains, 68; in Lower St. Francis River Valley, 347–49; in mining country, 41, 71, 284, 294–95, 295n23, 361, 362–63, 363n9; in Mississippi alluvial plain, 65–66; in Plattin Valley, 340–45; production statistics on, 433; in rough hills region, 69–70, 340–45; and rural property sizes, 460–61; in Saline County, 302n33; and soil, 31–33, 31n12, 33n15, 39, 340. *See also* Cattle; Food; Hogs; Horses; and specific crops
Ague, 58
Alabama territory, 90, 95n44
Allen, Thomas, 383, 383n39
Alley, William, 333
Alluvial plain, 41n37, 63–66, *64*, 65n85, 65n87
Alluvial rivers, 39–44
Alluvium, 31–32, 33, 47, 340, 455

Alta Luisiana (Upper Louisiana), 6–7
Alvord, Clarence, 20–21, 165n104
American Board of Land Commissioners. *See* Board of Land Commissioners
American Bottoms, 6, 63, 65, 389
American Fur Company, 431
American Indians. *See* Indians
American Mines, 332–34
American Revolution, 89, 341, 341n2, 370, 371, 386, 387n47
American settlements: compact settlement, *203*, 204, 220–21; compared with French villages, 209, 214, 216, 414, 452; discontinuous valley settlement, *203*, 204, 219–20, 345–50; farming in, 218, 219–21, 220n24; founding of, 216; linear valley settlement, *203*, 204, 216, 218–19, 218–19nn20–23, 340–45; map locations of, *215*; and Maxwell grant, 357–58; planned towns, *203*, 204, 209, 214, 216; population of, 205–6; property turnover rates of, 413; town plans of, *217*; valley settlements, 340–50. *See also* American settlers; and specific settlements
American settlers: and attempts to limit, 164–73, *168–70, 172*; in Bellevue Valley, 110, 343, 359–65; and bluff lands and karst plain, 66, 68, 221; definition of, 16n17; as entrepreneurs, 452–53; farming by, 29–30, 69–70, 105, 218–21, 220n24, 265, 340–45, 429; and fevers, 57; fishing by, 47; and foods, 36; and forests, 34, 35; and fur trapping, 29, 36; and impact of winds on health, 62; Indian attacks on and Indian fighting by, 275, 340, 347; land settlement regulations (1804–1818), 76, 114–73, 195–96, 458–59; land settlement regulations (1818–1830), 76, 174–97, 458–59; and lead manufacturing, 49, 50; in Lower Louisiana during Spanish period, 104–5; in Lower St. Francis River

521

522 Index

Valley, 345–50; mining by, 16, 105, 284, 287, 287–88n7; and Mississippi alluvial plain, 66; mobility of, 462; in Plattin Valley, 340–45; relations between French settlers and, 448–49, 448n1; and rough hills region, 69–70, 340–45; in Ste. Genevieve District, 16–17; settlement forms of, *203,* 204, 209, 214, 216, 218–21; in Spanish Louisiana, 16–17, 88, 88–89n29, 90–94, 92n36, 93n39, 96–106, 108, 110, 112, 287; trade skills of, 105; and trees, 35, 70. *See also* American settlements; Protestant immigrants; Ste. Genevieve District (American period); United States; and specific American settlers and administrators
Anderson, Hattie M., 18n19
Andreis, Father Felix de, 393
Animals. *See* Livestock; Wildlife
Antedating of Spanish concessions, 109–10, 126–30, 148, 291, 331–34
Appalachia, 16, 76, 97, 218, 340–41, 346, 360, 388, 398
Apple Creek, 7, 37, 67, 70, 95, 104, 193, 205, 213, 372–75, 374n13, 380, 381, 381n33, 385, 387, 387n47, 396, 397, 397n65, 422, 422n25
Appleton, 422n25
Aranda, count of, 113
Arcadia, 353
Archaeology, 21, 374, 374n13
Arensberg, Conrad M., 218n20
Arkansas, 146, 177n6, 186, 301, 348, 366, 426–27, 428, 442
Arkansas Post, 21–22, 161n92
Arkansas River, 175, 428
Arkansas village, 147n65
Armstrong Diggings, 334
Army, U.S., 187–88, 427
Arnold, Morris, 21–22, 161n92
Aron, Stephen, 19, 375n16
Arpents, 82, 103, 104, 313n49
Arundinaria gigantea, 66
Ashe, Thomas, 387n47
Ashley, William Henry, 22, 53, 184, 284, 296, 366n17
Astor, John Jacob, 434
Atherton, Lewis E., 429nn37–38, 438n48, 439n52
Austin, Emily, 288, 325
Austin, Horace, 444
Austin, James, 297
Austin, Maria (Mary) Brown, 288
Austin, Moses: arrival of, in Missouri, 269, 286, 288; and Bellevue Valley,

361–63; on Big River Bend, 329, 334, 336; confirmation process for land grant to, 139, 186; craftsmen accompanying, 101, 104, 288–89, 330, 362; death of, 286, 303, 326, 334; and earthquakes, 54; family of, 288, 297; financial problems of, 303, 325, 326, 366; food for family of, 363, 363n9; on fraudulent land claims, 126–27, 130; and French Creoles, 289, 290, 292, 320; on French land claims, 123; and Herculaneum, 287, 321–24, 321n64; homes of, 289, 289n12, *290, 292,* 300–301, *301, 323, 324,* 326; land grant to, 100–101, 104, 135, 139, 204, 223, 287–91, 288nn10–11, *290,* 303n34; and lead mining, 16, 22, 27, 47n50, 48n51, 49, 52, 54, 100–101, 101n9, 104, 105, 135, 138, 142, 189, 284, 286–91, 330, 430, 449, 457; and lead shipping, 320–21, 321n64; on limit of settlement of Missouri, 167, *168;* and map making, 456; and Mine à Breton, 286–91; on Mississippi River as settlement divide, 166; in Missouri legislature, 392n54; on natural resources, 27, 34; and Old Mines, 305, 306n40, 317, 318; and partnership with Delassus, Vallé, and Jones, 287, 287n9, 289, 321n64; and Potosi, 287, 295–97, 300–302, 426; on protection of frontier, 166; and Red River colony, 366n17; report by, on lead mines, 140, 286; roads and bridges built by, 287, 294, 302, 318, 425; and Rush's land claim, 153; sale of property by, 224n32, 303, 303n34, 325, 326, 326n72; and Schoolcraft, 326n72, 327n73; on settlement rights, 144; slaves of, 363; Smith T as adversary of, 142; in Ste. Genevieve village, 288n10; and Stoddard, 286; survey of land claim of, 288, 289, *290;* and Texas colony, 303, 326, 366, 366n17, 427; value of mining property of, 286; Virginia lead plantation of, 100, 287, 291, 362; and Washington County, 287, 300
Austin, Stephen, 288
Autumnal fever. *See* Malaria
Aux Vases River, 67, 68, 69, 100, 177, 204, 205, *229,* 261, 262, 264–65, 417, 422
Axes, 37, 37n27
Azor dit Breton, François, 284–85, 285n1, 359, 386

Backdating of Spanish concessions, 109–10, 126–30, 148, 291, 331–34
Baker, John, 333
Baker, Thomas, 110
Balesi, Charles J., 6n3, 20–21, 210n12, 222n27
Baltimore, 437, 439n51
Banner, Stuart, 21, 111n38, 457
Baptist church, 282, 336, 337n92, 349, 365
Barite, 48, 319n61
Barrens, 38–39, 39n33, 68
Barrens Settlement: businesses in, 394, 395; church in, 393, 397; European settlers in, 397; family size in, 391–92, 391–92n54; land sales in, 177; land subdivision by families in, 182, 391–92, 391–92n54, 461; map location of, *369;* map of landownership in, *118;* Maryland Catholic settlers in, 383, 388–95; naming of, 389n50; population of, 205, 390n51, 414; property turnover rate of, 413; prosperity of, 39; seminary at, 22, 393–94, 395; settlement type of, 204, 221; slavery in, 392, 394, 395
Barseloux, Jean-Baptiste, 346n12, 370
Barton, David, 186, 190
Basler, Lucille, 20, 444
Basse Louisiane, 4
Bastrop, baron de, 222n28
Bateaux, 417, 418
Bates, Frederick: and Bellevue Valley, 362, 364; and earthquakes, 54n67; on Herculaneum, 321n64; land purchase considered by, 364; and lead-land leasing program, 139, 141–43; as member of Board of Land Commissioners, 119, 125; on "mineral-mania," 294; and pending land claims after 1812, 148–49, 175; rejection of claim of Maria Tomar by, 277; on saltpeter, 52; on squatters, 161, 189; and U.S. rectangular survey, 175
Bates, Moses, 288, 297
Bat guano, 52–53
Bays, Rev. Joseph, 366n17
Bear, Thomas, 343
Beauvais, Vital, 308
Belgique, 397n66
Belle Fontaine, 422
Bellefontaine Mine, 315
Bellevue Valley: agriculture in, 41, 295, 361, 362–63, 365, 367, 437; American settlers in, 110, 343, 359–65; and Austin, 361–63; businesses in, 364, 365; Caledonia settlement in, 365–67; churches in, 365, 402; compact settlements in, 204, 221, 364–65; Detchmendy's concession in, 152–53, 361; discovery of, 360; economy of, 363, 365; historical studies on, 22; iron and ironworks in, 362, 366; land claims and landownership in, 110, *118,* 360–62, 364; land subdivision by families in, 182; map of landownership in, *118;* movement from, to other locations, 358, 366; naming and spelling of, 359, 359n1; population of, 205, 363, 364n10; and roads, 302, 365–66, 423, 426; salt supply for, 362; site and natural resources of, 359–60; slavery in, 361, 362, 363; surveying of, 152–53, 157, 361. *See also* Caledonia
Bellin maps, 225–26, 226n3
Benton, Thomas Hart, 186, 190, 191
Bent, Silas, 128, 175
Berquin-Duvallon, 432, 453
Berthiaume, François, 104, 387, 387n47
Berthold, Bartholomew, 244n37
Bettis, Elijah, Jr., 349
Bettis, Elijah, Sr., 348–49
Bettis family, 348–49
Bettis Ferry, 349, 427
Bettis Settlement, 348–49
Big Field (Le Grand Champ), 63, 227n6, 240, *242*
Big Lick. *See* Bellevue Valley
Big River, 37, 184, 187, 204, 205, 304, 320, 325, 329–31, 330n78, 359, 365, 412, 416, 422. *See also* Big River Bend
Big River Bend: agriculture in, 295, 329–30, 336; French Creoles in, 460; inhabitants of, 329–30, 336, 358; landownership for lead-mining purposes in, 330–34, *335;* lead mining in, 330–34, *335,* 464; map location of, *328;* maps of landownership in, *118, 335;* plain at, 327, 329; population of, 205, 336; and roads, 422, 426; settlement type of, 204, 221. *See also* Big River
Big River mines, 302
Big St. Francis River, 353
Bining, Arthur Cecil, 292n17
Birkbeck, Morris, 438–39n51
Black Legend (*Leyenda Negra*), 93
Black River, 36, 161, 220, 223, 345, 349, 352, 353, *355,* 356, 358, 358n14, 380, 426, 427, 456
Black River Valley, 414

Blacks. *See* Slavery
Bloom, Peter, 163
Blouin, Daniel, 85n20
Bluff lands and karst plain, *64,* 66–68, 221, 231–33, 241, 244, 244n38, 264, 387
Board of Land Commissioners: and Bellevue Valley, 364; and Boone's Spanish land grant, 98n2; and concepts of landownership, 119–20; creation of first board, 117; creation of second board, 151, 197; decisions of and testimony before second board, 151–52, 152n77, 222n29, 307, 318, 333–34, 350n21, 362, 364, 387n47, 388; and evidence of landownership, 124–26; and Fenwick Settlement, 388; final report (1812) of first board, 146–48, 175; and floating concessions, 135, 141, 149; and fraudulent claims, 126–30, 130–31n29, 144–45, 148, 151, 151–52nn76–77, 388; and French tripartite villages, 122–24, 147; function of, 117, 146; and Indian lands, 131–35, 373n9, 377–78; and large land claims, 121–24, 122n13, 146, 146n62, 151–52, 222n29; and lead-land leasing, 136, 138–43, 193; and map of landownership in 1804, *118;* and Maxwell grant, 356; membership of, 119, 472; and mineral lands, 135–43; number of claims handled by, 146–47; and Old Mines group concession of 1803, 318–19; and Pratte's land claim, 332–34; and reasons for population mobility, 412; and settlement rights, 120, 125–26, 143–45, 144n58, 146, 341, 348; on size limitations and actual occupancy of land claims, 120–21, 122; and Spanish land regulations and land grants, 98n2, 119–20, 125–26, 143, 273, 356; and St. Michel concession, 273, 280; and surveying of private lands, 153; and unconfirmed claims, 146–47, 151–52
Boats, 324, 417–18, 427–28
Boilvin, Nicholas, 307, 315
Bois Brule: and American Bottom, 63; American settlers in, 66, 99, 348, 360, 368, 370, 382–83; and bank caving, 44n41; concessions in, 370, 382; early settlers in, 370, 370n3; European settlers in, 397; French Creoles in, 460; *grand marais* in, 65; Indians in, 370, 371; land sales in, 177; map of landownership in, *118;* migration from, 414; population of, 205, 370; property turnover rate of, 413; settlement type of, 204; size of, 263; slaves in, 370; survey of, 153–54, *154;* trees on, 263
Bois Brule Creek, 160
Bolduc landholdings, 65n87
Bolon, Hypolite, 261
Boone, Daniel, 98, 98n2
Boonslick, 149, 167, 171, 183, 302, 302n33, 343, 424, 425n30, 426, 439, 442, 464
Borderlands, 18–19
Bornes de pierre, 102
Bottom or point, 40–43, *40, 41*n37
Bouis, Antoine Vincent, 244n37
Boyer, Charles, 333
Boyer, Joseph, 314
Boyer, Pierre, 284, 286
Boyer, Risin (or Risine or Rissene), 100
Boyle, Susan, xix, 20, 432n41, 452
Brackenridge, Henry Marie, 34n16, 58, 208n6, 218n21, 271, 417
Bradbury, John, 49, 52
Braudel, Fernand, 24, 209n8, 211n13, 401
Braun, John, 346, 346–47n12
Brazeau: founding of, 396; land sales in, 177; map location of, *369;* map of landownership in, *118;* migration from, 414; population of, 205; settlement type of, 204
Brazeau Creek, 385, 387n47, 388, 396, 397n65
Brazeau River, 386, 387
Brazeau Valley, 388, 417
Breton Creek, 286, 289, 294, 295
Brewer family, 389
Bridges, 287, 294, 321n66, 326, 327n73
Briggs, Winstanley, 20–21, 78, 210, 211
British territories and settlers: in Canada, 13; and French Canadians, 89; and fur trade, 81, 90; in interior of North America, 113, 162; and Perry County, 368, 370, 397; and proclamation line of 1763, 165n104; and roads, 419; and Spain, 75, 81, 90; and Treaty of Paris (1763), 13, 79
Brown, Burton L., 31n11
Brown, Joseph C., 177n6, 236, 242
Brown, Lionel, 296
Browne, Joseph, 378
Bryan, Emily Austin, 288, 325
Bryan, James, 284, 325, 334, 344, 364
Bryan's Mines, 334

Buford family, 357, 366
Buford Mountain, 359
Bull, Jacqueline P., 99n4
Bullitt report, 321n64
Burns, Barnabas, 382, 382n37
Burns, Michael, 382
Burns Settlement, 369
Burton. *See* Mine à Breton; Potosi
Butcher, Bartholomew, 163, 163n99
Butcher, Michael, 163, 163n99
Butcher, Sebastian, 163
Byrd family, 101

Cabanage, 213n16
Cabanes, 285, 308
Cabins: construction of, 309, 317, 331, 375; in Greenville, 350n20; of Indians, 374, 375; in Lower St. Francis River Valley, 346, 347; of miners, 212–13, 213n16, 285–86, 289, 308, 309, 316, 333; in Perry County, 368, 370; in Plattin Valley, 341; of slaves, 450; of Smith T, 218n21, 315
Cadastral patterns, 154, 156, 156n84, 159n88, 265
Cadillac, Antoine de Lamothe, 267, 419
Cahokia: and flooding, 44; founding of, 11, 77; harborage for, 46; historical studies on, 21; map location of, *8;* maximum size of, 209n9; and roads, 415, 419; settlement by seigneury in, 221, 222n27; town plan of, 238
Cahokia Creek, 46
Cahokia Indians, 11
Caillot dit Lachance family, 91, 248, 254 272, 285
Caillot dit Lachance *père*, Nicholas, 270n8, 272n11
Caledonia: businesses in, 439, 452; founding of, 216, 365, 402; historical studies on, 22; and land speculation, 366; limitation on growth of, 366; map location of, *215;* population of, 206, *410;* post office at, 442; and roads, 426; settlement type of, 204; town plan of, *217. See also* Bellevue Valley
California, 196n48
Callaway's Mill, 275n17
Call, Steven R., 20
Calvert, Lord, 255n62
Camino real (royal road), *420,* 421–22, 421nn23–24, 423, 427
Canada: Acadia and Acadians, 3, 79, 88, 210n11, 384; British control of, 13; cultural conservatism of French colonists in, 211n13; French aristocrats in, 453; French loss of, 13; large landholdings in, 460; Quebec land subdivision system, 10n6; seigneuries of, 209–10, 211, 221–22, 255n62; Spanish control of, 13; travel to, on Mississippi River, 418–19
Cane, 66
Cane Creek, 275, 275n17
Canoes, 417
Cape Girardeau: commandant's salary at, 109n33; earthquakes at, 54; and flooding, 44; Lorimier as commandant at, 83, 100, 347, 372, 375, 378; Lorimier's land at, 95, 372; map location of, *8, 215;* population of, *410;* and Protestant American settlers, 92n36, 100; and roads, 421, 422, 423, 423n28; rough hills at, 69; settlers to St. Francis Valley from, 347, 348; and steamboats, 428
Cape Girardeau County, 67, *178, 411,* 423n28
Cape Girardeau District, 21, 37, 100, 372, 396, 397n65
Capeheart (Kephart), John, 337n90
Carbonate bedrock, 30–31, 61
Carlin, Eliza, 343
Carlin, Thomas, 343
Carlos III, King, 82n14, 83, 240
Carolina, 16, 218, 346, 348–49. *See also* North Carolina; South Carolina
Caron, 231n11
Carondelet, Francisco Luis Hector, baron de: and agriculture, 432; and Catholic immigrants, 94, 95, 351, 384; and concessions for poorer people, 265–66; and contract for lead, 287; French background of, 83, 83n19, 94, 246; immigration policy of, 94–96, 98; and Indians, 372–73, 372n6, 376–78; land grants by, 100–101, 104, 122, 222n28, 246, 288–91, 288n11, 356, 376–78; and La Saline, 261–63; and New Bourbon, 253, 262
Carondelet settlement, *8,* 147n65, 212, *410*
Carpenter, Ephraim, 259
Carr, William C., 138, 140, 141, 146
Castor Creek, 275, 276, 278, 287n7
Castor Creek Valley, 277
Castor Hill Treaty, 381
Castor Mine, 268, 268n3, 275
Castor River, 268, 268n3
Castor Village, 279
Catholic church: and France, 82–83; in

La Saline, 261; Maryland (English) Catholics, 383–84, 385n43, 388–95; and Maxwell grant, 351–53, 356–58; in Perryville, 22; and Protestant immigrants to Spanish Louisiana, 91–93, 92n36, 97, 99–100, 103, 105–6; and slavery, 12n11; and Spain, 82–83; in Ste. Genevieve and New Bourbon, 230, 233, 237, 253, 271, 271n9; in St. Louis, 385; and St. Mary's of the Barrens seminary, 22, 393–94, 395; in St. Michel, 282. *See also* Catholic immigrants

Catholic immigrants: and Barrens Settlement, 388–95, 397; and Carondelet, 94, 95, 351; and Fenwick Settlement, 383–88; and Galvez, 89; and Gayoso, 103–4; and Maxwell grant, 351–53, 356, 358; and Miró, 91; in Perry County, 368; in Perryville, 395n61; Spanish policy on, 88, 89, 91, 94, 95, 103–6, 112, 195, 453; and Trudeau, 105

Cattle, 29, 33, 35, 38, 50, 53, 69, 87–88, 87n26, 207, 218, 233, 259, 309, 341, 347, 349, 375, 376, 390, 415, 416, 433

Cayton, Andrew, 457
Cedar Creek, 366
Cedar trees, 35, 69, 360
Cerré, Gabriel, 135n39
Cerré family, 308n43
Cession Treaty (1803), 114–17, 192–93, 193n43, 377
Champ (agricultural land), 207–8, 209, 212, 278–80
Charettes (carts), 415, 419
Chemin de roi (royal road), 421, 421n24
Cherokee Indians, 346, 368, 372n6, 374n11
Chert, 30, 71
Chert-free limestones, 31, 32, 329
Chevallier, Pierre, 272
Cheveux Blanc, Chief (White Hair, or Paw-Hiu-Skah), 133, 134
Chicago, Ill., 21
Chicago (Mo.) salt enterprise, 362
Chickasaw Indians, 11, 131n30
Childress, James D., 31n11
Chouteau, Auguste, 12, 79, 122n13, 402, 451
Chouteau, Jean Pierre, 122n13, 128n25, 134–35
Chouteau family, 297, 308n43, 375
Churches. *See* Baptist church; Catholic church; Lutheran church; Methodist church; Presbyterian church

Cinq Hommes, *118,* 395
Cinq Hommes Creek, 68, 69, 372, 374, 376–77, 385, 388, 390, 422
Cinq Hommes River, 95
Cissell family, 389, 391
Cisterns, 47
City planning. *See* Settlement forms
Civil War, 357n10
Clamorgan, Jacques, 94, 122n13
Clamorgan family, 308n43
Clark, Francis, Jr., 383
Clark, Francis, Sr., 383
Clark, George Rogers, 89
Clark, William, 111n39, 379, 381, 381n35, 403n3, *405*
Clark's Fork, 348
Clendenen, Harbert, 22, 27n1, 160n89, 219nn22–23, 345n10
Climate, 28–30, 28nn2–3, 61–62
Coleman dit Racola, François, 264, 308
Coleman dit Racola, François, II, 317n59
Coleman family, 315, 317–18
Collot, Georges-Victor, 28n3, 39, 102n12
Colluvial slopes, 234–35
Common fields, 208, 208n6, 209n9, 240–43, 241n35, *242,* 251–52, *276,* 278–80, 465
Commons, 208–9, 208–9nn7–8, 226–27, 227nn5–6, 233, *276,* 280
Commune (commons), 208, 208–9nn7–8, 226–27, 227nn5–6, 233
Community, 201–2, 201–2n2
Compact settlement, *203,* 204, 220–21
Concessions: backdating of, 109–10, 126–30, 148, 291, 331–34; definition of, 88; floating concessions, 135, 141, 149, 222n29, 293, 361, 385, 387; and Louisiana purchase to U.S., 109–11, 114–17, 136, 192–93; size of Spanish concessions, 122; in Spanish period, 88, 98–113, 104n15; written concession document as evidence of landownership, 125. *See also* Settlement regulations

Congress, U.S.: creation of first Board of Land Commissioners by, 117; creation of second Board of Land Commissioners, 151; and final report (1812) of Board of Land Commissioners, 146–48, 175; and Indian landownership, 131–32, 377; and Indian policy, 379; and land-confirmation process, 120–24, 145, 147–51, 196, 377; and large land claims, 123–24, 377; and Lead and Salt Leasing Act (1807), 138–43, 138n45, 197; and lead-leasing system, 138–43, 187, 190–92, 190–

91n38, 197; and Maxwell grant, 357; and Mine à Breton, 297–98; and mineral lands, 137, 163; Missouri senators and representative in, 186, 190; and New Madrid earthquakes, 149; and preemption rights, 163–64, 164n101; and rectangular land survey, 174–75, 175n1; and sale of lead lands, 192–93; and sale of public lands in Missouri, 142; and settlement-rights claims, 145; and trespassing on public domain, 137, 163, 164
Connecting tissues: administrative organization, 402–12; definition of, 401; economic activity, 429–40; and population mobility, 412–15, 461–62; and post offices, 440–43, 441; roads and river travel, 415–28
Conner, William, 342
Construction materials, 35–36
Cook, Daniel, 338n93
Cook, John, 338n93
Cook, Nathaniel, 281, 283, 338
Cook County, Ill., 338n93
Cook's Settlement: agriculture in, 338; development, 336, 338–39; inhabitants' move to Fredericktown from, 283; map location of, 327, 328; map of landownership in, 118; and roads, 277, 426; settlement type of, 204, 221; squatters on masons' tract in, 163, 163n99
Cooper, William, 247n44
Corn, 29, 29n6, 33, 69, 329, 340, 341, 346, 347, 349, 376, 387, 433, 437
Côte, 67n90
Cotton, 30, 347, 349, 433
County seats, 220, 281, 282, 283, 287, 295, 300, 325, 327, 339, 350n20, 394–95, 426, 460, 463
Courtois River, 187, 204
Coxe, Tench, 429n37
Cox, James, 344
Coyteux, Louis, 370
Craighead, Alexander, 141, 364, 365, 402
Craig, Seth, 314, 434
Cramer, R., 349n19
Crawford, Elizabeth Franserge, 254n60, 255
Crawford, Robert, 254n60, 255
Crawford, William H., 150
Creoles: definition of, 13n14. See also French Creoles
Crime, 62
Cronon, William, 21, 33n15, 454

Crow, Benjamin, 110, 144
Crow, Walter, 144
Cruzat, Francisco, 89, 91, 228
Cultural diversity, 18–20, 447–54. See also American settlers; French Creoles; French territories and settlers; Indians; Spanish territories and settlers; and specific groups, such as Scotch-Irish settlers
Cumberland River, 92
Currenton, 427
Current River, 22, 53, 219n23, 342n3, 348, 352, 380, 427
Current River Valley, 414

Dana, Edmund, 227n6
Darby, William, 427
Deblois, Joseph, 350, 350n21
Deblois family, 350
Dedham, Mass., 209n9, 212n14, 265n82, 461
Deer, 36, 68
De Gruy, Antoine Valentin, 257, 257n66, 304
Deguire, Paul, 275
Deguire dit Larose family, 248
Deguire family, 272
Delassus, Charles de Hault: and American settlers, 105, 360; backdating of concessions by, 109–10, 126–29, 148; as commandant at New Madrid, 253; French background of, 17, 83, 83n19; land claim of, 122n13, 192, 331; land grants by, 104, 107–10, 119n9, 126–27, 129, 134–35, 148, 333, 352, 430, 434; and lead mining, 287; as lieutenant governor generally, 17, 253; and Maxwell grant, 352; and militia, 347; move to New Orleans by, 128; and Old Mines concession, 307, 309–10, 313; and partnership with Austin, 287, 287n9; and request for concession around Mine la Motte, 269
Delassus, Philippe-François-Camille, 331–32
Delassus family, 291, 314n52, 422
Delaurière, Frémon, 246n41, 266
Delaware Indians: as early settlers in Bois Brule, 368, 370; farming by, 68, 70, 373, 376; and French colonials, 131n30, 450; mobility of, 462; at New Ste. Genevieve, 243; and Osage Indians, 450; population of, 373, 380, 381n33; removal and resettlement of, 193–94, 377n24, 379–82, 450–51; settlement form of, 213n17;

Shawnee-Delaware Indian grant, 95, 132–33, 133n33, 193–94, 204, 205, 213, 347, *369*, 371–82, 385, 450–51; and Treaty of Castor Hill, 381; villages of, 374–75, 374nn12–13, 381–82; white encroachments on land grant to, 377–79. *See also* Indians
DeLeyba, Fernando, 88
Demangeon, Albert, 211n13
Denman, David, xix, 20, 21, 210–11, 228
Derbigny, Pierre-Augustin, 253
Des Moines River, 176
Detchmendy, Pascal, 152–53, 305, 309, 361
Detroit, 117
Diamond, Sigmund, 255n62
Diphtheria, 55
Discontinuous valley settlement, *203*, 204, 219–20, 345–50
Diseases. *See* Health issues; and specific diseases
Dispersed French settlements, 264–66
Distilleries, 325, 365
Doctors, 100, 105, 384, 429
Dodge, Henry, 141
Dodge, Israel, 155n82, 157, *158*, 244, 254, 255, 384
Dodge, John, 91, 244
Dodge and Company, 141
Dodge's Creek, *229*
Dodson, John, 99
Doggett, Jacob, 331–32
Doggett, Susanna, 331–32
Dolomites, 30–31
Donaghoe, Jane, 330n77
Donahoe, Thomas, 391n53
Donaldson, Thomas, 117n6, 124n16
Donner, Jacob, 341
Doolin, John, 341, 343
Dorrance, Ward, 21, 115n3, 316n56, 412n8, 416n13
Dowd, Gregory Evans, 95n44
Drake, Daniel, 28n2, 46, 55n69, 57, 213n16
Drinking water, 46–47
Dry Fork, 348, 349
DuBourg, Bishop William Louis, 385, 393–94
Duclos, Alexandre, 314
Dueling, 293, 386
Dugan, 257
Dumville, Robert, 337n91
Duncan, Frances L. S., 99n4
Dunklin, Daniel, 296, 298n29
Duralde, Martin, 102n13

Durand, Father Marie Joseph, 392
Duval, John, 257

Eads, Abraham, 330, 333
Earle, Carville, 453
Earthquakes, 53–55, 54n67, 149, 348, 396
Easton, Rufus, 127–28, 130, 166, 334
Easton, Byrne, and Cook, 314n52
Economic activity: in American period, 433–40; in French period, 429–30; and post offices, 440–43, *441;* in Spanish period, 430–33, 431–32nn41–42; zones of economic development, 435–40, *436. See also* Agriculture; Fur trapping and fur trade; Mining; Salt and saltworks
Egers (or Eagen), George, 370n3
Ekberg, Carl: on Thomas Allen, 383n39; on American town layout, 214n18; on Big River region, 218n20; on Cahokia, 222n27; on economic life, 429n37; on French settlers, 459; historical focus of, 20–21; on illnesses, 55n69; on La Saline, 257n66; on militia at Ste. Genevieve, 162n95; on mining camps, 213n16; on slavery, 450; on Ste. Genevieve, 226n3, 227n5, 231n11, 243n37; and St. Michel, 267n1; on tripartite village, 202, 202n4, 208nn6–7, 209n9, 210, 211, 212n14
Emplacement (village lots), 107
England, 89, 90, 297. *See also* British territories
English settlers. *See* British territories and settlers
Entrepreneurial activity, 452–53. *See also* specific towns and villages
Environment. *See* Natural resources
Equisetum hyemale, 66
Establishment Creek, 419, 419n20, 422
Establishment River, 34, 34n16, 67, 68, 69, 204, 205, 223, 224n32, 264
Ethnicity. *See* American settlers; Cultural diversity; French Creoles; French territories and settlers; Indians; Spanish territories and settlers; and specific groups, such as Scotch-Irish settlers

Faragher, John Mack, 19, 21, 202n2, 239n28, 449, 451
Farm animals. *See* Livestock
Farming. *See* Agriculture
Farmington: American settlers near, 100; businesses in, 339; as county seat, 327, 339, 426, 460, 463; found-

ing of, 216; map location of, *215, 328;* population of, 206; post office at, 442; settlement type of, 204; surveying of, 157, 339n96; town plan of, *217,* 339
Farmington Plain, *327,* 336–39, 343
Farrell, James, 261
Featherstonhaugh, George W., 350n20
Fencing, 208n7, 226–27, 252
Fender, Christian, 370
Fenwick, Chloe, 385
Fenwick, Ezekiel, 385, 386
Fenwick, Joseph, 223, 244, 244–45n39, 247n44, 384–86, 389, 391
Fenwick, Julie Vallé, 386
Fenwick, Thomas, 385–86
Fenwick, Walter, 100, 151–52n76, 155n82, 244, 247, 247n44, 384, 386
Fenwick Settlement, *369,* 383–88, 387n47, 390n51, 396, 414
Ferries, 349, 419, 421
Festervand, Dorris F., 31n11
Fevers, 55–61, 55–57nn69–71, 60nn76–77, 253, 346, 348, 350n20, 360, 386, 389, 455
Finiels, Nicolas de, 34n16, 67, 250, 254n60, 259, 265, 374, 421–22, 432n42
Fires, 38
Fish and fishing, 47, 263
Fisher, James, 348
Flader, Susan, xix, 20, 210, 211, 225n1, 226
Flaget, Bishop Benoit Joseph, 393
Flanders, Robert, 22, 365n14, 367
Flatboats, 417–18
Flatrès, Pierre, 211n13
Flat River, 331
Flax, 433
Fleming, Jane Donaghoe, 330n77
Fleming, Patrick, Jr., 329, 329–30n77, 330, 334
Fleming, Patrick, Sr., 330n77
Flemings, 397, 397n66
Flemish Netherlands, 84n19
Flint, 30
Flint, Timothy A., 144n58, 358, 381–82, 416, 444, 448
Floating concessions, 135, 141, 149, 222n29, 293, 361, 385, 387
Floods, 14, 39–44, *40,* 44n41, 44–45n43, 60, 91, 230, 232–33, 264, 280, 386, 389, 454
Flora Creek, 372, 374
Flour mills, 301, 433n44
Fontainebleau Treaty (1762), 13, 79

Food: for Austin's family, 363, 363n9; Indians' sale of, 374, 374n11; for mining communities, 294, 302n33, 363, 363n9, 402; at seminary in Barrens Settlement, 394; in woods and grasslands, 36. *See also* Agriculture
Forests, 33–37, 34n17. *See also* Trees
Fort de Chartres: and bank caving, 43–44; climate of, 28; and Establishment River, 34n16; and fevers, 58; French Canadians in, 11; and lead mining, 48; map location of, *8;* marriage at, 304; in Mississippi alluvial plain, 65; Renaut's grant in, 78; and roads, 415, 419; transient population of, 162
Fort Osage, 132
Fort San Carlos, 239n29
Fourche à Courtois, 184, 375, 414
Fourche à Duclos, 264, 422
Fourche à Renault, 128, 303n35, 304
Four humors, 61–62
Fourneaux (fire pits to boil saltwater), 257, 257–58n67, 259
Fox Indians, 131, 167, 167n108
France: émigrés from, to Spanish Louisiana, 95, 101, 101n10, 246–48; and French Revolution, 50, 95; and Napoleonic Wars, 50; Normandy villages in, 315–16, 315–16n56; peasant culture of, 211, 211n13; relations between Spain and, 82–83; retrocession of Louisiana to, in 1800, 17, 102n12, 108–9. *See also* French territories and settlers; Ste. Genevieve District (French period)
Franklin, 180, *410,* 439
Fraudulent land claims, 126–30, 144–45, 148, 151, 151–52nn76–77, 388
Fredericktown: bicentennial celebration of, 278n21; businesses in, 439, 452; church in, 282; as county seat, 281, 282, 283, 463; founding of, 216, 281n27, 283; inhabitants of, 283; land sales in, 177; map location of, *215;* migration to, 414; naming of, 281; population of, 205–6, *410;* post office at, 442; and roads, 426, 427; settlement type of, 204, 221; and St. Michel, 281–82; town plan of, *217,* 282, 338, 465
French and Indian War, 12
French Canadians: Acadians, 3, 79, 88, 210n11, 384; intermarriage with Indians, 13, 371, 375; as settlers, 3, 11; and slavery, 12. *See also* French Creoles

French Creoles: and agriculture, 28–30; Americanization of, 451–52; and Austin, 289, 290, 292, 320; and Bellevue Valley, 360; cultural legacy of, as prerevolutionary French peasantry, 453; definition of Creole, 13n14; house construction by, 259, 278; and Indians, 277, 371; landownership by, 114–15, 115nn3–4, 123, 125, 459–60; and Louisiana purchase by U.S., 114–15, 115nn3–4; marriage with Indians, 13, 371, 375; marriage with mulattos, 277; mining by, 123, 272n11, 275n16, 281, 285–86, 288–93, 303–9; and Mississippi alluvial plain, 65; in New Bourbon, 248; in Old Mines, 293, 303–9, 313–14, 319–20, 449, 460, 465; in Perry County, 368; of St. Louis, 21; settlement system of, 195, 211; in Spanish territories, 13–14, 79, 88, 89, 91, 101, 448; in St. Michel, 269–73, 277, 283, 283n32. *See also* specific Creoles

French territories and settlers: and agriculture, 28–30, 207–9, 243n37, 245, 265, 429, 433; Americanization of, 451–52; and Arkansas Post, 21–22; and bluff lands and karst plain, 66–68, 264; cultural diversity among the French, 447–48; definition of French, 13n14; description and map of, 4–6, 5, 403, 404; and entrepreneurial activity, 452–53; and fevers, 57–58; and fishing, 47; and flooding, 43–44, 264, 280; and foods, 36; and fur trapping, 29, 36; and geographic linkages, 9; and horses, 415; and house construction, 259, 278; and Illinois country as term, 6; and Indians, 11, 77–78, 79, 85, 131n30, 210, 231, 448; and land division system, 10, 10n6, 65, 82; land settlement regulations before 1770, 76, 77–80, 82, 83, 114–15; and middle Mississippi Valley, 11; and mining, 12, 49–51, 70–71, 70n99, 78; mobility of French settlers, 461; and place names, 246n41; and prairies, 37, 37n26; relations between American and French settlers, 448–49, 448n1; and rough hills region, 69; and seigneuries, 209–10, 211, 221–22, 222n27, 255n62; and slavery, 12, 12n11, 77, 77n1, 82; and Treaty of Fontainebleau, 13, 79; and Treaty of Paris (1763), 12–13, 79; and Treaty of San Ildefonso, 17, 108, 109–10, 333; and trees, 34, 35, 37, 69; and tripartite villages, 122–24, 147, 147n65, 176–77, 195, 202, 203, 204, 207–12, 270, 277–80, 452, 465; and Upper Louisiana as term, 4, 6; and winds' impact on health, 62. *See also* Canada; French Creoles; French villages; Ste. Genevieve District (French period); and specific French settlers and administrators

French Village, 422

French villages: agriculture in, 207–9; as "closed" communities, 414; and colluvial slopes, 234–35; compared with American towns, 209, 214, 216, 414; components of, 207–8; dispersed settlements, 264–66; founding of, 216; La Saline, 255–64; long lots in, 207–8, 208n6, 226; map location of, 215; maximum size of, 209, 209n9; New Bourbon, 244–55; New Ste. Genevieve, 235–43; Old Ste. Genevieve, 225–35; planned villages, 202, 203, 204, 207–12; population of, 205–6; property turnover rates of, 413; St. Michel, 267–83; town plans of, 217; tripartite villages, 122–24, 147, 147n65, 176–77, 195, 202, 203, 204, 207–12, 270, 277–80, 452, 465; unplanned villages, 203, 204, 212–13. *See also* specific villages

Frontier: administrative settlement policy and traditional practices on, 75, 76, 457–59; Austin on protection of, 166; environmental encounter on, 454–57; of exclusion, 19, 451; Gitlin on, 437n47; of inclusion, 19, 451; multicultural, 18–20, 447–54; settlement pattern on, 459–61; settlement systems on, 461–65; stages of, 461; Turner on, 152n77, 161n94, 196, 202n2, 461, 462, 463

Fry, Henry, 358
Fryberg, 283n33
Furniture construction, 35, 36
Fur trapping and fur trade, 29, 36, 77, 81–82, 90, 108, 115, 161, 210, 253, 418, 429, 430–31, 434, 437

Gabouri Creek, 67, 229, 234, 235, 237, 239n28, 417
Gabouri River, 46, 160, 230, 234, 265, 330
Gabouri Valley, 460
Galena District, 303
Galena mines, 325–26

Index 531

Gallatin, Albert, 175
Gallipolis colony, 95, 247–48, 252
Gallman, Robert E., 392n55
Galvez, Bernardo, 86, 89–90, 92
Gambling, 271, 292
Gasconade River, 36, 53, 171n110, 380
Gates, Paul Wallace, 117n6, 152n77, 196, 196n48, 458
Gayoso de Lemos, Manuel: and backdating of Spanish concessions, 127; and Catholic immigrants, 103–4; and Luzières's plan for New Bourbon, 250, 250n49; settlement instructions of 1797, 86, 103–4, 106, 106n23, 107, 120, 145, 273
General Land Office, 171, 184, 187, 417n15
Georgia, 159n88, 160n89
German and German-American immigrants, 70n98, 100, 218, 341, 370, 370n3, 396, 397nn64–65, 464
Gerrard, Joseph, 155n82, 329, 330, 334
Gibson, Humphrey, 341, 342, 343
Giraud, Marcel, 37n26, 303n35
Gitlin, Jay, 18–19, 21, 437n47
Glacken, Clarence J., xvii
Glaise (shallow depression or marl), 33, 33n14
Glaize Creek, 341
Govereau, Etienne, 412
Gracy, David B. II, 101n9, 363n9
Graham, George, 184, 184n17
Grand Champ (Big Field), 63, 227n6, 240, *242*
Grand Marais, 68, 241
Grand Tower tract, 386
Granite, 31
Grasses and grasslands, 37–39, 65, 65n85, 68
Great Lakes, 418, 431
Greenville: as county seat, 220, 350n20, 463; Featherstonhaugh on, 350n20; founding of, 216; map location of, *215;* population of, 206, *410;* post office at, 442, 443n58; and roads, 427; settlement type of, 204; town plan of, *217*
Greenville Treaty, 349
Griffard, Alexis, 251
Gristmills, 271–72, 325, 349
Guibourd, Jacques, 305n38
Gunpowder, 52–53

Hagan family, 389
Hale's Crossing, 338–39
Hall, John W., 159n88
Halley's comet, 54–55
Hamlets versus villages, 211n13
Hammond, Samuel, 321
Harborages, 46, 323, 427, 428, 455
Harris, Cole, 211–12nn13–14, 218n20, 453, 454, 460
Harrison, William Henry, 126n21
Harrison family, 387n47
Harrisonville, Ill., 324
Harvey, Henry, 375n14, 377n24
Haute Louisiane, 4
Hawkins, John, 389n50
Hay, 39, 69
Hayden, Clement, 391
Hayden family, 389
Hazel Run, 364
Health issues: and doctors, 100, 105, 384; general healthfulness of environment, 55–62, 402; and lead production, 50–51, 61, 457; malaria and other fevers, 55–61, 55–57nn69–71, 60nn76–77, 346, 348, 350n20, 360, 386, 389, 455; and medicinal plants, 36; and saltpeter works, 53; and winds, 61–62
Hematite, 51
Hemp, 30, 105, 429, 429n38, 433
Henry, Andrew, 284, 296, 330n77, 364, 364n12, 365, 386, 392
Herculaneum: Austin's home in, 323, *324,* 326; Austin's plans for, 287, 321–24, 321n64, *322;* businesses in, 324–26, 439, 452; as county seat, 325, 327; decline and sale of, 326–27; and flooding, 44; founding of, 216; harborage for, 46, 323, 427; and lead shipping, 321n64, 323–26, 344, 344n9, 428, 464; map location of, *215;* naming of, 321n65; population of, 206, 324–25, *410;* post office at, 442; and roads, 302, 425; rough hills at, 69; settlement type of, 204; shot towers in, 323–24, *324;* site of, 321, 321n65; sketch of, *324;* as St. Joseph Lead Company town, 327; town plan of, *217,* 321–23, *322*
Hibbard, Benjamin, 171n111
Highways. *See* Roads
Hildebrand, John, 88, 88–89n29
Hilliard, Sam B., 159n88
Hillsboro, 327, 327n73
"Hinge" cities and towns, 462–64
Hinkston family, 387n47
Historic Ste. Genevieve District. *See* Ste. Genevieve District
Hix's Ferry, 427

Hofstra, Warren R., 221n26
Hogan, John, 356–57n10
Hogs, 29, 33, 53, 70, 218, 233, 341, 376, 429, 433
Holston River, 360
Holston Valley, 101
Homestead Act, 171n111
Horine, Frederick, 341
Horine brothers, 341, 344
Horse Island, 66
Horses, 50, 69, 207, 346, 347, 375, 376, 390, 415–16, 415n12, 419, 421, 433, 451
Houck, Louis, 88–89n29, 132n31, 268n5, 305n38, 330n78, 337n90, 387n47, 413, 415n12
House construction, 35–36, 259, 278, 309, 317, 375. *See also* Cabins
Hubardeau, Jean-Baptiste, 265
Hudson, John C., 201n1, 363n9
Hunt, Anne Lucas, 233, 418n17
Hunting and trapping, 36, 68, 70, 115, 161, 161n92, 162, 162n95, 218, 219, 345–48, 350, 376, 427. *See also* Fur trapping and fur trade; Wildlife
Hurricanes, 54n67
Hutchins, Thomas, 259

Icing of Mississippi River, 45, 45n44
Illinois Country: American immigrants in Spanish Illinois, 16–17; Austin on, 27; description of, as term, 6–7, 6n3; flooding in, 39–44; French land settlement policy in, 77–80; French period before 1770 in, 77–80, 82, 83, 403; historical studies on, 20–21, 202; Indians in, 78; mining in, 12; settlements in, 11; slavery in, 12, 77, 77n1, 93; Spanish period (1770–1795) in, 80–96; Spanish period (1796–1804) in, 97–113. *See also* Ste. Genevieve District; Upper Louisiana
Illinois immigrants, 382
Illinois mines, 187, 325–26
Illinois River, 418, 431, 431n40
Illinois tribes of Indians, 6, 11, 21, 85, 132, 368, 371. *See also* Indians; and specific tribes
Illnesses. *See* Health issues; and specific illnesses
Immigration policies of Spanish, 14, 16–17, 76, 81, 88, 89–106, 92n36, 93n39, 108, 112, 195, 458
Indemnity scrip, 155, 155n82
Indiana Territory, 7, 471–72
Indian Creek, 385
Indians: attacks on American settlers by and Americans' fighting against, 275, 340, 347; attacks on miners by, 231, 268–69, 268n5, 292, 305; in Bellevue Valley, 360, 361; in Bois Brule, 370, 371; boundaries of, 402–3, 402–3n2; as Christians, 85; defeat and expulsion of, from Northwest Territory, 349; farming by, 68, 70, 373, 376; and fevers, 56; fishing by, 47; and French colonials, 11, 77–78, 79, 85, 131n30, 210, 231, 448, 450; and French Creoles, 277, 371; historical studies on, 21; and horses, 346, 361, 415n12; in Illinois Country, 6; intermarriage with French and Americans, 13, 371, 375; Jefferson's policies on, 75, 166, 378; landownership by, 85, 131–35, 133n33, 134n35, 193–94, 451; Madison's policy on, 379; in Mississippi Valley generally, 18, 19; mobility of, 462; in Northwest Territory, 95, 349, 371, 378; "pass-through tract" for, 193, 193n44; population of, 17, 205–6, 373, 380, 381n33, 403, *409;* in Saline River Valley, 256, 256n64, 263; Shawnee-Delaware Indian grant, 95, 132–33, 193–94, 204, 205, 213, 347, 371–82, 385, 450–51; as slaves of French colonials, 77; and Spanish colonials, 85, 90, 94–95, 95n44, 132–33, 371–73, 372n6, 376–78, 450–51; in Ste. Genevieve District, 14, 17, 21, 55, 78; Stoddard on, 131n30; terms for, 131n30; trade with, 345, 346, 347, 370, 373–74, 374n11, 430, 431; U.S. government treaties with, 131–32, 134n35, 167, 167n108, 193–94, 194n46, 381; U.S. removal and resettlement of Delaware and Shawnee Indians, 193–94, 377n24, 379–82, 450–51; villages of, 213, 213n17, 374–75, 374nn12–13, 381–82, 385. *See also* specific tribes
Indigo, 30
Infant deaths, 59–60
Intermarriage, 13, 277, 371, 375
Iowa, 162–63n98, 177n6
Irish immigrants, 351–53, 356, 356–57n10, 368, 370, 383. *See also* Scotch-Irish settlers
Iron and ironworks, 47, 51–52, 105, 362, 366, 460, 464
Iron and lead plantations, 291–92, 292n17
Iron County, 409n6
Iron Mountain, 362, 460

Ironton, 353
Irregular polygon shapes in land surveys, 156
Isle du Bois, *118*, 204, 205
Italians, 243–44n37

Jackson: economy of, 439, 464; land office in, 180, 182, 182n13, 184, 186, 188–89, 396; map location of, *215;* population of, 206, *410;* post office at, 442
James (DeGimes), Mr., 260
James, James, 348
James, Thomas, 51–52
James, William, 99
James Fork, 380
Janet, Richard Joseph, 22
Janis family, 91
Jefferson, Thomas: and American settlers in Spanish Louisiana, 92–93; on climate of North American interior, 28n2; and fraudulent land claims in Upper Louisiana, 127–28; Indian policies of, 75, 166, 378; and limitation of settlement of Upper Louisiana, 166; on Louisiana, 423; and mineral lands, 138, 140; and town plan for Jeffersonville, Ind., 60n77; on unauthorized settlements on public domain, 163; and U.S. rights to navigate Mississippi River, 92
Jefferson Barracks, 60
Jefferson City, 442
Jefferson County, *178, 181,* 187, 325, 327, 327n73, 342, 343, 350, 409, *411,* 426
Jeffersonville, Ind., 60n77
Joachim Creek, 187, 341, 343
Joachim River, 46, 69, 204, 205, 262, 321, 323, 326, 416
Joachim River Valley, 320, 326, 342
Johnson, Hugh N., 214n18
Johnson Shut-Ins State Park, 353, 354
Jones, John A., 160
Jones, John Rice, 287, 287n9, 295, 296, 326n71
Jordan, Terry G., xv–xvi, 87n26, 220n24, 435n47, 452
Journal of a Tour into the Interior of Missouri (Schoolcraft), 3

Kansas River, 176, 381
Karst plain and bluff lands, *64,* 66–68, 100, 221, 231–33, 241, 244, 244n38, 264, 387
Kaskaskia: British in, 14; cattle market in, 349; church in, 271; earthquakes in, 54; as economic and administrative center, 11, 77, 79, 462; and fevers, 58; and flooding, 42, 91, 233, 454; founding of, 11, 77; and founding of Old Ste. Geneviève village, 209, 209n9, 225; General Clark's seizure of, in 1778, 89; historical studies on, 21; Indian trade with, 374n11; land sales in, 91; and La Saline, 256n64; and lead mining, 48, 268; long lots in, 226; loosening of ties between Ste. Geneviève and, 430; map location of, *8;* in Mississippi alluvial plain, 65; Moore as visitor to, 388; population of, 225; processing French land claims at, 117; relocation from, to Ste. Geneviève District, 91, 91n35, 248, 272; and roads, 419; town plan of, 238; travel between Ste. Geneviève village and, 39, 45, 225; and winter icing of Mississippi River, 45
Kaskaskia Indians, 11, 256
Kaskaskia Island, 63, 64
Kaskaskia River, 46, 262–63, 454
Kaups, Matti, 220n24, 452
Kedro, Milan J., 21
Keelboats, 417, 417–18nn17–18, 418
Kelly, Isaac, 347, 348
Kelly, Jacob, Jr., 347
Kelly, Jacob, Sr., 347
Kelly family, 255, 348–49
Kendal, Judather, 321
Kentucky: administrative settlement policy versus traditional practices, 457; agriculture in, 432–33, 432n43; barrens in, 39n33; from frontier to "civilized world," 19; hemp in, 429n38; immigrants to Upper Louisiana from, 16, 22, 34, 39n33, 68, 76, 92, 98–100, 99n4, 105, 116, 144–45, 183, 187, 218, 248, 282, 330, 340, 341, 346, 348, 358, 382, 383–89, 397; and Lancaster square plan, 282; land speculation and land claims in, 97–98, 116, 120n10; malaria and other fevers in, 58; Maryland Catholics in, 384, 388–89; ridge and valley landscapes of, 416; settlement of, 93
Kentucky River, 432n42
Kerlegand, Jean-René Guiho, sieur de, 101n10, 247, 250n48, 253–54
Keslin, Richard O., 256n64
Kiel, Henry, 160
King, Leslie J., 201n1
Kingshighway, 421n23
Koehler, Kevin J., 20, 247n43, 254nn58–59
Krytz (or Kreutz, Crites), William, 370n3

Labeaume family, 291
La Breche, Jean-Baptiste, 333
LaBriere, Julien, 232n13
Lachance, Benjamin, 272
Lachance, François, 275
Lachance, Nicholas, 283n32
Lachance family. *See* Caillot dit Lachance family
Laclède, Pierre de, 12, 79, 86, 239n27, 430
Lafoix, Vincent, 264
La Fontaine de la Prairie, 286
Lake Michigan, 186
Lake Wappapello, 350n20
Lami (Lamy), Michael, 244n37
Lancaster square plan, 282
Land-confirmation process: community well in, 124–26; and concepts of landownership, 119–20; and establishment of Board of Land Commissioners, 117; and Fenwick Settlement, 388; and floating concessions, 135, 141, 149; and fraudulent claims, 126–30, 144–45, 148, 151, 151–52nn76–77, 388; and French tripartite villages, 122–24, 147; and indemnity scrip, 155, 155n82; and Indian lands, 131–35; and large land claims, 121–24, 122n13, 146, 146n62, 150–52; and lead-land leasing, 136, 138–43, 193n43; and map of landownership in 1804, *108;* and membership of Board of Land Commissioners, 119; and Mine à Breton, 297–98; and mineral/lead lands, 135–43, 150, 183–93; and New Madrid certificates, 149–50, 150n70, 163; number of claims handled by first Board of Land Commissioners, 146–47; results of, 146–52, 196–97; and settlement rights, 120, 125–26, 143–45, 144n58, 146, 341, 348; and size limitations and actual occupancy of land claims, 120–21, 122; and Spanish land regulations and land grants, 98n2, 119–20, 125–26, 143; and surveying of private lands, 152–59; and unconfirmed claims, 146–52; U.S. government officials involved in, 471–72; in Upper Louisiana from 1804–1818, 117–45, 195–96. *See also* Landownership
Land division: American system of, 10; in Barrens Settlement, 391–92, 391–92n54; drawing lots for, 313n50; in eastern U.S., 459; within families, 182, 391–92, 391–92n54, 461; French-Spanish system of, 10, 10n6, 65, 459–60; in Northwest Territory, 221. *See also* Landownership
Landless. *See* Squatters
Land measurement units, 82
Landownership: and attempts to limit settlement during American period (1804–1818), 164–73, *168–70, 172;* backdating of Spanish concessions, 109–10, 126–30, 148, 291, 331–32; confirming of land claims by U.S., 117–45, 471–72; different concepts of, 119–20; evidence of, 124–26; by French Creoles, 114–15, 115nn3–4, 123, 125, 459–60; French system of, 76, 77–80, 82, 83, 114–15, 459–60; by Indians, 85, 131–35, 133n33, 193–94, 451; in Kentucky, 98n2; land sales in American period (1804–1818), 116–17; land sales in American period (1818–1830), 116, 177–83, 177n8, *178, 181;* and Louisiana purchase by U.S., 109–11, 114–17, 136; in Louisiana Territory, 17–18; maps of, *118, 181;* of mineral/lead lands, 135–43, 150, 183–93; and preemption rights, 163–64, 164n101, 177n7, 182, 197; results of land-confirmation process by U.S., 146–52; and settlement rights, 120, 125–26, 143–45, 144n58, 146, 341, 348; Spanish land-granting process, 88, 98–113, 98n2, 104n15, 114, 120; spatial pattern of, 459–61; and squatters in American period, 116, 159–64, 196–97, 346, 449; and surveying by Spanish, 102, 102n13, 109–10, 125–26; surveying of private lands by U.S. (1804–1818), 152–59; U.S. government officials involved in land-confirmation process, 471–72. *See also* Land-confirmation process; specific settlements
Land patents, 180n11, 188
Land routes. *See* Roads
Land sales: in American period (1804–1818), 116–17; in Kaskaskia, 91; of lead lands, 192–93; payment for, 180; price per acre, 179, 179n10, 192; of public lands (1818–1830), 116, 177–83, 177n8, *178,* 179n9, *181;* in Ste. Genevieve County, 412–13, 413n9; in St. Michel, 282; volume of sales, 182, 182n13
Land settlement regulations. *See* Settlement regulations
Land speculation, 97–98, 119, 120n10,

127–28, 138, 149–50, 163, 216, 224n32, 314–15, 343–44, 351, 366, 395, 434

Land surveys. *See* Surveys and surveyors

La Nouvelle Bourbon. *See* New Bourbon

Large tracts: and Congress, 123–24, 377; impact of, on settlement process, 460–61; and land-confirmation process, 121–24, 122n13, 146, 146n62, 150–52; as settlement form, *203*, 204, 221–24, 222–24nn27–32; in Spanish period, 222–23, 222nn28–29, 287–91, 362. *See also* specific landowners, such as Austin, Moses

Larimore, Ann E., 296n25

La Saline: and American immigrants, 261, 370n1; blacksmith at, 412; compared with lead-mining villages, 255–56; concessions at, 257–59, 257–58nn66–68, *260*, 456; founding of, 216; harborage for, 46; historical studies on, 21; history of site of, 256; houses in, 259; inhabitants of, 255–56, 260–61, 385; as lead-shipping point, 257, 268, 419; map location of, *8*; on maps by 1770, 80; and Peyroux, 254, 255, 257–58, 260–63, 456; population of, 80, 205, 260–61, 264; as proposed district post, 261–63; and relocation of Ste. Genevieve village, 234, 262, 263; and roads, 419; salt resources and saltworks at, 53, 256–59, 261, 264, 430; settlement type of, 204, 259; site of, 262–63; slavery in, 260

La Salle, Robert Cavelier, sieur de, 4

Lasource, Louis, 231n12

Laws of the Indies, 81, 83, 85, 86–87, 86n24, 157n86, 238–40, 239nn27–29

Lawrence County, 255

Layton, Bernard, 395

Layton family, 389

Lead, 47, 47–48nn50–51, 48–51, 70–71, 108, 135–43, 183–93, *185*, 326, 432, 457. *See also* Mining

Lead and iron plantations, 291–92, 292n17, 362

Lead and Salt Leasing Act (1807), 138–43, 138n45, 197

Lead lands, 135–43, 150, 163, 179, 183–93, *185*. *See also* Mining; Mining country

Lead shipping, 320–21, 321n64, 323–26, 344, 428, 463–64

Lead smelters/smelting, 36, 49–50, 457

Leasing of lead-bearing lands, 136, 138–43, 138n45, 183, 187–92, 190–91n38, 192n42, 197, 293, 458

Leclerc, François, 245

LeGrand, Jean-Marie, 264–65

Lemon, James T., 218n20

Lesterville, 353

LeSueur, Charles, *324*

Lewis, John, 144

Lewis, Meriwether, 111n39, 138n45, 167, 167n108, *169*, 292n19, 378, 403n3, *405*

Leyba, Fernando de, 370

Limestone, 30–31, 32, 47, 61, 67, 68, 71, 329, 455

Linear valley settlement, *203*, 204, 216, 218–19, 218–19nn20–23, 340–45

Line (or Lene), Joab, 337n91

Little Meramec River, 303–4

Little Prairie, 147n65

Little Saline Creek, 133

Little St. Francis River, 268n3, 275, 277, 353

Livestock, 29, 33, 33n15, 35, 38, 50, 66, 68, 69–70. *See also* Cattle; Hogs; Horses

Lockridge, Kenneth A., 209n9, 212n14, 265n82, 461

Loess, 31, 32, 67, 68, 387, 390, 455, 457

Logan, David, 348–49

Logan brothers, 348

Log cabins. *See* Cabins

Long, Maj. Stephen, 381n33

Long lots, 207–8, 208n6, 226, 241, 251, 265

Long-lot shapes in land surveys, 156, 156n84

Longtown, 396

Lorimier, Louis: and Berthiaume, 104; Cape Girardeau land of, 95, 372; as commandant at Cape Girardeau, 83, 100, 347, 372, 375, 378; and German American settlers, 100; as illiterate, 347, 372n6; at La Saline, 261; as Métis trader, 95; in Northwest Territory, 95; and Shawnee-Delaware Indian grant, 371–73, 376, 378; size of land claims of, 122n13; and Virginia Settlement, 348

Louisiana: American settlers in Lower Louisiana, 104–5; description and map of, as French territory, 4–6, *5;* fevers in Lower Louisiana, 58; flooding in Lower Louisiana, 43; French community of Lower Louisiana, 104–5; large land grants in Lower Louisiana, 222, 222nn27–28; lieutenant governors of Upper Louisiana, 17, 469; naming of, 4; retrocession of, from Spain to France, 17, 102n12, 108–9; Spanish government of Lower Louisiana, 84, 84n19, 89; terms pertaining to, 4, 6–7; U.S. purchase of, 17, 109–11, 113, 114–17, 136, 192–93. *See also* New Orleans; Upper Louisiana

Louisiana Purchase, 17, 109–11, 114–17, 136, 192–93

Louisiana Territory: date of, as term, 7; population of, 409; U.S. government officials involved in land-confirmation process in, 471–72; U.S. national policy on landownership in, 17–18

Louis XVI, King, 246

Louis XV, King, 83

Lower Louisiana. *See* Louisiana

Lower St. Francis River Valley, 345–50, 414, 442–43

Lucas, Anne. *See* Hunt, Anne Lucas

Lucas, John B. C.: on French Creoles, 115; on French settlements, 210; on Indian lands, 131–32; and lead-land leasing, 140, 142; on Old Mines, 316; and petition of 1803 for Old Mines concession, 308n43, 309; and settlement rights, 144; on tripartite villages, 123

Luna, Francisco, 288n10

Lutheran church, 396, 397, 397n65

Luzières, Camille, 253

Luzières, Pierre-Charles Delassus de: arrival of, in 1790s, 14, 245–26; on burning of woodlands, 457; children of, 253; and church for New Bourbon, 253; as commandant of New Bourbon District, 14, 95, 246–48, 250, 250n49, 253–54, 254n59, 453–54; death of, 255; elitism of, 246–47, 247n44, 289, 449; and Fenwick family, 387; financial problems of, 454; French background of, 246, 246n41; house of, 246, 246n42, 247, 253, 289, 422; illness of, 253; on inhabitants and subordinates, 453–54; land grant to, 192, 222–23, 244, 253, 339, 460; and lead mining, 137, 138; and naming of St. Michel for, 270; and recruitment of American settlers in Spanish Illinois, 98–100, 99n4, 127n22, 248, 287, 384

Luzières family, 122n13, 291, 297, 331, 454

Macagné, Joel, 329, 330

Macagné, Laurent, 329, 330

Macarty, Jean-Jacques, 225n1, 226, 227

Maclot, John, 323, 423

Maddin, Thomas: and Bois Brule settlers, 383; as deputy surveyor under Soulard, 102; and Detchmendy's concession, 152–53, 361; and Old Mines survey, 310, *311,* 313, 314; and St. Michel survey, 273, *274;* and Tayon survey, 128, *129*

Madison, James, 164, 379

Madison County, 48n50, *178, 181,* 277, 281, 283n32, 338, 345–50, 356, 409, *411*

Mahoney, Timothy R., 401n1, 412n7, 462

Mail service, 302

Maison Rouge, marquis de, 222n28

Malaria, 55–61, 55nn69–70, 346, 348, 350n20, 455

Manitto (black slave), 333

Manning, Joseph, 391

Maple sugar, 433

Map making, 455–46

Marais (pond), 30, 33, 65

Marais des Cygnes River, 134n35

Marais des Liards, 212n15, *410*

Maramec Iron Works, 51–52, 292n17, 363n9

Marbut, Curtis Fletcher, 31n11

Marquette, Father, 47

Marriage. *See* Intermarriage

Marshall, John, 192–93

Maryland, 255n62, 383–84, 453

Masonic lodge, 243

Massachusetts, 209n9, 212n14, 265n82, 461

Mathers, William, 138n45

Matis (or Matisse), Jérome, 251, 272, 275

Maxent, Gilbert Antoine, 430

Maxwell, Father James: arrival of, in Ste. Genevieve, 351; and Barrens church, 393; death of, 255, 356–57; land grant to, 122, 122n13, 223, 351–58, 352n4, *354, 355,* 366, 456; and map making, 456; and Maryland (or English) Catholics, 385n43, 393; as parish priest in Ste. Genevieve, 271, 271n9, 351; will of, 357n11

Maxwell, Hugh H., 357, 357n11

Maxwell, John P., 357, 357n11

Maxwell grant, 122, 122n13, 223, 351–58, 352n4, *354, 355,* 366

McBee, Margaret, 160

McCandless, Perry, 18n19

McClain family, 342

McClain House, 422n25

McCormack, Peter, 341, 342, 344
McDaniel, Chad, xix, 55n69, 59–60
McDermott, John Francis, 69n93
McLaughlin, James, 144
McLaughlin, Thomas, 144
McMullin family, 342
McNair, Alexander, 164
McQuillan, D. Aidan, 10n6
McWhiney, Grady, 160n89
Meade, Melinda, 55n69
Meinig, Donald W., 112n40, 119n7, 216n19, 255n62, 401n1, 435n47, 452, 459
Melom, Halvor Gordon, 439n52
Menard, Pierre, 193, 373, 387, 387n47, 431
Meramec River, 7, 36, 47, 48n50, 53, 68, 71, 88, 88n29, 94, 105, 138n45, 171n110, 184, 194, 213, 219n23, 220, 303n35, 325, 340, 370, 374n11, 375, 380, 421, 456
Meramec Valley, 345
Mercedes (graces), 111, 111n39, 128, 222n29, 347
Methodist church, 337n92, 338, 341, 342, 344, 349, 360, 365, 366, 368, 396, 402
Métis, 12n11, 85, 95, 133–34, 134n35, 371
Mexican cession, 124n16
Mexican territory, 442
Mikesell, Marvin, 19
Miliet, Jean-Baptiste, 305, 305n38
Militias, 161, 162n95, 347
Mill Creek, 230, 275, 275n17
Miller, Andrew, 139
Mills: flour mills, 301, 433n44; gristmills, 271–72, 325, 349; in Herculaneum, 324–25; in Mine à Breton, 285, 288; Owsley's Mill (or Callaway's Mill), 275n17, 277; sawmills, 35–36, 271, 301, 325; water mills, 244, 245, 247, 252, 253, 254, 263, 359, 433n44
Mine à Breton: agriculture in, 294–95, 295n23; attorney in, 385; Austin's home in, 289, 289n12, 290, 292, 300–301, 301, 326; Austin's land grant at, 287–91, 288nn10–11, 290, 303n34, 430; blacksmith at, 412; businesses in, 285–86, 291, 294, 301, 302, 402; class differences in, 449; *commissaire de police* for, 291–92; community well in, 124; confusion over land in, 293–94, 297–98; cost of transporting lead from, 320; discovery of lead ore at, 284–85, 285n1, 294, 304; disorder and lawlessness in, 292–93, 292n19; and earthquakes, 54; food suppliers for, 294–95; founding of, 216, 269, 284; French Creoles in, 272n11, 285–86, 288–93, 305; furnace and mill at, 288–89, 290, 336; houses in, 294; land claims at, 123, 124, 149, 176; and land-confirmation process, 297–98; lead mining at, 53, 54, 187, 284–85, 334, 433; map location of, 8, 215; map of landownership in, 118; migration to, 414; and miners from Old Mines, 305–6, 306n40; mining camps at, 213, 285–86, 289; naming of, 285; and Osage attacks on miners, 292, 305; Perry's property in, 142; population of, 205–6, 286, 291, 301–2; prescriptive rights to mining in, 137, 293; and roads, 302, 320, 419, 420, 422–23, 423n26, 425, 426; settlement type of, 204; sketch of, 301, 301; survey of, 298–300, 298n29, 299; survey of Austin's land grant at, 288, 289, 290; timber depletion at, 36–37; wages for miners in, 294; water pollution at, 305
Mine à Breton Creek, 286, 289, 294, 295
Mine à Gabouri (or Gerbore), 100, 192, 330, 337n90
Mine à Joe, 334
Mine à la Platte, 333
Mine à Maneto, 332–34
Mine à Renault, 71, 78, 302
Mine au Castor, 268, 268n3, 275
Mine la Motte: and Americans, 281; and French Creoles, 272n11, 275n16, 281; land claims in, 151–52n76; lead mining at, 71, 184, 187, 267, 334; map location of, 8; map of landownership in, 118; on maps by 1770, 80; and Maxwell, 353; meaning of term, 268n3; mining camps at, 213; mountains near, 31; population of, 205, 269; prairie location of, 37; Renaut's grant in, 78; request for concession to land around, 269, 269n6; and roads, 277, 419, 420, 426; settlement type of, 204; shipment of lead from, 257, 268; square shape of, during Spanish regime, 37; timber depletion at, 36
Mine la Motte Road, 281
Mineral Fork, 425n30
Minerals, 28, 47–53, 47–48nn50–51, 70–71, 135–43. *See also* Mining
Mining: by Americans generally, 16, 105, 284, 286–87n7, 287; and Austin, 16, 22, 27, 48n51, 49, 52, 54, 100–

101, 101n9, 104, 105, 135, 138, 142, 189, 284, 286–91, 330, 430, 457; Austin's report on, 140, 286; Carr's report on, 140, 141, 146; decline of, at end of 1820s, 182; description of lead mining, 48–51; eastern investments in, 314, 434; employment statistics in, 186–87; factors influencing lead production, 187; by French, 12, 49–51, 70–71, 70n99, 78, 135; by French Creoles, 115, 123, 272n11, 275n16, 281, 285–86, 288–93, 303–9; French definition of *mine,* 70n99; "French" log smelters or furnaces for lead smelting, 287, 315; Galena mines, 325–26; health hazard of lead production, 50–51, 61; historical studies of, 22, 23; and Indians' attacks on miners, 231, 268–69, 268n5, 292, 305; lead production statistics, 186, 188, 334; and lead shipping, 320–21, 321n64, 323–26, 344, 344n9, 428, 463–64; and leasing of lead-bearing lands, 136, 138–43, 138n45, 183, 187–92, 190–91n38, 192n42, 197, 293, 458; number of producing mines, 187; and ownership of mineral/lead lands, 135–43, 150, 183–93; pollution of creek water by, 305, 306, 449, 457; prescriptive rights systems for ownership of mines, 136–37, 195, 285, 293, 305; and retrocession of Louisiana to France, 108; reverberatory furnace for lead smelting, 287, 291, 457; and Scotch hearth furnace, 189–90; silver mining, 48n50; and slavery, 268, 285, 294, 303, 307, 314n52, 333; and Smith T, 141–42, 284, 315, 334, 336, 457; technological improvements for, 457, 464; and timber depletion, 36–37; value of lead in 1799, 432, 432n42; wages for miners, 294. *See also* Mining country; and headings beginning with Mine

Mining camps. *See* Mining country

Mining country: agriculture in, 41, 71, 284, 294–95, 295n23, 361, 362–63, 363n9; Big River Bend, 327–336; description of, as natural region, *64,* 70–71; Farmington, 336–39; healthfulness of, 60–61; Herculaneum, 320–27; map of, *64;* Mine à Breton, 284–95; and mining camps generally, 212–13, 213n16, 285–86, 289; Old Mines, 303–20; and population mobility, 414; Potosi, 295–303; roads in, 21, 188, 277, 286, 302, 419, *420, 422–26, 424,* 464. *See also* Mining; headings beginning with Mine; and specific settlements

Miró, Esteban, 91–93, 92n36, 94, 222

Mississippi alluvial plain, 63–66, *64,* 65n85, 65n87

Mississippi River: agriculture along, 437; alluvial loams on bottoms of, 31; and bank caving, 42–44, 44n41, 230, 231, 233, 454; changing nature of, 39–44, *40,* 263n77; cost of lead shipping down, 320, 325–26; drinking water from, 46; and ease of crossing, 39, 456; and European colonials, 9–10; ferries across, 419, 421; fishing in, 47; flooding of, 14, 39–44, *40,* 44–45n43, 60, 230, 232–33, 264, 280, 386, 389, 454; and fur trade, 431; as international boundary, 6, 13, 90, 113, 456; and Jefferson's Indian policy, 166; La Salle's journey down, 4; as limit of American settlement, 166; maps of, 455; and rectangular land survey (1815), 175; Spanish control over, 90; steamboats on, 324, 418, 427–28; time needed for travel on, 418–19; transportation on, 417–19; and tributary mouths, 45–46, 455; U.S. access to, 90, 92, 93, 96, 113; width and depth of, 39, 39n34; winter icing of, 45, 45n44

Mississippi territory, 90, 95n44

Mississippi Valley: American colony in, 93; and European colonials, 9–10, 90, 113; French settlements in, 18, 77; historical studies on, 20–23; as multicultural frontier, 18–20; significance of, to U.S., 9–11. *See also* Illinois Country; Ste. Genevieve District; Upper Louisiana

Missouri: constitutional convention in, 338; date of Missouri Territory as term, 7; first permanent European settlement in, 12; as "mother of the West", xv; population of, *410;* sale of public lands in, 142; senatorial elections in, 338; state capital of, 442; statehood for, 7, 19, 190; as term during Spanish administration, 7; Territorial Assembly of, 379, 395, 409n5; U.S. government officials involved in land-confirmation process in, 471–72. *See also* Ste. Genevieve District (American period)

Missouri Indians, 11, 132*n*31
Missouri River, 6, 7, 10, 79, 131–32, 167, 176, 212*n*15, 239*n*29, 302, 348, 381, 428, 430–31, 431*n*40, 442, 455
Missouri River Valley, 21
Mitchell, Robert D., 401*n*1, 460
Mitchigami Indians, 11
Mobility of population, 412–15, 461–62
Mongrain, Noel, 133–34, 134*n*35
Monroe, James, 176*n*5, 177
Mont Généreux, 235, 245, 245*n*40, 247, 248
Monticello, 327, 327*n*73
Montreal, 238
Moogk, Peter, 210*n*11, 211*n*13, 453*n*9
Moore, Isidore, 386, 386*n*45, 388–93, 392*n*54, 395
Moore, James, 255
Moore, Lewis, 392
Morales, Juan Bonaventure: and Austin's land grant, 288, 288*n*11; and backdating of Spanish concessions, 127; control of concessions by, 110; and halt to land granting, 108, 108*n*30; settlement regulations of 1799 by, 86, 106–8, 106*n*23, 120
Morgan, George, 58, 93, 222, 323, 387*n*47, 418
Morgan, Solomon, 154, *154*
Morrison, Joshua, 361
Morrow, Lynn, 21
Morse, Rev. Jedidiah, 380
Mosquitoes, 55–56
Mulattos, 12*n*11, 277
Multicultural frontier, 18–20, 447–54
Murphy, David, 337*n*91, 339
Murphy, Joseph, 337*n*91
Murphy, Sarah, 337, 337*nn*91–92
Murphy, William, Jr., 337
Murphy, William, Sr., 336–37, 337*n*92
Murphy's Settlement: agriculture in, 337, 460; development of, 336–37; land subdivision by families in, 182; map location of, 327, *328*, 337*n*90; map of landownership in, *118*; population of, 205; and roads, 277, 423; Schoolcraft on, 337; settlement type of, 204, 221; survey of, 154*n*81

Natchitoches, 277, 426–27
Native Americans. *See* Indians
Natural resources: Austin on, 27, 34; of bluff lands and karst plain, *64,* 66–68; climate, 28–30, 28*nn*2–3, 61–62; and environmental encounter, 454–57; healthfulness of environment, 55–62; minerals, 28, 47–53, 47–48*nn*50–51, 70–71, 135–43, 457; of mining country, *64,* 70–71; of Mississippi alluvial plain, 63–66, *64,* 65*n*85, 65*n*87; plants, 33–39; rivers and water, 39–47; rock, 30–31, 61; of rough hills region, *64,* 68–70, 70*n*98; soil, 31–33, 31–33*nn*11–15, 39, 51, 68, 71, 179, 329, 340, 359–60, 455, 457; trees, 33–37, 34*n*17, 68, 69, 70, 191, 231, 416, 457; wildlife, 36, 53. *See also* Mining
Neal, Jacob, 104
Neal, Samuel, 104
Neighborhood, 201–2
Nerinckx, Father Charles, 390*n*51
New Andalusia, 94
New Bourbon District: boundary of, *407;* establishment of, 14, 246; fevers in, 60; fires in, 38; founding of, 216; land-confirmation process in, 146; levees proposed for, 43; Luzières as commandant of, 14, 95, 246–48, 250, 250*n*49, 253–54, 254*n*58, 453–54; map of, *8, 407;* Peyroux's plan for, 262; population of, 205–6, 403, *410;* reincorporation of, into Ste. Genevieve District, 409; trees in, 36
New Bourbon village: blacksmith at, 412; bluffs at, 67, 243; church for, 253, 255, 271, 271*n*9; and class separations, 246–47, 247*n*44, 289, 449; commandant at, 95, 109, 109*n*33, 192; common field of, 244–45*n*39, 251–52, 465; commons in, 208, 245; community well in, 47; famine at, 44; farming in, 265; founding of, 244–45, 247, 247*n*43, 262; historical studies on, 20; houses in, 247, 247*n*44, 249–51, 251*n*51, 289; inhabitants of, 95, 101*n*10, 247–48, 255, 272, 384, 386; land claims in, 24, 147*n*65; map location of, *8, 215;* maps of landownership in, *118, 249;* migration from, 414; mill for, 47; and Mississippi alluvial plain, 63, 65; *place* in, 250, 251; population of, 205–6, 250; property descriptions for, 250–51; reasons for failure of, 252–55, 254*n*59, 449; relationship between St. Michel and, 254–55, 254*n*60; and roads, 250–51, 422; settlement type of, 204; site of, 244–45, 254, 255; slavery in, 250; town plan of, 86, *217,* 247*n*43, 248–50, *249*

New England, 33n15, 202n2, 209n9, 212n14, 265n82, 297, 313n50, 454, 461
New Hartford, 325n70
New Madrid: commandant at, 253; earthquakes at, 53–54, 149, 396; historical studies on, 21; land claims in, 147n65; Morgan's aborted planned town at, 323; population of, 166, 206, 410; Protestant American settlers at, 92n36, 93, 222; and roads, 421, 422, 423, 427; squatters in, 162n95
New Madrid earthquake certificates, 149–50, 150n70, 163, 395, 434
New Orleans: as agricultural market, 432–33; climate of, 28n3; overland route to, 427; Spanish administration of, 83; town plan of, 238, 239; travel to, 439n51; U.S. right of deposit at, 96
New Ste. Genevieve village. *See* Ste. Genevieve village (new)
New York, 439n51
Nicknames of French, 285, 285n1, 308
Nicolle, Gabriel, 272, 277, 277n19
Nigh, Tim, 27n1
Niter, 47, 52
Normandy villages, 315–16, 315–16n56
Norris, F. T., 230n9
Norris, J. D., 292n17
North American Cattle Ranching Frontiers (Jordan), xv
North Carolina, 348–49, 360, 392n55, 396, 397, 397n65. *See also* Carolina
North Dakota, 363n9
North Gabouri Valley, 422
North River, 132n31
Northwest Territory: as eastern part of Upper Louisiana, 6; Indian defeat and expulsion from, 95, 349, 371, 378; land policy in, 91, 116, 117, 119, 119n7, 120, 221; leasing program for mineral lands in, 138; rectangular land subdivision in, 221; slavery prohibited in, 91, 389

Oak trees, 69, 70, 360
Ohio, 159n87, 247
Ohio River, 6, 10, 53, 92, 93, 95, 171, 262, 264, 432n42, 437
Ohio River Valley, 432n43
Ohio Territory, 95
Ohio Valley, 19, 95
Old Mines: and absentee lead operators, 306–8; Austin's petition for mining concession at, 305; church in, 318, 318n60; founding of, 216, 275n15; French Creoles at, 293, 303–9, 313–14, 319–20, 449, 460, 465; gardens in, 316, 319; group concessions of 1803 for, 307–10, 318–19; and gunpowder production, 53; historical studies on, 21; houses in, 314, 316, 317; inhabitants of, 304–6, 308–9, 314; land claims in, 124; lead mining at, 187, 284, 303–6, 315, 334; log furnaces in, 315; map location of, 8, 215; map of landownership in, 118; migration to, 414; Millet's request for concession at, 305; name of, 304, 304n37; natural terrain at, 71; and Osage attacks on miners, 305; population of, 205–6, 305–6, 306n40, 308, 317, 410; prescriptive rights to mining at, 305; price for land in, 314; and Renaut's lead-mining grant, 303–4, 303n35, 305; residents' 1796–1797 petition for concession at, 305, 307; and roads, 302, 425n30; sale of land in, 314–15, 314n52; settlement type of, 204; Smith T in, 315; survey of, 310–14, 311, 312, 317–18, 317n59; timber depletion at, 37; unincorporated, amorphous villages in, 320n62; village at, in 1740s, 304; village plan for, 315–18, 315–16n56; water pollution at, 306, 449
Old Mines Creek, 128, 304, 306, 310, 317
Old Ste. Genevieve village. *See* Ste. Genevieve village (old)
O'Mara, Joseph, 401n1
Onuf, Peter S., 221n25
O'Reilly, Alejandro, 84n19, 85–88, 87n25, 120
Organization of space: administrative, 402–12; definition of connecting tissues, 401; economic activity, 429–40; by landowners, 402; and population mobility, 412–15, 461–62; and post offices, 440–43, 441; roads and river travel, 415–28
Osage Fork, 375
Osage Indians: attacks on miners by, 231, 268–69, 268n5, 292, 305, 451; in Bellevue Valley, 360, 361; and Chouteau land grant, 134; and Delaware Indians, 380, 450; and European colonials, 11, 85, 131n30, 371, 377, 449, 451; historical studies on, 21; horse theft by, 346, 361, 451; landownership by, 451; mobility of,

462; and Mongrain land grant, 133–34, 134n35; at New Ste. Genevieve, 55; and Shawnee Indians, 377, 450; survey of Osage Cession boundary, 177n6; trails of, 419, 427; U.S. policy on, 131n30, 194; U.S. treaty of 1808 with, 132, 132n32, 133; U.S. treaty of 1825 with, 134n35; villages of, 85; violence between American and French settlers and, 275, 340, 377, 451. *See also* Indians

Osage River, 133, 442

Osage Treaty: of 1808, 133; of 1825, 134n35

Overby, Osmund, xix, 20

Owen, Robert, 89n29a

Owsley, Jonathan, 271

Owsley's Mill, 275n17, 277

Owsley's Settlement: American settlers in, 271–72; Cook's purchase of land in, 338; founding of, 271, 348; map of landownership in, *118;* and Owsley's Mill, 275n17, 277; settlement type of, 204

Ozarks: historical studies of, 21–22. *See also* Ste. Genevieve District

Padgett, Henry, 358

Palmyra land office, 180

Panic of 1819, 180, 303, 326, 364

Parc (parque), 208, 208n7, 240–43, 240n33, 241n35, 251, 255

Parish, Joseph, 348

Parish family, 349

Paris Treaty: of 1763, 13, 79; of 1784, 90

Partenais dit Mason, Amable, 142, 190n37, 297, 305, 306, 314, 314n52

Pattison, William D., 159n87

Paul Wilhelm, Duke, 425

Pays des Illinois (Illinois Country), as term, 6

Pearceall, Samuel, 331–32

Peck, James H., 110n35, 151, 239n29

Peck, Rev. John Mason, 161, 282, 346, 358

Pénicaut, André, 256n64

Pennsylvania, 341, 382

Penn, William, 218n20

Penrose, Clement Biddle, 146, 147

Peoria, Ill., 119n7

Peoria Indians, 11, 79, 213, 371, 380

Perez, Manuel, 329, 372

Perry, Samuel, 142, 298n29

Perry, William, 296

Perry County: Abernathy Settlement in, *369,* 396; agriculture in, 263n77; Barrens Settlement in, 39, *118,* 177, 182, 204, 205, 221, 388–95; Bois Brule in, 63, 66, 99, *118,* 153–54, 177, 204, 205, 360, 382–83; Brazeau Settlement in, *118,* 177, 204, 205, *369, 396;* county seat of, 394–95; courthouse for, 395; Fenwick Settlement in, *369,* 383–88; and German immigrants after 1830, 464; Indians in generally, 368, 370; karst plain in, 67; Kentuckians as settlers in, 39, 39n33; map of landownership in 1830, *181;* map of public domain land sales in, *178;* maps of, *369, 411;* Perryville in, 395; prairie and grasslands in, 39; settlement complexity in, 368–70, 397–98; Shawnee-Delaware Indian grant in, 95, 132–33, 193–94, 204, 205, 213, 347, 368, 371–82

Perryville: Catholics in, 395n61; as county seat, 394–95, 463; founding of, 216, 395; historical studies on, 22; karst plain near, 67; map location of, *215;* population of, 206; post office at, 442; seminary at, 22, 393–94, 395, 439, 442; settlement type of, 204; town plan of, *217,* 395

Peterson, Charles E., 20, 89n29, 239

Petit Canada, 422

Petite Rivière, 46, *229,* 230

Petit Rocher (Little Rock), 323

Peyroux de la Coudrenière, Henri: and Big River Bend concession, 329; as commandant at New Madrid, 346; as commandant at Ste. Genevieve, 94, 262, 346; on farm crops, 29; on hunter-trapper-trader, 346; La Saline concession of, 254, 255, 257–58, 257–58nn67–68, 260–63, *260,* 263n77, 368, 449, 456; and New Bourbon District plan, 262; and New Ste. Genevieve, 238; proposal for new settlement by, 94; and Vallé, 262, 263

Philadelphia, 214n18, 437, 438, 439n51

Philipson brothers, 438

Physicians. *See* Doctors

Piankashaw Indians, 380

Piernas, Pedro, 87, 102n13, 161n94, 228

Pigs. *See* Hogs

Pinckney's Treaty, 95–96

Pineries, 35–36, 290

Pine trees, 34, 35–36, 35n20, 69, 360

Pinus echinata, 35–36

Pirogues, 417

Pittman, Philip, 39, 46, 228

Placet, Jean-Baptiste, 285
Planned American towns, 203, 204, 209, 214, 216, 465. *See also* specific towns
Planned French villages, 202, 203, 204, 207–12. *See also* specific villages
Plante, Nicolas, 264
Plants, 33–39, 50, 66
Platin, 69, 69n93
Plat of survey, 125–26
Plattin Creek, 187, 340–41, 343
Plattin mines, 344
Plattin River, 69, 204, 205
Plattin Valley, 326, 340–45
Plattin way station, 344
Pocahantas settlement, 396, 396–97n64
Pointe Basse, 12, 14, 63, 65n87, 66–68, 79, 95, 157, 225–27, 227n6, 230–34, 240–46, 245n40, 251, 252, 261, 264, 265, 419, 433
Point (or bottom), 40–43, 40, 41n37
Portage des Sioux, 8, 147n65, 410
Post oak trees, 50, 69
Post offices, 440–43, 441, 443n58
Poston, Henry, 339n96
Potassium nitrate, 52
Potel, Jean, 314
Potosi: Austin's plans for, 287, 295–97, 300–302, 426; businesses in and economy of, 301, 402, 439, 452, 464; community well in, 47; as county seat, 287, 295, 300, 463; courthouse in, 300–301, 301, 302; decline of, 303; food suppliers for, 295; founding of, 216; incorporation of, 302; land claims at, 149; land sales near, 177; lead mining in, 3, 53, 188; lot sales and American investors in, 296–97; map location of, 215; migration to, 414; naming of, 296, 296n25; population of, 206, 301–2, 410; post office at, 442; and roads, 302, 338–39, 424, 425, 426, 426n31, 427; Schoolcraft at, 3, 300–301, 301; settlement type of, 204; town plan of, 217, 295–96, 465
Prairie à Bollieur (Boyer?), 100
Prairie du Pont, 209n9
Prairie du Rocher, 8, 11, 65
Prairie Gauthier, 265
Prairies, 37–39, 37n26, 65, 327, 390
Prairie Salle, 327
Prairie Spring Creek, 385
Pratte, Antoine, 334
Pratte, Bernard, 133, 154n81
Pratte, Father, 271n9
Pratte, Jean-Baptiste, 269, 283n33, 332–34

Pratte, Joseph, 107, 305n38, 362, 460
Pratte, Pierre Auguste, 331, 334
Pratte family, 291, 297
Preemption rights, 163–64, 164n101, 177n7, 182, 197
Presbyterian church, 360–63, 365, 368, 396, 402
Prescriptive rights for mine ownership, 136–37, 195, 285, 293, 305
Price, Anna, xix, 452
Price, Edward, 459
Price's Spring, 422
Proclamation line of 1763, 165n104
Protestant immigrants: in Abernathy Settlement, 396; in Bellevue Valley, 360–62, 365–67; in Brazeau Settlement, 396; in Brazeau Valley, 388; conversion of, to Catholic church, 351; in La Saline, 261; in Perry County, 368; to Spanish Louisiana, 91–93, 92n36, 93n39, 97, 99–100, 105–6. *See also* American settlers; Baptist church; Lutheran church; Methodist church; Presbyterian church
Public lands: in early twentieth century, 464–65; lead-land leasing, 136, 138–43, 138n45, 183, 187–92, 190–91n38, 192n42, 197; and rectangular land survey, 174–77, 183; sale of, 116, 177–83, 177n8, 178, 179n9, 181; sale of lead lands, 192–93, 192n42; trespassing on, 137, 163, 164
Putnam, Rufus, 159n87

Quarles report, 321n64
Quebec, 10n6, 78, 209–10, 211n13, 222n27
Quick families, 397n64
Quin, Michael, 295n32

Racola, 317, 317n59
Rectangular land survey, 155–56, 158, 167, 171, 174–77, 175n1, 178, 181, 183, 184n16, 221n25
Rector, William, 155, 175–76
Red River, 100, 349, 366n17, 424, 426–27
Red River Country, 277, 301, 366n17, 428, 442
Red River Road, 277, 426–27
Reed, Andrew, 347–48
Reed, Joseph, 144
Reed, Robert, 144
Reed, William, Jr., 144, 360, 361, 365
Reed family, 366
Régistre d'Arpentage, 128, 128n26, 130
Renaudière (miner), 31, 304

Renault mines, 71, 78, 302, 304
Renaut, Philippe François: colonization by, 223n30; heirs of, 268n2; land and mining grants for, 78, 135, 156, 267, 419; Old Mines concession for, 303, 303n35, 305; spelling of name of, 78n5
Reps, John W., 209n9, 214n18, 238
Revolutionary War. See American Revolution
Reynolds County, 353, 356
Rhorer, David, 257
Richwoods, 49, *118*, 187, 204
Riddick, Thomas, 146
Ridge roads, 416–17
Ring, Thomas, 348, 350
Ring Creek, 348
Ripley County, 356–57n10
Ríu, Capt. Francisco, 82
River Aux Vases, 67, 68, 69, 100, 177, 204, 205, *229*, 261, 262, 264–65, 417, 422
River des Pères, 89n29, 111n39
Rivers and water, 39–47, 90, 113, 417–19
Rivière à Gaboury, 330
Rivière au Castor, 268, 268n3
Rivière aux Cannes, 275
Roads: in American period, 423–28, *424;* Austin's financing of, 287, 302, 318, 425; in Bellevue Valley, 365–66; *camino real* (royal road), *420*, 421–22, 421nn23–24, 423, 427; on Farmington Plain, 338–39; in French period, 419–21, *420;* and La Saline, 256n64; and linear valley settlements, 219; in mining country, 21, 188, 277, 286, 302, 320, 419, *420*, 422–26, *424*, 464; and New Bourbon village, 250–51; Red River Road, 277, 426–27; ridge roads, 416–17; Shawnee Trace, 375, 422; in Spanish period, *420*, 421–23; state roads, 425, 425n29; St. Louis–Ste Genevieve route, 39, 421–22, 423, *424*, 425, 425n29; and St. Michel, 277; terminology on, 415n11
Robbins, Roy M., 164
Robert, Jean, 314, 316
Robert, Polite, 314
Roberts, Edmund, 160
Rocheblave, Philippe-François Rastel de, 80
Rock, 30–31, 61
Rogers, Chief, 375, 377
Rogers Town, 375
Rosati, Bishop Joseph, 389, 393, 395
Rose, Allen Henry, 23

Rough hills region, *64,* 68–70, 70n98, 340–45
Routes. See Roads
Rozier, Ferdinand, 438
Rozier, Fermin, 55
Rush, Benjamin, 59n75
Rush, Thomas, 153
Russell, William, 162n98, 171, *172*

Sac Indians, 131, 167, 167n108
St. André, 212n15
St. Ange de Bellerive, Louis, 79–80, 82
St. Charles, *8,* 83, 111n39, 147n65, 421, 426
St. Cloud Creek, 244
St. Domingue, 28n3
Ste. Gemme Beauvais family, 91
Ste. Genevieve County, 263n77, 395n61, 409, *411,* 412–13, 464
Ste. Genevieve District: administrative settlement policy and traditional practices in, 75, 76, 457–59; administrative structure of, 402–12; agriculture in, 28–30, 29n6, 33; bibliographic essay for, 18–23; bluff lands and karst plain, *64,* 66–68, 100, 221, 231–33, 241, 244, 244n38, 264, 387; boundary of, *407;* cultural diversity in, 447–54; decline of, 464; definition of historic Ste. Genevieve District, 9; description of, as place, 4–11; economic activity in, 429–40; environmental encounter in, 454–57; first permanent European settlement in, 12; floods in, 14, 39–44, 60, 230, 232–33, 264, 280, 386, 389, 454; geographical perspectives on generally, 23–24; imprints of early settlers on, 465; Indians in, 14, 17, 21, 55, 78; land settlement under French before 1770 in, 76, 77–80, 114–15; land settlement under Spanish (1770–1795) in, 75, 76, 80–96, 114; land settlement under Spanish (1796–1804) in, 77, 97–113, 114; maps of, *5, 8, 18, 118, 181, 215, 406–8, 411;* mining country, 16, *64,* 70–71; natural regions of, 63–71, *64;* natural resources of, 27–62; need for geographical perspectives on, 23–24; overview of stages of land settlement in, 75–77; population mobility in, 412–15, 461–62; population of, 14, 15, 16–17, 18, 116, 162, 205–6, 403, 403nn3–4, *408,* 409; and post offices, 440–43, *441;* property turnover rate in, 412–13; Renault's grant in, 78; roads and

river travel in, 21, 415–28; rough hills region, *64,* 68–70, 70*n*98, 340–45; settlement of, after 1830, 464–65; settlement pattern in, 459–61; settlement systems in, 461–65; slavery in, 12*n*11, 14, 15, 17, 80, 91, 91*n*35, 221, 250, 260, 409, 430, 450; time period from eighteenth to early nineteenth century, 11–18. *See also* specific towns and villages

Ste. Genevieve District (American period): administrative structure of, 409–12; attempts to limit settlement (1804–1818), 164–73, *168–70, 172;* confirmation of land claims in, 117–45, 471–72; cultural diversity in, 449–54; description of, 7, 9; economic activity during, 433–40, *436;* and Indian lands, 131–35, 193–94; land sales (1804–1818), 116–17; land sales (1818–1830), 116, 177–83, 177*n*8, *178, 181;* land settlement (1804–1818), 114–73, 195–96; land settlement (1818–1830), 174–97; and map making, 456; maps of, *215, 406–8, 411;* maps of landownership in, *118, 181;* mineral/lead lands in, 135–43, 150, 179, 183–93, *185;* population of, 116, 162, 205–6, 403, *408,* 409; results of land-confirmation process by U.S., 146–52; roads during, 423–28, *424;* squatters in, 116, 159–64, 196–97, 346, 449; surveying of private lands (1804–1818), 102*n*13, 122*n*13, 126–30, 126*n*21, 152–59; and surveying of public lands, 174–83. *See also* American settlers; United States; and specific towns

Ste. Genevieve District (French period): economic activity during, 429–30; and map making, 455–56; population of, 206; regulating settlement before 1770, 76, 77–80, 82, 83, 114–15; roads during, 419–21, *420. See also* French territories and settlements; French villages; and specific towns and villages

Ste. Genevieve District (Spanish period): administrative structure of, 403, *405–8,* 409; American settlers in, 88, 88–89*n*29, 90–94, 92*n*36, 93*n*39, 96–106, 108, 110, 112, 287; and backdating of Spanish concessions, 109–10, 126–30, 148, 291, 331–34; Catholic immigrants to, 88, 89, 91, 94, 95, 103–6, 112, 195, 351–53, 356, 453; cultural diversity in, 447–51; description and map of, 7, *8;* economic activity during, 430–33, 431–32*nn*41–42; and fevers, 60, 60*n*76; and floods, 60; French administrators in, 83, 83–84*n*19; French Creoles in, 13–14, 79, 88, 89, 91, 101; French émigrés from revolutionary France to, 95, 101, 101*n*10, 246–48; governors-general of Louisiana, 467; immigration policies for, 14, 16–17, 76, 81, 88, 89–106, 92*n*36, 93*n*39, 112, 195, 458; and Indians, 85, 90, 94–95, 95*n*44, 132–33, 371–73, 372*n*6, 376–78, 450–51; lieutenant governors of Upper Louisiana, 17, 469; and Louisiana purchase by U.S., 109–11, 114–17; and map making, 455–56; maps of, *406–8;* population of, 14, 15, 16–17, 18, 206, 403*n*3; Protestant immigrants to, 91–93, 92*n*36, 97, 99–100, 103, 105–6; and retrocession of Louisiana to France in 1800, 17, 102*n*12, 108–9; roads during, *420,* 421–23; settlement regulations (1770–1795), 75, 76, 80–96, 114, 165*n*104, 195, 458–59; settlement regulations (1796–1804), 77, 97–113, 114, 165*n*104, 195, 356, 458–59; subdivisions of Upper Louisiana, 403, *405;* surveyors in, 102, 102*n*13, 109–10, 125–26. *See also* Spanish territories and settlers; and specific towns and villages

Ste. Genevieve Project, xix, 413*n*10, 473

Ste. Genevieve village (new): and agriculture, 243, 243*n*37; businesses in, 243, 320, 412, 438, 444, 452; cattle market in, 349; central *place* of, 237–38, 237*n*25; church in, 237, 253, 271, 271*n*9; commandants at, 83, 109*n*33; common fields at, 240–43, 241*n*35, *242,* 251, 252, 465; commons in, 208; decline of, 444, 464; as economic center, 402, 439; employment in, 243; fevers in, 59–60; harborage for, 46, 323, 427, 428; as "hinge" town, 462; historical studies on, 20; and Indians, 371, 374*n*11; land claims in, 123*n*15, 124, 147*n*65; map location of, *8, 215;* map of, 235*n*21, *236;* masonic lodge in, 243; migration from, 414; and Peyroux, 262; population of, 91, 205, 206, *410;* post office at, 443; relocation of original village, 14, 31, 91, 211–12, 216, *229,* 230–35, 231–32*nn*11–13, 262, 263, 454; settlement type of, 204; surveys of, 102n13, 155, 177n6, 235–37, *236;* town plan of, 62, 86, *217,* 235–40, *236,* 237n23, 239n28, 465

Ste. Geneviève village (old): and bank caving, 230, 231, 233, 454; cemetery in, 233, 233n20; church in, 230, 233, 237; and colluvial slopes, 234–35; commons in, 226–27, 227nn5–6, 233; fevers in, 59–60; and floods, 14, 42, 91, 230, 232–33; founding of, 12, 76, 79, 209, 209n9, 216, 225–26; harborage for, 46; historical studies on, 20; houses in, 227, 228, 228n8, 230, 235; and lead mining, 268; map location of, *8,* 225–26, 226n3; population of, 14, 80, 228; relocation of, 14, 31, 91, 211–12, *229,* 230–35, 231–32nn11–13, 262, 263, 454; Rocheblave as local commandant of, 80; slaves in, 80; town plan of, 20, *217,* 227–28, 230, 238; tripartite form of, 204, 210, 211, 226–27

St. Ferdinand, 147n65

St. Francis River, 48n50, 51, 71, 106, 176, 184, 192, 194, 204, 205, 220, 271, 275, 277, 330, 337n90, 345, 346, 348, 349, 358, 380, 383, 427

St. Francis River Valley, lower, 345–50

St. Francois County, 339, *411,* 464

St. Francois Mountains, 31

St. Gemme Beauvais, Jean-Baptiste, 269, 285n2

St. Joseph Lead Company, 327

St. Laurent River, 258, 352n4

Saint Lawrence River, 43, 209

St. Louis: and agriculture, 30, 433; American settlers in, 16, 93n39, 116; attack on, in 1780, 84; Board of Land Commissioners in, 117; businesses in, 364, 365, 437–38, 439n52, 464; Catholic church in, 385, 389, 393; cattle market in, 349; economy of, 430–34, 439, 439n52, 443–44, 462–64; founding of, 12, 76, 79, 210, 216, 239n27, 464; and France, 17; French Creoles in, 21; French government of, 79; and fur trade, 430–32, 434; as "hinge" city, 462–63; historical studies on, 21; Indians in, 132, 374n11; land claims in, 121, 122, 147n65; land office in, 177n7, 180, 182n13, 184, 186; and lead industry, 48, 430; and lead shipping, 464; map location of, *8, 215;* Moore as visitor to, 389; overland travel from Ste. Geneviève to, 39, 421–22, 423, *424,* 425, 425n29; population of, 206, *410;* post offices in, *441,* 442; settlement type of, 210, 212; significance of, 11; as Spanish administrative center, 13, 17, 80–83, 87, 443, 464; St. Ange de Bellerive as local commandant of, 79–80, 82; and steamboats, 428; Stoddard as administrator of, 17, 110–11, 116, 124, 148, 434, 451; surveys of, 102n13, 155n83, 237n24; town plan of, 86, *217,* 237n24, 238, 239; winter icing of Mississippi River at, 45

St. Louis County, *411*

St. Louis District, 212n15

St. Louis Missouri Fur Company, 434

St. Mary's (town), 352n4

St. Mary's of the Barrens seminary, 22, 393–94, 395, 439, 442

St. Mary's (Village Creek) Road, 281

St. Michael's. *See* St. Michel

St. Michel: American settlers in, 277, 282; background on founding of, 267–69; church in, 281, 282, 318; class differences in, 449; common field of, *276,* 278–80, 455, 465; commons in, *276,* 280; community well in, 47; concession for, 273–75, *276,* 313n50; and flooding, 280; founding of, 212, 216, 254, 269–73, 270n8, 348; and Fredericktown, 281–83; French Creoles in, 269–73, 277, 283, 283n32, 460; houses in, 278; inhabitants of, 272, 273, 275n16; land claims in, 124; land sales in, 282; and lead mining, 254; location of, in sedimentary basin, 275; map location of, *8, 215;* map of landownership in, *118;* migration to, 414; naming of, 270, 270n8; population of, 205–6, 272, 273, 281; reasons for founding of, 270–72; relationship between New Bourbon and, 254–55, 254n60; relocation to new village, *276,* 280–81, 280n26; renamed as St. Michael's, 282–83; and roads, 277, 427; site of original village, 278n21; slavery in, 272; survey of, 273–75, *274;* town plan of, 86, *276,* 277–78; township sale in, 177; tripartite pattern of, 124, 204, 212, 270, *276,* 277–80

St. Philippe, 11, 44

St. Pierre, Paul de, 238

St. Vrain, Jacques Ceran Delassus de, 122n13, 222n29, 253

Salaries. *See* Wages

Salée (white, sticky clay), 33

Sale of land. *See* Land sales

Saline County, 302n33
Saline Creek, 79, 88n29, 177, 272, 275, 278–82
Saline River, 46, 47, 53, 63, 68, 204, 205, 226, *229*, 256, 258, 262, 264–65, 352n4, 368, 422
Saline River Valley, 21, 255–59, 256n64, 263
Salt and saltworks, 22, 23, 27, 36, 46, 47, 52, 53, 79, 135, 139, 192n42, 234, 255–59, 261, 264, 362, 368, 429, 430, 433, 449
Saltpeter, 47, 52–53, 53n62, 138, 138n45
Salt River, 183
Samuel, Mr., 260
Sanchem, Juan Baptista. *See* St. Gemme Beauvais, Jean-Baptiste
Sandy Creek, 187, 341
San Ildefonso Treaty, 17, 108, 109–10, 333
San Lorenzo del Escorial Treaty, 95–96
Santa Fe, 442
Saucier, François, 226, 227
Sauer, Carl O., 27n1, 31n11, 39n33, 201–2n2, 453
Sauk Indians, 131n30
Sawmills, 35–36, 271, 301, 325
Schiavo, Giovanni E., 243–44n37
Schoolcraft, Henry Rowe: and Austin, 326n72, 327n73; on Farmington Plain, 338; on farms in river valleys, 219n23; at Fredericktown, 281, 281n27; on lead mines of Missouri, 47n50, 326n72; on Murphy's Settlement, 337; at Potosi, 3, 300–301, *301;* on rough hills, 69; on St. Michael's, 283
Schools, 394, 394n59. *See also* Seminary
Schroeder, Walter A., 27n1, 159n88, 390n52, 439n53
Schultz, C., 375n16
Scotch-Irish settlers, 360–62, 365–67, 368, 382–83, 396, 397n64
Scott, Andrew, 296
Scott, John, 190, 190n37, 193
Sebastian, Benjamin, 92
Seigneuries, 209–10, 211, 221–22, 222n27, 255n62
Selma, 344, 344n9, 428
Seminary, 22, 393–94, 395, 439, 442
Semple, Ellen Churchill, 439n52
Seth Craig and Company, 314, 434
Settlement forms: and community, 201–2, 201–2n2; compact settlement, *203,* 204, 220–21; county seats, 463; discontinuous valley settlement, *203,* 204, 219–20, 345–50; geometry of, 201; "hinge" cities and towns, 462–64; introduction to, 201–2; large tracts, *203,* 204, 221–24, 222–24nn27–32, 460–61; linear valley settlement, *203,* 204, 216, 218–19, 218–19nn20–23, 340–45; and neighborhood, 201–2; planned American towns, *203,* 204, 214, 216, 465; planned French villages, 202, *203,* 204, 207–12; Spanish city planning, 62; and spatial pattern of landholdings, 459–60; and topography, 460; tripartite villages of French settlers, 122–24, 147, 147n65, 176–77, 195, 202, *203,* 204, 207–12, 270, 277–80, 452, 465; unplanned French villages, *203,* 204, 212–13, 316–18; variety in, 398
Settlement regulations: in American period (1804–1818), 76, 114–73, 195–96, 458–59; in American period (1818–1830), 76, 174–97, 458–59; attempts to limit settlement during American period (1804–1818), 164–73, *168–70, 172;* and backdating of Spanish concessions, 109–10, 126–30, 148, 291, 331–34; confirming of land claims by U.S., 117–45, 471–72; in French period before 1770, 76, 77–80, 82, 83, 114–15; Galvez's policies of 1778, 86, 89–90; Gayoso's instructions of 1797, 86, 103–4, 106, 106n23, 107, 120, 145, 273; and Indian landownership, 85, 131–35, 133n33, 134n35, 193–94; land sales in American period (1804–1818), 116–17; land sales in American period (1818–1830), 116, 177–83, 177n8, *178, 181;* and Law of the Indies, 81, 83, 85, 86–87, 86n24, 157n86, 238–40, 239nn27–29; and Louisiana purchase by U.S., 109–11, 114–17, 136, 192–93; *mercedes* (graces) granted by Spanish authorities, 111, 111n39, 128; Morales's regulations of 1799, 86, 106–8, 106n23, 120; O'Reilly's regulations of 1770, 85–88, 87n25, 120; results of land-confirmation process by U.S., 146–52; and Spanish authorities' discretionary powers, 111–12; in Spanish period (1770–1795), 75, 76, 80–96, 114, 165n104, 195, 458–59; in Spanish period (1796–1804), 77,

97–113, 114, 165n104, 195, 458–59; and squatters in American period, 116, 159–64, 196–97; and surveying of private lands by U.S. (1804–1818), 152–59; versus traditional practices, 75, 76, 457–59

Settlement rights, 116, 120, 125–26, 143–45, 144n58, 146, 341, 348

Shannon, William, 438, 444

Shawnee-Delaware Indian grant, 95, 132–33, 193–94, 204, 205, 213, 347, 369, 371–82, 385, 450–51

Shawnee Indians: adoption of captives by, 375, 375–76n16; Americanization of, 375, 375n16; and Berthiaume concession, 104; as early settlers in Bois Brule, 368, 370; farming by, 68, 70, 373, 376; and French colonials, 131n30, 261, 450; independent subgroups of, 194n46; mobility of, 462; and Osage Indians, 377, 450; population of, 373, 380, 381n33; removal and resettlement of, 193–94, 377n24, 379–82, 450–51; settlement form of, 213n17; Shawnee-Delaware Indian grant, 95, 132–33, 133n33, 193–94, 204, 205, 213, 347, 369, 371–82, 385, 450–51; and Spanish colonials, 451; and Treaty of Castor Hill, 381; villages of, 374–75, 374nn12–13, 381–82, 385; white encroachments on land grant to, 377–79. *See also* Indians

Shawnee Trace, 375, 422

Shelbyville (Tenn.) plan, 339, 395

Shenandoah Valley, 221n26, 460

Shibboleth, 218n21, 344

Shibboleth Mine, 53, 141–42, 187, 315, 336, 344

Shoemaker, Floyd Calvin, 146n62

Shortleaf pine trees, 35–36

Shortridge, James R., 440n55

Shot towers, 323–24, 324

Shreve, H. M., 438

Sibley, George C., 325, 343

Sickle-cell trait, 56, 56n71

Silver and silver mining, 48–49, 48n50, 288n10, 304

Simpich, Frederick, xv

Sims, Robert, 337n91

Sinkholes, 30–31

Sitios (square tracts for cattle raising), 87, 269n6

Slavery: Austin's slaves, 363; in Barrens Settlement, 392, 394, 395; in Bellevue Valley, 361, 362, 363; in Bois Brule, 370; Caribbean slave revolts, 361; and Fenwick Settlement, 384, 385; fevers and slaves, 56, 56–57n71; and French and Spanish colonials, 12, 12n11, 77, 77n1, 82, 91–92, 93, 450; in Longtown, 396; in Lower St. Francis River Valley, 347, 348, 349; and mining, 268, 285, 294, 303, 306, 314n52, 333; in Murphy's Settlement, 337; Partenais dit Mason as large slave owner, 314n52; in Plattin Valley, 341; population of slaves, 14, 15, 80, 93, 250, 260, 409; prohibition of, in Northwest Territory, 91, 389; in Ste. Genevieve District, 12n11, 14, 15, 17, 80, 91, 93n35, 221, 250, 260, 409, 430; in St. Michel, 272; terms for slaves, 12n11

Smallpox, 55

Smith, Reuben, 394n59

Smith T, John: as adversary of Moses Austin, 142; and agriculture, 302n33; and Big River Bend, 334; biography of, 22; and floating concessions, 141–42, 293; houses of, 218n21, 315; intimidation methods of, 293; and lead-shipping port of Selma, 344; and mining, 141–42, 284, 315, 334, 336, 457; and Old Mines, 314, 314n52, 315, 318, 318n60; and purchase of land from Capeheart (Kephart), 337n90; size of land claims of, 122n13

Soil, 31–33, 31–33nn11–15, 39, 51, 68, 71, 179, 329, 340, 359–60, 455, 457

Soulard, Antoine: Austin survey by, 288; and backdating of Spanish concessions, 109–10, 126, 128; and Board of Land Commissioners, 130; collection of survey plats of, 125, 130; deputy surveyors under, 102, 171; French surveying practices used by, 102n13; Kerlegand survey by, 250n48; land claims by, 130; Luzières family survey by, 122n13; Maxwell survey by, 352n4; and Mine à Breton road, 423n26; and orientation of properties for surveys, 157, 159n87; St. Michel survey by, 273; as surveyor in American period, 126, 126n21, 130; as surveyor in Spanish period, 102, 109–10

South Carolina, 54n67, 358, 397. *See also* Carolina

Space, organization of. *See* Organization of space

Spain: Black Legend on, 93; lead mining

in, 50; and Pinckney's Treaty, 95–96; relations between England and, 89; relations between France and, 82–83; relations between U.S. and, 90, 95–96; and retrocession of Louisiana to France in 1800, 17, 102n12, 108–9. *See also* Ste. Genevieve District (Spanish period); Spanish territories and settlers

Spanish Illinois, 7

Spanish Netherlands, 50, 83n19

Spanish territories and settlers: and American settlers, 88, 88–89n29, 90–94, 92n36, 93n39, 96–106, 108, 110, 112, 287, 458; in Canada, 13; and Catholic immigrants, 88, 89, 91, 94, 95, 103–6, 112, 195, 351–53, 356, 358, 453; and cattle raising, 87–88, 87n26; and city planning, 62; and floods, 44n41; and French Canadians, 13–14, 79, 88, 89, 91; French émigrés from revolutionary France to, 95, 101, 101n10, 246–48; French local administrators in Spanish Upper Louisiana, 83, 83–84n19, 101; and fur trade, 81–82; and Galvez's policies of 1778, 86, 89–90; Gayoso's instructions of 1797, 86, 103–4, 106, 106n23, 107, 120, 145, 273; and geographic linkages, 9; governors-general of Louisiana, 467; and horses, 415n12; and Illinois country as term, 6–7; and immigration policies, 14, 16, 76, 81, 88, 89–106, 92n36, 93n39, 108, 112, 195, 458; and impact of winds on health, 62; and Indians, 85, 90, 94–95, 95n44, 132–33, 371–73, 372n6, 376–78; and land division system, 10; land settlement regulations (1770–1795), 75, 76, 80–96, 114, 165n104, 195, 458–59; land settlement regulations (1796–1804), 77, 97–113, 114, 165n104, 195, 356, 458–59; and large land grants, 222–23, 222nn28–29, 287–91, 362; and Law of the Indies, 81, 83, 85, 86–87, 86n24, 157n86, 238–40, 239nn27–29; lieutenant governors of Upper Louisiana, 17, 469; in Lower Louisiana, 84, 84n19, 89; and mining country, 71; Morales's regulations of 1799, 86, 106–8, 106n23, 120; and O'Reilly's regulations of 1770, 85–88, 87n25, 120; and Protestant immigrants, 91–93, 92n36, 93n39, 97, 99–100, 105–6; and rough hills region, 70; and slavery, 12n11, 17, 82, 91–92, 93, 450; and Treaty of Fontainebleau, 13, 79; and Treaty of San Ildefonso, 17, 108, 109–10, 333; and trees, 34; and Upper Louisiana as term, 6–7. *See also* Ste. Genevieve District (Spanish period); and specific Spanish settlers and administrators

Speculation. *See* Land speculation

Springfield iron furnace, 366

Square shape and land surveys, 156

Squatters, 97–98, 116, 159–64, 160n89, 162n95, 162–63nn98–99, 196–97, 346, 449

Stamps, George, 348

Steamboats, 324, 418, 427–28

Steel, Andrew, 348

Stevenson, William O., 366n17

Stevenson family, 366

Steward, Dick, 302n33, 318n60

Stewart, John, 360

Stilgoe, John R., 238–39

Stoddard, Amos: on agriculture, 65; and Austin, 286; and Bellevue Valley, 361–62n5; and evidence of landownership, 124; on fevers, 57, 60, 61; on fraudulent land claims, 127, 127n23, 130; on French settlers, 448; on icing of Mississippi River, 45; on Indians, 131n30, 374; on keelboats, 417–18nn17–18; on Lead and Salt Leasing Act (1807), 141; on limit of settlement of Missouri, 167, *170;* on Mississippi River, 39n34; and New Bourbon, 252; on population of Ste. Genevieve District, 403n4; on prairies, 37–38; on river bank caving, 44n41; on settlement rights, 144n58; on silver mine, 48n50; on Spanish land policy, 111; as territorial administrator in St. Louis, 17, 110–11, 116, 148, 434; on unity versus diversity of settlers, 451; and water pollution at Old Mines, 306, 449; on winds and health, 62

Stout's Creek, 363, 366, 426

Stout's Settlement, 353, 363–64, 426

Strickland family, 341

Stringtown, 281

Sturgus, John, 341, 341n2

Sugar camps (*sucreries*), 264

Sulphur, 52

Sundé, Jacques, 133

Supreme Court, U.S., 192–93, 222n28, 303n34, 332

Surveys and surveyors: in American period, 102n13, 122n13, 126–30, 126n21, 152–59, 155n82, 174–83, 176–77nn5–6, 472; and Bois Brule settlers, 383; and cadastral patterns, 154, 156, 156n84, 159n88; cost of, 176n5; demand for, 153n8; French measure for lot dimension, 313n49; of French tripartite villages, 176–77; and indemnity scrip, 155, 155n82; and irregular polygon shapes, 156; and lead lands, 183–86, *185;* and long-lot shapes, 156, 156n84; Mine à Breton survey, 298–300, 298n29, *299;* Old Mines survey, 310–14, *311, 312,* 317–18, 317n59; and orientations of properties, 156–59, 157n86, *158,* 159nn87–88; rectangular survey by U.S., 155–56, *158,* 167, 171, 174–77, 175n1, *178, 181,* 183, 184n16, 221n25; and Shawnee-Delaware Indian grant, 373; in Spanish period, 102, 102n13, 105, 109–10, 125–26, 152–53; and square shape, 156; and township subdivision, 176

Swine. *See* Hogs

Swiss immigrants, 397

Tardiveau, Barthélémi, 247, 250, 253
Taylor, Alan, 221n26, 247n44
Taylor, Peter J., 201n1
Tayon, Charles, 83, 128
Tayon, François, 111n39, 128, 128n25, *129,* 303n35
Tennessee, 16, 22, 92, 93, 97, 101, 102, 218, 282, 339, 340, 360, 396, 397, 416, 428
Tennessee River, 92
Terrains (lots), 207, 230, 233, 234, 240, 247–51, 255, 259, 270, 278, 280
Terre Bleue Creek, 286
Terre Bleue River, 327, 330, 334, 364
Terres, 208, 226, 233–35, 240, 241, 251–52, 255, 270, 278–80
Tesreau (Tesserot) family, 271
Tesserot, Elizabeth Levrard, 254n60
Tesserot, Joseph, 254n60, 271
Tesserot (Tesreau) family, 271
Texas, 277, 301, 303, 326, 349, 366, 366n17, 392n54, 424, 426–27, 428, 442
Thomas, Lt. Martin, 187–88, 189, 191, 321n64, 344n9
Thompson, H. C., II, 270n8, 280n26
Three-Notch Road, 277, 419
Tiff (*tuf blanc*), 47–48n50, 48
Tison, Albert, 246n41
Tobacco, 30, 33, 433, 437
Tomar, Maria, 277
Towns. *See* American settlements; Settlement forms; and specific towns

Townsend, John, 160
Traces on the Rhodian Shore (Glacken), xvii
Transportation. *See* Boats; Horses; Roads
Treasury Department, 187
Treat, Payson Jackson, 119n7
Treaty of Castor Hill, 381
Treaty of Cession (1803), 114–17, 192–93, 193n43, 377
Treaty of Fontainebleau (1762), 13, 79
Treaty of Greenville, 349
Treaty of Paris: of 1763, 12–13, 79; of 1784, 90
Treaty of San Ildefonso, 17, 108, 109–10, 333
Treaty of San Lorenzo del Escorial, 95–96
Trees, 33–37, 34n17, 50, 68, 69, 70, 191, 231, 416, 457
Trespassers. *See* Squatters
Trewartha, Glenn T., 213n16, 218n20
Trimble, Michael K., 20n23, 213n16, 257n67
Tripartite French villages, 122–24, 147, 147n65, 176–77, 195, 202, *203,* 204, 207–12, 270, 277–80, 452, 465. *See also* specific villages
Trudeau, Zenon: on agriculture, 432; and American or English settlers, 94, 98, 105, 265; and backdating of Spanish concessions by Delassus, 109, 127; on collapse of open-field system in French villages, 212n14; on flooding, 43n40; French background of, 83, 83n19; and fur traders, 431; on Illinois Country in Upper Louisiana, 6–7; and Indians, 373, 376; and Irish immigrants, 352; land grants issued by, 119n9, 273; and Luzières, 246; and New Bourbon, 253, 265; and New Ste. Genevieve, 262; and Pointe Basse, 252; and Protestant immigrants, 103; on stipends for commandants, 109n33; and St. Michel concession, 273
Tucker, James, 391
Tucker, Joseph, 385n43, 389n50, 391, 392
Tucker, Josephus, 389n50
Tucker, Michael, 391
Tucker families, 389, 389n50
Tucker Settlement. *See* Barrens Settlement
Turnbo, S. C., 160n89
Turner, Frederick Jackson, 152n77, 161n94, 196, 202n2, 461, 462, 463

Twelve Mile Creek, 348, 349
Twyman, James, 287n7
Tyler, Thomas, 370
Tywappity Bottoms, 372

Uhlig, Harald, 208n6
Ulloa, Antonio de, 82n14, 87n25, 239n29, 384
United States: and access to Mississippi River, 90, 92, 93, 96, 113; and American Revolution, 89; Indian policy of, 131–35, 132n32, 133n33, 134n35; and Louisiana purchase, 17, 109–11, 113, 114–17, 136, 192–93; and Northwest Territory, 91; and Pinckney's Treaty, 95–96; relations between Spain and, 90, 95–96; and right of deposit in New Orleans, 96; Treaty of Paris (1784), 90; and westward expansion, 92, 96, 112, 112n40. *See also* American settlers; Ste. Genevieve District (American period)
U.S. Public Land Commission, 117n6
Unplanned French villages, 203, 204, 212–13, 316–18. *See also* specific villages
Unzaga, Luis, 88
Upper Louisiana: American military presence in, 17; American plans for, 75–76; American settlers in, during Spanish period, 88, 88–89n29, 90–94, 96–106, 108, 110, 112, 287, 458; Austin on, 27; description of, as term, 4, 6–7, 6n3; first establishment of American family in, 88, 88–89n29; flooding in, 39–44; French period before 1770 in, 77–80, 82, 83, 403; large land grants in, 222–24, 222–23nn29–30, 287–91, 362, 460–61; Lewis and Clark map of, 403n3, 405; lieutenant governors of, 17, 469; and Louisiana purchase by U.S., 110–11, 112, 114–17, 136, 192–93; population of, 16–17, 89, 116, 403; Spanish period (1770–1795) in, 80–96; Spanish period (1796–1804) in, 97–113; Spanish subdivisions of, 403, 405. *See also* Illinois Country; Louisiana; Ste. Genevieve District
Usner, Daniel H., Jr., 19, 21, 57n71, 451

Vacheries (cattle ranches), 87–88, 87n26, 258, 269n6, 339, 422, 460
Vallé, Bazile, 245, 285–86
Vallé, François, I: and Bois Brule settlers, 370; concessions for, 244; death of son of, 268; and lead mining, 268

Vallé, François, II: and Bellevue Valley, 361; as commandant at Ste. Genevieve, 83, 287; compared with Luzières, 254; concessions for, 122, 122n13, 237n25, 244n38, 245, 258–59, 368, 456; and contract for lead, 287; and Indians, 373, 373n9; and Luzières's concession, 246; and Mine à Breton *commissaire de police,* 291–92; mine claims by, 269; and new Ste. Genevieve, 250; and Old Mines, 315, 449; and partnership with Austin, 287, 287n9, 289, 321n64; and Peyroux, 262, 263, 456; and petition of 1803 for Old Mines concession, 307; and town plan for New Ste. Genevieve, 238–40; and water mill, 244, 245, 247, 252, 253; and water pollution at Mine à Breton, 305
Vallé, Jean-Baptiste, 137, 223, 269, 306
Vallé, Joseph, 268, 268n5
Vallé, Julie, 386
Vallé family, 224n32, 283n33, 297, 431, 449, 454
Vanburken (or Vanburkelow), William, 370n3
Vance, James E., Jr., 218n20, 461
Van Zandt, Nicholas Biddle, 338
Vase (soft mud), 33, 33n14, 46
Vial, Pedro, 288n10
Vieilles Mines. *See* Old Mines
View of the Lead Mines of Missouri (Schoolcraft), 47n50, 300–301, 301
Village à Robert, 147n65, 212n15
Village Creek, 275
Villages: definition of, 202n3, 209, 213; hamlets compared with, 211n13. *See also* French villages; Settlement forms; and specific villages
Villars, Antoine, 331
Villars, Athanase, 331
Villars family, 422
Vincennes, 117
Violette, Eugene M., 152n77
Virginia, 16, 98n2, 183, 218, 221n26, 248, 287, 291, 330, 341, 348, 362, 382, 397
Virginia Settlement, 348, 349, 350
Viriat, Pierre, 272, 277
Volney, Count, 28n2, 61, 62

Wade, Richard C., 462
Wages: of commandants, 109n33; of laborers in American East, 294; of miners, 294
War Department, 187–88
War of 1812, 17, 19, 119n7, 145, 150,

175n1, 187, 296, 326, 348, 388, 391, 396
Washington County: barite mining in, 319n61; courthouse for, 300; creation of, 287, 295, 363, 409; farmers from, in Red River country, 366n17; land-ownership in, 162, *181;* maps of, *178, 411;* mining in, 48n50; naming of, 350; population of, 162; public domain land sales in, *178;* squatters in, 162; St. Joachim Church in, 318
Water. *See* Rivers and water; and specific rivers
Water for drinking, 46–47
Water mills, 244, 245, 247, 252, 253, 254, 263, 359, 433n44
Wayne, Anthony, 349
Wayne County, *178, 181,* 220, 277, 345–50, 349n19, 350n21, 409, *411,* 414, 442, 443n58
Wea Indians, 11
Webb, Benedict Joseph, 389n50
West Florida, 94
Westover, Job, 99–100
Wheat, 28–29, 329, 349, 432, 433, 437
White, Joseph M., 151n75
White River, 160n89, 219n23, 380
Whitewater Dutch community, 100, 397n65
Whitewater River, 373, 374n11
Whooping cough, 55
Wickliffe Mounds Research Center, 246n42
Wideman, Francis, 321, 327n73
Wideman family, 341
Wifvarenne, Marie, 304
Wifvarenne, Pierre, 304

Wilcox, David, 105
Wildlife, 36, 53, 68. *See also* Hunting
Wilkinson, James A., 126–28, 128n26, 130, 137
Williams, Michael, 292n17
Williamsen, Tora Lorraine, 20, 239, 239n28
Willkie, Duncan C., 374n13
Wills, David, 443n58
Wilson, N., 141
Wilson's Creek, 380
Wilt, Christian, 423, 437–38
Winds, 61–62
Wine, 36, 36n22
Wines family, 342
Winslow, Arthur, 22
Winter icing of Mississippi River, 45, 45n44
Wisconsin, 187, 192n42
Wish, Harvey, 115n3
Women's roles, 431–32n41
Wood, Joseph S., 202n2
Wood, M. M., 423n28, 426n31
Wood, Peter H., 56–57n56
Woods, 34–47, 34n17
Wyckoff, William, 221n26

Yazoo, 94
Yellow fever, 55, 253
Ylinoa (Illinois Country), 6–7
York Chapel, 396
Yosti, Emilian, 244n37

Zitomersky, Joseph, 21, 213n17, 296n25
Zones of economic development, 435–40, *436*